Spon's Landscape and External Works Price Book

2003

Also available from Spon Press:

Spon's Building Costs Guide for Educational Premises – *Derek Barnsley* 0419233105

Spon's Railways Construction Price Book – *Franklin + Andrews* 0419242104

Spon's Middle East Construction Costs Handbook – *Franklin + Andrews* 0415234387

Spon's Latin American Construction Costs Handbook – *Franklin + Andrews* 0415234379

Spon's African Construction Costs Handbook – *Franklin + Andrews* 0415234352

Spon's Asia Pacific Construction Costs Handbook, 3rd edition – *Davis Langdon & Seah* 0419254706

Spon's European Construction Costs Handbook, 3rd edition – *Davis Langdon & Everest* 0419254609

Spon's Irish Construction Price Book – *Franklin + Andrews* 0419242201

Spon's Estimating Costs Guide to Electrical Works – *Bryan Spain* 0415256429

Spon's Estimating Costs Guide to Plumbing and Heating – *Bryan Spain* 0415256410

Spon's Estimating Costs Guide to Minor Works, Refurbishment and Repairs – *Bryan Spain* 0415266459

Spon's Construction Resource Handbook – *Bryan Spain* 0419236805

Spon's House Improvements Price Book – *Bryan Spain* 0419241507

Spon's First Stage Estimating Handbook – *Bryan Spain* 0415234360

Guidelines for Landscape and Visual impact Assessment – *The Landscape Institute and the Institute of Environmental Management and Assessment, Sue Wilson* 041523185X

External Works, Roads and Drainage – *Phil Pitman* 0419257608

The Regeneration of Public Parks – *Jan Woudstra & Ken Fieldhouse* 0419259007

Design for Outdoor Recreation – *Simon Bell* 0419203508

Manufactured Sites – *Niall Kirkwood* 0415243653

To order or obtain further information on the above or receive a full catalogue please contact:
The Marketing Department, Spon Press, 11 New Fetter Lane, London, EC4P 4EE.
Tel: 020 7583 9855 Fax 020 7842 2298
www.sponpress.com

Spon's Landscape and External Works Price Book

Edited by

DAVIS LANGDON & EVEREST

in association with
LANDSCAPE PROJECTS
Landscape Surveyors

2003

Twenty-second edition

Spon Press
Taylor & Francis Group

LONDON AND NEW YORK

First edition 1978
Twenty-second edition published 2003
by Spon Press
11 New Fetter Lane, London EC4P 4EE

Simultaneously published in the USA and Canada
by Spon Press
29 West 35th Street, New York, NY 10001

Spon Press is an imprint of the Taylor & Francis Group

© 2003 Spon Press

Printed and bound in Great Britain by
TJ International Ltd, Padstow, Cornwall

All rights reserved. No part of this book may be reprinted or reproduced or utilised in any form or by any electronic, mechanical, or other means, now known or hereafter invented, including photocopying and recording, or in any information storage or retrieval system, without permission in writing from the publishers.

The publisher makes no representation, express or implied, with regard to the accuracy of the information contained in this book and cannot accept any legal responsibility or liability for any errors or omissions that may be made.

Publisher's note
This book has been produced from camera-ready copy supplied by the authors.

British Library Cataloguing in Publication Data
A catalogue record for this book is available from the British Library

Library of Congress Cataloging in Publication Data
A catalogue record for this book has been requested

ISBN 0-415-30120-3
ISSN 0267-4181

Contents

	Preface to the Twenty Second Edition	ix
	List of Manufacturers and Suppliers	xi
	Common Arrangement of Work Sections	xxvii
1	**How to use this Book**	**1**
	1.1 Introduction	1
	1.2 Important notes on the profit element of rates in this book	1
	1.3 Introductory notes on costing	2
	1.4 Adjustment and variation of the rates in this book	4
	1.5 Approximate Estimates	5
	1.6 How this book is updated each year	6
2	**Cost Information**	**7**
	2.1 Labour Rates used in this edition	7
	2.2 Rates of Wages	7
	2.3 Daywork and Prime Cost	11
	2.4 Computation of Labour Rates, Cost of Materials and Plant	14
	2.5 Cost Indices	19
	2.6 Regional Variations	21
3	**Preliminaries**	**23**
	3.1 Preliminaries/General Conditions	23
4	**Landscape Consultant's Appointment**	**31**
	4.1 Landscape Consultant's Appointment	31
5	**Fees for Professional Services**	**55**
	Landscape Architects' Fees	55
	Architects' Fees	58
	Quantity Surveyors' Fees	62
6	**The Aggregates Tax**	**103**
7	**Prices for Measured Works**	**111**
	New Items for 2003	111
	A **Preliminaries**	**123**
	A11 Tender and contract documents	123
	A32 Employers requirements: Management of works	124
	A34 Employers requirements: Security/safety/protection	124
	A41 Contractors general cost items: Site accommodation	124
	A44 Contractors general cost items: Temporary roads	125

B		**Complete buildings/structures/units**	**126**
B10		Prefabricated buildings/structures	126
D		**Groundwork**	**127**
D11		Soil stabilization	127
D20		Excavation and filling	134
E		**In situ concrete/large precast concrete**	**140**
E10		In situ concrete	140
E20		Formwork for in situ concrete	143
E30		Reinforcement for in situ concrete	143
F		**Masonry**	**144**
F10		Brick/block walling	145
F20		Natural stone rubble walling	148
F22		Cast stone walling	149
F30		Accessories/Sundry items for brick/block/stone walling	151
F31		Precast concrete sills/lintels/copings/features	151
G		**Structural/Carcassing metal/timber**	**152**
G31		Prefabricated timber unit decking	152
H		**Cladding/covering**	**153**
H51		Natural stone slab cladding/features	153
H52		Cast stone slab cladding/features	153
J		**Waterproofing**	**154**
J10		Specialist waterproof rendering	154
J20		Mastic asphalt tanking/damp proofing	154
J30		Liquid applied tanking/damp proofing	154
J40		Flexible sheet tanking/damp proofing	155
J50		Green Roof Systems	155
M		**Surface finishes**	
M10		Cement:sand/Concrete screeds/toppings	160
M20		Plastered/Rendered/Roughcast coatings	160
M40		Stone/Concrete/Quarry/Ceramic tiling/Mosaic	160
M60		Painting/clear finishing	161
P		**Building fabric sundries**	**163**
P30		Trenches/Pipeways/Pits for buried engineering services	163
Q		**Paving/planting/fencing/site furniture**	**164**
Q10		Stone/concrete/brick kerbs/edgings/channels	164
Q20		Bases and sub-bases to roads/pavings	169
Q21		In situ concrete roads/pavings/bases	170
Q22		Coated macadam/asphalt roads/pavings	172
Q23		Gravel/hoggin roads/pavings	175
Q24		Interlocking brick/block roads/pavings	178
Q25		Slab/brick/sett/cobble pavings	179
Q26		Special surfacings for sport	189
Q30		Seeding/turfing	192

		Contents		vii
	Q31	Planting	215	
	Q35	Landscape maintenance	251	
	Q40	Fencing and gates	258	
	Q50	Site/street furniture/equipment	273	
	R	**Disposal systems**	**292**	
	R12	Drainage below ground	292	
	R13	Land drainage	299	
	S	**Piped supply systems**	**314**	
	S10	Cold water	314	
	S14	Irrigation	314	
	S15	Fountains/water features	315	
	V	**Electrical supply/power/lighting systems**	**320**	
	V41	Street area floodlighting	320	
8		**Approximate Estimates**	**335**	
	8.1	Preliminaries	337	
	8.2	Demolition and site clearance	338	
	8.3	Groundwork	339	
	8.4	Ground Stabilization	341	
	8.5	Insitu Concrete	344	
	8.6	Brick\Block Walling	345	
	8.7	Roads and Pavings	347	
	8.8	Special surfaces for Sport\Playgrounds	363	
	8.9	Preparation for Planting\Turfing	366	
	8.10	Seeding and Turfing	368	
	8.11	Aftercare of Turf Areas	371	
	8.12	Planting	372	
	8.13	Fencing	378	
	8.14	Street Furniture	384	
	8.15	Drainage	385	
	8.16	Irrigation	392	
	8.17	Water features	393	
	8.18	Timber decking	394	
	8.19	Lighting	395	
9		**Tables and Memoranda**	**397**	
10		**Index**	**469**	

SPON'S PRICEBOOKS 2003

Spon's Architects' & Builders' Price Book 2003
Davis Langdon & Everest

The most detailed professionally relevant source of construction price information currently available.

New features for 2003 include:
- A new section on the Aggregates Tax.
- A new section on Part L of the Building Regulations and its impact upon costs.

HB & CD-ROM ♦ £115.00
0-415-30116-5 ♦ 1056pp

Spon's Landscape & External Works Price Book 2003
Davis Langdon & Everest

The only comprehensive source of information for detailed landscape and external works cost.

New features for 2003 include:
- Notes on the aggregate tax with examples of the impact of the tax on measured works.
- Costs of external wall block work, special bricks, clay drainage and much more!

HB & CD-ROM ♦ £85.00
0-415-30120-3 ♦ 512pp

Spon's Mechanical & Electrical Services Price Book 2003
Mott Green & Wall

Still the only annual services engineering price book available.

New features for 2003 include:
- Wide range of building types for both elemental and all-in m2 rates.
- The electrical section is now in line with the CAWS that SMM7 follow.

HB & CD-ROM ♦ £115.00
0-415-30122-X ♦ 704pp

Spon's Civil Engineering & Highway Works Price Book 2003
Davis Langdon & Everest

More than just a price book, it is a comprehensive, work manual that all those in the civil engineering, surveying and construction business will find it hard to work without.

New features for 2003 include:
- A revision and expansion of both the Outputs and the Tables and Memoranda sections with more useful data.

HB & CD-ROM ♦ £120.00
0-415-30118-1 ♦ 704pp

Receive a **free cd-rom** when you order any Spon 2003 Pricebook to help you produce tender documents, customise data, perform keyword searches and simple calculations.

To Order: Tel: +44 (0) 8700 768853, or +44 (0) 1264 343071 Fax: +44 (0) 1264 343005, or Post: Spon Press Customer Services, Thomson Publishing Services, Cheriton House, Andover, Hants, SP10 5BE, UK Email: book.orders@tandf.co.uk

For a complete listing of all our titles visit www.sponpress.com

Preface to the Twenty Second Edition

The landscape and external works sector of the construction industry continues to enjoy a period of growth. At the time of going to press the industry suffers from a skills shortage in the management sector. Although the colleges are now producing a future crop of skilled managers, as a result of the degree in Landscape Management, only a trickle of these potential managers have emerged in their intended managerial posts. The weekly Horticultural magazine which serves the industry offers no fewer than 9 senior management positions in the week that this edition is submitted for publication. The management skills shortage is further exacerbated by the desire of many capable managers viewing with optimism the possibilities of beginning their own companies operating in the domestic and small commercial sector. The possibilities of higher earnings for even small sole traders have placed much pressure on labour prices and availability in the industry.

Consultants and contractors are both finding profitable work in the domestic sector as well. This is in part due to the property boom and the availability of extra cash amongst householders who are spending increased amounts on their gardens. Specialist London based contractors report profit margins of 20% – 40% in this sector with lead in times of up to six months. Domestic customers indicate that they are prepared to wait for their favoured contractor to become available.

It is unfortunate that the opportunities for additional work and turnover are not proportionately translated to profit for the industry in general. This is directly due again to the ability of companies to carry out their secured works with their available labour and management.

The current rate of labour used in this book (£15.00 per man hour) reflects an 11% increase over last years' rate which in turn showed an increase of 8.7% over the previous year. It must be stressed that the rates used are base upon a hypothetical site in the South East and readers should apply the weightings as shown in the table on page 21.

The other area of impact in the current year is the aggregate tax. Full notes are provided in section 6. Our observations are that most major suppliers of aggregate based materials such as concrete paving or building materials have increased their prices by 4% - 7%. Our prices for ready mixed concrete show an increase in the supply price of concrete of between 10% - 15%.

x   Preface to the Twenty Second Edition

Prices for landscape plants have remained the same for the third year running. The nursery which supplies mature tree prices to this publication has decreased the prices of supply of British grown stock reporting a general trend for this in the market place.

Some directly compared examples between the current and the previous years book are as follows: The variation is shown in brackets
Worked example on page 4; Pre-cast flag £15.19 / m2 (21%); Site mixed concrete 1:3:6 to foundations £116.00 / m3 (13%) Class B Engineering bricks laid 1 brick thick £78.67 / m2 (6%)
Planting labours only of a 14 -16 cm bare root tree £15.00 (11%).

This book has a new section in the beginning of the measured works. It now shows as a separate section the items which are included for the first time. These items also appear in their relevant section.

We would like to thank the companies who supply us our basic cost data. We approach them each year as we consider them to be the leading suppliers of services or materials in their sector. These companies work closely with us to provide accurate and correct information and we encourage our readers to support them at the time of specification or purchase of materials or services. A directory of all contributors is listed at the front of this publication.

DAVIS LANGDON & EVEREST

Princes House

39 Kingsway,

London

WC2B 6TP

SAM HASSALL

LANDSCAPE PROJECTS

Landscape Surveyors

6 Sandrock Hill Road

Farnham

Surrey

GU10 4NS

Tel: 020 7497 9000

e-mail: data.enquiry@davis langdon-uk.com

Tel: 01252 725513

e–mail: info@landpro.co.uk

List of Manufacturers and Suppliers

This list has been compiled from the latest information available but as firms frequently change their names, addresses and telephone numbers due to re-organization, users are advised to check this information before placing orders.

A Plant Acrow
Colnbrook By-Pass
Colnbrook
Slough
Berkshire
Plant Hire
Tel: 01753 680420
Fax: 01753 681338

A Plant Hire Co Ltd
Coothamlea Workshop
Cootham
Storrington
West Sussex RH20 4JN
Plant Hire
Tel: 01903 742348
Fax: 01903 742351

Abacus Municipal Ltd
Oddicroft Lane
Sutton-in-Ashfield
Notts. NG17 5FT
Lighting & Street Furniture
Tel: 01623 511111
Fax: 01623 552133

Ace Minimix
British Rail Goods Yard
York Way
Kings Cross
London N1 0AU
Ready Mixed Concrete
Tel: 0345 697690
Fax: 020 7278 5333

Addagrip Surface Treatments UK Ltd
Addagrip House
Bell Lane Industrial Estate
Uckfield
East Sussex TN22 1QL
Epoxy bound surfaces
Tel: 01825 761333
Fax: 01825 768566

AKZO Coatings Plc
Crown House
Hollin Lane
Darwin
Lancashire BB3 0BG
Sandtex matt paint
Tel: 01254 704951
Fax: 01254 774414

Alpha Rail Ltd
Alpha House
Urban Road
Kirkby-in-Ashfield
Nottingham NG17 8AP
Steel railings and fencing
Tel: 01623 750214
Fax: 01623 756596

Alumasc Exterior Building Products Ltd
White House Works
Bold Road
Sutton, St.Helens
Merseyside WA9 4JG
Green Roof Systems
Tel: 01744 648400
Fax: 01744 648401

List of Manufacturers and Suppliers

Amberol Ltd
The Plantation
Spencer Road
Belper
Derbyshire DE56 1JW
Street furniture
Tel: 01773 830930
Fax: 01773 834191

Amenity and Horticultural Services Ltd
Coppards Lane
Northiam
East Sussex TN31 6QP
Horticultural composts & fertilizers
Tel: 01797 252728
Fax: 01797 252724

Anderton Concrete Products Ltd
Anderton Wharf
Soot Hill
Anderton
Northwich
Cheshire CW9 6AA
Concrete fencing
Tel: 01606 79436
Fax: 01606 871590

Anglo Aquarium Plant Co Ltd
Strayfield Road
Enfield
Middlesex EN2 9JE
Aquatic plants
Tel: 0800 376 8001
Fax: 020 8363 8547
Website   www.anglo-aquarium.co.uk
e-mail: office@anglo-aquarium.co.uk

Arrow Chemicals Ltd
Stanhope Road
Swadlincote
Derbyshire DE11 9BE
Specialist coatings
Tel: 01283 221044
Fax: 01283 225731

Atlas Stone Products
Westington Quarry
Chipping Campden
Gloucestershire GL55 6EG
Paving manufacturer
Tel: 01386 840226
Fax: 01386 841356
Website: www.atlasstone.co.uk
e-mail: sales@atlasstone.co.uk

Autopa Ltd
Cottage Leap off Butlers Leap
Rugby
Warwickshire CV21 3XP
Street furniture
Tel: 01788 550556
Fax: 01788 550265
Website: www.autopa.co.uk
e-mail: info@autopa.co.uk

AVS Fencing Supplies Ltd
Unit 5F, Six Acre Works
Old Portsmouth Road
Peasmarsh
Guildford, Surrey GU3 1NE
Fencing
Tel: 01483 575212
Fax: 01483 306194
Website: www.avsfencing.co.uk
e-mail: sales@avsfencing.co.uk

Baylis Landscape Contractors
Hartshill Nursery
Thong Lane
Gravesend
Kent DA12 4AD
Sportsfield Constructon
Tel: 01474 569576
Fax: 01474 321587

BDC Concrete Products Ltd
Corporation Road
Newport
Gwent NP9 0WT
Precast concrete paving
Tel: 01633 244181
Fax: 01633 216655

List of Manufacturers and Suppliers

Belwood Trees Ltd
Brigton of Ruthven
Meigle
Perthshire PH12 8RQ
Mature tree grower
Tel: 01828 640219
Fax: 01828 640623
Website: www.belwoodtrees.co.uk
e-mail: belwood@belwoodtrees.co.uk

Blanc De Bierges
Eastrea Road
Whittlesey
Peterborough
Cambridgeshire PE7 2AG
Concrete paving supplier
Tel: 01733 202566
Fax: 01733 205405
Website: www.blancedebierges.com
e-mail: mail@blancedebierges.com

Board Walk
366 Upper Richmond Road West
East Sheen
London SW14 7JU
Timber Decking
Tel: 020 8392 2662
Fax: 020 8876 2420

Boughton Loam Ltd
Telford Way
Telford Way Industrial Estate
Kettering, Northants. NN16 8UN
Loams and Topsoils
Tel: 01536 510515
Fax: 01536 510691
Website: www.boughton-loam.co.uk
e-mail: enquiries@boughton-loam.co.uk

Breedon Plc
Breedon-on-the-Hill
nr. Melbourne
Derby DE73 1AP
Specialist gravels
Tel: 01332 862254
Fax: 01332 863149
Website: www.breedon.co.uk
e-mail: breedon@breedon.co.uk

British Seed Houses Ltd
Camp Road
Swinderby
Lincoln
LN6 9QJ
Grass and wildflower seed
Tel: 01522 868714
Fax: 01522 868095
Website: www.bshlincoln.co.uk
e-mail: simontaylor@bshlincoln.co.uk

Broxap Mawrob
121a - 125a Sefton Street
Southport
Merseyside
PR8 5DR

Tel: 01704 501011
Fax: 01704 541403
Website: www.broxap.com
e-mail: sales@mawrob.fsnet.co.uk

Broxap Streetscene
Rowhurst Industrial Estate
Chesterton
Newcastle-under-Lyme
Staffordshire ST5 6BD
Street furniture
Tel: 01782 564411
Fax: 01782 565357
Website: www.broxap.com

Burnham & Co (Onyx) Ltd
Burnham Way
Lower Sydenham
London SE26 5AG
Street furniture
Tel: 020 8659 1525
Fax: 020 8659 4707

Capital Garden Products Ltd
Gibbs Reed Barn
Pashley Road
Ticehurst
East Sussex TN5 7HE
Plant Containers
Tel: 01580 201092
Fax: 01580 201093
e-mail: sales@capital-garden.com

List of Manufacturers and Suppliers

CED Ltd
728 London Road
West Thurrock
Grays,
Essex RM20 3LU
Natural stone
Tel: 01708 867237
Fax: 01708 867230
Website: www.ced.ltd.uk
e-mail: sales@ced.ltd.uk

Champion Timber
205-209 Burlington Road
New Malden
Surrey KT3 4NB
Fencing
Tel: 020 8949 1621
Fax: 020 8942 9752
Website: www.championtimber.com

Charcon Hard Landscaping
Hulland Ward
Ashbourne
Derbyshire DE6 3ET
Paving and street furniture
Tel: 01335 372222
Fax: 01335 370074
Website: www.charcon.com
e-mail: charcon.sales@aggregate.com

Cirencester Civil Engineering
9 Rendcomb Drive
Cirencester
GL7 1YN
Macadam contractor
Tel: 01285 652020
Fax: 01285 651007

Clifton Nurseries Ltd
5a Clifton Villas
Little Venice
London W9 2PH
Nurserymen
Tel: 020 7286 6622
Fax: 020 7286 5655

Coblands Nurseries Ltd
Trench Road
Tonbridge
Kent TN10 3HQ
Nursery stock supplier
Tel: 01732 770999
Fax: 01732 770271

Colas Ltd
Rowfant
Crawley
West Sussex RH10 4NF
Road repair
Tel: 01342 711000
Fax: 01342 711199

Colourpave Ltd
Laymore
Forest Vale Industrial Estate
Cinderford
Gloucestershire GL14 2PH
Macadam Contractor
Tel: 01594 826768
Fax: 01594826598

Columbia Cascade Ltd
Business Development Centre
Amanford Road
Tychroes
Amanford
Carmarthenshire SA18 3QN
Street furniture
Tel: 01269 596396
Fax: 01269 596395

Cooper Clarke Group
Special Products Division
Bloomfield Road
Farnworth
Bolton BL4 9LP
Wall drainage and erosion control
Tel: 01204 862222
Fax: 01204 793856
Website: www.cooperclarke.com
e-mail: ccgmarketing@btinternet.com

List of Manufacturers and Suppliers

Coverite Ltd
Palace Gates
Bridge Road
Wood Green
London N22 4SP
Waterproofing specialist
Tel: 020 8888 7821
Fax: 020 8889 0731

Crowders Nurseries
Lincoln Road
Horncastle
Lincolnshire LN9 5LZ
Plant protection
Tel: 01507 525000
Fax: 01507 524000

CU Phosco Ltd
Lower Road
Great Amwell
Ware
Hertfordshire SG12 9TA
Lighting
Tel: 01920 462272
Fax: 01920 461370
Website: www.cuphosco.co.uk
e-mail: sales@cuphosco.co.uk

D W Windsor Ltd
Marsh Lane
Ware
Hertfordshire SG12 9QL
Lighting
Tel: 01992 474600
Fax: 01992 474601

Dee-Organ Ltd
5 Sandyford Road
Paisley
Renfrewshire PA3 4HP
Street furniture
Tel: 0141 889 7000
Fax: 0141 889 7764
Website: www.dee-organ.co.uk
E-mail: signs@dee-organ.co.uk

DLS Perryfields Ltd
Thorn Farm
Inkberrow
Worcestershire WR7 4LJ
Grass and wild flower seed
Tel: 01386 793135
Fax: 01386 792715
Website: www.perryfields.co.uk
e-mail: amenity@perryfields.co.uk

Duracourt (Spade Oak) Ltd
Town Lane
Wooburn Green
High Wycombe
Buckinghamshire HP10 0PD
Tennis courts
Tel: 01628 528421
Fax: 01628 810509
Website: www.spadeoak.co.uk
e-mail: duracourt@spadeoak.co.uk

E.T. Clay Products
7 Fowler Road
Hainault
Essex IT6 3UT
Brick supplier
Tel: 020 8501 2100
Fax: 020 8500 9990
e-mail: eddie@etclay.freeserve.co.uk

Earth Anchors Ltd
15 Campbell Road
Croydon
Surrey CR0 2SQ
Anchors for site furniture
Tel: 020 8684 9601
Fax: 020 8684 2230
Website: www.earth-anchors.com

Edwards Sports Products Ltd
North Mills
Bridport
Dorset DT6 3AH
Sports equipment
Tel: 01308 424111
Fax: 01308 455800

List of Manufacturers and Suppliers

Elliot Workspace
The Fen
Baston
Peterborough
Cambridgeshire PE6 9PT
Site Offices
Tel: 01778 560891
Fax: 01778 560881

English Woodlands
Burrow Nursery
Cross in Hand
Heathfield
East Sussex TN21 0UG
Plant protection
Tel: 01435 862992
Fax: 01435 867742
Website: www.ewburrownursery.co.uk
e-mail: sales@ewburrownursery.co.uk

Ensor Rylance Ltd
Blackamoor Road
Guide
Blackburn
Lancashire BB1 2LQ
Drainage suppliers
Tel: 01254 52244
Fax: 01254 682371

Erisco-Bauder Ltd
Broughton House
Broughton Road
Ipswich
Suffolk IP1 3QR
Green Roof Systems
Tel: 01473 257671
Fax: 01473 230761

Eve Trakway
Coxmoor Road
Sutton-in-Ashfield
Nottinghamshire NG17 5LA
Portable roads
Tel: 01623 515333
Fax: 01623 440154

Exclusive Leisure Ltd
28 Cannock Street
Leicester LE4 7HR
Artificial sports surfaces
Tel: 0116 233 2255
Fax: 0116 246 1561
Website: www.exclusiveleisure.co.uk
e-mail: info@exclusiveleisure.co.uk

Exxon Chemical Geopolymers Ltd
Mamhilad Park Industrial Estate
Pontypool
Gwent NP4 0YR
Geofabrics
Tel: 01495 757722
Fax: 01495 762383

Fairwater Water Gardens
Lodge Farm
Malthouse Lane
Ashington
West Sussex RH20 3BU
Water Feature Contractors
Tel: 01903 892522
Fax: 01903 892228

Farmura Ltd
Stone Hill
Egerton
Ashford,
Kent, TN27 9DU
Organic fertilizer suppliers
Tel: 01233 756241
Fax: 01233 756419
Website: www.farmura.com
www.alginure.co.uk
e-mail: info@farmura.com

Fleet (Line Markers) Limited
Fleet House
Spring Lane Industrial Estate
Malvern Link
Worcestershire WR14 1AT
Sports line marking
Tel: 01684 573535
Fax: 01684 892784

List of Manufacturers and Suppliers

Forticrete Ltd
Bridal Way
Bootle
Merseyside L30 4UA
Retaining wall systems
Tel: 0151 521 3545
Fax: 0151 521 5696

Furnitubes International
Seager Buildings
Brookmill Road
London SE8 4HL
Street furniture
Tel: 020 8694 9333
Fax: 020 8694 8315
Website: www.furnitubes.com
e-mail: sales@furnitubes.com

Geometric Furniture Ltd
Birch Mill
Heywood Old Road
Heywood
Lancashire OL10 2QQ
Street furniture
Tel: 0161 653 2233
Fax: 0161 653 2299

Grace Construction Products
Ajax Avenue
Slough
Berkshire SL1 4BH
Bitu-thene tanking
Tel: 01753 692929
Fax: 01753 691623

Grass Concrete Ltd
Walker House
22 Bond Street
Wakefield
West Yorkshire WF1 2QP
Grass block paving
Tel: 01924 379443
Fax: 01924 290289
Website: www.grasscrete.com
e-mail: info@grasscrete.com

Greenfix Ltd
Laverham House
77 St Georges Place
Cheltenham
Glos. GL50 3PP
Seeded erosion control mats
Tel: 01242 700092
Fax: 01242 700093
Websit: www.greenfix.co.uk
e-mail: cheltenham@greenfix.co.uk

Greenkeeper Ltd
Shelford Manor
Shelford
Nottingham NG12 1ER
Specialist turfs
Tel: 01949 21144
Fax: 01949 21144
Website: www.greenkeeperltd.co.uk

H.E. Services Limited
Whitewall Road
Strood
Kent ME2 4DZ
Plant hire
Tel: 01634 291491
Fax: 01634 295626

H.S. Jackson & Son (Fencing) Ltd
Stowting Common
Ashford
Kent TN25 6BN
Fencing
Tel: 01233 750393
Fax: 01233 750403
Website: www.jacksons-fencing.co.uk
e-mail: sales@jacksonsfencing.co.uk

H.S.S. Hire Shops
Group Office
25 Willow Lane
Mitcham
Surrey CR4 4TS
Tool and plant hire
Tel: 020 8260 3100
Fax: 020 8687 5005

List of Manufacturers and Suppliers

Haddonstone Ltd
The Forge House
East Haddon
Northampton NN6 8DB
Architectural stonework
Tel: 01604 770711
Fax: 01604 770027
Website: www.haddonstone.co.uk
e-mail: info@haddonstone.co.uk

Hanson Brick
Unicorn House
Wellington Street
Ripley
Derbyshire DE5 3DZ
Brick manufacturer
Tel: 08705 258258
Fax: 01773 514041
Website: www.hanson-brickseurope.com
e-mail: info@hansonbrick.com

Harrison Flagpoles
Borough Road
Darlington
Co Durham DL1 1SW
Flagpoles
Tel: 01325 355433
Fax: 01325 461726

Hepworth Plc
Hazlehead
Stocksbridge
Sheffield S30 5HG
Drainage
Tel: 01226 763561
Fax: 01226 764827

Hill & Smith Ltd
Springvale Business & Industrial Park
Bilston
Wolverhampton WV14 0QZ
Safety barriers
Tel: 01902 499400
Fax: 01902 499419

Hills Industries Ltd
Pontygwindy Industrial Estate
Caerphilly
Mid Glamorgan CF8 3HU
Industrial washing lines
Tel: 029 2088 3951
Fax: 029 2088 6102
Website: www.hills-industries.co.uk
e-mail: info@hills-industries.co.uk

Hodkin & Jones (Sheffield) Ltd
Callywhite Lane
Dronfield
Sheffield S18 6XP
Drainage
Tel: 01246 290890
Fax: 01246 290292

Inturf
The Chestnuts
Wilberfoss
York YO4 5NT
Turf
Tel: 01759 321000
Fax: 01759 380130
Website: www.inturf.co.uk
e-mail: info@inturf.co.uk

J Toms Ltd
Grigg Lane
Headcorn
Ashford
Kent TN27 9XT
Plant Protection
Tel: 01622 891111
Fax: 016922 890783

John Anderson Mobile Toilets
Smallford Lane
St. Albans
Herts. AL4 0LL
Mobile toilet facilities
Tel: 07000 822485
Fax: 01727 822886

List of Manufacturers and Suppliers

Johnsons Wellfield Quarries Ltd
Crosland Hill
Huddersfield
West Yorkshire HD4 7AB
Natural Yorkstone pavings
Tel: 01484 652311
Fax: 01484 460007
Website: www.johnsons-wellfield.co.uk
e-mail: sales@johnsons-wellfield.co.uk

Jonathan James Ltd
15-17 New Road
Rainham
Essex RM13 8DJ
Specialist screeds and tanking
Tel: 01708 556921
Fax: 01708 520751

Jones of Oswestry
Whittington Road
Oswestry
Shropshire SY11 1HZ
Channels, gulleys, manhole covers
Tel: 01691 653251
Fax: 01691 658222

Keith Banyard Landscape Services
Nettletree Farm
Horton Heath
Wimborne
Dorset BH21 7JN
Grounds maintenance contractors
Tel: 01202 828800
Fax: 01202 820128

Keller Comtec
Barham Business Court
Teston, Maidstone
Kent ME18 5BZ
Hydro seeding specialist
Tel: 01622 618799
Fax: 01622 618790

Kompan Ltd
Unit 20, Denbigh Hall
Bletchley
Milton Keynes
Buckinghamshire MK3 7QT
Play equipment
Tel: 01908 642466
Fax: 01908 270137
Website: www.kompan.com

Land and Water Services
3 Weston Farm Yard
The Street
Albury
nr. Guildford
Surrey
Specialist plant hire
Tel: 01483 202733
Fax: 01483 202510

Landline Ltd
1 Bluebridge Industrial Estate
Halstead
Essex CO9 2EX
Pond and lake installation
Tel: 01787 476699
Fax: 01787 472507

Landscapes By Design
The Studio
17 Church Road
Farnborough
Kent BR6 7DB
Trellis
Tel: 01689 851570
Fax: 01689 861151

Lappset UK Ltd
Lappset House
Henson Way
Telford Way Industrial Estate
Kettering
Northamptonshire NN16 8PX
Play equipment
Tel: 01536 412612
Fax: 01536 521703

List of Manufacturers and Suppliers

LDC Limited
Loampits Farm
99 Westfield Road
Woking
Surrey GU22 9QR
Willow walling
Tel: 01483 767488
Fax: 01483 764293

Leaky Pipe Systems Ltd
Frith Farm
Dean Street
East Farleigh
Maidstone, Kent ME15 0PR
Irrigation systems
Tel: 01622 746495
Fax: 01622 745118

Lister Lutyens Co Ltd
6 Alder Close
Eastbourne
East Sussex BN23 6QF
Street furniture
Tel: 01323 431177
Fax: 01323 639314
Website: www.listerteak.com
e-mail: sales@listerteak.com

Louis Poulsen UK Ltd
Surrey Business Park
Weston Road
Epsom
Surrey KT17 1JG
Outdoor lighting
Tel: 01372 848800
Fax: 01372 848801

Maccaferri Ltd
7400 The Quorum
Oxford Business Park North
Garsington Road
Oxford   OX4 2JZ
Gabions
Tel: 01865 770555
Fax: 01865 774550

Malcolm, Ogilvie & Co
Constable Works
31 Constitution Street
Dundee DD3 6NL
Scotland
Geofabrics
Tel: 01382 322974
Fax: 01382 202123

Marlin Lighting
Hanworth Trading Estate
Hampton Road West
Feltham TW13 6DR
Lighting
Tel: 020 8894 5522
Fax: 020 8898 8480

Marshalls Mono Ltd
Southowram
Halifax
West Yorkshire HX3 9SY
Hard landscape materials and street furniture
Tel: 01422 306000
Fax: 01422 306185

Melcourt Industries Limited
Boldridge Brake
Long Newnton
Tetbury
Gloucestershire GL8 8RT
Mulch and compost
Tel: 01666 502711
Fax: 01666 504398
Website: www.melcourt.co.uk
e-mail: mail@melcourt.co.uk

Milton Pipes Ltd
Milton Regis
Sittingbourne
Kent ME10 2QF
Soakaway Rings
Tel: 01795 425191
Fax: 01795 420360

List of Manufacturers and Suppliers

Monarflex Geomembranes Ltd
Lyon Way
St Albans
Hertfordshire AL4 0LB
Lake liners
Tel: 01727 830116
Fax: 01727 858998

Neptune Outdoor Furniture Ltd
Thompsons Lane
Marwell
Winchester
Hampshire SO21 1JH
Street furniture
Tel: 01962 777799
Fax: 01962 777723
Website: www.nofl.co.uk
E-mail: sales@nofl.co.uk

Netlon Professional
New Wellington Street
Blackburn
Lancashire BB2 4PJ
Erosion control, soil stabilisation, plant protection
Tel: 01254 262431
Fax: 01254 680008

Noral Ltd
26 Vincent Avenue
Crownhill
Milton Keynes MK8 0AB
Lighting
Tel: 01908 561818
Fax: 01908 569785
Website: www.noral.net

Norbury Ltd
28 Marshgate Drive
Hertford
Herts. SG13 7AJ
Hardwood gates
Tel: 01992 554327
Fax: 01992 505978

Norris and Gardener
Reay House
Elm Hill, Normandy
Guildford
Surrey GU3 2HU
Grounds maintenance contractors
Tel: 01252 323734
Fax: 01252 331818

Notcutts Nurseries Ltd
Woodbridge
Suffolk IP12 4AF
Nursery stock supplier
Tel: 01394 445440
Fax: 01394 383344

Orchard Street Furniture Ltd
Castle House
Castle Square
Benson
Oxfordshire OX10 6SD
Street furniture
Tel: 01491 642123
Fax: 01491 642126

Orsogril UK Ltd
c/o McArthur Group Ltd
Geddings Road
Hoddesdon
Hertfordshire EN11 0NZ
Fencing
Tel: 01992 470407
Fax: 01992 470407

Outdoor Lighting
Surrey Business Park
Weston Road
Epsom
Surrey KT17 1JG
Outdoor lighting
Tel: 01372 848800
Fax: 01372 848801

PHI Group
13 Royal Crescent
Cheltenham
Gloucestershire GL50 3DA
Soil Stabilisation, Erosion Control
Tel: 0870 333 4126
Fax: 0870 333 4127
Website: www.phigroup.co.uk
e-mail: info@phigroup.co.uk

Platipus Anchors Ltd
Kingsfield Business Centre
Philanthropic Road
Redhill
Surrey RH1 4DP
Tree anchors
Tel: 01737 762300
Fax: 01737 773395
Website: www.platipus-anchors.com
e-mail: info@platipus-anchors.com

Polybau Ltd
PO Box 58
Trafford Park Road
Newbridge, Trafford Park
Manchester M17 1JD
Drainage channels
Tel: 0161 872 1472
Fax: 0161 877 6592

Polypipe Civils Ltd
Union Works
Bishop Meadow Road
Loughborough
Leicestershire LE11 5RE
Drainage materials
Tel: 01509 615100
Fax: 01509 610215

Rawell Water Control Systems Ltd
Carr Lane
Hoylake, Wirral
Merseyside CH47 4FE
Lake liners
Tel: 0151 632 5771
Fax: 0151 632 4363

RCC
Barholm Road
Tallington
Stamford
Lincolnshire PE9 4RL
Precast concrete retaining units
Tel: 01778 344460
Fax: 01778 345949
Website: www.tarmacprecast.co.uk
e-mail: rcc@tarmac.co.uk

Recycled Materials Ltd
PO Box 519
Surbiton
Surrey KT6 4YL
Disposal Contractors
Tel: 020 8390 7010
Fax: 020 8390 7020
e-mail: rml@demon.co.uk

Rigby Taylor Ltd
The Riverway Estate
Portsmouth Road
Peasmarsh, Guildford
Surrey GU3 1LZ
Horticultural supply
Tel: 0800 424919
Fax: 01483 534058
Website: www.rigbytaylor.com
e-mail: sales@rigbytaylor.com

RIW Ltd
Arc House
Terrace Road South
Binfield, Bracknell
Berkshire RG42 4PZ
Waterproofing products
Tel: 01344 861988
Fax: 01344 862010
Website: www.riw.co.uk
e-mail: enquiries@riw.co.uk

Rolawn (Turf Growers) Ltd
Elvington
York YO41 4XR
Industrial turf
Tel: 01904 608661
Fax: 01904 608272
Website: www.rolawn.co.uk
e-mail: ian.elwick@rolawn.co.uk

List of Manufacturers and Suppliers

Rom Ltd
Wheaton Road
Witham
Essex CM8 3BU
Reinforcement steel bars
Tel: 01376 514321
Fax: 01376 518628

Ruberoid Building Products Ltd
Tewin Road
Welwyn Garden City
Hertfordshire AL7 1BP
Synthaprufe waterproofing
Tel: 01707 822222
Fax: 01707 333763

Russell Leisure Ltd
Roddinglaw
Gogar
Edinburgh EH12 9DW
Scotland
Play equipment
Tel: 0131 335 5400
Fax: 0131 335 5401
Website: www.russell-leisure.co.uk
e-mail: sales@russell-leisure.co.uk

Sarena Plastics Ltd
Beechings Way
Gillingham
Kent ME8 6PT
Street furniture
Tel: 01634 370887
Fax: 01634 370915
Website: www.sarena.co.uk
e-mail: info@sarena.freeserve.co.uk

Scotts UK Professional Ltd
Paper Mill Lane
Bramford
Ipswich
Suffolk IP8 4BZ
Fertilizers and chemicals
Tel: 01473 830492
Fax: 01473 830386

Siddall and Hilton Mesh Ltd
Birds Royd Lane
Brighouse
West Yorkshire HD6 1LT
Fencing
Tel: 01484 401610
Fax: 01484 721028
Website: www.shmesh.com
e-mail: sales@shmesh.com

SMP Playgrounds Ltd
Ten Acre Lane
Thorpe
Egham
Surrey KT16 8EJ
Playground equipment
Tel: 01784 489100
Fax: 01784 431079
Website: www.smp.co.uk
e-mail: sales@smp.co.uk

Spade Oak Construction Co Ltd
Town Lane
Wooburn Green
High Wycombe
Bucks. HP10 0PD
Macadam Contractors
Tel: 01628 529421
Fax: 01628 810509
Website: www.spadeoak.co.uk
e-mail: email@spadeoak.co.uk

Sportsmark Group Ltd
Ealing Road
Brentford
Middlesex TW8 0LH
Sports surfaces
Tel: 020 8560 2010
Fax: 020 8568 2177

St Gobain Pipelines
Cray Avenue
St Mary Cray
Orpington
Kent BR5 3RH
Drainage Systems
Tel: 01689 891900
Fax: 01689 822372

List of Manufacturers and Suppliers

Steelway-Fensecure
Parkside Industrial Estate
Hickman Avenue
Wolverhampton
West Midlands WV1 2EN
Fencing
Tel: 01902 490919
Fax: 01902 490929

Sugg Lighting Ltd
Sussex Manor Business Park
Gatwick Road
Crawley
West Sussex RH10 2GD
Lighting
Tel: 01293 540111
Fax: 01293 540114
Website: www.sugglighting.co.uk
E-mail: sales@sugglighting.co.uk

Tarmac Topmix
British Rail Goods Yard
York Way
Kings Cross
London N1 0AU
Ready mixed concrete
Tel: 020 7837 7011
Fax: 020 7278 5333

Tensar International (Netlon Group)
New Wellington Street
Blackburn
Lancashire BB2 4PJ
Erosion control, soil stabilisation
Tel: 01254 262431
Fax: 01254 266868
Website: www.tensar-international.com
E-mail: sales@tensar.co.uk

Terram Limited
Mamhilad Park Industrial Estate
Pontypool
Gwent NP4 0YR
Geofabrics
Tel: 01495 767402
Fax: 01495 767436
Website: www.terram.com
E-mail: info@terram.co.uk

Tinsley Wire - Sheffield
PO Box 119
Shepcote Lane
Sheffield S9 1TY
Security Fencing
Tel: 01142 561561
Fax: 01142 619351

Townscape Products Ltd
Fulwood Road South
Sutton-in-Ashfield
Nottinghamshire NG17 2JZ
Hard landscaping
Tel: 01623 513355
Fax: 01623 440267
Website: www.townscape-products.co.uk
E-mail: sales@townscape-products.co.u

Tubex Ltd
Aberaman Park
Aberdare
Mid Glamorgan CF44 6DA
Plant Protection, tree guards
Tel: 01685 88800
Fax: 01685 888001
Website: www.tubex.com
E-mail: plantcare@tubex.com

Turf Management Systems
Dromenach Farm
Seven Hills Road
Iver Heath
Bucks. SL0 0PA
Specialist turf systems
Tel: 01895 834411
Fax: 01895 834892

Urban Soils Limited
High Legh Estate Office,
Knutsford,
Cheshire
WA16 0QS
Structural Tree Soil
Tel: 01925 757800
Fax: 01925 757800

List of Manufacturers and Suppliers

Visqueen Agri
Yarm Road
Stockton-on-Tees
Cleveland TS18 3GE
Lake liners
Tel: 01642 677228
Fax: 01642 664325

Wavin Building Products
Parsonage Way
Chippenham
Wiltshire SN15 5PN
Drainage products
Tel: 01249 766600
Fax: 01249 443286

White Horse Contractors Ltd
Blakes Oak Farm
Lodge Hill
Abingdon
Oxfordshire OX14 2JD
Drainage
Tel: 01865 736272
Fax: 01865 326176
Website: www.whitehorsecontractors.co.uk
e-mail: whc@whitehorsecontractors.co.uk

Wicksteed Leisure Ltd
Digby Street
Kettering
Northamptonshire NN16 8YJ
Play equipment
Tel: 01536 517028
Fax: 01536 410633
Website: www.wicksteed.co.uk
e-mail: sales@wicksteed.co.uk

Woodhouse UK Plc
Spartan Close
Tachbrook Park
Warwick
Warwickshire CV34 6RR
Lighting
Tel: 01926 314313
Fax: 01926 883778

Woodscape Ltd
Upfield, Pike Lowe
Brinscall
Chorley
Lancashire PR6 8SP
Street furniture
Tel: 01254 830886
Fax: 01254 831846
Website: www.woodscape.co.uk
e-mail: sales@woodscape.co.uk

Wybone Ltd
Mason Way
Platts Common Industrial Estate
Hoyland, Barnsley
South Yorkshire S74 9TF
Street furniture
Tel: 01226 744010
Fax: 01226 350105
Website: www.wybone.co.uk
e-mail: sales@wybone.co.uk

Yeoman Aggregates Ltd
Stone Terminal
Horn Lane
Acton
London W3 9EH
Aggregates
Tel: 020 8896 6820
Fax: 020 8896 6829
e-mail: yeoman.sales@ukonline.co.uk

ESSENTIAL READING FROM SPON PRESS

Guidelines for Landscape and Visual Impact Assessment

Second Edition

Edited by The Landscape Institute and the Institute of Environmental Management and Assessment, Co-ordinated by Sue Wilson, Landscape consultant, UK

'For any self-respecting landscape consultant, *Guidelines for Landscape and Visual Impact Assessment* is an essential reference book. The book is well-constructed, concise, compact and well-made and illustrated.'
The Architects' Journal on the first edition

This new edition of these good practice guidelines provides advice on assessing the landscape and the visual impacts of development projects. Issues covered by the guidelines include:

- integration of landscape and visual issues into the development process
- the need for a transparent approach to landscape and visual impact assessment;
- describes the baseline conditions;
- determining the magnitude and significance of impacts;
- reviews the landscape and visual components of an EIS (Environmental Impact Statement).

The guidelines will enable developers and decision makers to be aware of the issues to be addressed by this type of assessment, as well as providing environmental consultants with a framework for good practice.

April 2002: 246x189: 176pp
75 colour illustrations and 6 line drawings
Hb: 0-415-23185-X: £35.00

To Order: Tel: +44 (0) 8700 768853, or +44 (0) 1264 343071 Fax: +44 (0) 1264 343005, or Post: Spon Press Customer Services, Thomson Publishing Services, Cheriton House, Andover, Hants, SP10 5BE, UK Email: book.orders@tandf.co.uk

For a complete listing of all our titles visit:
www.sponpress.com

Common Arrangement of Work Sections

The main work sections relevant to landscape work and their grouping:

A Preliminaries/General Conditions

- A10 Project particulars
- A11 Tender and contract documents
- A12 The site/existing buildings
- A13 Description of the work
- A20 The contract/sub-contract
- A30 Employers requirements: tendering/sub-letting/supply
- A31 Employers requirements: provision, content and use of documents
- A32 Employers requirements: management of the works
- A33 Employers requirements: quality standards/control
- A34 Employers requirements: security/safety/protection
- A35 Employers requirements: specific limitations on method/sequence/timing/ use of site
- A36 Employers requirements: facilities/temporaryworks/services
- A37 Employers requirements: operation/maintenance of the finished building
- A40 Contractors general cost items: management and staff
- A41 Contractors general cost items: site accommodation
- A42 Contractors general cost items: services and facilities
- A43 Contractors general cost items: mechanical plant
- A44 Contractors general cost items: temporary works
- A50 Works/products by/on behalf of the employer
- A51 Nominated sub-contractors
- A52 Nominated suppliers
- A53 Work by statutory authorities/undertakers
- A54 Provisional work
- A55 Dayworks
- A60 Preliminaries/general conditions for demolition contract
- A61 Preliminaries/general conditions for investigation/survey contract
- A62 Preliminaries/general conditions for piling/embedded retaining wall contract
- A63 Preliminaries/general conditions for landscape contract
- A70 General specification requirements for work package

B Complete buildings/structures/units

- B10 Prefabricated buildings/structures
- B11 Prefabricated building units

C Existing site/buildings/services

- C10 Site survey
- C11 Ground investigation
- C12 Underground services survey

D Groundwork

- D11 Soil stabilisation
- D20 Excavating and filling
- D41 Crib walls/gabions/reinforced earth

Common Arrangement of Work Sections

E   In situ concrete/large precast concrete

- E05 In situ concrete construction generally
- E10 Mixing/casting/curing in situ concrete
- E20 Formwork for in situ concrete
- E30 Reinforcement for in situ concrete

F   Masonry

- F10 Brick/block walling
- F20 Natural stone rubble walling
- F21 Natural stone ashlar walling/dressings
- F22 Cast stone ashlar walling/dressings
- F30 Accessories/sundry items for brick/block/stone walling
- F31 Precast concrete sills/lintels/copings/features

G   Structural/carcassing metal/timber

- G31 Prefabricated timber unit decking

H   Cladding/covering

- H51 Natural stone slab cladding/features
- H52 Cast stone slab cladding/features

J   Waterproofing

- J10 Specialist waterproof rendering
- J20 Mastic asphalt tanking/damp proofing
- J21 Mastic asphalt roofing/insulation/finishes
- J22 Proprietary roof decking with asphalt finish
- J30 Liquid applied tanking/damp proofing
- J31 Liquid applied waterproof roof coatings
- J40 Flexible sheet tanking/damp proofing
- J44 Sheet linings for pools/lakes/waterways

M   Surface finishes

- M10 Cement:sand/concrete screeds
- M20 Plastered/rendered/roughcast coatings
- M40 Stone/concrete/quarry/ceramic tiling/mosaic
- M60 Painting/clear finishing

P   Building fabric sundries

- P30 Trenches/pipeways/pits for buried engineering services

Q   Paving/planting/fencing/site furniture

- Q10 Kerbs/edgings/channels/paving accessories
- Q20 Granular sub-bases to roads/pavings
- Q21 In situ concrete roads/pavings
- Q22 Coated macadam/asphalt roads/pavings
- Q23 Gravel/hoggin/woodchip roads/pavings
- Q24 Interlocking brick/block roads/pavings
- Q25 Slab/brick/sett/cobble pavings
- Q26 Special surfacings/pavings for sport/general amenity
- Q30 Seeding/turfing
- Q31 Planting
- Q32 Planting in special environments
- Q35 Landscape maintenance
- Q40 Fencing
- Q50 Site/street furniture/equipment

R Disposal systems

R12 Drainage below ground
R13 Land drainage

S Piped supply systems

S10 Cold water
S14 Irrigation
S15 Fountains/water features

V Electrical supply/power/lighting systems

V41 Street/area floodlighting

ESSENTIAL READING FROM SPON PRESS

External Works Roads and Drainage
A Practical Guide
Phil Pitman

External Works Roads and Drainage: A Practical Guide bridges the gap between theory and practice in building, construction and civil engineering, providing practical guidance, and the knowledge required 'on the job'.

This comprehensive book includes sections on legislation, environmental issues, surface water, highway and foul drainage design, road and pavement design, and external works. The book is well illustrated, with flow charts to help guide you through the design process, bullet points and tables.

Written in an accessible style, this guide is an essential reference for engineers, architects, designers, technicians and students looking to increase their design and project planning knowledge.

'The text provides practical input which can be of benefit to both student and professional alike.' **Building Engineer**

2001: 234x156: 328pp:
7 line drawings, 30 tables
Pb: 0419257608: £29.99

To Order: Tel: +44 (0) 8700 768853, or +44 (0) 1264 343071 Fax: +44 (0) 1264 343005, or Post: Spon Press Customer Services, Thomson Publishing Services, Cheriton House, Andover, Hants, SP10 5BE, UK Email: book.orders@tandf.co.uk

For a complete listing of all our titles visit:
www.sponpress.com

DAVIS LANGDON & EVEREST

LONDON
Princes House
39 Kingsway
London WC2B 6TP
Tel: (020) 7497 9000
Fax: (020) 7497 8858
Email: rob.smith@davislangdon-uk.com

BIRMINGHAM
29 Woodbourne Road
Harborne
Birmingham
B17 8BY
Tel: (0121) 4299511
Fax: (0121) 4292544
Email: richard.d.taylor@davislangdon-uk.com

BRISTOL
St Lawrence House
29/31 Broad Street
Bristol
BS1 2HF
Tel: (0117) 9277832
Fax: (0117) 9251350
Email: alan.trolley@davislangdon-uk.com

CAMBRIDGE
36 Storey's Way
Cambridge
CB3 0DT
Tel: (01223) 351258
Fax: (01223) 321002
Email: stephen.bugg@davislangdon-uk.com

CARDIFF
4 Pierhead Street
Capital Waterside
Cardiff CF10 4QP
Tel: (029) 20497497
Fax: (029) 20497111
Email: paul.edwards@davislangdon-uk.com

EDINBURGH
39 Melville Street
Edinburgh
EH3 7JF
Tel: (0131) 240 1350
Fax: (0131) 240 1399
Email: ian.mcandie@davislangdon-uk.com

GLASGOW
Cumbrae House
15 Carlton Court
Glasgow G5 9JP
Tel: (0141) 4296677
Fax: (0141) 4292255
Email: hugh.fisher@davislangdon-uk.com

LEEDS
No 4 The Embankment
Victoria Wharf
Sovereign Street
Leeds LS1 4BA
Tel: (0113) 2432481
Fax: (0113) 2424601
Email: tony.brennan@davislangdon-uk.com

LIVERPOOL
Cunard Building
Water Street
Liverpool L3 1JR
Tel: (0151) 2361992
Fax: (0151) 2275401
Email: john.davenport@davislangdon-uk.com

MANCHESTER
Cloister House
Riverside
New Bailey Street
Manchester M3 5AG
Tel: (0161) 8197600
Fax: (0161) 8191818
Email: paul.stanion@davislangdon-uk.com

MILTON KEYNES
Everest House
Rockingham Drive
Linford Wood
Milton Keynes MK14 6LY
Tel: (01908) 304700
Fax: (01908) 660059
Email: kevin.sims@davislangdon-uk.com

NEWCASTLE
2nd Floor
The Old Post Office Building
St Nicholas Street
Newcastle upon Tyne NE1 1RH
Tel: (0191) 221 2012
Fax: (0191) 261 4274
Email: garey.lockey@davislangdon-uk.com

NORWICH
63 Thorpe Road
Norwich
NR1 1UD
Tel: (01603) 628194
Fax: (01603) 615928
Email: michael.ladbrook@davislangdon-uk.com

NOTTINGHAM
Hamilton House
Clinton Avenue
Nottingham NG5 1AW
Tel: (0115) 962 2511
Fax: (0115) 969 1079
Email: rod.exton@davislangdon-uk.com

OXFORD
Avalon House
Marcham Road
Abingdon
Oxford OX14 1TZ
Tel: (01235) 555025
Fax: (01235) 554909
Email: paul.coomber@davislangdon-uk.com

PETERBOROUGH
Clarence House
Minerva Business Park
Lynchwood
Peterborough PE2 6FT
Tel: (01733) 362000
Fax: (01733) 230875
Email: stuart.bremner@davislangdon-uk.com

PLYMOUTH
3 Russell Court
St Andrew Street
Plymouth PL1 2AX
Tel: (01752) 668372
Fax: (01752) 221219
Email: gareth.steventon@davislangdon-uk.com

PORTSMOUTH
St Andrews Court
St Michaels Road
Portsmouth PO1 2PR
Tel: (023) 92815218
Fax: (023) 92827156
Email: chris.tremellen@davislangdon-uk.com

SOUTHAMPTON
Brunswick House
Brunswick Place
Southampton SO15 2AP
Tel: (023) 80333438
Fax: (023) 80226099
Email: richard.pitman@davislangdon-uk.com

DAVIS LANGDON CONSULTANCY
Princes House
39 Kingsway
London WC2B 6TP
Tel: (020) 7379 3322
Fax: (020) 7379 3030
Email: jim.meikle@davislangdon-uk.com

MOTT GREEN & WALL
Africa House
64-78 Kingsway
London WC2B 6NN
Tel: (020) 7836 0836
Fax: (020) 7242 0394
Email: barry.nugent@mottgreenwall.co.uk

SCHUMANN SMITH
14th Flr, Southgate House
St Georges Way
Stevenage
Hertfordshire SG1 1HG
Tel: (01438) 742642
Fax: (01438) 742632
Email: nschumann@schumannsmith.com

with offices throughout Europe, the Middle East, Asia, Australia, Africa and the USA which together form

DAVIS LANGDON & SEAH INTERNATIONAL

DAVIS LANGDON & EVEREST
Authors of Spon's Price Books

Davis Langdon & Everest is an independent practice of Chartered Quantity Surveyors, with some 1092 staff in 20 UK offices and, through Davis Langdon & Seah International, some 2,300 staff in 82 offices worldwide.

DLE manages client requirements, controls risk, manages cost and maximises value for money, throughout the course of construction projects, always aiming to be - and to deliver - the best.

TYPICAL PROJECT STAGES, DLE INTEGRATED SERVICES AND THEIR EFFECT:

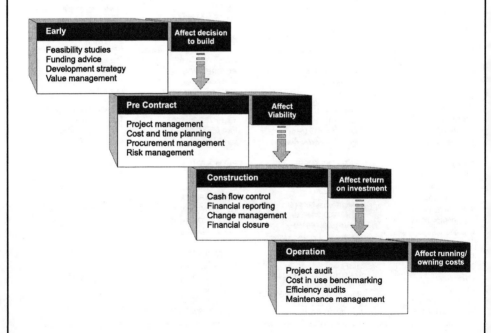

EUROPE ⇨ ASIA - AUSTRALIA - AFRICA - AMERICA
GLOBAL REACH - LOCAL DELIVERY

1
How to use this Book

1.1 INTRODUCTION

First-time users of *Spon's Landscape and External Works Price Book* and others who may not be familiar with the way in which prices are compiled may find it helpful to read this section before starting to calculate the costs of landscape works.
The cost of an item of landscape construction or planting is made up of many components:
- the cost of the product;
- the labour and additional materials needed to carry out the job;
- the cost of running the contractor's business.

These are described more fully below.

1.2 IMPORTANT NOTES ON THE PROFIT ELEMENT OF RATES IN THIS BOOK

The rates shown in the Measured Works and Approximate Estimates sections of this book do not generally contain a profit element unless the rate has been provided by a subcontractor.

Analysed Rates versus Sub-Contractor Rates
As a general rule if a rate is shown as an analysed rate in the Measured Works section, i.e. it has figures shown in the columns other than the "Unit" and "Total Rate" column, it can be assumed that this has no profit or overhead element and that this calculation is shown as a direct labour/material supply/plant, task being performed by a contractor.

On the other hand if a rate is shown as a "Total Rate" only, this would normally be a sub-contractors rate and would contain the profit and overhead for the subcontractor.

The foregoing applies for the most part to the Approximate Estimates section. However in some items there may be an element of sub-contractor rates within direct works build-ups.

As an example of this, to excavate, lay a base and place a macadam surface; a general landscape contractor would normally perform the earthworks and base installation. The macadam surfacing would be performed by a specialist subcontractor to the landscape contractor.

The Approximate Estimate for this item uses rates from the Measured Works section to combine the excavation, disposal, base material supply and installation with all associated plant. There is no profit included on these elements. The cost of the surfacing, however, as supplied to us in the course of our annual price enquiry from the macadam subcontractor, would include the subcontractors profit element.

The landscape contractor would add this to his section of the work but would normally apply a mark up at a lower percentage to the subcontract element of the composite task.
Users of this book should therefore allow for the profit element at the prevailing rates. Please see the worked examples below and the notes on "Overheads" for further clarification.

1.3 INTRODUCTORY NOTES ON COSTING

The Prices for Measured Work are intended to apply to a medium-sized contract of about 6000 m2 in the outer London area, and assume that 50% is hard surfacing and 50% is soft landscaping and planting. Similarly it has been necessary to assume that the work is undertaken as a main contract, but if the work is let as a sub-contract, consideration should be given to the need for the addition of main contractor's discount and profit.

As explained in more detail later the prices are generally based on wage rates and material costs current at Spring 2002. They do not allow for preliminary items, which are dealt with in Section 2.3, or for any Value Added Tax which may be payable.

Adjustments should be made to standard rates for time, location, local conditions, site constraints and any other factors likely to affect the costs of a specific scheme.

Term contracts for general maintenance of large areas should be executed at rates somewhat lower than those given in this section.

There is now a facility available to readers that enables a comparison to be made between the level of prices given and those for projects carried out in regions other than outer London; this is dealt with in Section 2.6.

The units of measurement have been varied to suit the type of work and care should be taken when using any prices to ascertain the basis of measurement adopted.

The prices per unit of area for executing various mechanical operations are for work in the following areas and under the following conditions.

Prices per m^2 relate to areas not exceeding 100 m^2
(any plan configuration)

Prices per 100 m^2 relate to areas exceeding 100 m^2 but not exceeding 1/4 ha
(generally clear areas but with some sub-division)

Prices per ha relate to areas over 1/4 ha
(clear areas suitable for the use of tractors and tractor-operated equipment)

The prices per unit area for executing various operations by hand generally vary in direct proportion to the change in unit area.

Measured Works

Prime Cost: Commonly known as the 'PC'. Prime Cost is the actual price of the material item being addressed such as paving, shrubs, bollards or turves, as sold by the supplier. Prime Cost is given 'per square metre', 'per 100 bags' or 'each' according to the way the supplier sells his product. In researching the material prices for the book we requested that the suppliers price for full loads of their product delivered to a site close to the M25 in London. Spon's rates do not include VAT. Some companies may be able to obtain greater discounts on the list prices than those shown in the book. Prime Cost prices for those products and plants which have a wide cost range will be found under the heading of Market Prices in the main sections of this book, so that the user may select the product most closely related to his specification.

Materials: The PC material plus the additional materials required to fix the PC material.
Every job needs materials for its completion besides the product bought from the supplier. Paving needs sand for bedding, expansion joint strips and cement pointing; fencing needs concrete for post setting and nails or bolts; tree planting needs manure or fertilizer in the pit, tree stakes, guards and ties. If these items were to be priced out separately, Spon's Landscape and External Works Price Book (and the Bill of Quantities) would be impossibly unwieldy, so they are put together under the heading of Materials.

Labour. This figure covers the cost of planting shrubs or trees, laying paving, erecting fencing etc. and is calculated on the wage rate (skilled or unskilled) and the time needed for the job. Extras such as highly skilled craft work, difficult access, intermittent working and the need for labourers to back up the craftsman all add to the cost. Large regular areas of planting or paving are cheaper to install than smaller intricate areas, since less labour time is wasted moving from one area to another.

Labour Rates used in this edition

Based on surveys carried out on a cross section of external works contractors, the following rates for labour have been used in this edition.
These rates include for company overheads such as employee administration, transport insurance, and on costs such as National Insurance. The rates do not include for profit.

The rates for all labour used in this edition is as follows

General contracting	£ 15.00 / hour
Maintenance contracting	£ 12.00 / hour

Plant, consumable stores and services: This rather impressive heading covers all the work required to carry out the job which cannot be attributed exactly to any one item. It covers the use of machinery ranging from JCB's to shovels and compactors, fuel, static plant, water supply (which is metered on a construction site), electricity and rubbish disposal. The cost of transport to site is deemed to be included elsewhere and should be allowed for elsewhere or as a preliminary item. Hired plant is calculated on an average of 36 hours working time per week

Overheads. An allowance for this is included in the labour rates which are described above. The general overheads of the contract such as insurance, site huts, security, temporary roads and the statutory health and welfare of the labour force are not directly assignable to each item, so they are distributed as a percentage on each, or as a separate preliminary cost item. The contractor's and sub-contractor's profits are not included in this group of costs. Site overheads, which will vary from contract to contract according to the difficulties of the site, labour shortages, inclement weather or involvement with other contractors, have not been taken into account in the build up of these rates, while Overhead (or profit) may have to take into account losses on other jobs and the cost to the contractor of unsuccessful tendering.

Sub-contract rates
Where there is no analysis against an item, this is deemed to be a rate supplied by a sub-contractor. In most cases these are specialist items where most external works contractors would not have the expertise or the equipment to carry out the task described. It should be assumed that sub-contract rates include for the sub contractors profit.
An example of this may be found for the Tennis court rates in section Q26

1.4   ADJUSTMENT AND VARIATION OF THE RATES IN THIS BOOK
It will be appreciated that a variation in any one item in any group will affect the final Measured Work price.
Any cost variation must be weighed against the total cost of the contract and a small variation in Prime Cost where the items are ordered in thousands may have more effect on the total cost than a large variation on a few items, while a change in design that necessitates the use of earth-moving equipment which must be brought to the site for that one job will cause a dramatic rise in the contract cost. Similarly, a small saving on multiple items will provide a useful reserve to cover unforeseen extras.

Worked examples
A variation in the Prime Cost of an item can arise from a specific quotation.
For example:,
The PC of 900 x 600 x 50mm precast concrete flags is given as £2.78 each which equates to £5.14/m^2; and to which the costs of bedding and pointing are added to give a total material cost of £9.19/m^2.
Further costs for labour and mechanical plant give a resultant price of £15.19/m^2.
If a quotation of £5.00/m^2 was received from a supplier of the flags the resultant price would be calculated as £15.19 less the original cost (£5.14) plus the revised cost (£5.00) to give £15.05/m^2.

A variation will also occur if, for example, the specification changes from 50 light standard trees to 50 extra heavy standard trees. In this case the Prime Cost will increase due to having to buy extra heavy standard trees instead of light standard stock.

Original price: The prime cost of an Acer platanoides light standard is given as £7.80, to which the cost of handling and planting, etc is added to give an overall cost of £13.05. Further costs for labour and mechanical plant for mechanical excavation of a 600 x 600 x 600 mm pit (£2.97), a single tree stake (£7.18), and importing "Topgrow" compost in 80 litre bags (£0.62) and backfilling (£5.25) give a resultant price of £29.07. Therefore the total cost of 50 light standard trees is £1453.50.

How To Use This Book 5

Revised price: The prime cost of an Acer platanoides extra heavy standard is given as £34.00, however, taking into account the additional staking, extra labour for handling and planting, a larger tree pit, a larger stake and more compost, and this cost gives a total unit price of £73.38. This is significantly more than the light standard version. Therefore the total cost of 50 extra heavy trees is £3669.00 and the additional cost is £2215.50

This example of the effects of changing the tree size, illustrates that caution is needed when revising tender prices as merely altering the prime cost of an item is will not accurately reflect the total cost of the revised item.

1.5 APPROXIMATE ESTIMATES

These are combined Measured Work prices which give an approximate cost for a complete section of landscape work. For example, the construction of a car park comprises excavation, levelling, road-base, surfacing and marking parking bays. Each of these jobs is priced separately in the Measured Works section, but a comprehensive price is given in the Approximate Estimates section, which is intended to provide a quick guide to the cost of the job. It will be seen that the more items that go to make up an approximate estimate, the more possibilities there are for variations in the PC prices and the user should ensure that any PC price included in the estimate corresponds to his specification.

Worked example

In many instances a modular format has been use in order to enable readers to build up a rate for a required task. The following table describes the trench excavation pipe laying and backfilling operations as contained in section R12 of this book.

Pipe laying 100 m	£
Excavate for drain 150 mm wide x 450 mm deep inclusive of disposal on site; by machine	87.43
Lay flexible plastic pipe 110 mm wide	94.06
Backfilling with gravel rejects, blinding with sand and topping with 150 mm topsoil	218.35
Total	399.84

The figures given in Spon's Landscape and External Works Price Book are intended for general guidance, and if a significant variation appears in any one of the cost groups, the Price for Measured Work should be re-calculated.

1.6 HOW THIS BOOK IS UPDATED EACH YEAR

The basis for this book is a database of Material, Labour, Plant and Sub-Contractor resources each with its own area of the database.

Material, Plant and Sub-Contract Resources

Each year the suppliers of each material, plant or subcontract item are approached and asked to update their prices to those that will prevail in September of that year.

These resource prices are individually updated in the database. Each resource is then linked to one or many tasks. The tasks in the task library section of the database is automatically updated by changes in the resource library. A quantity of the resource is calculated against the task.

The calculation is generally performed once and the links remain in the database.
On occasions where new information or method or technology are discovered or suggested these calculations would be revisited. A further source of information is simple time and production observations made during the course of the last year.

Labour Resource Update

Most tasks except those shown as sub-contractor rates (see above) employ an element of labour. The Data Department at Davis Langdon and Everest conducts ongoing research into the costs of labour in various parts of the country.

Tasks or entire sections would then be re-examined and recalculated. Comments on the rates published in this book are welcomed and may be submitted to the contact addresses shown in the preface

2
Cost Information

2.1 LABOUR RATES USED IN THIS EDITION

Based on surveys carried out on a cross section of external works contractors, the following rates for labour have been used in this edition.

These rates include for company overheads such as employee administration, transport insurance, and on costs such as National Insurance. The rates do not include for profit.

The rates for all labour used in this edition is as follows

General contracting	£ 15.00 / hour
Maintenance contracting	£ 12.00 / hour

2.2 RATES OF WAGES

2.2.1 Building industry

Authorized rates of wages, etc., in the building industry in England and Wales and Scotland agreed by the Construction Industry Joint Council. Effective from 25 June 2001. Basic pay. Weekly rate based on 39 hours.

Craft Rate	Skill Rate 1	£271.05	General Operative	£214.11
£284.70	Skill Rate 2	£261.30		
	Skill Rate 3	£244.53		
	Skill Rate 4	£230.49		

The Prices for Measured Works used in this book are based upon the commercial wage rates currently used in the landscaping industry and are typical of the London area.
However, it is recognized that some contractors involved principally in soft landscaping and planting works may base their wages on the rates determined by the Agricultural Wages Board (England and Wales) and therefore the following information is given to assist readers to adjust the prices if necessary.

2.2.2 Agricultural Wages, England and Wales

Minimum payments agreed by the Agricultural Wages Board for standard 39 hour, five day week. Effective 1 October 2001.

Age	Unit	Basic pay	Craft grade		Appointment Grade	
			Certificate Grade	NVQ3 Grade	Grade 2	Grade 1
Weekly pay:						
19+	£/ week	186.03	214.11	219.57	232.44	251.16
18	£/ week	157.95				
17	£/ week	130.26				
16	£/ week	111.54				
15 & under	£/ week	93.21				
Overtime pay:						
19+	£/ week	7.16	8.24	8.45	8.94	9.66
18	£/ week	6.08				
17	£/ week	5.01				
16	£/ week	4.29				
15 & under	£/ week	3.59				

Since 1997 there have been three main types of worker:
- Full time workers (standard and flexible)
- Part time workers (standard and flexible)
- Casual workers

Weekly rates apply to Full Time Standard Workers for 39 hours work.

Hourly rates apply to the hours worked in a week by a Part Time Standard worker.

Overtime rates apply to all overtime hours worked by Standard Workers -
- more than 39 hours in any week
- or more than 8 hours on any day
- or Saturday or Sunday
- or Public Holiday

From 1 June 2000 the minimum rate for holiday pay will be the same as the minimum rate for basic pay.

Copies of the Agricultural Wages Order 2001 (Number 1) may be obtained from
The Agricultural Wages Board for England and Wales

Nobel House
17 Smith Square
London SW1P 3JR   Telephone: 020 7238 5862

Grades and responsibilities

Certificate Craft Grade
Must have worked in agriculture for at least 3 out of last 4 years and hold a valid craft certificate issued by the National Proficiency Tests Council in an approved subject.

Craftsman (NVQ3)
Must have worked in agriculture for at least 3 out of last 4 years and hold a valid Level 3 National Vocational Qualification (NVQ) in an approved subject and have a valid NVQ Level 3 Certificate issued by the National Proficiency Tests Council.

Appointment Grade 2
Day to day responsibility for supervising the work on a farm or part of it, and implementing management decisions, or responsibility for the instruction and supervision of staff.

Appointment Grade 1
A management responsibility for an entire farm or part of it run as a separate operation or business, or responsibility for employing and disciplining staff.

2.2.3 Agricultural Wages, Scotland

New rules relating to agricultural wages and other conditions of service took effect from 1 January 2002. The new rules are contained in the Agricultural Wages (Scotland) Order No. 49 2001. The Order contains the detailed legal requirements for the calculations of minimum pay etc, but the Scottish Office has produced 'A Guide for Workers and Employers' which attempts to explain the new rules in simpler terms.

Both 'The Agricultural Wages (Scotland) Order No. 49 2001' and 'A Guide for Workers and Employers' may be obtained from:

Scottish Agricultural Wages Board
Pentland House
47 Robb's Loan EDINBURGH EN14 1TY   Telephone: 0131 244 6195

The new Order applies equally to all workers employed in agriculture in Scotland. It makes no distinction between full-time employees, part-time employees and students.

The old classifications of general farm worker, shepherd, stockworker, tractorman and supervisory grade have been removed and replaced by a single category covering all agricultural workers. The one exception to this is that special arrangements have been made for hill-shepherds.

Instead of minimum weekly rates of pay, the new Order works on minimum hourly rates of pay (with the exception hill-shepherds). It is very important to note that although wages are to be calculated on an hourly basis, a worker is still entitled to be paid for the full number of hours for which he is contracted (unless he is unavailable for work). In particular, this means that an employer cannot reduce the number of hours for which he is to be paid simply by e.g. sending him home early.

Cost Information

Minimum rates of wages

There are two main pay scales. One is for workers who have been with their present employer for not more than 13 weeks and the other for workers who have been with their present employer for more than 13 weeks.

Within these two scales, there are varying minimum hourly rates of pay depending on the age of the worker.

The minimum hourly rates of pay are as follows:

Age	Minimum Hourly Rates (£)	
	Up to 13 weeks	After 13 weeks
	£	£
Under 16	1.72	2.30
16 and under 17	2.08	2.77
17 and under 18	2.43	3.23
18 and under 19	3.59	3.93
19 and over	4.20	4.62

Workers who have been with the same employer for more than 13 weeks and hold either:

(a) a Scottish, or National, Vocational Qualification in an agricultural subject at Level III or above, or
(b) an apprenticeship certificate approved by Lantra, NTO (formerly ATB Landbase), or a certificate of acquired experience issued by the ATB Landbase.

shall be paid, for each hour worked, an additional sum of not less than £0.69.

Minimum hourly overtime rate

The minimum hourly rate of wages payable to a worker :

(a) for each hour worked in excess of 8 hours on any day, and
(b) for each hour worked in excess of 39 hours in any week.

shall be calculated in accordance with the following formula:

M x 1.65 (where M = the minimum hourly rate of pay to which the worker is entitled)

2.3 DAYWORK AND PRIME COST

The tables below analyse rates for daywork. The calculations used in the measured works and the derived approximate estimates use current commercial rates for the build-up of all items and do not employ the rates used below. Users should refer to Section 2.1 " Labour Rates used in this edition" at the beginning of this section for information on the actual hourly rates used.

When work is carried out which cannot be valued in any other way it is customary to assess the value on a cost basis with an allowance to cover overheads and profit. The basis of costing is a matter for agreement between the parties concerned, but definitions of prime cost for the building industry have been prepared and published jointly by The Royal Institution of Chartered Surveyors and the Construction Confederation (formerly the Building Employers Confederation, formerly the National Federation of Building Trades Employers) for the convenience of those who wish to use them. Reference should be made to *Spon's Architects' and Builders' Price Book* where these documents are reproduced by kind permission of the publishers together with the Schedule of Basic Plant Charges (January 1990 revision).

The calculations on page 15 shows an example of the calculation of typical standard hourly base rates for craft and general operatives prepared in accordance with the Definition of Prime Cost of Daywork carried out under a Building Contract. A building contract will normally provide for a contractor to tender a percentage adjustment for the base rates for incidental costs, overheads and profit.

Cost Information

Daywork Rates - Building Operatives

Rates effective from 24 June 2002

			Craft operative		General operative	
			Rate	*Annual cost*	*Rate*	*Annual cost*
			£	£	£	£
Guaranteed minimum weekly earnings						
Standard basic rate	46.2 wks	284.70		13153.14	214.11	9891.88
Guaranteed minimum bonus	46.2 wks	0.00		0.00	0.00	0.00
				13153.14		9891.88
Employer's National Insurance contribution (11.8% after the first £89.00 per week)				1050.07		665.25
				14203.21		10557.13
Employer's Contribution to:						
CITB levy (0.5% of payroll)				74.02		55.67
Holiday Pay	4.2 wks	284.70		1195.74	214.11	899.26
Public Holidays	1.6 wks	284.70		455.52	214.11	342.58
EasyBuild Stakeholder Pension (Death and accident cover is provided free)	52 wks	5.00		260.00	5.00	260.00
Annual cost of labour as defined in Section 3				16188.50		12114.64

Hourly base rates as defined in section 3, clause 3.2 (annual cost of labour / hours per annum	8.98	6.72

Notes

1. Calculated following Definition of Prime Cost of Daywork carried out Under a Building Contract, published by the Royal Institution of Chartered Surveyors and the Construction Confederation.
2. Standard basic rates effective from 24 June 2002
3. Standard working hours per annum calculated as follows:

52 weeks @ 39 hours	=	2028.0
Less		
4.2 weeks holiday @ 39 hours	=	163.8
8 days public holidays @ 7.8 hours	=	62.4
		226.2
		1801.8

All labour costs incurred by the contractor in his capacity as an employer other than those contained in the hourly base rate, are to be taken into account under Section 6.

5. The above example is for guidance only and does not form part of the Definition; all the basic costs are subject to re-examination according to the time when and in the area where the daywork is executed.
6. National Insurance payments are at not-contracted out rates applicable from 5 April 2002
7. Basic rate and GMB number of weeks =

52.0	weeks
- 4.2	weeks annual holiday
- 1.6	weeks public holiday
46.2	weeks

2.4 COMPUTATION OF LABOUR RATES, COST OF MATERIALS AND PLANT

Different organizations will have varying views on rates and costs, which will in any event be affected by the type of job, availability of labour and the extent to which mechanical plant can be used. However this information should assist the reader to:
(1) compare the prices to those used in his own organization;
(2) calculate the effect of changes in wage rates or prices of materials;
(3) calculate analogous prices for work similar to but differing in detail from the examples given.

Computation of labour rates

From 24 June 2002 basic weekly rates of pay for craft and general operatives are £284.70 and £214.11 respectively; to these rates have been added allowances for the items below in accordance with the recommended procedure of the Chartered Institute of Building in its Code of Estimating Practice. The resultant hourly rates are £10.05 and £7.53 for craft operatives and general operatives respectively.

The items referred to above for which allowances have been made are:

- Lost time
- Non-productive overtime
- Sick pay
- Construction Industry Training Board Levy
- Employer's contribution towards holidays with pay
- Employer's contribution towards pension and death benefit scheme
- Employer's contribution towards National Insurance
- Severance pay and sundry costs
- Employer's liability and third party insurance

Note: For travelling allowances and site supervision see Preliminaries, Section 2.3.

The tables which follow illustrate how the hourly rates referred to above have been calculated. Productive time has been based on a total of 1801.8 hours worked per year for daywork calculations above and for 1953.54 hours worked per year for the all-in labour rates (including 5 hours per week average overtime) for the all in labour rates below.

Cost Information

Building Craft and General Operatives

Rates effective from 24 June 2002

		Craft operative		General Operative	
		Rate	*Annual cost*	*Rate*	*Annual cost*
		£	£	£	£
Wages at standard basic rate, productive time:	44.30 wks	284.70	12611.04	214.11	9484.19
Lost time allowance:	0.904 wks	284.70	257.37	214.11	193.56
Non-productive overtime	5.80 wks	427.05	2476.89	321.17	1862.76
			15345.30		11540.51
Extra payments under National Working Rules	45.20 wks		-		-
Sick pay:	1.00 wk		-		-
CITB Allowance (0.50% of payroll):	1 year		86.57		65.11
Holiday Pay	4.20 wks	339.50	1425.89	255.32	1072.35
Public Holiday	1.60 wks	339.50	543.20	255.32	408.51
Employers' contribution to: EasyBuild Stakeholder Pension (Death and accident cover is provided free)	52.00 wks	5.00	260.00	5.00	260.00
National Insurance (average weekly payment)	47.8 wks	27.38	1308.75	17.99	859.78
			18969.71		14206.25
Severance pay and sundry costs	Plus	1.5%	284.55	1.5%	213.09
			19254.26		14419.34
Employers' liability and third-party insurance	Plus	2.0%	385.09	2.0%	288.39
Total cost per annum			19639.35		14707.73
Total cost per hour			10.05		7.53

Notes:

1. Absence due to sickness has been assumed to be for periods not exceeding 3 days for which no payment is due. (Working Rule 20.7.3)
2. EasyBuild Stakeholder Pension effective from 1 July 2002. Death and accident benefit cover is provided free of charge.
3. All N.I. Payments are at not-contracted out rates applicable from 6 April 2002

- National Insurance paid for 47.8 weeks (52wks - 4.2wks) is based on employer making regular monthly payments into the Template holiday pay scheme and by doing so the employer achieves National Insurance savings on holiday wages.

COMPUTATION OF LABOUR RATES IN PRICES FOR MEASURED WORK

A survey of typical landscape/ external works companies indicates that they are currently paying above average wages for multi skilled operatives regardless of specialist or supervisory capability. In our current overhaul of labour constants and costs we have departed from our previous policy of labour rates based on national awards and used instead a gross rate per hour for all rate calculations. This rate is tabled at the beginning of this section. Estimators can readily adjust this rate if they feel it inappropriate for their work.

COMPUTATION THE COST OF MATERIALS

Percentages of default waste are placed against material resources within the supplier database. These range from 2.5% (for bricks) to 20% for topsoil

An allowance for the cost of unloading, stacking etc should be added to the cost of materials.

The following are typical hours of labour for unloading and stacking some of the more common building materials.

Material	Unit	Labourer hour
Cement	tonne	0.67
Lime	tonne	0.67
Common bricks	1000	1.70
Light engineering bricks	1000	2.00
Heavy engineering bricks	1000	2.40

COMPUTATION OF MECHANICAL PLANT COSTS

Plant used within resource calculations in this book have been used according to the following principles:

1. Plant which is non-pedestrian plant is fueled by the Plant Hire company.
2. Operators of excavators above the size of 5 tonnes are provided by the Plant hire company and no labour calculation is allowed for these operators within the rate.
3. Three to Five tonne excavators and dumpers are operated by the landscape contractor and additional labour is shown within the calculation of the rate.
4. Small plant is operated by the landscape contractor.
5. No allowance for delivery or collection has been made within the rates shown.
6. Downtime is shown against each plant type either for non productive time or mechanical down time

The following tables lists the plant resources used in this years book

Plant Supplied by H.E Services

Prices shown reflect 36 working hours per week

Description	Unit	Supply Quantity	Rate Used £	Down Time %
Excavator 360 Tracked 21 tonne- *fueled and operated*	hour	1	29.73	10
Excavator 360 Tracked 5 Ton *fueled only*	hour	1	12.42	10
Excavator 360 Tracked 7 Ton *fueled and operated*	hour	1	22.40	10
Mini Excavator JCB 803 Rubber Tracks Self Drive *fueled only*	hour	1	8.95	10
Mini Excavator JCB 803 Steel Tracks Self Drive *fueled only*	hour	1	8.62	10
JCB 3Cx 4X4 Sitemaster, *fueled and operated*	hour	1	27.60	10
JCB 3Cx 4 X 4 Sitemaster + Breaker *fueled and operated*	hour	1	27.60	10
Dumper 5 Ton, Thwaites Self Drive *fueled only*	hour	1	6.71	10
Manitou Telehandler *fueled only*	hour	1	34.00	50

Cost Information

Plant Supplied by HSS Hire Shops

Description	Unit	Supply Quantity	Supply Cost £	Rate Used £	Down Time %
Auger Mechanical 2 Man Weekly Rate	hour	25	98.00	3.92	37.5
Engine Poker Vibrator	hour	25	89.00	3.56	37.5
Vibrating Plate Compactor	hour	25	45.00	2.32	37.5
Petrol Masonry Saw Bench	hour	25	112.00	4.80	35.5
Diamond Blade Consumable, 230 mm	mm	1	25.00	25.00	-
Access Platform To 5.2 M	day	5	140.00	28.00	-
Cutting Torch Oxy Acetylene	week	1	78.00	78.00	-
Cutting Torch Oxy Acetylene	hour	1	5.35	5.75	33.3
Bolt Croppers, 900 mm	hour	1	1.02	1.42	-
Hydraulic Breaker Diesel 5Hrs /Day	hour	1	5.28	7.92	37.5
Rotavator Howard Gem (Based On Weekly Rate) 25 Hrs/Week	hour	1	5.00	8.00	37.5
Rotavator Tractor Mounted 1200mm + Operator	hour	1	23.50	25.85	-
Compactor Bomag 136Kg/ 10.1Kn 4 Hrs /Day	hour	1	4.80	5.76	50
Compactor Bomag 470 Kg /18.7Kn 4 Hrs /Day	hour	1	6.00	7.20	50

2.5 COST INDICES

The purpose of this section is to show changes in the cost of carrying out landscape work (hard surfacing and planting) since 1976. It is important to distinguish between costs and tender prices: the following table reflects the change in cost to contractors but does not necessarily reflect changes in tender prices. In addition to changes in labour and material costs, which are reflected in the indices given below, tender prices are also affected by factors such as the degree of competition at the time of tender and in the particular area where the work is to be carried out, the availability of labour and materials, and the general economic situation. This can mean that in a period when work is scarce tender prices may fall despite the fact that costs are rising, and when there is plenty of work available, tender prices may increase at a faster rate than costs.

The Constructed Cost Index

A Constructed Cost Index based on PSA Price Adjustment Formulae for Construction Contracts (Series 2). Cost indices for the various trades employed in a building contract are published monthly by HMSO and are reproduced in the technical press.

The indices comprise 49 Building Work indices plus seven 'Appendices' and other specialist indices. The Building Work indices are compiled by monitoring the cost of labour and materials for each category and applying a weighting to these to calculate a single index.

Although the PSA indices are prepared for use with price-adjustment formulae for calculating reimbursement of increased costs during the course of a contract, they also present a time series of cost indices for the main components of landscaping projects. They can therefore be used as the basis of an index for landscaping costs.

The method used here is to construct a composite index by allocating weightings to the indices representing the usual work categories found in a landscaping contract, the weightings being established from an analysis of actual projects. These weightings totalled 100 in 1976 and the composite index is calculated by applying the appropriate weightings to the appropriate PSA indices on a monthly basis, which is then compiled into a quarterly index and rebased to 1976 = 100.

Constructed Landscaping (Hard Surfacing and Planting) Cost Index
Based on approximately 50% soft landscaping area and 50% hard external works.
1976 = 100

Year	First Quarter	Second Quarter	Third Quarter	Fourth quarter	Annual average
1990	330	334	353	357	344
1991	356	356	364	364	360
1992	364	365	376	378	371
1993	380	381	383	384	382
1994	387	388	394	397	391
1995	399	404	413	413	407
1996	417	419	428	430	424
1997	430	432	435	442	435
1998	442	445	466	466	455
1999	466	468	488	490	478
2000	493	496	513	514	504
2001	512	514	530	529	521
2002	529*				

* *Provisional*

This index is updated every quarter in Spon's Price Book Update. The updating service is available, free of charge, to all purchasers of Spon's Price Books. (Complete the reply-paid card enclosed.)

2.6 REGIONAL VARIATIONS

Prices in Spon's Landscape and External Works Price Book are based upon conditions prevailing for a competitive tender in the outer London area. For the benefit of readers, this edition includes regional variation adjustment factors which can be used for an assessment of price levels in other regions.

Special further adjustment may be necessary when considering city centre or very isolated locations.

Region	Adjustment Factor
Outer London	1.00
Inner London	1.10
South East	0.96
South West	0.87
Midlands	0.82
East Anglia	0.87
Northern	0.78
North West	0.83
Scotland	0.82
Wales	0.78
Northern Ireland	0.60
Channel Islands	1.25

The following example illustrates the adjustment of prices for regions other than outer London, by use of regional variation adjustment factors.

£

A. Value of items priced using Spon's Landscape and External Works Price Book   100 000

B. Adjustment to value of A. to reflect Midlands Region price level 100 000 x 0.82   82 000

ESSENTIAL READING FROM SPON PRESS

The Regeneration of Public Parks

Edited by Jan Woudstra and Ken Fieldhouse, University of Sheffield, UK

'I feel this book will be welcome in the library of every landscape designer, every park manager and every garden historian. I would recommend it to anyone working in the landscape, urban design and parks profession and it is essential reading for anyone with an interest or active involvement in the care and refurbishment of public parks.' - *Journal of Environmental Planning and Management*

The Regeneration of Public Parks highlights the new enthusiasm for Britain's public parks and provides a valuable overview of the public park design, sustainability, management and regeneration. It will be essential to all those involved with or interested in public parks. The expanding interest in public parks in the UK has been encouraged by the grants of the Heritage Lottery Fund (Urban Parks Programme), but the renaissance is an international phenomenon, fired by general concerns about the urban environment.

The book has been prepared by a team of experts involved with, and responsible for public parks, all are specialists in their fields. The first part of the book deals with general issues relating to public parks, while the second part relates to specific design elements within the parks and their management: ironwork, walks and paved surfaces, planting and bedding, lakes and water bodies.

June 2000: 276x219: 200pp
132 b+w photos; 8 pages colour illustrations
Pb: 0-419-25900-7: £30.00

To Order: Tel: +44 (0) 8700 768853, or +44 (0) 1264 343071 Fax: +44 (0) 1264 343005, or Post: Spon Press Customer Services, Thomson Publishing Services, Cheriton House, Andover, Hants, SP10 5BE, UK Email: book.orders@tandf.co.uk

For a complete listing of all our titles visit:
www.sponpress.com

3
Preliminaries

3.1 PRELIMINARIES/GENERAL CONDITIONS

The number of items priced in the Preliminaries section of Tender Documents and the manner in which they are priced vary considerably between Contractors. Some Contractors, by modifying their percentage factor for overheads and profit, attempt to cover the costs of Preliminary items in their Prices for Measured Work. However, the cost of Preliminaries will vary widely according to job size and complexity, site location, accessibility, degree of mechanization practicable, position of the Contractor's head office and relationships with local labour/domestic sub-contractors. It is therefore usually far safer to price Preliminary items separately on their merits according to the job.

Reference should be made to *Spon's Architects' and Builders'* Price Book for the normal clause descriptions from the Preliminaries section of a bill of quantities where the *JCT Standard Form of Building Contract* 1980 Edition is to be used.

For the convenience of readers the clause descriptions which are likely to be found in the Preliminaries section of the tender documents for a project which consists of solely or mainly soft landscaping works and is therefore to be based on the *JCLI Form of Agreement* for Landscape Works January 1992 Edition, June 1996 revision, are given below together with further details against those items which are usually priced in tenders.

Note. The term 'Not priced' where used throughout this section means either that the cost implication is negligible or that it is usually included elsewhere in the tender.

1.	**Project, parties and consultants.**	Not priced
2.	**Description of site.**	Not priced
3.	**Drawings and other documents.**	Not priced
4.	**Form, type and conditions of contract.**	

Clause No.
1.0	***Intentions of the parties***	
1.1	Contractor's obligation.	Not priced
1.2	Landscape Architect's duties.	Not priced
1.3	Contract Bills and SMM	Not priced
1.4	Reappointment of Planning Supervisor or Principal Contractor - notification to Contractor.	Not priced
1.5	Alternative B2 in the 5th recital - notification by Contractor - regulation 7(5) of the CDM Regulations.	Not priced

2.0 Commencement and completion

2.1	Commencement and completion.	Not priced
2.2	Extension of contract period.	Not priced
2.3	Damages for non-completion.	Not priced
2.4	Practical completion.	Not priced
2.5	Defects liability. Inevitably some defects will arise and an allowance will often be made to cover this either here or against Clause 2.7.	
2.6	Partial practical completion.	Not priced
2.7	Failures of plants A or B. Inevitably some replacements will be required and an allowance should be made here or in the rates. Plant replacements up to and including 10% loss are usually covered in the planting measured work prices.	

3.0 Control of the Works

3.1	Assignment.	Not priced
3.2	Sub-contracting.	Not priced
3.3	Contractor's representative. Any allowance for site supervision/administration which is not included in the rates should be priced here together with the cost of setting out the works.	
3.4	Exclusion from the Works.	Not priced
3.5	Landscape Architect's instructions.	Not priced
3.6	Variations.	Not priced
3.7	PC and Provisional sums. An amount should be added to the PC sum items, if required, for profit and a further sum, where applicable, for attendance.	
3.8	Objections to a Nomination.	Not priced

4.0 Payment

4.1	Correction of inconsistencies.	Not priced
4.2	Progress payments and retention.	Not priced
4.3	Penultimate certificate	Not priced
4.4	Final certificate.	Not priced
4.5	Contribution, levy and tax changes.	Not priced
4.6 A	Fixed price. The additional cost of providing a fixed price can be included here or in the rates.	
B	Fluctuations. An allowance can be made here for the estimated shortfall in reimbursement (if any) under a fluctuating price contract.	

5.0 Statutory obligations

5.1	Statutory obligations, notices, fees and charges. The cost of complying with this clause should be included here.	
5.2	Value Added Tax.	Not priced
5.3	Statutory tax deduction scheme.	Not priced
5.4	Number not used.	Not priced
5.5	Prevention of corruption.	Not priced

Preliminaries 25

5.6	Employer's obligation - Planning Supervisor - Principal Contractor	Not priced
5.7	Duty of Principal Contractor	Not priced
5.8	Successor appointed to the Contractor as Principal Contractor	Not priced
5.9	Health and Safety file	Not priced

6.0 *Injury, damage and insurance*
6.1 Injury to or death of persons. The cost of maintaining insurance should be included here if not allowed for elsewhere.
6.2 Injury or damage to property. As above in 6.1.
6.3 A Insurance of the Works by Contractor - Fire, etc. If at the Contractor's risk the cost of maintaining insurance cover must be sufficient to include the full cost of reinstatement, all increases in cost, professional fees and any consequential costs such as clearing debris.
   B Insurance of the Works and any existing structures by Employer - Fire, etc. As above in 6.3 A.
6.4 Evidence of insurance.   Not priced
6.5 Malicious damage or theft A or B. If at the Contractor's risk the cost of complying with this clause should be included here.

7.0 *Determination*
7.1	Notices	Not priced
7.2	Determination by Employer.	Not priced
7.3	Determination by Contractor.	Not priced

8.0 *Supplementary Memorandum*
8.1 Meaning of references in 4.5, 4.6, 5.2 and 5.3.   Not priced

9.0 *Settlement of Disputes - Arbitration*   Not priced

5. **Contractor's liability.**   Not priced
(If the JCLI Form of Agreement is used this item is covered under Item 6. If a different Form of Contract is used it may be necessary to include for Contractor's insurances, etc., under this item.)

6. **Employer's liability.**   Not priced
(If additional insurances are to be allowed for in the contract sum such cost will be given as a provisional sum.)

7. **Local Authorities' fees and charges.**   Not priced
(If the JCLI Form of Agreement is used this item is covered under Item 5.
If a different Form of Contract is used it may be necessary to include for the fees and charges which the Contractor is required to pay under this Item unless given as provisional sums.)

8. **Obligations and restrictions imposed by the employer.**
   These include the following items and costs can only be assessed in the light of circumstances on a particular job.
   - (a) Access to and possession of use of the site.
   - (b) Limitations of working space.
   - (c) Limitations of working hours.
   - (d) The use or disposal of any materials found on site.
   - (e) Hoardings, fences, and the like, temporary name boards and advertising rights.
   - (f) The maintenance of existing live drainage, water, gas and other main or power services on or over the site.
   - (g) The execution or completion of the work in any specific order or in sections or phases.
   - (h) Temporary accommodation and facilities for the use of the Employer including heating, lighting, furnishing and attendance.
     This will include an office for the Clerk of Works if there is to be one on the site. Against this item the following should be priced:
     - i. Hire of Clerk of Works Office
     - ii. Transport to and from the site
     - iii. Erecting on a suitable base and later dismantling
     - iv. Lighting, heating and attendance on office
     - v. Local Authority rates
   - (i) The installation of telephones for the use of the Employer. (The cost of his telephone calls should be covered by a provisional sum.)
   - (k) Any other obligation or restriction.

     Additional obligations may include the provision of a performance bond. If the Contractor is required to provide sureties for the fulfilment of the work the usual method of providing this is by a bond provided by one or more insurance companies. The cost of a performance bond depends largely on the financial standing of the applying contractor. Figures tend to range from 0.25 to 0.5% of the contract sum.

Works by nominated sub-contractors, goods and materials from nominated suppliers and works by public bodies

9. **Works by nominated sub-contractors.** Not priced
   (Work to be carried out by a nominated sub-contractor and given as a prime cost sum to which an amount should be added, if required, for profit and a further sum for attendance.)

10. **Goods and materials from nominated suppliers.** Not priced
    (Goods and materials which are required to be obtained from a nominated supplier and given as a prime cost sum to which an amount should be added, if required, for profit.)

11. **Works by Public Bodies.** Not priced
    (Works which are to be carried out by a Local Authority or public undertaking given as a provisional sum.)

12. **Works by others directly engaged by the Employer.** Not priced
    (A description given for works by others directly engaged by the employer, and any attendance that is required shall be priced in the same way as works by nominated sub-contractors.)

General facilities and obligations

13. **Pricing**
    For convenience in pricing items will be listed which may include the following. The contractor should include maintaining any temporary works in connection with the items, adapting, clearing away and making good, and all notices and fees to local authorities and public undertakings.

    a) *Plant, tools and vehicles*

    The sixth edition of the Standard Method of Measurement of Building Works requires that items for plant be given at the beginning of each section, whereas the seventh edition provides for these items to be covered under A42 and A43. However, Contractors often include the cost of their own plant within the measured rates e.g. earthmoving.

    b) *Site administration and security*

    If the JCLI Form of Agreement is used the cost of administrative staff is often included against Item 3. When required allow for the provision of a watchman or inspection by a security organization.

c) *Transport for workpeople*

The labour rates per hour on which Prices for Measured Work have been based do not cover travel and lodging allowances which must be assessed according to the particular circumstances.

d) *Protecting the works from inclement weather*

In areas likely to suffer particularly inclement weather, some nominal allowance should be included for tarpaulins, polythene sheeting, etc., and the effect of any delays in concreting or brickwork by such weather.

e) *Water for the works*

Charges should properly be ascertained from the local Water Authority. If these are not readily available, an allowance of 0.33% of the value of the contract is probably adequate, providing water can be obtained directly from the mains. Failing this, each case must be dealt with on its merits. In all cases an allowance should also be made for temporary plumbing including site storage of water if required.

f) *Lighting and power for the works*

The Contractor is usually responsible for providing all temporary lighting and power for the works and all charges involved.

g) *Temporary roads, hardstandings, crossings and similar items*

Quite often consolidated bases of eventual site roads are used throughout a contract to facilitate movement of materials around the site. However, during the initial setting up of a site, with drainage works outstanding this is not always possible and occasionally temporary roadways have to be formed and ground levels later reinstated.

h) *Temporary accommodation for the use of the Contractor*

This includes all temporary offices and sheds for the Contractor and his domestic sub-contractors' use (temporary office for a Clerk of Works is covered under obligations and restrictions imposed by the employer).

Allowances must also be made for transport, erection, dismantling, etc., as previously shown under the item for Clerk of Works office.

j) *Temporary telephones for the use of the Contractor*

Against this item should be included the cost of installation, rental and an assessment of the cost of calls made during the contract.

k) *Traffic regulations*

Waiting and unloading restrictions can occasionally add considerably to costs, resulting in forced overtime or additional weekend working. Any such restrictions must be carefully assessed for the job in hand.

l) *Safety, health and welfare of workpeople*

The Contractor is required to comply with all relevant codes, regulations, agreements and statutes and in particular should allow for provision of the following:

1. Shelter from inclement weather
2. Accommodation for clothing
3. Accommodation and provision for meals
4. Provision of drinking water
5. Sanitary conveniences
6. Washing facilities
7. First aid
8. Site conditions

A variety of self-contained mobile or jack-type units are available for hire and allowance must be made in addition for transport costs to and from site, setting up costs, connections to mains, fuel supplies and attendance. A general provision to comply with the above code is often 0.50 to 0.75% of the contract value. The cost of safety supervisors (required for firms employing more than 20 people) is usually part of head office overhead costs.

m) *Disbursements arising from the employment of workpeople*

Travelling and lodging allowances have been dealt with under Transport for Workpeople and usually all other on-costs and disbursements are included in the all-in hourly rate used in the calculation of Prices for Measured Work. However, it is as well to check that all such disbursements have been included elsewhere.

n) *Maintenance of public and private roads*

Some additional insurance or value may be required against this item to insure against damage to entrance gates, kerbs or bridges caused by extraordinary traffic in the execution of the works.

o) *Removing rubbish, protective casings and coverings and cleaning the works on completion*

This includes removing surplus materials and final cleaning of the site prior to handover. Allow for sufficient 'bins' for the site throughout contract duration and for some operative time at the end of the contract for final clearing and cleaning ready for handover.

A general allowance of 0.20% of contract value is probably sufficient.

p) *Temporary fencing, hoardings and similar items*

This item must be considered in some detail as it is dependent on site perimeter, phasing of the work, etc.

q) *Control of noise, pollution and all other statutory obligations*

The Local Authority may impose restrictions on the timing of certain operations, particularly noisy or dust-producing operations, and may necessitate the carrying out of these outside normal working hours or using special tools and equipment.

The situation is most likely to occur in built up areas such as city centres, etc., where the site is likely to be in close proximity to offices or residential property.

14. Contingencies

Generally. Provision for contingencies is normally given as a provisional sum.

4
Landscape Consultant's Appointment

FOREWORD

This document has been designed to advise Landscape Consultants and their Clients in the execution of landscape commissions. It has been registered with the Office of Fair Trading under the terms of the Restrictive Trade Practices Act 1976.

The Landscape Institute is the Chartered Institute in the UK for Landscape Architects, incorporating Designers, Managers and Scientists. One of its objects is to promote the highest standard of professional service in the application of the arts and sciences of landscape architecture and management. All members of the Institute, collectively referred to in this document as Landscape Consultants, are governed by the Institute's Code of Professional Conduct, but the roles of the members of the three different landscape disciplines vary in accordance with their education and experience.

Landscape Architects (Design) identify and solve problems using design techniques based on an understanding of the external environment, knowledge of the functional and aesthetic characteristics of landscape materials, and of the organization of landscape elements, external spaces and activities.

Landscape Architects (Management) are concerned with the long term care and development of new and existing landscapes and also with policy and planning for future management and use. This involves them in the organization of manpower, machinery and materials and requires a working knowledge of statutory measures and Grant Aid Schemes in order to preserve and enhance the quality of the landscape.

Landscape Architects (Science) are specialists in the physical and biological aspects of landscape design and management. Relating scientific expertise to the practical problems of designers and managers, they may work in any of the many scientific subjects that are relevant to landscape, especially botany, ecology and soil surveys.

The Landscape Institute has produced three inter-related documents which assist clients and landscape consultants to reach clear agreement on the terms and conditions of an appointment. The documents are: **The Landscape Consultants Appointment** which sets out the Memorandum of Agreement as well as the standard and additional services provided, conditions of service which apply and a schedule of services and fees.

Engaging a Landscape Consultant - Guidance to Clients on Fees which shows the various bases on which fees are calculated, and the fees typically charged for the Standard services for projects of a range of sizes and complexity. It provides a basis against which the fee for a particular project can be compared or a starting point for negotiation.

Guide to Procedure for Competitive Tendering which guides clients through the initial considerations and the procedures needed to obtain comparable and valid fee tenders for projects.

Any question on or arising out of the information contained herein may be referred to the Director General, The Landscape Institute, 6/8 Barnard Mews, London SW11 1QU. Tel: 020 7738 9166. Fax: 020 7738 9134.

0 INTRODUCTION

0.1 The Landscape Consultant's professional responsibility is to act as the Client's adviser based upon a clear understanding of the Client's requirements and to administer any contract between Client and Contractor impartially and fairly. A good working relationship between the Client and the Landscape Consultant is therefore essential when proceeding with a commission. The Clients also have an important role. They must provide adequate information on the project, site and budget and fully understand and approve the Landscape Consultant's proposals at various stages of the work as it proceeds.

0.2 The most successful relationships are those which proceed in an atmosphere of mutual trust and goodwill. An understanding of the Client's and the Consultant's respective obligations by both parties is fundamental to the creation of such an atmosphere. The Landscape Institute therefore advises its members that before accepting a commission for professional services, they should agree with the Client the terms of the commission, including the scope of the services; the allocation of responsibilities and any limitation of liability; the payment of fees including the rates and methods of calculation and the provision for termination.

0.3 This document consists of the following parts:

0.3.1 **Memorandum of Agreement**
The Institute advises its members to use the Memorandum of Agreement and Schedule of Services and Fees (Appendix II). Alternatively, letters of appointment could also serve the purpose, provided that the services, responsibilities and fee basis are fully defined. If the agreement is not comprehensive it is likely to create uncertainties for either of both parties as the commission progresses.

0.3.2 **The Landscape Consultant's Appointment** (Appendix I) in three parts:
Part 1 - Landscape Consultant's Services. The work which a Landscape Consultant normally undertakes after accepting a commission is primarily concerned with landscape design, construction (including planting), and management. It comprises Preliminary Services and Standard Services, sub-divided into work stages as follows:

Preliminary Services	Work Stage A - Inception
	Work Stage B - Feasibility
Standard Services	Work Stage C - Outline Proposals
	Work Stage D - Sketch Scheme Proposals
	Work Stage E - Detailed Proposals
	Work Stage F - Production Information
	Work Stage G - Bills of Quantities
	Work Stage H - Tender Action
	Work Stage J - Contract Preparation
	Work Stage K - Operations on site
	Work Stage L - Completion

The Landscape Consultant may also provide services outside the design, construction and management process, for example public inquiry commissions, landscape planning issues, environmental impact assessment, landscape appraisal and evaluation.

Part 2 - Other Services. Many commissions will require Other Services additional to the Preliminary and Standard Services, and it may be necessary to obtain and incorporate advice from various different consultants.

Part 3 - Conditions of Appointment. This part describes conditions which normally apply when a Landscape Consultant is commissioned.

0.4 The Parts are described in detail in succeeding sections of this document. In addition, the samples of the Schedule of Services and Memorandum of Agreement, including details of the arrangements for their separate publication, can be found in the Appendices.

0.5 Guidance on the various types of fee arrangements and their appropriate use is currently the subject of further discussion with the Office of Fair Trading.

APPENDIX I

PART 1	1	**LANDSCAPE CONSULTANT'S SERVICES**
	1.1	This Part describes the Preliminary and Standard Services which a Landscape Consultant will normally provide. These services are common to large and small commissions and none should be omitted if the commission is to be completed successfully. However, it may be prudent to vary the sequence of the Work Stages or to combine two or more stages to suit the particular circumstances. Where, for any reason, partial services only are to be provided, the agreement between the Client and the Consultant should indicate precisely the extent of those services. Preliminary Services comprise Work Stages A and B; Standard Services comprise Work Stages C to L inclusive. Other Services are described in Part 2.
Work Stage A	1.2	**Preliminary Services - Work Stage A: Inception**
Brief	1.2.1	Discuss and assess the Client's requirements including the timescale and financial limits; advise the Client on how to proceed; advise the Client of the Employer=s duties under the Construction (Design and Management) Regulations 1994 ("the CDM Regulations"); agree the Landscape Consultant's services and the terms of engagement and fee payment basis, confirming these in writing with the Client.
Information which may be provided by the Client	1.2.2	Obtain from the Client information on ownership, other legal interests in the site, existing features, including underground services, and any other matters which may influence the development or management requirements.
Site Appraisal	1.2.3	Visit the site and carry out an initial appraisal.
Advice on other Consultants, Specialist Firms and Site Staff	1.2.4	Advise on the need for other consultants' services and the extent of these services; advise on the need for specialist contractors or suppliers to execute the works; advise on the need for site staff.
Programme	1.2.5	Advise on outline programme and fee basis for further services, and obtain the Client=s agreement thereto.

Work Stage B	**1.3**	**Preliminary Services - Work Stage B: Feasibility**
Feasibility Studies	1.3.1	Undertake such studies as are necessary to determine the feasibility of the Client=s requirements; discuss with the Client alternative solutions and their technical and financial implications, advise on the need to obtain planning permissions and other statutory requirements.
Landscape Consultants Range of Services	1.3.2	In the light of the feasibility studies, agree with the Client the detailed extent of Standard and Other Services as required.
Work Stage C	**1.4**	**Standard Services - Work Stage C: Outline Proposals**
Outline Proposals	1.4.1	Broadly analyze the Client=s requirements, prepare outline proposals and approximate estimate of the cost of executing the proposals for the Client=s approval with other Consultants where appointed.
CDM Designers Duties	1.4.2	Perform the following duties of the Designer as defined in the CDM Regulations at the appropriate stages of the commission: - co-operate with the Planning Supervisor, if appointed; - pass relevant information to the Planning Supervisor, if appointed, for incorporation in the initial Health and Safety File.
Work Stage D	**1.5**	**Standard Services - Work Stage D: Sketch Scheme Proposals**
Sketch Scheme Proposals	1.5.1	Develop the sketch scheme proposals from those agreed in outline, taking into account any changes requested by the Client, prepare cost estimates and programme for implementation with other Consultants where appointed. The sketch scheme proposals should indicate the size and character of the project in sufficient detail to enable the Client to agree the spatial arrangements, materials and appearance.

Changes in Scheme Proposals	1.5.2	Advise the Client of the implications of any changes in the cost and timing for executing the proposals and obtain approval for such changes.
Outline Planning Application	1.5.3	If appropriate, consult with the Planning Authority and submit any necessary application for outline planning permission using sketch scheme proposals. See also 2.2.8.
Other Approvals	1.5.4	Similarly, make application for any other approvals from statutory bodies, using sketch scheme proposals, where these approvals are not dependent on detailed proposals being available. See also 2.2.8.
Work Stage E	**1.6**	**Standard Services - Work Stage E: Detailed Proposals**
Detailed Proposals	1.6.1	Develop the proposals in sufficient detail to obtain the Client's approval of the proposed materials, techniques and standards of workmanship. When acting as design team leader, co-ordinate the proposals made by other consultants, specialist contractors or suppliers; obtain quotations and other information in relation to specialist work.
Cost Checks and Changes in Detailed Proposals	1.6.2	Carry out cost checks where necessary and advise the Client of the consequences of any changes to the estimated cost and programme. Obtain the Client's consent to proceed.
Detailed Statutory Approvals	1.6.3	Make detailed applications for approvals under planning and building legislation where necessary, using detailed scheme proposals. See also 2.2.8.
Work Stages F & G	**1.7**	**Standard Services - Work Stages F and G: Production Information and Bills of Quantities**
Production Information	1.7.1	Prepare all production drawings, schedules and specification of materials and workmanship required for the execution of the work
Bills of Quantities	1.7.2	Provide information for bills of quantities to be prepared by others. All information to be supplied in sufficient detail to enable a contract to be negotiated or competitive tenders to be invited. See also 2.6.1.

1.8 Standard Services - Work Stages H and J: Tender Action and Contract Preparation

Work Stages H & J

Other Contracts — 1.8.1 Where necessary, arrange for other contracts to be let in advance of the Main Contractor starting work.

Tender Lists — 1.8.2 With the Client's participation advise on suitable contractors and obtain approval of a final list of tenderers.

Tender Action or Negotiation — 1.8.3 Invite tenders from approved contractors; appraise and advise on tenders submitted. Alternatively, arrange for a price to be negotiated with a contractor by the Quantity Surveyor. See also 2.6.1.

Contract Document Preparation — 1.8.4 Advise the Client on the appointment of the Contractor and on the responsibilities of the Client, the Contractor and the Landscape Consultant under the terms of the contract document; prepare the contract and arrange for it to be signed by the Client and the Contractor; provide production information as required by the contract.

1.9 Standard Services - Work Stage K: Operations on Site During Construction and 12 Months' Maintenance

Work Stage K

Contract Administration — 1.9.1 Administer the contract during operations on site including control of the Clerk of Works where appointed.

Inspections — 1.9.2 Visit the site at intervals appropriate to the Contractor's programmed activities to inspect the progress and quality of the works. Frequency of the inspections shall be agreed with the Client.

Accounts — 1.9.3 Check and certify the authenticity of accounts.

Financial Appraisal and Programme — 1.9.4 Make periodic financial reports to the Client; with other Consultants where appointed identify any variation in the cost of the works or in the expected duration of the contract.

1.10 Standard Services - Work Stage L: Completion

Work Stage L

Completion of Works — 1.10.1 Administer the terms of the contract relating to the completion of the works and give general guidance on activities after completion of contract.

PART 2   2   OTHER SERVICES

2.1 This part describes services which may be provided by the Landscape Consultant by prior agreement with the Client to augment the Preliminary and Standard Services described in Part 1 or which may be the subject of a separate appointment. The list of services is not exhaustive.

2.2 Surveys and Investigations

Site Evaluation	2.2.1	Advise on the selection and suitability of sites; conduct negotiations concerned with sites and their features.
Measured Surveys	2.2.2	Make measured surveys, take levels and prepare plans of sites and their features, including any existing buildings.
Site Investigation	2.2.3	Undertake investigations into, prepare reports and schedules, and give advice on the nature and condition of the vegetation, soil or other features of the site. Investigate failures, arrange and supervise exploratory work by contractors or specialists.
Maintenance and Management Cost in Use	2.2.4	Survey and analyze the usage, management and maintenance of a site, undertake cost in use studies, and analyze the need for additional design work.
Environmental Assessment	2.2.5	Undertake environmental assessment studies of the impact of development proposals and land use changes.
Development Plans	2.2.6	Prepare development plans for sites, where development of part of the whole site will not be immediate. Prepare Masterplans showing general scheme principles and layout, and integration with surrounding land uses or proposals.
Demolition and Clearance	2.2.7	Provide services in connection with demolition and clearance works.
Special Drawings and Models	2.2.8	Prepare special drawings, models or technical information for the use of the Client for applications under planning or building regulations, or other statutory requirements, or for negotiations with ground landlords, adjoining owners, public authorities, licensing authorities, mortgagees and

others; prepare plans for conveyancing, Land Registry and other legal or record purposes.

Prototype	2.2.9	Develop prototype proposals for repetitive use by the client, but only when such repetitive use is specifically agreed by the Landscape Consultant and Client as appropriate.
Site Furniture and Equipment	2.2.10	Design or advise on the selection of site furniture and equipment; arrange and inspect the fabrication of site furniture, arrange trials and training in use of equipment.
Multi-Disciplinary Meetings	2.2.11	Attendance at multi-disciplinary meetings held to discuss projects or elements outside the Landscape Consultant=s appointment or fee basis (where attendance would not otherwise be required for whole or part of meeting).
Public Meetings	2.2.12	Prepare and organise material for public consultation and liaison; attend public meetings.
Works of Art	2.2.13	Advise on the commissioning or selection of works of art in connection with landscape commissions.
Scientific Developments	2.2.14	Undertake research and conduct trials if necessary; especially where technical problems indicate that traditional solutions are inadequate. It is recommended that a full brief, fee basis and funding be separately agreed.
Visits to Nurseries	2.2.15	Visit horticultural nurseries to ascertain the quantity, quality and cost of stock available for purchase.

2.3 Cost Estimating and Financial Advisory Services

Cost Plans and Cash Flow Requirements	2.3.1	Carry out cost planning for a project, including the cost of professional fees; advise on cash flow requirements for fees, works and cost in use.
Schedules of Rates and Quantities	2.3.2	Prepare schedules of rates or bills of quantities for tendering purposes; measure work executed.
Cost of Replacement and Reinstatement of Damaged Landscape	2.3.3	Carry out inspections and surveys, prepare estimates for the replacement and reinstatement of damaged landscapes; submit and negotiate claims for compensation.

Grant Applications	2.3.4	Provide information; make applications for and conduct negotiations for grants.
	2.4	**Planning Negotiations**
Planning Applications; Exceptional Negotiations	2.4.1	Conduct detailed negotiations with a planning authority that become prolonged because of complexity.
Planning Appeals and Public Inquiries	2.4.2	Prepare and submit an appeal under planning acts; advise on other work in connection with planning appeals. Prepare and submit a proof of evidence for a public inquiry, appear as an expert witness at a public inquiry.
Royal Fine Art Commission	2.4.3	Make submissions to the Royal Fine Art Commission.
Building Regulations; Exceptional Negotiations	2.4.4	Conduct detailed negotiations for approvals under the building regulations; negotiate waivers or relaxations; all of which may become prolonged because of complexity.
Landlords= Approvals	2.4.5	Submit plans of proposed works for the approval of landlords, mortgagees, freeholders or others.
Rights of Owners and Lessees	2.4.6	Advise on the rights and responsibilities of owners or lessees including right of ways, rights of support, boundary, drainage and wayleave responsibilities etc.; such advice shall always be subject to confirmation by the Client's legal adviser.
Statutory Bodies	2.4.7	Liaise with, submit plans or technical information and conduct negotiations with statutory bodies e.g. Environment Agency, highway authority, statutory undertakers.
	2.5	**Additional Administration of Projects**
Site Staff	2.5.1	Provide or recruit site staff with the Client=s agreement, for frequent or constant inspection of the works.
Extended Administration	2.5.2	Administer and inspect contract aftercare and maintenance works extending beyond 12 months from completion of the capital works.

Suspension	2.5.3	Suspend and recommence work on projects where progress is unexpectedly delayed by more than 3 months due to financial constraints, delayed approvals or other reasons.
Project Management	2.5.4	Provide management from inception to completion; prepare briefs, appoint and co-ordinate consultants, implementation managers, agents, suppliers and contractors; monitor time cost and agreed targets; monitor progress of the works; handover projects on completion.
Record Drawings	2.5.5	Provide the Client with a set of drawings showing the main elements of the scheme; arrange for drawings of other services to be provided as appropriate.
Landscape Management Plans	2.5.6	Prepare management plans and maintenance schedules; prepare drawings, schedules and operational manuals, assess cost and staffing implications of proposals.
Contract Claims	2.5.7	Carry out the administration, evaluation and settlement of contract claims.
Litigation and Arbitration	2.5.8	Prepare and give evidence, settle proofs, confer with solicitors and counsel; attend court and arbitration hearings; appear before tribunals; act as arbitrator or adjudicator.
Health and Safety Plan	2.5.9	Prepare pre-tender Health and Safety Plan; notify the Health and Safety Executive of the project; assemble the Health and Safety File; prepare >as built= drawings for the Health and Safety File.
Planning Supervisor	2.5.10	Provide the services of Planning Supervisor under the CDM Regulations. The terms and fee for this service should be agreed under a separate appointment.
	2.6	**Services normally provided by other Consultants**
Other Consultants' Services	2.6.1	Where services such as quantity surveying, architecture, civil, structural, mechanical or electrical engineering, town planning or graphic design are provided from within the Landscape Consultant's own office or by other Consultants in association with the Landscape Consultant, it is recommended that the fees be separately agreed.

2.7 Additional Design Services

2.7.1 Alter or modify any design, specification, drawing or other document as a result of new or modified instructions from the Client or for other reasons which could not reasonably have been foreseen. Make and issue reproductions of such altered documents.

PART 3

3 CONDITIONS OF APPOINTMENT

3.1 This Part describes the conditions which apply to a Landscape Consultant's appointment. If different or additional services are required they should be set out in the Schedule of Services, Memorandum of Agreement or letter of appointment.

Duty of Care

3.2 The Landscape Consultant will use reasonable skill, care and diligence in accordance with the normal standards of the profession

Contamination or Pollution

3.3 Unless this clause is expressly stated not to apply, nothing in the agreement shall require the Consultant to provide advice or services in connection with the presence of or risk of contamination or pollution by harmful substances. The Client shall retain sole responsibility for determining what, if any, investigations and actions shall be taken in relation to such substances, and shall commission such professional advice as he considers necessary.

Landscape Consultant's Authority

3.4 The Landscape Consultant will act on behalf of the Client in the matters set out or implied in the Landscape Consultant's appointment. The Landscape Consultant will obtain the authority of the Client before initiating any service or Work Stage.

Modifications

3.5 The Landscape Consultant shall not significantly alter an approved proposal without Client approval. Should changes be found to be necessary during implementation, the Client shall be informed and consent obtained without delay.

Terms for Payment

3.6 The Client shall pay the Landscape Consultant for his services in accordance with paragraph 2 of the Memorandum of Agreement, save that in the event of termination or suspension of the commission the

Consultant shall be entitled to payment of a reasonable proportion of the interim payment or at the agreed hourly rate, whichever is applicable, for all work carried out to the date of termination or suspension and not previously the subject of an invoice for interim payment.

Revisions to the Conditions of Appointment	3.7	The agreement between a Client and a Landscape Consultant is deemed to allow for revisions due to changing circumstances. In long-term commissions such changes will probably be due to unforeseen factors or matters beyond the control of the Landscape Consultant at the date of the appointment.
Project Control	3.8	The Landscape Consultant will report any significant variations in authorised expenditure or contract period.
Impartiality	3.9	The Landscape Consultant will be impartial in administering the terms of a contract between Client and Contractor.
Appointment of other Consultants	3.10	Consultants may be appointed either by the Client direct or by the Landscape Consultant subject to acceptance by each party.
Sub-contracting	3.11	The Landscape Consultant shall not sub-contract any part of the commission without notifying the Client and receiving formal agreement on the division of responsibilities that will apply.
Liability of other Consultants	3.12	Where a Consultant is appointed under clause 3.10, the Landscape Consultant shall not be held liable for the other Consultant's work, provided that in relation to the execution of such work under a contract between Client and Contractor nothing in this clause shall affect any responsibility of the Landscape Consultant to perform his duties under the terms of that contract.
Consultant Co-ordination	3.13	The Landscape Consultant will have the responsibility to co-ordinate and integrate into the overall design the services provided by any consultant, however employed.
Design work by Contractors/ Suppliers	3.14	A specialist contractor, sub-contractor or supplier who is employed by the Client and who supplies design drawings to the Landscape Consultant for

		incorporation in the works may be appointed by agreement. The Landscape Consultant shall not be held liable for the execution and performance of this work. The Landscape Consultant will have the authority to integrate and co-ordinate this design information into the overall design.
Contractor Responsibility	3.15	The Client will employ a contractor under a separate agreement to undertake construction or other works not undertaken by the Landscape Consultant. The Client will hold the contractor, and not the Landscape Consultant, responsible for the contractor's operational methods and for the proper execution of the works.
Site Inspections	3.16	The Landscape Consultant will visit the site at intervals appropriate to the progress of the works. As these intervals will vary depending on the nature of the work, the Landscape Consultant will explain to the Client at the outset when inspections will be made and agree these with the Client. If more inspections/visits to the site are required by the Client, details, as an extension to the Standard Services, will be agreed in writing with the Client.
Site Staff and Resident Landscape Staff	3.17	Where frequent or constant professional inspection is agreed to be required, a resident professional shall be appointed on a full or part-time basis by the Consultant under specific terms of appointment and remuneration.
Site Staff/Clerk of Works	3.18	Where frequent or constant inspection of the works is required a Clerk of Works suitably qualified in the supervision of landscape operations shall be employed. The Clerk of Works may be employed by the Client, or the Landscape Consultant, but in either case will be under the control and direction of the Landscape Consultant
Information from Client	3.19	The Client is required to provide the Landscape Consultant with such information and make such decisions as are necessary for the proper performance of the agreed service. The requirement and reasons for such timely action shall be explained to the Client by the Landscape Consultant so that the implications of delay are clearly understood by both parties.

Client Representative	3.20	The Client, if a firm or other body of persons, will, when requested by the Landscape Consultant, nominate a responsible representative through whom all instructions will be given.
Copyright	3.21	Copyright in all documents and drawings prepared by the Landscape Consultant shall unless otherwise agreed remain the property of the Consultant. Where so agreed, copyright shall be passed to the Client only after all fees due to the Landscape Consultant have been paid.
Copyright Entitlement	3.22	The Client will be entitled to use documents and drawings in executing the works for which they were prepared by the Landscape Consultant provided that: a) All fees due to the Landscape Consultant have been submitted or paid. b) The entitlement relates only to that site or part of the site for which the design was prepared. This entitlement applies to the design, maintenance and management of the works.
Assignment	3.23	Neither the Client nor the Landscape Consultant shall assign the appointment in whole or in part without prior written agreement as to the division of responsibilities that apply.
Suspension	3.24	The length of notice for suspension should be agreed in writing at the outset. The Client may suspend the Landscape Consultant's appointment in whole or in part, the notice given being in accordance with the agreed timing and in writing.
Suspension by Consultant	3.25	The Landscape Consultant will give immediate notice in writing to the Client of any situation arising from force majeure which makes it impractical to carry out any of the agreed services, and agree with the Client a suitable course of action.
Resumption of Service	3.26	Following the notice in accordance with clause 3.24 if no instruction has been received within 6 months, the Landscape Consultant shall make a written request for instructions. If no instruction is received within 30 days, the appointment shall be treated as terminated.

Suspension of Obligations	3.27	If the Client fails to make payment in accordance with Clause 2 of the Memorandum of Agreement and no effective notice to withhold payment has been given, the Landscape Consultant may, following 7 days= notice setting out the grounds for suspension, suspend the performance of his obligations under the agreement until payment is received.
Termination	3.28	The Landscape Consultant's appointment may be terminated by either party by following the procedure in clause 3.24, where this is permitted by the appointment.
Death or Incapacity	3.29	If death or incapacity of a sole practitioner stops the Landscape Consultant from carrying out the agreed duties under this appointment, it shall be terminated. As soon as all outstanding fees have been submitted or paid, the Client will be entitled to use all data prepared on the project subject to the provisions in respect of copyright in accordance with clause 3.21 and 3.22.
Adjudication	3.30	In the event of a dispute arising under the agreement, either party may give notice at any time to the other of his intention to refer the dispute to adjudication.
Appointment of Adjudicator	3.31	An adjudicator shall be appointed by agreement between the parties within 2 working days of receipt of notice under 3.30 or, failing agreement, within 7 days of the said notice by the President or a Vice-President of the Landscape Institute. The adjudicator shall conduct the adjudication in such manner as he considers fit, having regard to the Construction Industry Council's Model Adjudication Procedure, and subject to the following matters:

within 7 working days of notice under 3.30 the parties shall agree and provide to the adjudicator a joint statement of undisputed facts (so as to reduce the area of dispute to a minimum);

the adjudicator shall act impartially and shall reach a decision within 28 days of referral of the dispute to him or such longer period as the parties may agree; the adjudicator shall be entitled to extend the period of 28 days by up to 14 days with the consent of the party giving notice under 3.30;

the adjudicator shall be entitled to take the initiative in ascertaining the facts and the law;

the parties shall accept the adjudicator's decision as binding upon them until such time as the dispute is finally resolved in accordance with a ruling under 3.32, arbitration under 3.33 or by agreement;

alternatively the parties may agree to accept the decision of the adjudicator as finally determining the dispute;

the adjudicator, or any employee or agent of his, shall not be liable to the parties for anything done or omitted in the discharge or purported discharge of his functions as adjudicator unless the act or omission is in bad faith.

Ruling on a Joint Statement	3.32	Any difference or dispute arising from a written appointment under this document may, by agreement, be referred to the Landscape Institute for a ruling by the President. The parties must agree: to prepare and submit with their submission a joint statement of undisputed facts to reduce the area of dispute to a minimum; and to accept the ruling as final and binding.
Arbitration	3.33	Subject to the parties' rights under 3.30 to 3.32, any difference or dispute arising out of the appointment shall be referred to arbitration by a person to be agreed between the parties or, failing agreement within 21 days after either party has given to the other a written request to concur in the appointment of an arbitrator, a person to be appointed at the request of either party by the President or a Vice-President for the time being of the Chartered Institute of Arbitrators.
Alternative for Scotland	3.34	In Scotland, any difference or dispute arising out of the appointment which cannot be resolved in accordance with clause 3.32 or 3.33 shall be referred to arbitration by a person to be agreed between the parties or, failing agreement within 21 days after either party has given to the other a written request to concur in the appointment of an arbiter, a person to be nominated at the request of either party by the Chairman for the time being of the Scottish Branch of the Chartered Institute of Arbitrators.

Settlement of Disputes by Agreement	3.35	Nothing herein shall prevent the parties agreeing to settle any difference or dispute arising out of the appointment without recourse to arbitration or adjudication.
Governing laws England and Wales	3.36	The application of these conditions shall be governed by the laws of England and Wales. or
Alternative for Scotland		The application of these conditions shall be governed by the laws of Scotland. or
Alternative for Northern Ireland		The application of these conditions shall be governed by the laws of Northern Ireland.
CDM Regulations 1994	3.37	Where the Client is required under the CDM Regulations to appoint a Planning Supervisor, the following Additional Conditions CDM.1 - CDM.6 shall apply to this agreement.
	CDM1	The Landscape Consultant shall, where the Client is required by CDM Regulations to appoint a Planning Supervisor, co-operate with and pass relevant information to the Planning Supervisor (whether within the same firm of landscape consultants or otherwise).
	CDM2	The Client shall, where required by the CDM Regulations, appoint as soon as reasonably practicable a competent Planning Supervisor. The Client shall procure that the appointment remains filled at all times until construction is completed.
	CDM3	The Landscape Consultant shall, where the Client is required by the CDM Regulations to have appointed a Planning Supervisor and where after the design information is complete the Client orders changes necessitating re-work, continue to co-operate with and pass relevant information to the Planning Supervisor.
	CDM4	The Client shall, where required by the CDM Regulations, appoint as soon as reasonably practicable a competent Principal Contractor.

CDM5 The Client shall, where required by the CDM Regulations to appoint a Planning Supervisor, and where consultants are appointed, procure the consultants' co-operation in the exchange of information relating to health and safety aspects of their work, and shall co-operate with the Planning Supervisor.

CDM6 The Client shall, where required by the CDM Regulations to appoint a Planning Supervisor, and where specialists are appointed, procure the specialists' co-operation in the exchange of information relating to health and safety aspects of their work, and shall co-operate with the Planning Supervisor.

APPENDIX I
MEMORANDUM OF AGREEMENT

Between Client and Landscape Consultant for use with the Landscape Consultant's Appointment.

This Agreement

is made on the ....day of ....(month and year)

Between ....
(insert name of Client)

Of ....

....
(hereinafter called the Client)

And ....
(insert name of Landscape Consultant or firm of Landscape Consultants)

Of ....

....
(hereinafter called the Consultant)

NOW IT IS HEREBY AGREED
that upon the Conditions of the Landscape Consultant's Appointment (Parts 1,2 and 3) (Revision) attached hereto as Appendix I

save as excepted or varied by the parties hereto in the Schedule of Services and Fees, hereinafter called the 'Schedule', attached hereto as Appendix II

and subject to any special conditions set out or referred to in the Schedule:

1. The Consultant will perform for the Client the services listed on the Schedule in respect of

....
(insert general description of the Project)

at ....
(insert location of the Project)

2.	The Client will pay the Consultant on demand for the services, fees and expenses as indicated in the Schedule; in accordance with the following provisions:
2.1	The Consultant shall deliver invoices in the instalments and on the dates or in the circumstances specified in the Schedule. Fees so invoiced shall be due to the Consultant on the date of receipt by the Client of the invoice. If sent by post, the invoice shall be deemed to have been received two working days after posting.
2.2	Not later than 5 days after the date of receipt of an invoice in accordance with clause 2.1 hereof, the client shall acknowledge receipt of the invoice and give notice to the Consultant specifying the amount (if any) of the payment to be made and the basis on which that amount is calculated.
2.3	The Final Date for Payment of any sum due from the Client to the Consultant shall be 17 days* [...] after the sum becomes due. *insert other period if required
2.4	Any notice given by the Client of intention to withhold payment pursuant to section 111(1) of the Housing Grants, Construction and Regeneration Act 1996 shall be given not later than 7 days before the Final Date for Payment of any sum due to the Consultant.

AS WITNESS the hands of the parties the day and year first above written

Signatures: Client..

Landscape Consultant...............................

Witnesses: Name............   Name............
Address............................   Address........................ .
Description........   Description........

APPENDIX II

SCHEDULE OF SERVICES AND FEES

Referred to in the Memorandum of Agreement dated ....

Between

....
(insert name of Client)

And

....
(insert name of Landscape Consultant)

For

....
(insert description of project)

Unless otherwise stated the services listed in the conditions of appointment will be as described in The Landscape Consultant's Appointment (Parts 1,2 and 3) (........ Revision), issued by The Landscape Institute. Clause references relate to that document.

S1 SERVICES

Service	Clause	Fee basis
		(State whether percentage, time or lump sum)
Preliminary Services		
Standard Services		
Other Services		

S2 SPECIAL CONDITIONS

CONDITIONS WHICH SHALL APPLY — Insert any special conditions which are to apply to the appointment.

CONDITIONS NOT TO APPLY — Insert any clauses which are not to apply to this appointment.

FEES

PERCENTAGE FEES — Fees based on a percentage of the total construction cost shall be calculated as follows:

LUMP SUM FEES — The total fee for all services shown under S1 Services to be on a lump sum basis shall be ,..........................

TIME CHARGE FEES — Rates for fees charged on a time basis shall be:

,....../hour
,....../hour
.1 for Principals ,....../hour
,....../hour
.2 for other staff grade......... ,....../hour

.3 for other staff grade............

.4 for other staff grade.........

.5 for other staff grade.........

INTERIM PAYMENTS — The Consultant shall be entitled to (and shall render invoices for) interim payment of fees:
.1 for work on a percentage or lump sum basis, at the completion of work stages and / or additional interim stages, as follows:

* insert stages and proportions to cover <u>all</u> interim payments

Work stage	Proportion of percentage or lump sum fee	Cumulative total

EXPENSES AND DISBURSEMENTS — * The fees charged are inclusive of all expenses and disbursements

Landscape Consultants Appointment

* delete where not applicable and insert location of rates schedule.

or

* Expenses and disbursements shall be charged in addition, in accordance with rates specified

....

* delete where not applicable and insert commencement date

.2 for work on a time-charge basis *monthly / quarterly / half-yearly from the commencement of work on

....

Signed: ....
              Client   Consultant

Date: ....

5
Fees for Professional Services

LANDSCAPE ARCHITECTS' FEES

The Landscape Institute does not set any fees scales for its members. A publication which offers guidance in determining fees for different types of project is available from

The Landscape Institute
6/8 Barnard Mews London SW11 1QY   Telephone: 020 7350 5200

The publication is entitled
"Engaging a Landscape Consultant – Guidance for Clients on Fees.

The document refers to suggested fee systems on the following basis:
- Time Charged Fee Basis
- Lump Sum Fee Basis
- Percentage Fee Basis

The publication lists various project types and relates them to complexity ratings for works valued at £20.000.00 and above. The suggested fee scales are dependent on the complexity rating of the project.

The following are samples of the Complexity Categories

Category 1:   Golf Courses, Country Parks and Estates, Planting Schemes

Category 2:   Coastal, River and Agricultural Works, Roads, Rural recreation schemes.

Category 3:   Hospital grounds, Sport stadia, Urban offices and Commercial properties, Housing

Category 4 :   Domestic/Historic Garden Design, Urban rehabilitation and Environmental improvements

Table showing suggested fees for various categories of complexity of Landscape Design or consultancy

	Complexity			
Project Value	1	2	3	4
20,000	3,000	3,250	3,600	4,200
30,000	3,900	4,275	4,725	5,460
50,000	5,600	6,125	6,750	7,875
100,000	9,250	10,200	11,200	13,000
160,000	12,800	14,400	15,840	18,400
200,000	15,600	17,500	21,000	24,200
300,000	21,300	24,000	26,400	31,200
500,000	33,000	37,000	45,000	52,500
750,000	48,750	52,500	57,750	67,500
1,000,000	62,000	69,000	75,000	88,000
10,000,000	610,000	680,000	710,000	830,000

Fees for Professional Services

Fees for Other Professional Services

Extracts from the scales of fees for architects, quantity surveyors and consulting engineers are given together with extracts from the Town and Country Planning Regulations 1993 and Building Regulation Charges. These extracts are reproduced by kind permission of the bodies concerned, in the case of Building Regulation Charges, by kind permission of the London Borough of Ealing. Attention is drawn to the fact that the full scales are not reproduced here and that the extracts are given for guidance only. The full authority scales should be studied before concluding any agreement and the reader should ensure that the fees quoted here are still current at the time of reference.

ARCHITECTS' FEES

Standard Form of Agreement for the Appointment of an Architect (SFA/99),	Pages 58 and 59
Conditions of Engagement for the Appointment of an Architect (CE/99),	Page 58 (brief notes only)
Small Works (SW/99),	Page 58 (brief notes only)
Employer's Requirements (DB1/99),	Page 58 (brief notes only)
Contractor's Proposals (DB2/99),	Page 58 (brief notes only)
Form of Appointment as Planning Supervisor (PS/99),	Page 59 (brief notes only)
Form of Appointment as Sub-Consultant (SC/99),	Page 59 (brief notes only)
Form of Appointment as Project Manager (PM/99),	Page 59 (brief notes only)

QUANTITY SURVEYORS' FEES

Scale 36,	inclusive scale of professional charges, page 62
Scale 37,	itemised scale of professional charges, page 67
Scale 40,	professional charges for housing schemes for Local Authorities, page 87
Scale 44,	professional charges for improvements to existing housing and environmental improvement works, page 93
Scale 45,	professional charges for housing schemes financed by the Housing Corporation, page 96

ARCHITECTS' FEES

RIBA Forms of Appointment 1999
- include forms to cover all aspects of an Architect's practice;
- identify the traditional roles of an Architect;
- reflect or take into account recent legislation such as the Housing Grants, Construction and Regeneration Act 1996, the Arbitration Act 1996 (*applies to England and Wales only*), the Unfair Terms in Consumer Contracts Regulations 1994, Latham issues etc;
- update the Outline Plan of Work Stage titles and descriptions;
- are sufficiently flexible for use with English or Scottish Law [1]; the building projects of any size (except small works); other professional services [2]; different project procedures.

[1] The Royal Incorporation of Architects in Scotland will publish appropriate replacement pages for use where the law of Scotland applies to the Agreement.
[2] Excluding acting as adjudicator, arbitrator, expert witness, conciliator, party wall surveyor etc.

All forms require the Architect to agree with the Client the amount of professional indemnity insurance cover for the project.

Standard Form of Agreement for the Appointment of an Architect (SFA/99)
The core document from which all other forms are derived. Suitable for use where an Architect is to provide professional services for a fully designed building project in a wide range of size or complexity and/or to provide other professional services. Used with Articles of Agreement. Includes optional Services Supplement.

Conditions for Engagement for the Appointment of an Architect (CE/99)
Suitable for use where an Architect is to provide services for a fully designed building project and/or to provide other professional services where a Letter of Appointment is preferred in lieu of Articles of Agreement. Includes optional (modified) Services Supplement and draft Model letter.

Small Works (SW/99)
Suitable for the provision of professional services where the cost of construction is not expected to exceed £150,000 and use of the JCT Agreement for Minor Works is appropriate. The appointment is made by a specially drafted Letter of Appointment, the Conditions and an optional Schedule of Services for Small Works.

Employer's Requirements (DB1/99)
A supplement to amend SFA/99 and CE/99 where an Architect is appointed by the Employer Client to prepare Employer's Requirements for a Design and Build contract. Includes replacement Services Supplement and notes on completion for initial appointment, where a change to Design and Build occurs later or where "consultant switch" is contemplated.

Contractor's Proposals (DB2/99)
A supplement to amend SFA/99 where an Architect is appointed by the Contractor Client to prepare Contractor's Proposals under a Design and Build contract. Includes replacement Services Supplement and notes on completion for initial appointment and for "consultant switch"

Form of Appointment as Planning Supervisor (PS/99)
Used with Articles of Agreement. Provides for the Planning Supervisor to prepare the Health and Safety File and the pre-tender Health and Safety Plan.

Form of Appointment as Sub-Consultant (SC/99)
Suitable for use where a Consultant wishes (another Consultant (Sub-Consultant) or Specialist) to perform a part of his responsibility but not for use where the intention is for the Client to appoint Consultants or Specialists directly. Used with Articles of Agreement. Includes draft form of Warranty to the Client.

Form of Appointment as Project Manager (PM/99)
Suitable for a wide range of projects where the Client wishes to appoint a Project Manager to provide a management service and/or other professional service. Used with Articles of Agreement. Does not duplicate or conflict with the Architect's services under other RIBA forms.

Guides

Matters connected with using and completing the forms are covered in "The Architect's Contract: A guide to RIBA Forms of Appointment 1999 and other Architect's appointments". The guidance covers the Standard Form (SFA/99) including a worked example; the options for calculating fees i.e. percentages, calculated or fixed lump sums, time charges and other methods; the other forms in the suite; the Design and Build amendments including advice on 'consultant switch' and 'novation' agreements; and topics for other appointments connected with dispute resolution, etc. Notes on appointments for Historic Buildings and Community Architecture projects are also included.

A series of guides "Engaging an Architect", addressed directly to clients, is also published on topics associated with the appointment of an Architect. These guides include graphs showing indicative percentage fees for the architect's normal services in stages C-L under SFA/99, CE/99 or SW/99 for works to new and existing buildings. Indicative hourly rates are also shown. Tables 2, 3 and 4 on page 8 show by interpolation of the graphs the indicative fees for different classes of building adjusted to reflect current tender price indices. Table 1 gives the classification of buildings. Indicative fees do not include expenses, disbursements or VAT fees.

The appointing documents and guides are published by:
RIBA Publications, Construction House, 56-64 Leonard Street, London. EC2A 4LT.
Telephone: 0207 251 0791   Fax: 0207 608 2375

The Standard Form of Agreement for the Appointment of an Architect (SFA/99)
SFA/99 is the core document, which is used as the basis for all the documents in the RIBA 1999 suite. It should be suitable for any Project to be procured in the 'traditional' manner. Supplements are available for use with Design and Build procurement. It is used with Articles of Agreement and formal attestation underhand or as a deed.
There are some changes from SFA/92 in presentation, in some of the responsibilities and

liabilities of the parties, to some definitions and clauses and their arrangement for greater clarity or flexibility, for compliance with the Construction Act; and to ensure that each role of the Architect is separately identified.

It comprises:

- the formal declaration of intent - Recitals and Articles;
- Appendix to the Conditions;
- Schedule 1 - Project Description;
- Schedule 2 - Services indicating the roles the Architect is to perform, which Work Stages apply and any other services required. A table of 'Other activities' identifies items not included in 'Normal Services';
- Schedule 3 - Fees and Expenses;
- Schedule 4 - Other appointments;
- the Conditions governing performance of the contract and obligations of the parties;
- a Services Supplement describing the Architect's roles and activities' which can be edited to suit the requirements of the Project or removed if it is not suitable for the Project, and if appropriate, another inserted in its place; and
- notes on completion.

TABLE 1   INDICATIVE PERCENTAGE FEES FOR NEW WORKS

Construction Cost £	Class 1 %	Class 2 %	Class 3 %	Class 4 %	Class 5 %
50,000	7.90	8.70	-	-	-
75,000	7.25	7.80	8.40	-	-
100,000	7.10	7.60	8.20	8.90	9.60
250,000	6.20	6.70	7.20	7.80	8.40
500,000	5.75	6.25	6.75	7.25	7.90
1,000,000	5.40	5.90	6.20	6.80	7.50
2,500,000	5.15	5.60	6.10	6.60	7.10
5,000,000	-	-	5.97	6.50	7.00
over 10,000,000	-	-	5.95	6.45	6.97

TABLE 2   INDICATIVE PERCENTAGE FEES FOR WORKS TO EXISTING BUILDINGS

Construction Cost £	Class 1 %	Class 2 %	Class 3 %	Class 4 %	Class 5 %
50,000	11.60	12.60	-	-	-
75,000	10.70	11.60	12.40	-	-
100,000	10.40	11.30	12.20	13.15	14.10
250,000	9.30	10.10	10.85	11.75	12.55
500,000	8.70	9.45	10.20	11.05	11.80
1,000,000	8.25	9.00	9.70	10.55	11.30
2,500,000	-	-	9.25	10.00	10.75
5,000,000	-	-	9.10	9.85	10.55
over 10,000,000	-	-	9.00	9.75	10.45

TABLE 3   INDICATIVE HOURLY RATES

Type of work	General	Complex	Specialist
Partner/Director or equivalent	£95	£140	£180
Senior Architect	£75	£105	£140
Architect	£55	£75	£95

QUANTITY SURVEYORS' FEES

Author's Note:

The Royal Institution of Chartered Surveyors formally abolished the standard Quantity Surveyors fee scales with effect from 31 December 1998. To date the Institution has not published a guidance booklet to assist practitioners in compiling fee proposals and we believe that they are unlikely to do so. The last standard Quantity Surveyors fee scales have therefore been reproduced here in order to assist the reader. Information considered to be relevant to Landscape and External works projects has been extracted from the published document. Readers should refer to the full document for full information.

SCALE 36
INCLUSIVE OF PROFESSIONAL CHARGES FOR QUANTITY SURVEYING SERVICES FOR BUILDING WORKS ISSUED BY THE ROYAL INSTITUTION OF CHARTERED SURVEYORS.
See Author's note above.

EFFECTIVE FROM JULY 1988 TO DECEMBER 1998

1.0. GENERALLY

1.1 This scale is for the use when an inclusive scale of professional charges is considered to appropriate by mutual agreement between the employer and the quantity surveyor.

1.2 This scale does not apply to civil engineering works, housing schemes financed by local authorities and the Housing Corporation and housing improvement work for which separate scales of fees have been published.

1.3 The fees cover quantity surveying services as may be required in connection with a building project irrespective of the type of contract from initial appointment to final certification of the contractor's account such as:

(a) Budget estimating; cost planning and advice on tendering procedures and contract arrangements.
(b) Preparing tendering documents for main contract and specialist sub-contracts; examining tenders received and reporting thereon or negotiating tenders and pricing with a selected contractor and/or sub-contractors.
(c) Preparing recommendations for interim payments on account to the contractor; preparing periodic assessments of anticipated final cost and reporting thereon; measuring work and adjusting variations in accordance with the terms of the contract and preparing final account, pricing same and agreeing totals with the contractor.
(d) Providing a reasonable number of copies of bills of quantities and other documents; normal travelling and other expenses. Additional copies of documents, abnormal travelling and other expenses (e.g. in remote areas or overseas) and the provision of checkers on site shall be charged in addition by prior arrangement with the employer.

Fees for Professional Services

1.4. If any of the materials used in the works are supplied by the employer or charged at a preferential rate, then the actual or estimated market value thereof shall be included in the amounts upon which fees are to be calculated.

1.5. If the quantity surveyor incurs additional costs due to exceptional delays in building operations or any other cause beyond the control of the quantity surveyor then the fees may be adjusted by agreement between the employer and the quantity surveyor to cover the reimbursement of these additional costs.

1.6. The fees and charges are in all cases exclusive of value added tax which will be applied in accordance with legislation.

1.7. Copyright in bills of quantities and other documents prepared by the quantity surveyor is reserved to the quantity surveyor.

2.0. INCLUSIVE SCALE

2.1. The fees for the services outlined in para.1.3, subject to the provision of para. 2.2, shall be as follows:

(a) Category A: Relatively complex works and/or works with little or no repetition.

Examples: Ambulance and fire stations; banks; cinemas; clubs; computer buildings; council offices; crematoria; fitting out of existing buildings; homes for the elderly; hospitals and nursing homes; laboratories; law courts; libraries; "one off" houses; petrol stations; places of religious worship; police stations; public houses, licensed premises; restaurants; sheltered housing; sports pavilions; theatres; town halls; universities, polytechnics and colleges of further education (other than halls of residence and hostels); and the like.

Value of work £	Category A fee		
	£		£
Up to 150 000	380 + 6.0%	(Minimum fee £3 380)	
150 000 - 300 000	9 380 + 5.0%	on balance over	150 000
300 000 - 600 000	16 880 + 4.3%	on balance over	300 000
600 000 - 1 500 000	29 780 + 3.4%	on balance over	600 000
1 500 000 - 3 000 000	60 380 + 3.0%	on balance over	1 500 000
3 000 000 - 6 000 000	105 380 + 2.8%	on balance over	3 000 000
Over 6 000 000	189 380 + 2.4%	on balance over	6 000 000

(b) Category B: Less complex works and/or works with some element of repetition.

Examples:
Adult education facilities; canteens; church halls; community centres; departmental stores; enclosed sports stadia and swimming baths; halls of residence; hostels; motels; offices other than those included in Categories A and C; railway stations; recreation and leisure centres; residential hotels; schools; self-contained flats and maisonettes; shops and shopping centres; supermarkets and hypermarkets; telephone exchanges; and the like.

Value of work £	Category B fee		
	£		£
Up to 150 000	360 + 5.8%	(Minimum fee £3 260)	
150 000 - 300 000	9 060 + 4.7%	on balance over	150 000
300 000 - 600 000	16 110 + 3.9%	on balance over	300 000
600 000 - 1 500 000	27 810 + 2.8%	on balance over	600 000
1 500 000 - 3 000 000	53 010 + 2.6%	on balance over	1 500 000
3 000 000 - 6 000 000	92 010 + 2.4%	on balance over	3 000 000
Over 6 000 000	164 010 + 2.0%	on balance over	6 000 000

(c) Category C: Simple works and/or works with a substantial element of repetition.

Examples:
Factories; garages; multi-storey car parks; open-air sports stadia; structural shell offices not fitted out; warehouses; workshops; and the like.

Value of work £	Category C fee		
	£		£
Up to 150 000	300 + 4.9%	(Minimum fee £2 750)	
150 000 - 300 000	7 650 + 4.1%	on balance over	150 000
300 000 - 600 000	13 800 + 3.3%	on balance over	300 000
600 000 - 1 500 000	23 700 + 2.5%	on balance over	600 000
1 500 000 - 3 000 000	46 200 + 2.2%	on balance over	1 500 000
3 000 000 - 6 000 000	79 200 + 2.0%	on balance over	3 000 000
Over 6 000 000	139 200 + 1.6%	on balance over	6 000 000

(d) Fees shall be calculated upon the total of the final account for the whole of the work including all nominated sub-contractors' and nominated supplier's accounts. When work normally included in a building contract is the subject of a separate contract for which the quantity surveyor has not been paid fees under any other clause hereof, the value of such work shall be included in the amount upon which fees are charged.

(e) When a contract comprises buildings which fall into more than one category, the fee shall be calculated as follows:

(i) The amount upon which fees are chargeable shall be allocated to the categories of work applicable and the amounts so allocated expressed as percentages of the total amount upon which fees are chargeable.

(ii) Fees shall then be calculated for each category on the total amount upon which fees are chargeable.

(iii) The fee chargeable shall then be calculated by applying the percentages of work in each category to the appropriate total fee and adding the resultant amounts.

(iv) A consolidated percentage fee applicable to the total value of the work may be charged by prior agreement between the employer and the quantity surveyor. Such a percentage shall be based on this scale and on the estimated cost of the various categories of work and calculated in accordance with the principles stated above.

(v) When a project is subject to a number of contracts then, for the purpose of calculating fees, the values of such contracts shall not be aggregated but each contract shall be taken separately and the scale of charges (paras. 2.1 (a) to (e)) applied as appropriate.

3.0. ADDITIONAL SERVICES

3.1. For additional services not normally necessary, such as those arising as a result of the termination of a contract before completion, liquidation, fire damage to the buildings, services in connection with arbitration, litigation and investigation of the validity of contractors' claims, services in connection with taxation matters, and all similar services where the employer specifically instructs the quantity surveyor, the charges shall be in accordance with para. 4.0 below.

4.0. TIME CHARGES

4.1. (a) For consultancy and other services performed by a principal, a fee by arrangement according to the circumstances including the professional status and qualifications of the quantity surveyor.

(b) When a principal does work which would normally be done by a member of staff, the charge shall be calculated as para. 4.2 below.

4.2. (a) For services by a member of staff, the charges for which are to be based on the time involved, such charges shall be calculated on the hourly cost of the individual involved plus 145%.

(b) A member of staff shall include a principal doing work normally done by an employee (as para. 4.1 (b) above), technical and supporting staff, but shall exclude secretarial staff or staff engaged upon general administration.

(c) For the purpose of para. 4.2 (b) above, a principal's time shall be taken at the rate applicable to a senior assistant in the firm.

(d) The supervisory duties of a principal shall be deemed to be included in the addition of 145% as para. 4.2 (a) above and shall not be charged separately.

(e) The hourly cost to the employer shall be calculated by taking the sum of the annual cost of the member of staff of:

(i) Salary and bonus but excluding expenses;
(ii) Employer's contributions payable under any Pension and Life Assurance Schemes;
(iii) Employer's contributions made under the National Insurance Acts, the Redundancy Payments Act and any other payments made in respect of the employee by virtue of any statutory requirements; and
(iv) Any other payments or benefits made or granted by the employer in pursuance of the terms of employment of the member of staff;
- and dividing by 1,650.

5.0. INSTALMENT PAYMENTS

5.1 In the absence of agreement to the contrary, fees shall be paid by instalments as follows:

(a) Upon acceptance by the employer of a tender for the works, one half of the fee calculated on the amount of the accepted tender.

(b) The balance by instalments at intervals to be agreed between the date of the first certificate and one month after final certification of the contractor's account.

5.2 (a) In the event of no tender being accepted, one half of the fee shall be paid within three months of completion of the tender documents. The fee shall be calculated upon the basis of the lowest original bona fide tender received. In the event of no tender being received, the fee shall be calculated upon a reasonable valuation of the works based upon the tender documents.

(b) In the event of the project being abandoned at any stage other than those covered by the foregoing, the proportion of fee payable shall be by agreement between the employer and the quantity surveyor.

NOTE: In the foregoing context "bona fide tender" shall be deemed to mean a tender submitted in good faith without major errors of computation and not subsequently withdrawn by the tenderer.

SCALE 37
ITEMISED SCALE OF PROFESSIONAL CHARGES FOR QUANTITY SURVEYING SERVICES FOR BUILDING WORK ISSUED BY THE ROYAL INSTITUTION OF CHARTERED SURVEYORS.
See Author's note on page 62.

EFFECTIVE FROM JULY 1988 TO DECEMBER 1998

1.0. GENERALLY

1.1. The fees are in all cases exclusive of travelling and other expenses (for which the actual disbursement is recoverable unless there is some prior arrangement for such charges) and of the cost of reproduction of bills of quantities and other documents, which are chargeable in addition at net cost.

1.2. The fees are in all cases exclusive of services in connection with the allocation of the cost of the works for purposes of calculating value added tax for which there shall be an additional fee based on the time involved (see paras. 19.1 and 19.2).

1.3. If any of the materials used in the works are supplied by the employer or charged at a preferential rate, then the actual or estimated market value thereof shall be included in the amounts upon which fees are to be calculated.

1.4. The fees are in all cases exclusive of preparing a specification of the materials to be used and the works to be done, but the fees for preparing bills of quantities and similar documents do include for incorporating preamble clauses describing the materials and workmanship (from instructions given by the architect and/or consulting engineer)

1.5. If the quantity surveyor incurs additional costs due to exceptional delays in building operations or any other cause beyond the control of the quantity surveyor then the fees may be adjusted by agreement between the employer and the quantity surveyor to cover the reimbursement of these additional costs.

1.6. The fees and charges are in all cases exclusive of value added tax which will be applied in accordance with legislation.

1.7. Copyright in bills of quantities and other documents prepared by the quantity surveyor is reserved to the quantity surveyor.

CONTRACTS BASED ON BILLS OF QUANTITIES: PRE-CONTRACT SERVICES

2.0. BILLS OF QUANTITIES

2.1. Basic scale

For preparing bills of quantities and examining tenders received and reporting thereon.

(a) Category A: Relatively complex works and/or works with little or no repetition.

Examples:
Ambulance and fire stations; banks; cinemas; clubs; computer buildings; council offices; crematoria; fitting out of existing buildings; homes for the elderly; hospitals and nursing homes; laboratories; law courts; libraries; "one off" houses; petrol stations; places of religious worship; police stations; public houses; licensed premises; restaurants; sheltered housing; sports pavilions; theatres; town halls; universities, polytechnics and colleges of further education (other than halls of residence and hostels); and the like.

Value of work £	Category A fee		
	£		£
Up to 150 000	230 + 3.0%	(Minimum fee £1730)	
150 000 - 300 000	4 730 + 2.3%	on balance over	150 000
300 000 - 600 000	8 180 + 1.8%	on balance over	300 000
600 000 - 1 500 000	13 580 + 1.5%	on balance over	600 000
1 500 000 - 3 000 000	27 080 + 1.2%	on balance over	1 500 000
3 000 000 - 6 000 000	45 080 + 1.1%	on balance over	3 000 000
Over 6 000 000	78 080 + 1.0%	on balance over	6 000 000

(b) Category B: Less complex works and/or works with some element of repetition.

Examples:
Adult education facilities; canteens; church halls; community centres; departmental stores; enclosed sports stadia and swimming baths; halls of residence; hostels; motels; offices other than those included in Categories A and C; railway stations; recreation and leisure centres; residential hotels; schools; self-contained flats and maisonettes; shops and shopping centres; supermarkets and hypermarkets; telephone exchanges; and the like.

Value of work	Category B fee		
	£		£
Up to 150 000	210 + 2.8%	(Minimum fee £1 680)	
150 000 - 300 000	4 410 + 2.0%	on balance over	150 000
300 000 - 600 000	7 410 + 1.5%	on balance over	300 000
600 000 - 1 500 000	11 910 + 1.1%	on balance over	600 000
1 500 000 - 3 000 000	21 810 + 1.0%	on balance over	1 500 000
3 000 000 - 6 000 000	36 810 + 0.9%	on balance over	3 000 000
Over 6 000 000	63 810 + 0.8%	on balance over	6 000 000

(c) Category C: Simple works and/or works with a substantial element of repetition

Examples:
Factories; garages; multi-storey car parks; open-air sports stadia; structural shell offices not fitted out; warehouses; workshops and the like.

Value of work £	Category C fee		
	£		£
Up to 150 000	180 + 2.5%	(Minimum fee £1 430)	
150 000 - 300 000	3 930 + 1.8%	on balance over	150 000
300 000 - 600 000	6 630 + 1.2%	on balance over	300 000
600 000 - 1 500 000	10 230 + 0.9%	on balance over	600 000
1 500 000 - 3 000 000	18 330 + 0.8%	on balance over	1 500 000
3 000 000 - 6 000 000	30 330 + 0.7%	on balance over	3 000 000
Over 6 000 000	51 330 + 0.6%	on balance over	6 000 000

(d) The scales of fees for preparing bills of quantities (paras. 2.1 (a) to (c)) are overall scales based upon the inclusion of all provisional and prime cost items, subject to the provision of para. 2.1 (g). When work normally included in a building contract is the subject of a separate contract for which the quantity surveyor has not been paid fees under any other clause hereof, the value of such work shall be included in the amount upon which fees are charged.

(e) Fees shall be calculated upon the accepted tender for the whole of he work subject to the provisions of para. 2.6. In the event of no tender being accepted, fees shall be calculated upon the basis of the lowest original bona fide tender received. In the event of no such tender being received, the fees shall be calculated upon a reasonable valuation of the works based upon the original bills of quantities.

NOTE: In the foregoing context "bona fide tender" shall be deemed to mean a tender submitted in good faith without major errors of computation and not subsequently withdrawn by the tenderer.

(f) In calculating the amount upon which fees are charged the total of any credits and the totals of any alternative bills shall be aggregated and added to the amount described above. The value of any omission or addition forming part of an alternative bill shall not be added unless measurement or abstraction from the original dimension sheets was necessary.

(g) When a contract comprises buildings which fall into more than one category, the fee shall be calculated as follows:
 (i) The amount upon which fees are chargeable shall be allocated to the categories of work applicable and the amounts so allocated expressed as percentages of the total amount upon which fees are chargeable.
 (ii) Fees shall then be calculated for each category on the total amount upon which fees are chargeable.
 (iii) The fee chargeable shall then be calculated by applying the percentages of work in each category to the appropriate total fee and adding the resultant amounts.

(h) When a project is the subject of a number of contracts then, for the purpose of calculating fees, the values of such contracts shall not be aggregated but each contract shall be taken separately and the scale of charges (paras. 2.1 (a) to (h)) applied as appropriate.

(i) Where the quantity surveyor is specifically instructed to provide cost planning services the fee calculated in accordance with paras. 2.1 (a) to (h) shall be increased by a sum calculated in accordance with the following table and based upon the same value of work as that upon which the aforementioned fee has been calculated:

Categories A & B: (as defined in paras. 2.1 (a) and (b)).

Value of work £	Fee		
	£		£
Up to 600 000	0.70%		
600 000 - 3 000 000	4 200 + 0.40%	on balance over	600 000
3 000 000 - 6 000 000	13 800 + 0.35%	on balance over	3 000 000
Over 6 000 000	24 300 + 0.30%	on balance over	6 000 000

Category C: (as defined in paras. 2.1 (c))

Value of work £	Fee		
	£		£
Up to 600 000	0.50%		
600 000 - 3 000 000	3 000 + 0.30%	on balance over	600 000
3 000 000 - 6 000 000	10 200 + 0.25%	on balance over	3 000 000
Over 6 000 000	17 700 + 0.20%	on balance over	6 000 000

2.2. Works of alteration
On works of alteration or repair, or on those sections of the works which are mainly works of alteration or repair, there shall be a fee of 1.0% in addition to the fee calculated in accordance with paras. 2.1.

2.3. Works of redecoration and associated minor repairs,
On works of redecoration and associated minor repairs, there shall be a fee of 1.5% in addition to the fee calculated in accordance with paras. 2.1.

2.4. Bills of quantities prepared in special forms
Fees calculated in accordance with paras. 2.1, 2.2 and 2.3 include for the preparation of bills of quantities on a normal trade basis. If the employer requires additional information to be provided in the bills of quantities or the bills to be prepared in an elemental, operational or similar form, then the fee may be adjusted by agreement between the employer and the quantity surveyor.

2.5. Reduction of tenders
(a) When cost planning services have been provided by the quantity surveyor and a tender, when received, is reduced before acceptance, and if the reductions are not necessitated by amended instructions of the employer or by the inclusion in the bills of quantities of items which the quantity surveyor has indicated could not be contained within the approved estimate, then in such a case no charge shall be made by the quantity surveyor for the preparation of bills of reductions and the fee for the preparation of the bills of quantities shall be based on the amount of the reduced tender.
(b) When cost planning services have not been provided by the quantity surveyor and if a tender, when received, is reduced before acceptance, fees are to be calculated upon the amount of the unreduced tender. When the preparation of bills of reductions is required, a fee is chargeable for preparing such bills of reductions as follows:
(i) 2.0% upon the gross amount of all omissions requiring measurement or abstraction from original dimensional sheets.
(ii) 3.0% upon the gross amount of all additions requiring measurement.
(iii) 0.5% upon the gross amount of all remaining additions.

NOTE: The above scale for the preparation of bills of reductions applies to work in all categories.

2.6 Generally:
If the works are substantially varied at any stage or if the quantity surveyor is involved in an excessive amount of abortive work, then the fees shall be adjusted by agreement between the employer and the quantity surveyor.

3.0. NEGOTIATING TENDERS

3.1.
(a) For negotiating and agreeing prices with a contractor:

Value of work £	Fee	
	£	£
Up to 150 000	0.5%	
150 000 - 600 000	750 + 0.3% on balance over	150 000
600 000 - 1 200 000	2 100 + 0.2% on balance over	600 000
Over 1 200 000	3 300 + 0.1% on balance over	1 200 000

(b) The fee shall be calculated on the total value of the works as defined in paras. 2.1 (d), (e), (f), (g) and (j).

4.0. CONSULTATIVE SERVICES AND PRICING BILLS OF QUANTITIES

4.1. Consultative services
Where the quantity surveyor is appointed to prepare approximate estimates, feasibility studies or submissions for the approval of financial grants or similar services, then the fee shall be based on the time involved (see paras. 19.1 and 19.2) or alternatively, on a lump sum or percentage basis agreed between the employer and the quantity surveyor.

4.2. Pricing bills of quantities
(a) For pricing bills of quantities, if instructed, to provide an estimate comparable with tenders, the fee shall be one-third (33.33%) of the fee for negotiating and agreeing prices with a contractor, calculated in accordance with paras. 3.1 (a) and (b).

Fees for Professional Services

CONTRACTS BASED ON BILLS OF QUANTITIES: POST-CONTRACT SERVICES

Alternative scales (I and II) for post-contract services are set out below to be used at the quantity surveyor's discretion by prior agreement with the employer.

5.0. ALTERNATIVE I: OVERALL SCALE OF CHARGES FOR POST-CONTRACT SERVICES

5.1. If the quantity surveyor appointed to carry out the post-contract services did not prepare the bills of quantities then the fees in paras. 5.2 and 5.3 shall be increased to cover the additional services undertaken by the quantity surveyor.

5.2. Basic scale
For taking particulars and reporting valuations for interim certificates for payments on account to the contractor, preparing periodic assessments of anticipated final cost and reporting thereon, measuring and making up bills of variations including pricing and agreeing totals with the contractor, and adjusting fluctuations in the cost of labour and materials if required by the contract.

(a) Category A: Relatively complex works and/or works with little or no repetition.

Examples:
Ambulance and fire stations; banks; cinemas; clubs; computer buildings; council offices; crematoria; fitting out existing buildings; homes for the elderly; hospitals and nursing homes; laboratories; law courts; libraries; "one-off" houses; petrol stations; places of religious worship; police stations; public houses; licensed premises; restaurants; sheltered housing; sports pavilions; theatres; town halls; universities, polytechnics and colleges of further education (other than halls of residence and hostels); and the like.

Value of work £	Category A fee		
	£		£
Up to 150 000	150 + 2.0%	(Minimum fee £1 150)	
150 000 - 300 000	3 150 + 1.7%	on balance over	150 000
300 000 - 600 000	5 700 + 1.6%	on balance over	300 000
600 000 - 1 500 000	10 500 + 1.3%	on balance over	600 000
1 500 000 - 3 000 000	22 200 + 1.2%	on balance over	1 500 000
3 000 000 - 6 000 000	40 200 + 1.1%	on balance over	3 000 000
Over 6 000 000	73 200 + 1.0%	on balance over	6 000 000

(b) Category B: Less complex works and/or works with some element of repetition.

Examples:
Adult education facilities; canteens; church halls; community centres; departmental stores; enclosed sports stadia and swimming baths; halls of residence; hostels; motels; offices other than those included in Categories A and C; railway stations; recreation and leisure centres; residential hotels; schools; self-contained flats and maisonettes; shops and shopping centres; supermarkets and hypermarkets; telephone exchanges; and the like.

Value of work £	Category B fee		
	£		£
Up to 150 000	150 + 2.0%	(Minimum fee £1 150)	
150 000 - 300 000	3 150 + 1.7%	on balance over	150 000
300 000 - 600 000	5 700 + 1.5%	on balance over	300 000
600 000 - 1 500 000	10 200 + 1.1%	on balance over	600 000
1 500 000 - 3 000 000	20 100 + 1.0%	on balance over	1 500 000
3 000 000 - 6 000 000	35 100 + 0.9%	on balance over	3 000 000
Over 6 000 000	62 100 + 0.8%	on balance over	6 000 000

(c) Category C: Simple works and/or works with a substantial element of repetition.

Examples:
Factories; garages; multi-storey car parks; open-air sports stadia; structural shell offices not fitted out; warehouses; workshops; and the like.

Value of work £	Category C fee		
	£		£
Up to 150 000	120 + 1.6%	(Minimum fee £920)	
150 000 - 300 000	2 520 + 1.5%	on balance over	150 000
300 000 - 600 000	4 770 + 1.4%	on balance over	300 000
600 000 - 1 500 000	8 970 + 1.1%	on balance over	600 000
1 500 000 - 3 000 000	18 870 + 0.9%	on balance over	1 500 000
3 000 000 - 6 000 000	32 370 + 0.8%	on balance over	3 000 000
Over 6 000 000	56 370 + 0.7%	on balance over	6 000 000

(d) The scales of fees for post-contract services (paras. 5.2 (a) to (c)) are overall scales based upon the inclusion of all nominated sub-contractors' and nominated suppliers' accounts, subject to the provision of para. 5.2 (g). When work normally included in a building contract is the subject of a separate contract for which the quantity surveyor has not been paid fees under any other clause hereof, the value of such work shall be included in the amount on which fees are charged.

(e) Fees shall be calculated upon the basis of the account for the whole of the work, subject to the provisions of para. 5.3.

(f) In calculating the amount on which fees are charged the total of any credits is to be added to the amount described above.

(g) When a contract comprises buildings which fall into more than one category, the fee shall be calculated as follows:

   (i) The amount upon which fees are chargeable shall be allocated to the categories of work applicable and the amounts so allocated expressed as percentages of the total amount upon which fees are chargeable.

   (ii) Fees shall then be calculated for each category on the total amount upon which fees are chargeable.

   (iii) The fee chargeable shall then be calculated by applying the percentages of work in each category to the appropriate total fee and adding the resultant amounts.

(h) When a project is the subject of a number of contracts then, for the purposes of calculating fees, the values of such contracts shall not be aggregated but each contract shall be taken separately and the scale of charges (paras. 5.2 (a) to (h)), applied as appropriate.

(j) When the quantity surveyor is required to prepare valuations of materials or goods off site, an additional fee shall be charged based on the time involved (see paras. 19.1 and 19.2).

(k) The basic scale for post-contract services includes for a simple routine of periodically estimating final costs. When the employer specifically requests a cost monitoring service which involves the quantity surveyor in additional or abortive measurement an additional fee shall be charged based on the time involved (see paras. 19.1 and 19.2), or alternatively on a lump sum or percentage basis agreed between the employer and the quantity surveyor.

(l) The above overall scales of charges for post-contract services assume normal conditions when the bills of quantities are based on drawings accurately depicting the building work the employer requires. If the works are materially varied to the extent that substantial remeasurement is necessary then the fee for post contract services shall be adjusted by agreement between the employer and the quantity surveyor.

6.0. ALTERNATIVE II: SCALE OF CHARGES FOR SEPARATE STAGES OF POST-CONTRACT SERVICES

6.1.
If the quantity surveyor appointed to carry out the post-contract services did not prepare the bills of quantities then the fees in paras. 6.2 and 6.3 shall be increased to cover the additional services undertaken by the quantity surveyor.

6.2. Valuations for interim certificates

(a) For taking particulars and reporting valuations for interim certificates for payments on account to the contractor.

Total of valuations £	Fee	
	£	£
Up to 300 000	0.5%	
300 000 - 1 000 000	1 500 + 0.4% on balance over	300 000
1 000 000 - 6 000 000	4 300 + 0.3% on balance over	1 000 000
Over 6 000 000	19 300 + 0.2% on balance over	6 000 000

NOTES:

1. Subject to note 2 below, the fees are to be calculated on the total of all interim valuations (i.e. the amount of the final account less only the net amount of the final valuation).

(b) When the quantity surveyor is required to prepare valuations of materials or goods off site, an additional fee shall be charged based on the time involved (see paras. 19.1 and 19.2).

6.3. Preparing accounts of variation upon contracts
For measuring and making up bills of variations including pricing and agreeing totals with the contractor:

(a) An initial lump sum of £600 shall be payable on each contract.
(b) 2.0% upon the gross amount of omissions requiring measurement or abstraction from the original dimension sheets.

Fees for Professional Services

(c) 3.0% upon the gross amount of additions requiring measurement and upon dayworks.

(d) 0.5% upon the gross amount of remaining additions which shall be deemed to include all nominated sub-contractors' and nominated suppliers' accounts which do not involve measurement or checking of quantities but only checking against lump sum estimates.

(e) 3.0% upon the aggregate of the amounts of the increases and/or decreases in the cost of labour and materials in accordance with any fluctuations clause in the conditions of contract, except where a price adjustment formula applies.

(f) On contracts where fluctuations are calculated by the use of a price adjustment formula method the following scale shall be applied to the account for the whole of the work:

Value of work £	Fee		
	£		£
Up to 300 000	300 + 0.5%		
300 000 - 1 000 000	1 800 + 0.3%	on balance over	300 000
Over 1 000 000	3 900 + 0.1%	on balance over	1 000 000

6.3. Cost monitoring services

The fee for providing all approximate estimates of final cost and/or a cost monitoring service shall be based on the time involved (see paras. 19.1 and 19.2), or alternatively on a lump sum or percentage basis agreed between the employer and the quantity surveyor.

7.0. BILLS OF APPROXIMATE QUANTITIES, INTERIM CERTIFICATES AND FINAL ACCOUNTS

7.1. Basic scale

For preparing bills of approximate quantities suitable for obtaining competitive tenders which will provide a schedule of prices and a reasonably close forecast of the cost of the works, but subject to complete remeasurement, examining tenders and reporting thereon, taking particulars and reporting valuations for interim certificates for payments on account to the contractor, preparing periodic assessments of anticipated final cost and reporting thereon, measuring and preparing final account, including pricing and agreeing totals with the contractor and adjusting fluctuations in the cost of labour and materials if required by the contract:

(a) Category A: Relatively complex works and/or works with little or no repetition.

Examples:
Ambulance and fire stations; banks; cinemas; clubs; computer buildings; council offices; crematoria; fitting out existing buildings; homes for the elderly; hospitals and nursing homes; laboratories; law courts; libraries; "one-off" houses; petrol stations; places of religious worship; police stations; public houses; licensed premises; restaurants; sheltered housing; sports pavilions; theatres; town halls; universities, polytechnics and colleges of further education (other than halls of residence and hostels); and the like.

Value of work £	Category A fee		
	£		£
Up to 150 000	380 + 5.0%	(Minimum fee £2 880)	
150 000 - 300 000	7 880 + 4.0%	on balance over	150 000
300 000 - 600 000	13 880 + 3.4%	on balance over	300 000
600 000 - 1 500 000	24 080 + 2.8%	on balance over	600 000
1 500 000 - 3 000 000	49 280 + 2.4%	on balance over	1 500 000
3 000 000 - 6 000 000	85 280 + 2.2%	on balance over	3 000 000
Over 6 000 000	151 280 + 2.0%	on balance over	6 000 000

(b) Category B: Less complex works and/or works with some element of repetition

Examples:
Adult education facilities; canteens; church halls; community centres; departmental stores; enclosed sports stadia and swimming baths; halls of residence; hostels; motels; offices other than those included in Categories A and C; railway stations; recreation and leisure centres; residential hotels; schools; self-contained flats and maisonettes; shops and shopping centres; supermarkets and hypermarkets; telephone exchanges; and the like.

Value of work £	Category B fee		
	£		£
Up to 150 000	360 + 4.8%	(Minimum fee £2 760)	
150 000 - 300 000	7 560 + 3.7%	on balance over	150 000
300 000 - 600 000	13 110 + 3.0%	on balance over	300 000
600 000 - 1 500 000	22 110 + 2.2%	on balance over	600 000
1 500 000 - 3 000 000	41 910 + 2.0%	on balance over	1 500 000
3 000 000 - 6 000 000	71 910 + 1.8%	on balance over	3 000 000
Over 6 000 000	125 910 + 1.6%	on balance over	6 000 000

(c) Category C: Simple works and/or works with a substantial element of repetition.

Examples:
Factories; garages; multi-storey car parks; open air sports stadia; structural shell offices not fitted out; warehouses; workshops; and the like.

Value of work £	£	Category C fee £
Up to 150 000	300 + 4.1%	(Minimum fee £2 350)
150 000 - 300 000	6 450 + 3.3% on balance over	150 000
300 000 - 600 000	11 400 + 2.6% on balance over	300 000
600 000 - 1 500 000	19 200 + 2.0% on balance over	600 000
1 500 000 - 3 000 000	37 200 + 1.7% on balance over	1 500 000
3 000 000 - 6 000 000	62 700 + 1.5% on balance over	3 000 000
Over 6 000 000	107 700 + 1.3% on balance over	6 000 000

(d) The scales of fees for pre-contract and post-contract services (paras. 7.1 (a) to (c)) are overall scales based upon the inclusion of all nominated sub-contractors' and nominated suppliers' accounts, subject to the provision of para. 7.1. (g). When work normally included in a building contract is the subject of a separate contract for which the quantity surveyor has not been paid fees under any other clause hereof, the value of such work shall be included in the amount on which fees are charged.

(e) Fees shall be calculated upon the basis of the account for the whole of the work, subject to the provisions of para. 7.2.

(f) In calculating the amount on which fees are charged the total of any credits is to be added to the amount described above.

(g) When a contract comprises buildings which fall into more than one category, the fee shall be calculated as follows.

    (i) The amount upon which fees are chargeable shall be allocated to the categories of work applicable and the amount so allocated expressed as percentages of the total amount upon which fees are chargeable.

    (ii) Fees shall then be calculated for each category on the total amount upon which fees are chargeable.

    (iii) The fee chargeable shall then be calculated by applying the percentages of work in each category to the appropriate total fee adding the resultant amounts.

(h) When a project is the subject of a number of contracts then, for the purpose of calculating fees, the values of such contracts shall not be aggregated but each contract shall be taken separately and the scale of charges (paras. 7.1(a) to (h)) applied as appropriate.

(j) Where the quantity surveyor is specifically instructed to provide cost planning services, the fee calculated in accordance with paras. 7.1 (a) to (j) shall be increased by a sum calculated in accordance with the following table and based upon the same value of work as that upon which the aforementioned fee has been calculated:

Categories A & B: (as defined in paras. 7.1 (a) and (b))

Value of work £			Fee
	£		£
Up to 600 000	0.70%		
600 000 - 3 000 000	4 200 + 0.40%	on balance over	600 000
3 000 000 - 6 000 000	13 800 + 0.35%	on balance over	3 000 000
Over 6 000 000	24 300 + 0.30%	on balance over	6 000 000

Category C: (as defined in para. 7.1 (c))

Value of work £			Fee
	£		£
Up to 600 000	0.50%		
600 000 - 3 000 000	3 000 + 0.30%	on balance over	600 000
3 000 000 - 6 000 000	10 200 + 0.25%	on balance over	3 000 000
Over 6 000 000	17 700 + 0.20%	on balance over	6 000 000

(k) When the quantity surveyor is required to prepare valuations of materials or goods off site, an additional fee shall be charged based on the time involved (see paras. 19.1 and 19.2).

(l) The basic scale for post-contract services includes for a simple routine of periodically estimating final costs. When the employer specifically requests a cost monitoring service which involves the quantity surveyor in additional or abortive measurement an additional fee shall be charged based on the time involved (see paras. 19.1 and 19.2), or alternatively on a lump sum or percentage basis agreed between the employer and the quantity surveyor.

7.2. Works of alteration.
On works of alteration or repair, or on those sections of the work which are mainly works of alteration or repair, there shall be a fee of 1.0% in addition to the fee calculated in accordance with paras. 7.1 and 7.2

7.3. Works of redecoration and associated minor repairs.
On works of redecoration and associated minor repairs, there shall be a fee of 1.5% in addition to the fee calculated in accordance with paras. 7.1 and 7.2.

Fees for Professional Services 81

7.4 Bills of quantities and/or final accounts prepared on special forms.
Fees calculated in accordance with paras. 7.1, 7.2, 7.3 and 7.4 include for the preparation of bills of quantities and/or final accounts on a normal trade basis. If the employer requires additional information to be provided in the bills of quantities and/or final accounts or the bills and/or final accounts to be prepared in an elemental, operational or similar form, then the fee may be adjusted by agreement between the employer and the quantity surveyor.

7.5 Reduction of tenders.

(a) When cost planning services have been provided by the quantity surveyor and a tender, when received, is reduced before acceptance and if the reductions are not necessitated by amended instructions of the employer or by the inclusion in the bills of approximate quantities of items which the quantity surveyor has indicated could not be contained within the approved estimate, then in such a case no charge shall be made by the quantity surveyor for the preparation of bills of reductions and the fee for the preparation of bills of approximate quantities shall be based on the amount of the reduced tender.

(b) When cost planning services have not been provided by the quantity surveyor and if a tender, when received, is reduced before acceptance, fees are to be calculated upon the amount of the unreduced tender. When the preparation of bills of reductions is required, a fee is chargeable for preparing such bills of reductions as follows:

(i) 2.0% upon the gross amount of all omissions requiring measurement or abstraction from original dimension sheets.
(ii) 3.0% upon the gross amount of all additions requiring measurement.
(iii) 0.5% upon the gross amount of all remaining additions.

NOTE: The above scale for the preparation of bills of reductions applies to work in all categories.

7.6. Generally
If the works are substantially varied at any stage or if the quantity surveyor is involved in an excessive amount of abortive work, then the fees shall be adjusted by agreement between the employer and the quantity surveyor.

8.0. NEGOTIATING TENDERS

8.1
(a) For negotiating and agreeing prices with a contractor:

Value of work £		Fee
	£	£
Up to 150 000	0.5%	
150 000 - 600 000	750 + 0.3% on balance over	150 000
600 000 - 1 200 000	2 100 + 0.2% on balance over	600 000
Over 1 200 000	3 300 + 0.1% on balance over	1 200 000

(b) The fee shall be calculated on the total value of the works as defined in paras. 7.1 (d), (e), (f), (g) and (j)

9.0. CONSULTATIVE SERVICES AND PRICING BILLS OF APPROXIMATE QUANTITIES

9.1. Consultative services
Where the quantity surveyor is appointed to prepare approximate estimates, feasibility studies or submissions for the approval of financial grants or similar services, then the fee shall be based on the time involved (see paras. 19.1 and 19.2) or alternatively, on a lump sum or percentage basis agreed between the employer and the quantity surveyor.

9.2. Pricing bills of approximate quantities
For pricing bills of approximate quantities, if instructed, to provide an estimate comparable with tenders, the fees shall be the same as for the corresponding services in paras. 4.2 (a) and (b).

10.0. INSTALMENT PAYMENTS

10.1. For the purpose of instalment payments the fee for preparation of bills of approximate quantities only shall be the equivalent of forty per cent (40%) of the fees calculated in accordance with the appropriate sections of paras. 7.1 to 7.5, and the fee for providing cost planning services shall be in accordance with the appropriate sections of para. 7.1 (k); both fees shall be based on the total value of the bills of approximate quantities ascertained in accordance with the provisions of para. 2.1 (e).

10.2. In the absence of agreement to the contrary, fees shall be paid by instalments as follows:

(a) Upon acceptance by the employer of a tender for the works the above defined fees for the preparation of bills of approximate quantities and for providing cost planning services.

(b) In the event of no tender being accepted, the aforementioned fees shall be paid within three months of completion of the bills of approximate quantities.

(c) The balance by instalments at intervals to be agreed between the date of the first certificate and one month after certification of the contractor's account.

10.3. In the event of the project being abandoned at any stage other than those covered by the foregoing, the proportion of fee payable shall be by agreement between the employer and the quantity surveyor.

11.0. SCHEDULES OF PRICES

11.1. The fee for preparing, pricing and agreeing schedules of prices shall be based on the time involved (see paras. 19.1 and 19.2). Alternatively, the fee may be on a lump sum or percentage basis agreed between the employer and the quantity surveyor.

12.0. COST PLANNING AND APPROXIMATE ESTIMATES

12.1. The fee for providing cost planning services or for preparing approximate estimates shall be based on the time involved (see paras. 19.1 and 19.2). Alternatively, the fee may be on a lump sum or percentage basis agreed between the employer and the quantity surveyor.

CONTRACTS BASED ON SCHEDULES OF PRICES: POST-CONTRACT SERVICES

13.0. FINAL ACCOUNTS

13.1. Basic Scale

(a) For taking particulars and reporting valuations for interim certificates for payments on account to the contractor, preparing periodic assessments of anticipated final cost and reporting thereon, measuring and preparing final account including pricing and agreeing totals with the contractor, and adjusting fluctuations in the cost of labour and materials if required by the contract, the fee shall be equivalent to sixty per cent (60%) of the fee calculated in accordance with paras. 7.1 (a) to (j).

(b) When the quantity surveyor is required to prepare valuations of materials or goods off site, an additional fee shall be charged on the basis of the time involved (see paras. 19.1 and 19.2).

(c) The basic scale for post-contract services includes for a simple routine of periodically estimating final costs. When the employer specifically requests a cost monitoring service which involves the quantity surveyor in additional or abortive measurement an additional fee shall be charged based on the time involved (see paras. 19.1 and 19.2), or alternatively on a lump sum or percentage basis agreed between the employer and the quantity surveyor.

13.2. Works of alterations
On works of alteration or repair, or on those sections of the work which are mainly works of alteration or repair, there shall be a fee of 1.0% in addition to the fee calculated in accordance with paras. 13.1 and 13.3.

13.3. Works of redecoration and associated minor repairs
On works of redecoration and associated minor repairs, there shall be a fee of 1.5% in addition to the fee calculated in accordance with paras. 13.1

13.4 Final accounts prepared in special forms
Fees calculated in accordance with paras. 13.1,13.2,13.3 include for the preparation of final accounts on a normal trade basis. If the employer requires additional information to be provided in the final accounts or the accounts to be prepared in an elemental, operational or similar form, then the fee may be adjusted by agreement between the employer and the quantity surveyor.

PRIME COST CONTRACTS: PRE-CONTRACT AND POST-CONTRACT SERVICES

14.0. COST PLANNING

14.1. The fee for providing a cost planning service shall be based on the time involved (see paras. 19.1 and 19.2). Alternatively, the fee may be on a lump sum or percentage basis agreed between the employer and the quantity surveyor.

15.0. ESTIMATES OF COST

15.1.
(a) For preparing an approximate estimate, calculated by measurement, of the cost of work, and, if required under the terms of the contract, negotiating, adjusting and agreeing the estimate:

Value of work £		£		Fee £
Up to	30 000	1.25%		
30 000 -	150 000	375 + 1.00%	on balance over	30 000
150 000 -	600 000	1 575 + 0.75%	on balance over	150 000
Over	600 000	4 950 + 0.50%	on balance over	600 000

(b) The fee shall be calculated upon the total of the approved estimates.

16.0. FINAL ACCOUNTS

16.1
(a) For checking prime costs, reporting for interim certificates for payment on account to the contractor and preparing final accounts:

Value of work £		£		Fee £
Up to	30 000	2.50%		
150 000 -	150 000	750 + 2.00%	on balance over	30 000
150 000 -	600 000	3 150 + 1.50%	on balance over	150 000
Over	600 000	9 900 + 1.25%	on balance over	600 000

(b) The fee shall be calculated upon the total of the final account with the addition of the value of credits received for old materials removed and less the value of any work charged for in accordance with para. 16.1 (c).
(c) On the value of any work to be paid for on a measured basis, the fee shall be 3%.
(d) When the quantity surveyor is required to prepare valuations of materials or goods off site, an additional fee shall be charged based on the time involved (see paras. 19.1 and 19.2).
(e) The above charges do not include the provision of checkers on the site. If the quantity surveyor is required to provide such checkers an additional charge shall be made by arrangement.

17.0. COST REPORTING AND MONITORING SERVICES

17.1. The fee for providing cost reporting and/or monitoring services (e.g. preparing periodic assessments of anticipated final costs and reporting thereon) shall be based on the time involved (see paras. 19.1 and 19.2) or alternatively, on a lump sum or percentage basis agreed between the employer and the quantity surveyor.

18.0. ADDITIONAL SERVICES

18.1. For additional services not normally necessary, such as those arising as a result of the termination of a contract before completion, liquidation, fire damage to the buildings, services in connection with arbitration, litigation and investigation of the validity of contractors' claims, services in connection with taxation matters and all similar services where the employer specifically instructs the quantity surveyor, the charges shall be in accordance with paras. 19.1 and 19.2.

19.0. TIME CHARGES

19.1.
- (a) For consultancy and other services performed by a principal, a fee by arrangement according to the circumstances including the professional status and qualifications of the quantity surveyor.
- (b) When a principal does work which would normally be done by a member of staff, the charge shall be calculated as para. 19.2 below.

19.2.
- (a) For services by a member of staff, the charges for which are to be based on the time involved, such charges shall be calculated on the hourly cost of the individual involved plus 145%.
- (b) A member of staff shall include a principal doing work normally done by an employee (as para. 19.1 (b) above), technical and supporting staff, but shall exclude secretarial staff or staff engaged upon general administration.
- (c) For the purpose of para. 19.2 (b) above, a principal's time shall be taken at the rate applicable to a senior assistant in the firm.
- (d) The supervisory duties of a principal shall be deemed to be included in the addition of 145% as para. 19.2 (a) above and shall not be charged separately.
- (e) The hourly cost to the employer shall be calculated by taking the sum of the annual cost of the member of staff of:

    - (i) Salary and bonus but excluding expenses;
    - (ii) Employer's contributions payable under any Pension and Life Assurance Schemes;
    - (iii) Employer's contributions made under the National Insurance Acts, the Redundancy Payments Act and any other payments made in respect of the employee by virtue of any statutory requirements; and
    - (iv) Any other payments or benefits made or granted by the employer in pursuance of the terms of employment of the member of staff; and dividing by 1,650.

19.3. The foregoing Time Charges under paras. 19.1 and 19.2 are intended for use where other paras. of the Scale (not related to Time Charges) form a significant proportion of the overall fee. In all other cases an increased time charge may be agreed.

20.0. INSTALMENT PAYMENTS

20.1. In the absence of agreement to the contrary, payments to the quantity surveyor shall be made by instalments by arrangement between the employer and the quantity surveyor.

SCALE 40
PROFESSIONAL CHARGES FOR QUANTITY SURVEYING SERVICES IN CONNECTION WITH HOUSING SCHEMES FOR LOCAL AUTHORITIES
See Author's note on page 62

EFFECTIVE FROM FEBRUARY 1983 TO DECEMBER 1998

1.0 GENERALLY

1.1 The scale is applicable to housing schemes of self-contained dwellings regardless of type (e.g. houses, maisonettes, bungalows or flats) and irrespective of the amount of repetition of identical types or blocks within an individual housing scheme and shall also apply to all external works forming part of the contract for the housing scheme. This scale does not apply to improvement to existing dwellings.

1.2 The fees set out below cover the following quantity surveying services as may be required:

(a) Preparing bills of quantities or other tender documents; checking tenders received or negotiating tenders and pricing with a selected contractor; reporting thereon.

(b) Preparing recommendations for interim payments on account to the contractor; measuring work and adjusting variations in accordance with the terms of the contract and preparing the final account; pricing same and agreeing totals with the contractor; adjusting fluctuations in the cost of labour and materials if required by the contract.

(c) Preparing periodic financial statements showing the anticipated final cost by means of a simple routine of estimating final costs and reporting thereon, but excluding cost monitoring (see para. 1.4).

1.3 Where the quantity surveyor is appointed to prepare approximate estimates to establish and substantiate the economic viability of the scheme and to obtain the necessary approvals and consents, or to enable the scheme to be designed and constructed within approved cost criteria an additional fee shall be charged based on the time involved (see para. 7.0) or, alternatively, on a lump sum or percentage basis agreed between the employer and the quantity surveyor. (Cost planning services, see para. 3.0).

1.4 When the employer specifically requests a post-contract cost monitoring service which involves the quantity surveyor in additional or abortive work an additional fee shall be charged based on the time involved (see para. 7.0) or, alternatively, on a lump sum or percentage basis agreed between the employer and the quantity surveyor.

1.5 The fees are in all cases exclusive of travelling and other expenses (for which the actual disbursement is recoverable unless there is some prior arrangement for such charges) and of the cost of reproduction of bills of quantities and other documents, which are chargeable in addition at net cost.

1.6 The fees are in all cases exclusive of services in connection with the allocation of the cost of the works for purposes of calculating value added tax for which there shall be an additional fee based on the time involved (see para. 7.0).

1.7 When work normally included in a building contract is the subject of a separate contract for which the quantity surveyor has not been paid fees under any other clause thereof, the value of such work shall be included in the amount upon which fees are charged.

1.8 If any of the materials used in the works are supplied by the employer or charged at a preferential rate, then the estimated or actual value thereof shall be included in the amount upon which fees are to be calculated.

1.9 The fees are in all cases exclusive of preparing a specification of the materials to be used and the works to be done, but the fees for preparing bills of quantities and similar documents do include for incorporating preamble clauses describing the materials and workmanship (from information given by the architect and/or consulting engineer).

1.10 If the quantity surveyor incurs additional costs due to exceptional delays in building operations or any other cause beyond the control of the quantity surveyor, then the fees shall be adjusted by agreement between the employer and the quantity surveyor to cover the reimbursement of these additional costs.

1.11 When a project is the subject of a number of contracts then for the purposes of calculating fees, the values of such contracts shall not be aggregated but each contract shall be taken separately and the scale of charges applied as appropriate.

1.12 The fees and charges are in all cases exclusive of value added tax which will be applied in accordance with legislation.

1.13 Copyright in bills of quantities and other documents prepared by the quantity surveyor is reserved to the quantity surveyor.

2.0 BASIC SCALE

2.1 The basic fee for the services outlined in para. 1.2 shall be as follows:-

Value of work £	Fee		
	£		£
Up to 75 000	250 + 4.6%		
75 000 - 150 000	3 700 + 3.6%	on balance over	75 000
150 000 - 750 000	6 400 + 2.3%	on balance over	150 000
750 000 - 1 500 000	20 200 + 1.7%	on balance over	750 000
Over 1 500 000	32 950 + 1.5%	on balance over	1 500 000

2.2 Fees shall be calculated upon the total of the final account for the whole of the work including all nominated sub-contractors' and nominated suppliers' accounts.

2.3 For services in connection with accommodation designed for the elderly or the disabled or other special category occupants for whom special facilities are required an addition of 10% shall be made to the fee calculated in accordance with para. 2.1.

2.4 When additional fees under para. 2.3 are chargeable on a part or parts of a scheme, the value of basic fee to which the additional percentages shall be applied shall be determined by the proportion that the values of the various types of accommodation bear to the total of those values.

2.5 When the quantity surveyor is required to prepare an interim valuation of materials or goods off site, an additional fee shall be charged based on the time involved (see para. 7.0).

2.6 If the works are substantially varied at any stage and if the quantity surveyor is involved in an excessive amount of abortive work, then the fee shall be adjusted by agreement between the employer and the quantity surveyor.

2.7 The fees payable under paras. 2.1 and 2.3 include for the preparation of bills of quantities or other tender documents on a normal trade basis. If the employer requires additional information to be provided in bills of quantities, or bills of quantities to be prepared in an elemental, operational or similar form, then the fee may be adjusted by agreement between the employer and the quantity surveyor.

3.0 COST PLANNING

3.1 When the quantity surveyor is specifically instructed to provide cost planning services, the fee calculated in accordance with paras. 2.1 and 2.3 shall be increased by a sum calculated in accordance with the following table and based upon the amount of the accepted tender.

Value of work £	Fee		£
Up to 150 000	0.45%		
150 000 - 750 000	675 + 0.35%	on balance over	150 000
Over 750 000	2 775 + 0.25%	on balance over	750 000

3.2 Cost planning is defined as the process of ascertaining a cost limit, where necessary, within the guidelines set by any appropriate Authority, and thereafter checking the cost of the project within that limit throughout the design process. It includes the preparation of a cost plan (based upon elemental analysis or other suitable criterion) checking and revising it where required and effecting the necessary liaison with other consultants employed.

3.3 (a) When cost planning services have been provided by the quantity surveyor and bills of reductions are required, then no charge shall be made by the quantity surveyor for the bills of reductions unless the reductions are necessitated by amended instructions of the employer or by the inclusion in the bills of quantities of items which the quantity surveyor has indicated could not be contained within the approved estimate.

(b) When cost planning services have not been provided by the quantity surveyor and bills of reductions are required, a fee is chargeable for preparing such bills of reductions as follows:

   (i) 2.0% upon the gross amount of all omissions requiring measurement or abstraction from original dimension sheets.
   (ii) 3.0% upon the gross amount of all additions requiring measurement.
   (iii) 0.5% upon the gross amount of all remaining additions.

5.0 INSTALMENT PAYMENTS

5.1 In the absence of agreement to the contrary, fees shall be paid by instalments as follows:

(a) Upon receipt by the employer of a tender for the works sixty per cent (60%) of the fees calculated in accordance with paras. 2.0 and 4.0 in the amount of the accepted tender plus the appropriate recoverable expenses and the full amount of the fee for cost planning services if such services have been instructed by the employer.

(b) The balance of fees and expenses by instalments at intervals to be agreed between the date of the first certificate and one month after final certification of the contractor's account.

5.2 In the event of no tender being accepted, sixty per cent (60%) of the fees, plus the appropriate recoverable expenses, and the full amount of the fee for cost planning services if such services have been instructed by the employer, shall be paid within three months of the completion of the tender documents. The fee shall be calculated on the amount of the lowest original bona fide tender received. In the event of no tender being received, the fee shall be calculated on a reasonable valuation of the work based upon the tender documents.

NOTE: In the foregoing context "bona fide tender" shall be deemed to mean a tender submitted in good faith without major errors of computation and not subsequently withdrawn by the tenderer.

5.3 In the event of the project being abandoned at any stage other than those covered by the foregoing, the proportion of fee payable shall be by agreement between the employer and the quantity surveyor.

5.4 When the quantity surveyor is appointed to carry out post-contract services only and has not prepared the bills of quantities then the fees shall be agreed between the employer and the quantity surveyor as a proportion of the scale set out in paras. 2.0 and 4.0 with an allowance for the necessary familiarisation and any additional services undertaken by the quantity surveyor. The percentages stated in paras. 5.1 and 5.2 are not intended to be used as a means of calculating the fees payable for post-contract services only.

6.0 ADDITIONAL SERVICES

6.1 For additional services not normally necessary such as those arising as a result of the termination of a contract before completion, liquidation, fire damage to the buildings, services in connection with arbitration, litigation and investigation of the validity of contractors' claims, services in connection with taxation matters, and all similar services where the employer specifically instructs the quantity surveyor, the charge shall be in accordance with para. 7.0.

7.0 TIME CHARGES

7.1   (a)   For consultancy and other services performed by a principal, a fee by arrangement according to the circumstances, including the professional status and qualifications of the quantity surveyor.

       (b)   When a principal does work which would normally be done by a member of staff, the charge shall be calculated as para. 7.2.

7.2   (a)   For services by a member of staff, the charges for which are to be based on the time involved, such hourly charges shall be calculated on the basis of annual salary (including bonus and any other payments or benefits previously agreed with the employer) multiplied by a factor of 2.5, plus reimbursement of payroll costs, all divided by 1600. Payroll costs shall include inter alia employer's contributions payable under any Pension and Life Assurance Schemes, employer's contributions made under the National Insurance Acts, the Redundancy Payments Act and any other payments made in respect of the employee by virtue of any statutory requirements. In this connection it would not be unreasonable in individual cases to take account of the cost of providing a car as part of the "salary" of staff engaged on time charge work when considering whether the salaries paid to staff engaged on such work are reasonable.

       (b)   A member of staff shall include a principal doing work normally done by an employee (as para. 7.1 (b) above), technical and supporting staff, but shall exclude secretarial staff or staff engaged upon general administration.

       (c)   For the purpose of para. 7.2 (b) above a principal's time shall be taken at the rate applicable to a senior assistant in the firm.

       (d)   The supervisory duties of a principal shall be deemed to be included in the multiplication factor as para. 7.2 (a) above and shall not be charged separately.

7.3   The foregoing Time Charges under paras. 7.1 and 7.2 are intended for use where other paras. of the scale (not related to Time Charges) form a significant proportion of the overall fee. In all other cases an increased Time Charge may be agreed.

SCALE 44
PROFESSIONAL CHARGES FOR QUANTITY SURVEYING SERVICES IN CONNECTION WITH IMPROVEMENTS TO EXISTING HOUSING AND ENVIRONMENTAL IMPROVEMENT WORKS

See Author's note on page 62.

EFFECTIVE FROM FEBRUARY 1973 TO DECEMBER 1998

1. This scale of charges is applicable to all works of improvement to existing housing for local authorities, development corporations, housing associations and the like and to environmental improvement works associated therewith or of a similar nature.

2. The fees set out below cover such quantity surveying services as may be required in connection with an improvement project irrespective of the type of contract or contract documentation from initial appointment to final certification of the contractor's account such as:

    (a) Preliminary cost exercises and advice on tendering procedures and contract arrangements.
    (b) Providing cost advice to assist the design and construction of the project within approved cost limits.
    (c) Preliminary inspection of a typical dwelling of each type.
    (d) Preparation of tender documents; checking tenders received and reporting thereon or negotiating tenders and agreeing prices with a selected contractor.
    (e) Making recommendations for and, where necessary, preparing bills of reductions except in cases where the reductions are necessitated by amended instructions of the employer or by the inclusion in the bills of quantities of items which the quantity surveyor has indicated could not be contained within the approved estimate.
    (f) Analysing tenders and preparing details for submission to a Ministry or Government Department and attending upon the employer in any negotiations with such Ministry or Government Department.
    (g) Recording the extent of work required to every dwelling before work commences.
    (h) Preparing recommendations for interim payments on account to the contractor; preparing periodic assessments of the anticipated final cost of the works and reporting thereon
    (j) Measurement of work and adjustment of variations and fluctuations in the cost of labour and materials in accordance with the terms of the contract and preparing final account, pricing same and agreeing totals with the contractor.

3. The services listed in para. 2 do not include the carrying out of structural surveys.

4. The fees set out below have been calculated on the basis of experience that all of the services described above will not normally be required and in consequence these scales shall not be abated if, by agreement, any of the services are not required to be provided by the quantity surveyor.

IMPROVEMENT WORKS TO HOUSING

5. The fee for quantity surveying services in connection with improvement works to existing housing and external works in connection therewith shall be calculated from a sliding scale based upon the total number of houses or flats in a project divided by the total number of types substantially the same in design and plan as follows:

Total number of houses or flats divided by total number of types substantially the same in design and plan	Fee
not exceeding 1	see note below
exceeding 1 but not exceeding 2	7.0%
exceeding 2 but not exceeding 3	5.0%
exceeding 3 but not exceeding 4	4.5%
exceeding 4 but not exceeding 20	4.0%
exceeding 20 but not exceeding 50	3.6%
exceeding 50 but not exceeding 100	3.2%
exceeding 100	3.0%

and to the result of the computation shall be added 12.5%

NOTE: For schemes of only one house or flat per type an appropriate fee is to be agreed between the employer and the quantity surveyor on a percentage, lump sum or time basis.

ENVIRONMENTAL IMPROVEMENT WORKS

6. The fee for quantity surveying services in connection with environmental improvement works associated with improvements to existing housing or environmental improvement works of a similar nature shall be as follows:

Value of work				Fee
£		£		£
Up to	50 000	4.5%		
50 000 -	200 000	2 250 + 3.0%	on balance over	50 000
200 000 -	500 000	6 750 + 2.1%	on balance over	200 000
Over	500 000	13 050 + 2.0%	on balance over	500 000

and to the result of that computation shall be added 12.5%

GENERALLY

7. When tender documents prepared by a quantity surveyor for an earlier scheme are re-used without amendment by the quantity surveyor for a subsequent scheme or part thereof for the same employer, the percentage fee in respect of such subsequent scheme or the part covered by such reused documents shall be reduced by 20%.

8. The foregoing fees shall be calculated upon the separate totals of the final account for improvement works to housing and environmental Government works respectively including all nominated sub-contractors' and nominated suppliers' accounts and (subject to para. 5 above) regardless of the amount of repetition within the scheme. When environmental improvement works are the subject of a number of contracts then for the purpose of calculating fees, the values of such contracts shall not be aggregated but each contract shall be taken separately and the scale of charges in para. 6 above applied as appropriate.

9. In cases where any of the materials used in the works are supplied by the employer, the estimated or actual value thereof is to be included in the total on which the fee is calculated.

10. In the absence of agreement to the contrary, fees shall be paid by instalments as follows:

    (a) Upon acceptance by the employer of a tender for the works, one half of the fee calculated on the amount of the accepted tender.

    (b) The balance by instalments at intervals to be agreed between the date of the first certificate and one month after final certification of the contractor's account.

11. (a) In the event of no tender being accepted, one half of the fee shall be paid within three months of completion of the tender documents. The fee shall be calculated on the amount of the lowest original bona fide tender received. If no such tender has been received, the fee shall be calculated upon a reasonable valuation of the work based upon the tender documents.

    (b) In the event of the project being abandoned at any stage other than those covered by the foregoing, the proportion of fee payable shall be by agreement between the employer and the quantity surveyor.

12. If the works are substantially varied at any stage or if the quantity surveyor is involved in an excessive amount of abortive work, then the fee shall be adjusted by agreement between the employer and the quantity surveyor.

13. When the quantity surveyor is required to perform additional services in connection with the allocation of the costs of the works for purposes of calculating value added tax there shall be an additional fee based on the time involved.

13. For additional services not normally necessary such as those arising as a result of the termination of the contract before completion, liquidation, fire damage to the buildings, services in connection with arbitration, litigation and claims on which the employer

specifically instructs the surveyor to investigate and report, there shall be an additional fee to be agreed between the employer and the quantity surveyor.

15. Copyright in the bills of quantities and other documents prepared by the quantity surveyor is reserved to the quantity surveyor.

16. The foregoing fees are in all cases exclusive of travelling expenses and lithography or other charges for copies of documents, the net amount of such expenses and charges to be paid for in addition. Subsistence expenses, if any, to be charged by arrangement with the employer.

17. The foregoing fees and charges are in all cases exclusive of value added tax which shall be applied in accordance with legislation current at the time the account is rendered.

SCALE 45
PROFESSIONAL CHARGES FOR QUANTITY SURVEYING SERVICES IN CONNECTION WITH HOUSING SCHEMES FINANCED BY THE HOUSING CORPORATION
See Author's note on page 62

EFFECTIVE FROM JANUARY 1982 TO DECEMBER 1998

1. (a) This scale of charges has been agreed between The Royal Institution of Chartered Surveyors and the Housing Corporation and shall apply to housing schemes of self-contained dwellings financed by the Housing Corporation regardless of type (e.g. houses, maisonettes, bungalows or flats) and irrespective of the amount of repetition of identical types or blocks within an individual housing scheme.
   (b) This scale does not apply to services in connection with improvements to existing dwellings.

2. The fees set out below cover the following quantity surveying services as may be required in connection with the particular project:

   (a) Preparing such estimates of cost as are required by the employer to establish and substantiate the economic viability of the scheme and to obtain the necessary approvals and consents from the Housing Corporation but excluding cost planning services (see para. 10)
   (b) Providing pre-contract cost advice (e.g. approximate estimates on a floor area or similar basis) to enable the scheme to be designed and constructed within the approved cost criteria but excluding cost planning services (see para. 10).
   (c) Preparing bills of quantities or other tender documents; checking tenders received or negotiating tenders and pricing with a selected contractor; reporting thereon.
   (d) Preparing an elemental analysis of the accepted tender (RICS/BCIS Detailed Form of Cost Analysis excluding the specification notes or equivalent).

Fees for Professional Services

(e) Preparing recommendations for interim payments on account to the contractor; measuring the work and adjusting variations in accordance with the terms of the contract and preparing the final account, pricing same and agreeing totals with the contractor; adjusting fluctuations in the cost of labour and materials if required by the contract.

(f) Preparing periodic post-contract assessments of the anticipated final cost by means of a simple routine of periodically estimating final costs and reporting thereon, but excluding a cost monitoring service specifically required by the employer.

3. The fees set out below are exclusive of travelling and of other expenses (for which the actual disbursement is recoverable unless there is some special prior arrangement for such charges) and the cost of reproduction of bills of quantities and other documents, which are chargeable in addition at net cost.

4. Copyright in the bills of quantities and other documents prepared by the quantity surveyor is reserved to the quantity surveyor.

5. (a) The basic fee for the services outlined in para. 2 (regardless of the extent of services described in para. 2) shall be as follows:

Value of work		Fee		
	£	£		£
Up to 75 000		210 + 3.8%		
75 000 - 150 000		3 060 + 3.0%	on balance over	75 000
150 000 - 750 000		5 310 + 2.0%	on balance over	150 000
750 000 - 1 500 000		17 310 + 1.5%	on balance over	750 000
Over 1 500 000		28 560 + 1.3%	on balance over	1 500 000

(b) (i) For services in connection with Categories 1 and 2 Accommodation designed for Old People in accordance with the standards described in Ministry of Housing and Local Government Circulars 82/69 and 27/70 (Welsh Office Circulars 84/69 & 30/70), there shall be a fee in addition to that in accordance with para. 5 (a), calculated as follows:

    Category 1 An addition of five per cent (5%) to the basic fee calculated in accordance with para. 5 (a)
    Category 2 An addition of twelve and a half per cent (12.5%) to the basic fee calculated in accordance with para. 5 (a).

(ii) For services in connection with Accommodation designed for the Elderly in Scotland in accordance with the standards described in Scottish Housing Handbook Part 5, Housing for the Elderly, the fee shall be calculated as follows:

Mainstream and Amenity Housing	Basic fee in accordance with para. 5 (a)
Basic Sheltered Housing (i.e. Amenity Housing plus Warden's accommodation and alarm system)	An addition of five per cent (5%) to the basic fee calculated in accordance with para. 5 (a)
Sheltered Housing, including optional facilities	An addition of twelve and a half per cent (12.5%) of the basic fee calculated in accordance with para. 5 (a)

(c) (i) For services in connection with Accommodation designed for Disabled People in accordance with the standards described in Department of Environment Circular 92/75 (Welsh Office Circular 163/75), there shall be an addition of fifteen per cent (15%) to the fee calculated in accordance with paragraph 5 (a).

(ii) For services in connection with Accommodation designed for the Disabled in Scotland in accordance with the standards described in Scottish Housing Handbook Part 6, Housing for the Disabled, there shall be an addition of fifteen per cent (15%) to the fee calculated in accordance with para. 5 (a).

(d) For services in connection with Accommodation designed for Disabled Old People, the fee shall be calculated in accordance with para. 5 (c).

(e) For services in connection with Subsidised Fair Rent New Build Housing, there shall be a fee, in addition to that in accordance with paras. 5 (a) to (d), calculated as follows:

Value of work £	Category A fee		
	£		£
Up to 75 000	20 + 0.40%		
75 000 - 150 000	320 + 0.20%	on balance over	75 000
150 000 - 500 000	470 + 0.07%	on balance over	150 000
Over 500 000	715		

6. (a) Where additional fees under paras. 5 (b) to (d) are chargeable on a part or parts of a scheme, the value of basic fee to which the additional percentages shall be applied shall be determined by the proportion that the values of the various types of accommodation bear to the total of those values.

(b) Fees shall be calculated upon the total of the final account for the whole of the work including all nominated sub-contractors' and nominated suppliers' accounts.

Fees for Professional Services

- (c) If any of the materials used in the works are supplied free of charge to the contractor, the estimated or actual value thereof shall be included in the amount upon which fees are to be calculated.

- (d) When a project is the subject of a number of contracts then, for the purpose of calculating fees, the values of such contracts shall not be aggregated but each contract shall be taken separately and the scale of charges applied as appropriate.

7. If bills of quantities and final accounts are prepared by the quantity surveyor for the heating, ventilating or electrical services, there shall be an additional fee by agreement between the employer and the quantity surveyor subject to the approval of the Housing Corporation.

8. In the absence of agreement to the contrary, fees shall be paid by instalments as follows:
   - (a) Upon receipt by the employer of a tender for the works, or when the employer certifies to the Housing Corporation that the tender documents have been completed, a sum on account representing ninety per cent (90%) of the anticipated sum under para. 8 (b) below.
   - (b) Upon acceptance by the employer of a tender for the works, sixty per cent (60%) of the fee calculated on the amount of the accepted tender, plus the appropriate recoverable expenses.
   - (c) The balance of fees and expenses by instalments at intervals to be agreed between the date of the first certificate and one month after final certification of the contractor's account.

9. (a) In the event of no tender being accepted, sixty per cent (60%) of the fee and the appropriate recoverable expenses shall be paid within six months of completion of the tender documents. The fee shall be calculated on the amount of the lowest original bona fide tender received. In the event of no tender being received, the fee shall be calculated upon a reasonable valuation of the work based upon the tender documents.

   NOTE: In the foregoing context "bona fide tender" shall be deemed to mean a tender submitted in good faith without major errors of computation and not subsequently withdrawn by the tenderer.

   - (b) In the event of part of the project being postponed or abandoned after the preparation of the bills of quantities or other tender documents, sixty per cent (60%) of the fee on this part shall be paid within three months of the date of postponement or abandonment.
   - (c) In the event of the project being postponed or abandoned at any stage other than those covered by the foregoing, the proportion of fee payable shall be by agreement between the employer and the quantity surveyor.

10. (a) Where with the approval of the Housing Corporation the employer instructs the quantity surveyor to carry out cost planning services there shall be a fee additional to that charged under para. 5 as follows:

Value of work £	Category A fee		
	£		£
Up to 150 000	0.45%		
150 000 - 750 000	675 + 0.35%	on balance over	150 000
Over 750 000	2 775 + 0.25%	on balance over	750 000

    (b)   Cost planning is defined as the process of ascertaining a cost limit where necessary, within guidelines set by any appropriate Authority, and thereafter checking the cost of the project within that limit throughout the design process. It includes the preparation of a cost plan (based upon elemental analysis or other suitable criterion) checking and revising it where required and effecting the necessary liaison with the other consultants employed.

11.   If the quantity surveyor incurs additional costs due to exceptional delays in building operations or any other cause beyond the control of the quantity surveyor, then the fees shall be adjusted by agreement between the employer and the quantity surveyor to cover the reimbursement of these additional costs.

12.   When the quantity surveyor is required to prepare an interim valuation of materials or goods off site, an additional fee shall be charged based on the time involved (see paras. 15 and 16) in respect of each such valuation.

13.   If the Works are materially varied to the extent that substantial remeasurement is necessary, then the fee may be adjusted by agreement between the employer and the quantity surveyor.

14.   For additional services not normally necessary, such as those arising as a result of the termination of a contract before completion, fire damage to the buildings, cost monitoring (see para. 2 (f)), services in connection with arbitration, litigation and investigation of the validity of contractors' claims, services in connection with taxation matters and similar all services where the employer specifically instructs the quantity surveyor, the charges shall be in accordance with paras. 15 & 16.

15.   (a)   For consultancy and other services performed by a principal, a fee by arrangement according to the circumstances, including the professional status and qualifications of the quantity surveyor.
        (b)   When a principal does work which would normally be done by a member of staff, the charge shall be calculated as para. 16.

16.   (a)   For services by a member of staff, the charges for which are to be based on the time involved, such hourly charges shall be calculated on the basis of annual salary (including bonus and any other payments or benefits previously agreed with the employer) multiplied by a factor of 2.5, plus reimbursement of payroll costs, all divided by 1600. Payroll costs shall include inter alia employer's contributions payable under any Pension and Life Assurance Schemes, employer's contributions made under the National Insurance Acts, the Redundancy Payments Act and any other payments made in respect of the

Fees for Professional Services

(b) A member of staff shall include a principal doing work normally done by an employee (as para. 15 (b) above), technical and supporting staff, but shall exclude secretarial staff or staff engaged upon general administration.

(c) For the purpose of para. 16 (b) above a principal's time shall be taken at the rate applicable to a senior assistant in the firm.

(d) The supervisory duties of a principal shall be deemed to be included in the multiplication factor as para. 16 (a) above and shall not be charged separately.

17. The foregoing Time Charges under paras. 15 and 16 are intended for use where other paras. of the scale (not related to Time Charges) form a significant proportion of the overall fee. In all other cases an increased time charge may be agreed.

18. (a) In the event of the employment of the contractor being determined due to bankruptcy or liquidation, the fee for the services outlined in para. 2, and for the additional services required, shall be recalculated to the aggregate of the following:

    (i) Fifty per cent (50%) of the fee in accordance with paragraphs 5 and 6 calculated upon the total of the Notional Final Account in accordance with the terms of the original contracts.

    (ii) Fifty per cent (50%) of the fee in accordance with paragraphs 5 and 6 calculated upon the aggregate of the total value (which may differ from the total of interim valuations) of work up to the date of determination in accordance with the terms of the original contract plus the total of the final account for the completion contract;

    (iii) A charge based upon time involved (in accordance with paragraphs 15 and 16) in respect of dealing with those matters specifically generated by the liquidation (other than normal post-contract services related to the completion contract), which may include (inter alia):

> Site inspection and (where required) security (initial and until the replacement contractor takes possession);
> Taking instructions from and/or advising the employer;
> Representing the employer at meeting(s) of creditors;
> Making arrangements for the continued employment of sub-contractors and similar related matters;
> Preparing bills of quantities or other appropriate documents for the completion contract, obtaining tenders, checking and reporting thereon;
> The additional cost (over and above the preparation of the final account for the completion contract) of pre-paring the Notional Final Account; pricing the same;
> Negotiations with the liquidator (trustee or receiver).

(b) In calculating fees under para. 18 (a) (iii) above, regard shall be taken of any services carried out by the quantity surveyor for which a fee will ultimately be chargeable under para. 18 (a) (i) and (ii) above in respect of which a suitable abatement shall be made from the fee charged (e.g. measurement of variations for purposes of the completion contract where such would contribute towards the preparation of the contract final account).

(c) Any interim instalments of fees paid under para. 8 in respect of services outlined in para. 2 shall be deducted from the overall fee computed as outlined herein.

(d) In the absence of agreement to the contrary fees and expenses in respect of those services outlined in para. 18 (a) (iii) above up to acceptance of a completion tender shall be paid upon such acceptance; the balance of fees and expenses shall be paid in accordance with para. 8 (c).

(e) For the purpose of this Scale the term "Notional Final Account" shall be deemed to mean an account indicating that which would have been payable to the original contractor had he completed the whole of the works and before deduction of interim payments to him.

19. The fees and charges are in all cases exclusive of Value Added Tax which will be applied in accordance with legislation.

EXPLANATORY NOTE:

(Source: Chartered Quantity Surveyor, August 1986)

For rehabilitation projects the basic fee set out in paragraph 5 (a) of the scale will apply with the addition of a further 1% fee calculated upon the total of the final account for rehabilitation works including all nominated sub-contractors' and nominated suppliers' accounts.
In the case of special housing categories (e.g., elderly people) the additional percentage should be applied before the application of the additional percentage set out in paragraph 5 (b). The provisions of paragraph 6 (a) of the scale will also apply.
There is no longer any distinction between "hostel" and "cluster dwellings" which now have a single category of shared housing.
For shared housing new build projects other than those specified below the fee should be calculated in accordance with paragraph 5 (a) plus an enhancement of 10%.
For shared housing rehabilitation projects other than those specified below the fee should be calculated in accordance with paragraph 5 (a) of the scale plus 1% plus an enhancement of 10%.
For shared housing projects comprising wheelchair accommodation (as described in the Housing Corporation's Design and Contract Criteria) or frail elderly accommodation (as described in Housing Corporation circular HCO1/85) the fee should be calculated in accordance with paragraph 5 (a), (plus 1% for rehabilitation schemes where applicable) plus an enhancement of 15%.
The additional percentage set out in paragraph 5 (b) does not apply to shared housing projects, but the provisions of paragraph 6 (a) are applicable.

6
The Aggregates Tax

The Tax

The Aggregates Tax came into operation on 1 April 2002 in the UK, except for Northern Ireland where it will be phased in over five years commencing in 2003.

It is currently levied at the rate of £1.60 per tonne on anyone considered to be responsible for commercially exploiting 'virgin' aggregate in the UK and should naturally be passed by price increase to the ultimate user.

All materials falling within the definition of 'aggregate' is liable for the levy unless it is specifically exempted.

It does not apply to clay, soil, vegetable or other organic matter.

The hope is that this will :-

- Encourage the use of alternative materials that would otherwise be disposed of to landfill sites.
- promote development of new recycling processes, such as using waste tyres and glass
- promote greater efficiency in the use of virgin aggregates
- reduce noise + vibration, dust + other emissions to air, visual intrusion, loss of amenity + damage to wildlife habitats

The intention is for part of the revenue from the levy to be recycled to business and communities affected by aggregates extraction, through :-

- a 0.1 percentage point cut in employers' national insurance contributions
- a new £35 million per annum 'Sustainability Fund' to reduce the need for virgin materials and to limit the effects of extraction on the environmental where it takes place. A number of key priorities have been identified with a promise to give effect to them through existing programmes. The intention was to publish the results early in 2003.

Definition

'Aggregate' means any rock, gravel or sand which is extracted or dredged in the UK for aggregates use. It includes whatever substances are for the time being incorporated in it or naturally occur mixed with it.

'Exploitation' is defined as involving any one or a combination of any of the following :-

- being removed from its original site
- becoming subject to a contract or other agreement to supply to any person
- being used for construction purposes
- being mixed with any material or substance other than water, except in permitted circumstances

Aggregates Tax

Incidence

It is a tax on primary aggregate production – i.e. 'virgin' aggregate won from a source and used in a location within the UK territorial boundaries (land or sea). The tax is not levied on aggregate which is exported nor on aggregate imported from outside the UK territorial boundaries.

It is levied at the point of sale.

Exemption from tax

An 'aggregate' is exempt from the levy if it is :-

- material which has previously been used for construction purposes
- aggregate that has already been subject to a charge to the aggregates levy
- aggregate which was previously removed from its originating site before the commencement date of the levy
- aggregate which is being returned to the land from which it was won
- aggregate won from a farm land or forest where used on that farm or forest
- rock which has not been subjected to an industrial crushing process
- aggregate won by being removed from the ground on the site of any building or proposed building in the course of excavations carried out in connection with the modification or erection of the building and exclusively for the purpose of laying foundations or of laying any pipe or cable
- aggregate won by being removed from the bed of any river, canal or watercourse or channel in or approach to any port or harbour (natural or artificial), in the course of carrying out any dredging exclusively for the purpose of creating, restoring, improving or maintaining that body of water
- aggregate won by being removed from the ground along the line of any highway or proposed highway in the course of excavations for improving, maintaining or constructing the highway + otherwise than purely to extract the aggregate
- drill cuttings from petroleum operations on land and on the seabed
- aggregate resulting from works carried out in exercise of powers under the New Road and Street Works Act 1991, the Roads (Northern Ireland) Order 1993 or the Street Works (Northern Ireland) Order 1995
- aggregate removed for the purpose of cutting of rock to produce dimension stone, or the production of lime or cement from limestone.
- Aggregate arising as a waste material during the processing of the following industrial minerals:
    - ball clay
    - barytes
    - calcite
    - china clay
    - coal, lignite, slate or shale
    - feldspar
    - flint
    - fluorspar
    - fuller's earth
    - gems and semi-precious stones
    - gypsum

Aggregates Tax

- any metal or the ore of any metal
- muscovite
- perlite
- potash
- pumice
- rock phosphates
- sodium chloride
- talc
- vermiculite

However, the levy is still chargeable on any aggregate arising as the spoil or waste from or the by-products of the above exempt processes. This includes quarry overburden.

Anything that consists 'wholly or mainly' of the following is exempt from the levy (note that 'wholly' is defined as 100% but 'mainly' as more than 50%, thus exempting any contained aggregates amounting to less than 50% of the original volumes :

- clay, soil, vegetable or other organic matter
- coal, slate or shale
- china clay waste and ball clay waste

Relief from the levy either in the form of credit or repayment is obtainable where :

- it is subsequently exported from the UK in the form of aggregate
- it is used in an exempt process
- where it is used in a prescribed industrial or agricultural process
- it is waste aggregate disposed of by dumping or otherwise, e.g. sent to landfill or returned to the originating site

BACKGROUND

Environment Act 1995

Provides for planning permissions for minerals extraction (which includes aggregates) to be reviewed and updated on a regular basis by the mineral planning authorities.

Minerals planning guidance notes were to be updated to ensure that all mineral operators provide local benefits that clearly outweigh the likely impacts of mineral extraction.

The Rural White Paper published in November 2000

This set out the government's 'vision of a countryside that is sustainable economically, socially and environmentally' apart from economic support, action was proposed to deal with a number of environmental threats, one is to reduce the environmental impact (such as damage to biodiversity and visual intrusion) of aggregate extraction by imposing an aggregates levy.

Finance Act 2001

This provides for a levy being charged on aggregate which is subject to commercial exploitation. It places the levy under the care and management of the Commissioners of Customs & Excise.

Anticipated impact

The British Aggregates Association has suggested that the additional cost imposed by quarries is more likely to be in the order of £2.65 per tonne on mainstream products, applying an above average rate on these in order that by-products and low grade waste products can be held at competitive rates, as well as making some allowance for administration and increased finance charges.

With many gravel aggregates costing in the region of £7.50 to £9.50 per tonne, there will be a significant impact on construction costs.

The Civil Engineering Contractors Association has predicted that the tax will push up total construction prices in the civil sector by 1.3%. Other models have come up with figures much higher than this, but the level for any particular project will of course depend on the constituents and type of work involved. It has been estimated that between 20 and 25% of aggregate production is used for road maintenance and construction. Obviously there will be a substantial cost impact on civil engineering schemes involving large quantities of aggregates as fill materials etc. in schemes such as roads, railways and airports, but there will also be a significant impact on any building work which employs large quantities of aggregates not only in their primary state but also as concrete or concrete based products. An average building project could attract an increase of approximately 0.8 to 1%.

Individual typical materials could be affected as follows :-

- Concrete: given that a cubic metre contains 1.9 - 2.4 tonnes of sand and aggregate, the levy could push up prices by £ 3.04 - 3.84 per m³ at the £1.60 rate, were this to be £2.65 as reasoned above, the resulting increase would be £5.04 - 6.36 per m³. On a ready mix supply price of say £65 per m³, this would represent an average rise of 5.29% or 8.77% respectively.
- Precast concrete blocks: anticipated at 10 p per m² on dense aggregate
- Concrete roof tiles: anticipated at 10 p to 50 p per m² of roofing
- Precast concrete pavings: anticipated at 2 to 5% on the cost of concrete pavings, flags, blocks, kerbs etc.
- Bitumen macadam paving: anticipated to be in the region of 2.5% on the cost of materials

Avoidance

An alternative to using new aggregate in filling operations is to crush and screen rubble which may become available during the process of demolition and site clearance as well as removal of obstacles during the excavation processes.

Aggregates Tax

Example : Assuming that the material would be suitable for fill material under buildings or roads, a simple cost comparison would be as follows (note that for the purpose of the exercise, the material is taken to be 1.80 t/m³ and the total quantity involved less than 1,000 m³) :-

Importing fill material :

	£/m³	£/tonne
Cost of 'new' aggregate delivered to site	17.10	9.50
Addition for Aggregates Tax	2.88	1.60
Total cost of importing fill materials	19.98	11.10

Disposing of site material :

	£/m³	£/tonne
Cost of removing materials from site	9.50	5.28
Total cost of disposal of site material	9.50	5.28

Crushing site materials :

	£/m³	£/tonne
Transportation of material from excavations or demolition places to temporary stockpiles	1.25	.69
Transportation of material from temporary stockpiles to the crushing plant	1.00	.56
Establishing plant and equipment on site; removing on completion	.75	.42
Maintain and operate plant	3.75	2.08
Crushing hard materials on site	5.50	3.05
Screening material on site	.75	.42
Total cost of crushing site materials	13.00	7.22

From the above it can be seen that potentially there is a great benefit in crushing site materials for filling rather than importing fill materials.

Setting the cost of crushing against the import price would produce a saving of £6.98 per m³. If the site materials were otherwise intended to be removed from the site, then the cost benefit increases by the saved disposal cost to £16.48 per m³.

Even if there is no call for any or all of the crushed material on site, it ought to be regarded as a useful asset and either sold on in crushed form or else sold with the prospects of crushing elsewhere.

Specimen Unit rates

Establishing plant and equipment on site; removing on completion
  Crushing plant   £500
  Screening plant   £250

Maintain and operate plant
  Crushing plant   £3,000 per week
  Screening plant   £750 per week

Transportation of material from excavations or demolition places   £1.25 per m³
to temporary stockpiles

Transportation of material from temporary stockpiles to the   £1.00 per m³
crushing plant

Breaking up material on site using impact breakers
  mass concrete   £5.50 per m³
  reinforced concrete   £6.50 per m³
  brickwork   £2.50 per m³

Crushing material on site
  mass concrete ne 1000m³   £5.50 per m³
  mass concrete 1000 - 5000m³   £5.00 per m³
  mass concrete over 5000m³   £4.50 per m³
  reinforced concrete ne 1000m³   £6.50 per m³
  reinforced concrete 1000 - 5000m³   £6.00 per m³
  reinforced concrete over 5000m³   £5.50 per m³
  brickwork ne 1000m³   £5.00 per m³
  brickwork 1000 - 5000m³   £4.50 per m³
  brickwork over 5000m³   £4.00 per m³

Screening material on site   £0.75 per m³

7
Prices for Measured Works

ESSENTIAL READING FROM SPON PRESS

Design for Outdoor Recreation

Simon Bell, Forestry Commission, UK

'Very successful ... nicely presented and easy to read ... will provide a useful reference and stimulus to design professions, country managers and planners.'
Journal of Environmental Planning and Management

This book takes a fresh look at all aspects of design of facilities needed by visitors to outdoor recreation destinations. The book follows the thought processes and physical requirements of visitors, from their perspective. This idea is followed step-by-step from setting off for a destination to the arrival and the visit itself, broken down into a range of different aspects, such as information needs, car parking, picnicking, hiking a trail, recreation by water, wildlife watching and camping. The book is fully illustrated with sketches, diagrams and photographs, many in colour, taken from a wide range of locations in the United Kingdom, Europe, North America and elsewhere.

Contents: Introduction. Recreation Planning. Design Concepts for Outdoor Recreation. The Journey to the Destination. Providing Visitor Information. Parking the Car. Toilet Facilities. Picnicking. Children's Play. Trails. Water Based Recreation. Wildlife Viewing. Design for Overnight Visitors. Interpretation. Comprehensive Site Design. Bibliography. Index.

January 1997: 276x219
200 line illustrations, 60 b+w photos, 45 colour illustrations
Pb: 0-419-20350-8: £49.00

To Order: Tel: +44 (0) 8700 768853, or +44 (0) 1264 343071 Fax: +44 (0) 1264 343005, or Post: Spon Press Customer Services, Thomson Publishing Services, Cheriton House, Andover, Hants, SP10 5BE, UK Email: book.orders@tandf.co.uk.

For a complete listing of all our titles visit:
www.sponpress.com

NEW ITEMS FOR 2003 Excluding overheads and profit	PC £	Labour hours	Labour £	Plant £	Material £	Unit	Total rate £
D11 GROUND STABILIZATION: ANCHORS							
Anchoring Systems; Surface stabilization; Platipus Engineering Systems; Centres shown should be verified with a design engineer. They may vary either way dependent on circumstances							
Concrete Revetments; Hard anodised stealth anchors with stainless steel accessories installed to a depth of 1 m							
"SO4" anchors to a depth of 1.00 -1.50 m at 1.00 m ccs	15.00	0.33	5.00	0.89	15.00	m²	20.89
Loadlocking and crimping tool; for stressing and crimping the anchors							
Purchase price	350.00	-	-	-	350.00	nr	350.00
Hire rate	60.00	-	-	-	60.00	nr	60.00
Surface erosion; geotextile anchoring on slopes; Aluminium alloy Stealth anchors							
"SO4", "SO6" to a depth of 1.0-2.0 m at 2.00 m ccs	10.00	0.20	3.00	0.89	10.00	m²	13.89
Brick block or in-situ concrete, distressed retaining walls; anchors installed in two rows at varying tensions between 18 kN - 50 kN along the length of the retaining wall. Core drilling of wall not included							
"SO8", "BO6", "BO8" bronze Stealth and Bat anchors installed in combination with stainless steel accessories to a depth of 6 m. Average cost base on 1.50 m ccs; long term solution	300.00	0.50	7.50	0.89	300.00	m²	308.39
"SO8", "BO6", "BO8" bronze Stealth and Bat anchors installed in combination with stainless steel accessories to a depth of 6 m. Average cost base on 1.50 m ccs; short term solution	150.00	0.50	7.50	0.89	112.50	m²	120.89
Timber retaining wall support; anchors installed 19 kN along the length of the retaining wall. "SO6", "SO8", bronze Stealth anchors installed in combination with stainless steel accessories to a depth of 3.0 m. 1.50 centres; Average cost;	45.00	0.50	7.50	0.71	45.00	m²	53.21
Gabions; anchors for support and stability to moving, overturning or rotating gabion retaining walls							
"SO8", "BO6" anchors including base plate at 1.00 m centres to a depth of 4.00 m	100.00	0.33	5.00	0.89	100.00	m²	105.89
Soil nailing of geofabrics; anchors fixed through surface of fabric or erosion control surface treatment (not included)							
"SO8", "BO6" anchors including base plate at 1.50 m centres to a depth of 3.00 m	100.00	0.17	2.50	0.59	100.00	m²	103.09

NEW ITEMS FOR 2003 Excluding overheads and profit	PC £	Labour hours	Labour £	Plant £	Material £	Unit	Total rate £
D20 EXCAVATING/FILLING							
Market Prices of Topsoil; Charles Morris Fertilizers Ltd							
Multiple source screened topsoil	-	-	-	-	8.33	m³	8.33
Single source screened topsoil	-	-	-	-	11.66	m³	11.66
"P30" high grade topsoil	-	-	-	-	22.33	m³	22.33
Filling to make up levels; mechanical							
Arising from the excavations							
average thickness 100 mm maximum thickness	-	-	0.06	0.12	-	m²	0.18
average thickness 150 mm maximum thickness	-	0.01	0.09	0.18	-	m²	0.27
average thickness 200 mm maximum thickness	-	0.01	0.12	0.24	-	m²	0.36
average thickness 250 mm maximum thickness	-	0.01	0.15	0.30	-	m²	0.45
average thickness 300 mm maximum thickness	-	0.01	0.20	0.40	-	m²	0.60
average thickness 400 mm maximum thickness	-	0.02	0.24	0.49	-	m²	0.73
average thickness 500 mm maximum thickness	-	0.02	0.30	0.61	-	m²	0.91
average thickness 750 mm maximum thickness	-	0.03	0.45	0.91	-	m²	1.36
Obtained from on site spoil heaps; average 25 m distance; multiple handling							
average thickness 250 mm;	-	0.02	0.30	0.74	-	m²	1.04
average thickness 300 mm;	-	0.02	0.36	0.89	-	m²	1.25
average thickness 400 mm;	-	0.03	0.48	1.19	-	m²	1.67
average thickness 500 mm;	-	0.04	0.63	1.55	-	m²	2.17
average thickness 750 mm;	-	0.06	0.94	2.32	-	m²	3.26
Obtained off site; soil PC £11.66 / m³							
average thickness 100 mm	1.17	0.01	0.10	0.20	1.40	m²	1.70
average thickness 200 mm	2.33	0.01	0.20	0.40	2.80	m²	3.40
average thickness 250 mm	2.92	0.02	0.25	0.51	3.50	m²	4.26
average thickness 300 mm	3.50	0.02	0.23	0.46	4.20	m²	4.88
average thickness 400 mm	4.66	0.02	0.30	0.61	5.60	m²	6.50
average thickness 500 mm	5.83	0.03	0.38	0.76	7.00	m²	8.13
average thickness 750 mm	8.74	0.04	0.56	1.14	10.49	m²	12.20
Obtained off site; hardcore; PC £16.00 / m³							
average thickness 100 mm	1.60	0.01	0.10	0.20	1.76	m²	2.06
average thickness 200 mm	3.20	0.01	0.20	0.40	3.52	m²	4.12
average thickness 250 mm	4.00	0.02	0.25	0.51	4.40	m²	5.16
average thickness 300 mm	4.80	0.02	0.23	0.46	5.28	m²	5.96
average thickness 400 mm	6.40	0.02	0.30	0.61	7.04	m²	7.95
average thickness 500 mm	8.00	0.03	0.38	0.76	8.80	m²	9.93
average thickness 750 mm	12.00	0.04	0.56	1.14	13.20	m²	14.90
Disposal by skip; 7 yd3 (5.35 m³)							
Excavated material; off site; to tip							
by machine	-	0.13	1.88	1.29	21.80	m³	24.97
by hand	-	1.00	15.00	-	21.80	m³	36.80

NEW ITEMS FOR 2003 Excluding overheads and profit	PC £	Labour hours	Labour £	Plant £	Material £	Unit	Total rate £
F10 BRICK/BLOCK WALLING							
Dense aggregate concrete blocks; "Tarmac Topblock" or other equal and approved; in gauged mortar (1:2:9)							
Walls							
Solid Blocks 10 n/mm²							
440 x 215 x 100 mm thick	6.30	1.20	18.00	-	7.25	m²	**25.25**
440 x 215 x 140 mm thick	9.09	1.30	19.50	-	10.16	m²	**29.66**
440 x 215 x 215 mm thick	10.24	1.50	22.50	-	11.90	m²	**34.40**
Solid Blocks 10N/mm² laid flat							
440 x 100 x 215 mm thick	12.85	3.22	48.30	-	15.40	m²	**63.70**
Hollow Concrete blocks							
440 x 215 x 215 mm thick	10.25	1.30	19.50	-	11.32	m²	**30.82**
Filling of Hollow concrete blocks with concrete as work proceeds; tamping and compacting							
440 x 215 x 215 mm thick	6.70	0.20	3.00	-	6.70	m²	**9.70**
H51 NATURAL STONE/SLAB CLADDING FEATURES							
Sawn Yorkstone cladding; Johnsons Stone Ltd;							
Six sides sawn stone; rubbed face ; sawn and jointed edges; fixed to blockwork (not included) with stainless steel fixings "Ancon Ltd" grade 304 stainless steel frame cramp and dowel 7mm; cladding units drilled 4 x to receive dowelless							
440 x 200 x 50 mm thick	74.95	1.87	28.05	-	77.11	m²	**105.16**
H52 CAST STONE SLAB CLADDING FEATURES							
Cast Stone cladding; Haddonstone Ltd;							
Reconstituted stone in Portland Bath or Terracotta; fixed to blockwork or concrete (not included) with stainless steel fixings "Ancon Ltd" grade 304 stainless steel frame cramp and dowel M6 mm; cladding units drilled 4 x to receive dowelless							
440 x 200 x 50 mm thick	85.73	1.87	28.05	-	87.89	m²	**115.94**

NEW ITEMS FOR 2003 Excluding overheads and profit	PC £	Labour hours	Labour £	Plant £	Material £	Unit	Total rate £
Q22 COATED MACADAM/ASPHALT ROADS/PAVINGS							
Preamble; Users should note the new terminology for the surfaces described below which is to European standard descriptions. The now redundant descriptions for each course are shown in brackets.							
Macadam Surfacing; Spadeoak Construction Co Ltd; Surface (Wearing) Course; 20 mm of 6 mm dense bitumen macadam to BS4987-1 2001 ref 7.5							
Machine lay; areas 1000 m² and over							
limestone aggregate	-	-	-	-	-	m²	3.92
granite aggregate	-	-	-	-	-	m²	3.98
red	-	-	-	-	-	m²	7.33
Hand Lay; areas 400 m² and over							
limestone aggregate	-	-	-	-	-	m²	5.02
granite aggregate	-	-	-	-	-	m²	5.08
red	-	-	-	-	-	m²	9.10
Macadam Surfacing; Spadeoak Construction Co Ltd; Surface (Wearing) Course; 30 mm of 10 mm dense bitumen macadam to BS4987-1 2001 ref 7.4							
Machine lay; areas 1000 m² and over							
limestone aggregate	-	-	-	-	-	m²	4.69
granite aggregate	-	-	-	-	-	m²	4.77
red	-	-	-	-	-	m²	10.25
Hand Lay; areas 400 m² and over							
limestone aggregate	-	-	-	-	-	m²	5.83
granite aggregate	-	-	-	-	-	m²	5.92
red	-	-	-	-	-	m²	12.21
Macadam Surfacing; Spadeoak Construction Co Ltd; Surface (Wearing) Course; 40 mm of 10 mm dense bitumen macadam to BS4987-1 2001 ref 7.4							
Machine lay; areas 1000 m² and over							
limestone aggregate	-	-	-	-	-	m²	5.94
granite aggregate	-	-	-	-	-	m²	6.05
red	-	-	-	-	-	m²	11.62
Hand Lay; areas 400 m² and over							
limestone aggregate	-	-	-	-	-	m²	7.17
granite aggregate	-	-	-	-	-	m²	7.29
red	-	-	-	-	-	m²	13.68

NEW ITEMS FOR 2003 Excluding overheads and profit	PC £	Labour hours	Labour £	Plant £	Material £	Unit	Total rate £
Macadam Surfacing; Spadeoak Construction Co Ltd; Binder (Base) Course; 50 mm of 20 mm dense bitumen macadam to BS4987-1 2001 ref 6.5							
Machine lay; areas 1000 m² and over							
limestone aggregate	-	-	-	-	-	m²	6.32
granite aggregate	-	-	-	-	-	m²	6.45
Hand Lay; areas 400 m² and over							
limestone aggregate	-	-	-	-	-	m²	7.57
granite aggregate	-	-	-	-	-	m²	7.71
Macadam Surfacing; Spadeoak Construction Co Ltd; Binder (Base) Course; 60 mm of 20 mm dense bitumen macadam to BS4987-1 2001 ref 6.5							
Machine lay; areas 1000 m² and over							
limestone aggregate	-	-	-	-	-	m²	6.95
granite aggregate	-	-	-	-	-	m²	7.10
Hand Lay; areas 400 m² and over							
limestone aggregate	-	-	-	-	-	m²	8.25
granite aggregate	-	-	-	-	-	m²	8.41
Macadam Surfacing; Spadeoak Construction Co Ltd; Base (Roadbase) Course; 75 mm of 28 mm dense bitumen macadam to BS4987-1 2001 ref 5.2							
Machine lay; areas 1000 m² and over							
limestone aggregate	-	-	-	-	-	m²	8.34
granite aggregate	-	-	-	-	-	m²	8.54
Hand Lay; areas 400 m² and over							
limestone aggregate	-	-	-	-	-	m²	9.73
granite aggregate	-	-	-	-	-	m²	9.94
Macadam Surfacing; Spadeoak Construction Co Ltd; Base (Roadbase) Course; 100 mm of 28 mm dense bitumen macadam to BS4987-1 2001 ref 5.2							
Machine lay; areas 1000 m² and over							
limestone aggregate	-	-	-	-	-	m²	10.52
granite aggregate	-	-	-	-	-	m²	10.78
Hand Lay; areas 400 m² and over							
limestone aggregate	-	-	-	-	-	m²	12.51
granite aggregate	-	-	-	-	-	m²	12.78
Macadam Surfacing; Spadeoak Construction Co Ltd; Base (Roadbase) Course; 150 mm of 28 mm dense bitumen macadam in two layers to BS4987-1 2001 ref 5.2							
Machine lay; areas 1000 m² and over							
limestone aggregate	-	-	-	-	-	m²	16.59
granite aggregate	-	-	-	-	-	m²	16.97
Hand Lay; areas 400 m² and over							
limestone aggregate	-	-	-	-	-	m²	20.26
granite aggregate	-	-	-	-	-	m²	20.67

NEW ITEMS FOR 2003 Excluding overheads and profit	PC £	Labour hours	Labour £	Plant £	Material £	Unit	Total rate £
Q22 COATED MACADAM/ASPHALT ROADS/PAVINGS - cont'd							
Macadam Surfacing; Spadeoak Construction Co Ltd; Base (Roadbase) Course; 200 mm of 28 mm dense bitumen macadam in two layers to BS4987-1 2001 ref 5.2							
Machine lay; areas 1000 m² and over							
limestone aggregate	-	-	-	-	-	m²	21.08
granite aggregate	-	-	-	-	-	m²	21.60
Hand Lay; areas 400 m² and over							
limestone aggregate	-	-	-	-	-	m²	25.06
granite aggregate	-	-	-	-	-	m²	25.61
Q24 GRASS BLOCKS RECYCLED PLASTIC							
Recycled polyethylene grassblocks; Netlon Ltd; interlocking units laid to prepared base or rootzone (not included); "Netpave 50"; load bearing 50 tonnes per m²; 500 x 500 x 50 mm deep							
minimum area 40 m²	15.00	0.20	3.00	0.21	16.68	m²	19.89
40 -1039 m²	10.40	0.20	3.00	0.21	12.08	m²	15.29
over 1039 m²	9.50	0.20	3.00	0.21	11.18	m²	14.39
"Netpave 25"; load bearing :light vehicles and pedestrians; 500 x 500 x 25 mm deep; laid onto established grass surface							
minimum area 40 m²	14.00	0.10	1.50	-	14.00	m²	15.50
40 -1039 m²	9.40	0.10	1.50	-	9.40	m²	10.90
over 1039 m²	8.50	0.10	1.50	-	8.50	m²	10.00
"Grassroad"; Cooper Clarke Group; heavy duty for car parking and fire paths verge hardening and shallow embankments Honeycomb cellular polypropylene interconnecting paviors with integral downstead anti-shear cleats including topsoil but excluding edge restraints; to granular sub-base (not included)							
635 x 330 x 42 overall laid to a module of 622 x 311 x 32	-	-	-	-	-	m²	21.23
Extra over for green colour	-	-	-	-	-	m²	0.26

NEW ITEMS FOR 2003 Excluding overheads and profit	PC £	Labour hours	Labour £	Plant £	Material £	Unit	Total rate £
Q30 SEEDING/TURFING							
Reinforced turf; Netlon Advanced Turf; Netlon Ltd; Blended mesh fibre elements incorporated into root zone; Rootzone spread and levelled over cultivated, prepared and reduced and levelled ground (not included); compacted with light roller.							
"ATS 300" with "R300" topping, seed and fertilizer; 100 thick							
100 - 199 m²	16.15	0.03	0.38	0.60	16.15	m²	17.13
200 - 299 m²	14.65	0.03	0.38	0.60	14.65	m²	15.63
300 - 499 m²	14.10	0.03	0.38	0.60	14.10	m²	15.08
over 500 m²	13.70	0.03	0.38	0.60	13.70	m²	14.68
"ATS 300" with "R300" topping, seed and fertilizer; 150 thick							
100 - 199 m²	21.10	0.03	0.50	0.80	21.10	m²	22.40
200 - 299 m²	19.60	0.03	0.50	0.80	19.60	m²	20.90
300 - 499 m²	18.95	0.03	0.50	0.80	18.95	m²	20.25
over 500 m²	18.50	0.03	0.50	0.80	18.50	m²	19.80
"ATS 300" with washed turf and fertilizer ("R300" topping not included) 100 thick							
100 -199 m²	19.70	0.03	0.50	0.80	19.70	m²	21.00
200 -499 m²	18.10	0.03	0.50	0.80	18.10	m²	19.40
over 500 m²	17.15	0.03	0.50	0.80	17.15	m²	18.45
"ATS 300" with washed turf and fertilizer ("R300" topping not included) 150 thick							
100 - 199 m²	24.55	0.04	0.60	0.90	24.55	m²	26.05
200 - 499 m²	22.95	0.04	0.60	0.90	22.95	m²	24.45
over 500 m²	21.85	0.04	0.60	0.90	21.85	m²	23.35
Q31 PLANTING							
Structural Soils; Amsterdam Tree Sand; Urban Soils Ltd; Non compressive soil mixture for tree planting in areas to receive compressive surface treatments.							
backfilling and light compacting in layers; excavation, disposal, moving of material from delivery position and surface treatments not included; by machine							
individual treepits	45.38	0.50	7.50	2.59	45.38	m³	55.46
in trenches	45.38	0.42	6.25	2.15	45.38	m³	53.78
backfilling and light compacting in layers; excavation, disposal, moving of material from delivery position and surface treatments not included; by hand							
individual treepits	45.38	1.60	24.00	-	45.38	m³	69.38
in trenches	45.38	1.33	19.95	-	45.38	m³	65.33

NEW ITEMS FOR 2003 Excluding overheads and profit	PC £	Labour hours	Labour £	Plant £	Material £	Unit	Total rate £
Q40 FENCING							
Metal Estate Fencing; Broxap Ltd; Mild steel flat bar angular fencing; galvanized; main posts at 5.00 m centres; fixed with 1:3:6 concrete 430 deep; intermediate posts fixed with fixing claw at 1.00 m centres.							
"Cheshire"; curved top posts and 5 nr flat rails							
1200 high	29.60	0.28	4.17	-	32.00	m	**36.17**
"Prestbury" 5 nr plain flat or round bar							
1200 high	27.20	0.28	4.17	-	29.60	m	**33.77**
Kissing gate; Mild steel construction, flat or angular steel;							
Galvanized	271.00	2.00	30.00	-	273.40	m	**303.40**
R12 CLAY DRAINAGE							
Excavating trenches; using 3 tonne tracked excavator; to receive pipes; grading bottoms; earthwork support; filling with excavated material to within 150 mm of finished surfaces and compacting; completing fill with topsoil; disposal of surplus soil							
Services not exceeding 200 mm nominal size							
average depth of run not exceeding 0.50 m	0.52	0.12	1.80	1.15	0.63	m	**3.58**
average depth of run not exceeding 0.75 m	0.52	0.16	2.44	1.59	0.63	m	**4.67**
average depth of run not exceeding 1.00 m	0.52	0.28	4.25	2.78	0.63	m	**7.66**
average depth of run not exceeding 1.25 m	0.52	0.38	5.75	3.74	0.52	m	**10.01**
Granular beds to trenches; lay granular material, to trenches excavated separately, to receive pipes (not included)							
300 wide x 100 thick							
reject sand	1.90	0.05	0.75	0.26	2.09	m	**3.10**
reject gravel	2.27	0.05	0.75	0.26	2.38	m	**3.39**
shingle 40 mm aggregate	2.61	0.05	0.75	0.26	2.74	m	**3.75**
sharp sand	2.62	0.05	0.75	0.26	2.88	m	**3.89**
300 wide x 150 thick							
reject sand	2.85	0.08	1.13	0.39	3.13	m	**4.65**
reject gravel	3.40	0.08	1.13	0.39	3.57	m	**5.08**
shingle 40 mm aggregate	3.92	0.08	1.13	0.39	4.11	m	**5.62**
sharp sand	3.93	0.08	1.13	0.39	4.32	m	**5.84**
Excavating trenches; using 3 tonne tracked excavator; to receive pipes; grading bottoms; earthwork support; filling with imported granular material type 2 and compacting; disposal of surplus soil							

NEW ITEMS FOR 2003 Excluding overheads and profit	PC £	Labour hours	Labour £	Plant £	Material £	Unit	Total rate £
Services not exceeding 200 mm nominal size							
average depth of run not exceeding 0.50 m	4 37	0.09	1.30	0.80	6.17	m	8.28
average depth of run not exceeding 0.75 m	6 56	0.11	1.62	1.01	9.26	m	11.89
average depth of run not exceeding 1.00 m	8.75	0.14	2.10	1.31	12.35	m	15.76
average depth of run not exceeding 1.25 m	10 94	0.23	3.43	2.23	15.44	m	21.10
Excavating trenches, using 3 tonne tracked excavator, to receive pipes; grading bottoms; earthwork support; filling with concrete, ready mixed ST2 ; disposal of surplus soil							
Services not exceeding 200 mm nominal size							
average depth of run not exceeding 0.50 m	10 13	0.11	1.60	0.41	12.19	m	14.20
average depth of run not exceeding 0.75 m	15.20	0.13	1.95	0.52	18.28	m	20.74
average depth of run not exceeding 1.00 m	20 27	0.17	2.50	0.69	24.37	m	27.56
average depth of run not exceeding 1.25 m	25 33	0.23	3.38	1.03	30.47	m	34.88
Earthwork Support; providing support to opposing faces of excavation; moving along as work proceeds; A Plant Acrow							
Maximum depth not exceeding 2.00 m							
distance between opposing faces not exceeding 2 00 m	-	0.80	12.00	15.39	-	m	27.39
Clay pipes and fittings: to BS EN295:1:1991:Hepworth Plc; Supersleve							
100 mm clay pipes; polypropylene slip coupling; in trenches (trenches not included)							
laid straight	1 83	0.25	3.75	-	3.18	m	6.93
short runs under 3.00 m	1 83	0.31	4.69	-	3.18	m	7.86
Extra over 100 mm clay pipes for							
bends; 15-90 degree; single socket	5.82	0.25	3.75	-	5.82	nr	9.57
junction; 45 or 90 degree; double socket	12 71	0.25	3.75	-	12.71	nr	16.46
slip couplings polypropylene	2 16	0.08	1.25	-	2.16	nr	3.41
gully with "P" trap; 100 mm; 154 mm x 154 mm plastic grating	47.29	1.00	15.00	-	68.23	nr	83.23
150 mm vitrified clay pipes; polypropylene slip coupling; in trenches (trenches not included)							
laid straight	7.30	0.30	4.50	-	7.30	m	11.80
short runs under 3.00 m	5.62	0.33	5.00	-	5.62	m	10.62
Extra over 150 mm vitrified clay pipes for							
bends; 15 - 90 degree	8 13	0.28	4.20	-	12.05	nr	16.25
junction; 45 or 90 degree; 100 x 150	15.56	0.40	6.00	-	23.40	nr	29.40
junction; 45 or 90 degree; 150 x 150	17 08	0.40	6.00	-	24.92	nr	30.92
slip couplings polypropylene	3 92	0.05	0.75	-	3.92	nr	4.67
taper pipe 100 -150 mm	12 24	0.50	7.50	-	12.24	nr	19.74
taper pipe 150 - 225 mm	30.05	0.50	7.50	-	30.05	nr	37.55
socket adaptor; connection to traditional pipes and fittings	7.89	0.33	4.95	-	11.81	nr	16.76
Accessories in clay							
150 mm access pipe	41.91	0.50	7.50	-	49.75	nr	57.25
150 mm rodding eye	39.98	0.50	7.50	-	44.43	nr	51.93
gully with "P" traps; 150 mm; 154 mm x 154 mm plastic grating	47.29	1.00	15.00	-	52.99	nr	67.99

R13 SOAKAWAYS

Soakaway design based on BS EN 752-4. Flat rate hourly rainfall = 50 mm /hr and assumes 100 impermeability of the run-off area. A storage capacity of the soakaway should be 1/3 of the hourly rainfall; Formulae for calculating soakaway depths are provided in the publications mentioned below and in the memoranda section of this publication. The design of soakaways is dependent on amongst other factors, soil conditions, permeability, groundwater level and runoff. The definitive documents for design of soakaways are CIRIA 156 and BRE Digest 365 dated September 1991; The suppliers of the systems below will assist through their technical divisions. Excavation earthwork support of pits and disposal not included.

NEW ITEMS FOR 2003 Excluding overheads and profit	PC £	Labour hours	Labour £	Plant £	Material £	Unit	Total rate £
Excavating; mechanical							
To reduce levels							
maximum depth not exceeding 1.00 m; JCB sitemaster	-	0.05	0.75	1.52	-	m³	2.27
maximum depth not exceeding 1.00 m; 360 Tracked excavator	-	0.04	0.60	1.19	-	m³	1.79
maximum depth not exceeding 2.00 m; 360 Tracked excavator	-	0.06	0.90	1.78	-	m³	2.68
Disposal; mechanical							
Excavated material; off site; to tip not exceeding 13 km; mechanically loaded.							
inert material	-	0.03	0.50	1.01	10.00	m³	11.51
Insitu concrete ring beam foundations to base of soakaway; 300 mm wide x 250 deep; poured on or against earth or unblinded hardcore							
internal diameters of rings							
900 mm	18.88	4.00	60.00	-	20.77	nr	80.77
1200 mm	25.17	4.50	67.50	-	27.69	nr	95.19
1500 mm	31.46	5.00	75.00	-	34.60	nr	109.60
2400 mm	50.34	5.50	82.50	-	55.37	nr	137.87
Concrete soakaway rings; Milton Pipes Ltd; perforations and step irons to concrete rings at manufacturers recommended centres; placing of concrete ring soakaways to insitu concrete ring beams (1:3:6) (not included); filling and surrounding base with gravel 225 deep (not included);							

NEW ITEMS FOR 2003 Excluding overheads and profit	PC £	Labour hours	Labour £	Plant £	Material £	Unit	Total rate £
Ring diameter 900 mm							
1 00 deep; volume 636 litres	31.00	1.25	18.75	13.01	99.95	nr	131.71
1 50 deep; volume 954 litres	46.50	1.65	24.75	34.34	147.85	nr	206.94
2 00 deep; volume 1272 litres	62.00	1.65	24.75	34.34	195.75	nr	254.84
Ring diameter 1200 mm							
1.00 deep; volume 1131 litres	46.00	3.00	45.00	31.22	125.77	nr	201.99
1 50 deep; volume 1696 litres	69.00	4.50	67.50	46.82	184.97	nr	299.30
2 00 deep, volume 2261 litres	92.00	4.50	67.50	46.82	244.17	nr	358.50
2 50 deep; volume 2827 litres	115.00	6.00	90.00	62 43	276.77	nr	429.21
Ring diameter 1500 mm							
1 00 deep; volume 1767 litres	84.00	4.50	67.50	31.22	183 13	nr	281.85
1 50 deep; volume 2651 litres	126.00	6.00	90.00	43.70	268.93	nr	402.63
2 00 deep; volume 3534 litres	168.00	6.00	90.00	43.70	354.73	nr	488.43
2 50 deep; volume 4418 litres	115.00	7.50	112.50	78.04	345.53	nr	536.07
Ring diameter 2400 mm							
1 00 deep; volume 4524 litres	214.00	6.00	90.00	31.22	347.33	nr	468.55
1 50 deep; volume 6786 litres	321.00	7.50	112.50	43.70	513.33	nr	669.53
2 00 deep; volume 9048 litres	428.00	7.50	112.50	43.70	683.13	nr	839.33
2 50 deep, volume 11310 litres	535.00	9.00	135.00	78.04	849.13	nr	1062.17
Cover slabs to soakaways							
heavy duty precast concrete							
900 diameter	53.00	1.00	15.00	15.61	53.00	nr	83.61
1200 diameter	68.00	1 00	15.00	15.61	68.00	nr	98.61
1500 diameter	108.00	1.00	15.00	15.61	108.00	nr	138.61
2400 diameter	383.00	1.00	15.00	15.61	383.00	nr	413.61
Step irons to concrete chamber rings.	19.20	-	-	-	19.20	m	19.20
Geofabric surround to soakaways; **Terram Ltd** "Terram" synthetic fibre filter fabric; to face of concrete rings (not included); anchoring whilst backfilling (not included)							
"Terram 1000", 0.70 mm thick; mean water flow 50 litre/m²/s	0.43	0.05	0.75	-	0.52	m²	1.27
Gravel surrounding to concrete ring soakaway 40 mm aggregate backfilled to vertical face of soakaway wrapped with geofabric (not included)							
250 thick	26.10	0.20	3.00	6.24	26.10	m³	35.34
Backfilling to face of soakaway; **Carefully compacting as work proceeds** arising from the excavations							
average thickness exceeding 0.25 m; depositing in layers 150 mm maximum thickness	-	0.03	0.50	3.12	-	m³	3.62

NEW ITEMS FOR 2003 Excluding overheads and profit	PC £	Labour hours	Labour £	Plant £	Material £	Unit	Total rate £
R13 SOAKAWAYS - cont'd							
"Aquacell" soakaway; Wavin Plastics Ltd; . Preformed polypropylene soakaway infiltration crate units; to trenches; surrounded by geotextile and 40 mm aggregate laid 100 thick. in trenches (Excavation, disposal and backfilling not included)							
1 00 x 500 x 400; internal volume 190 litres							
4 crates; 2.00 x 1.00 x 400; 760 litres	108.00	1.60	24.00	12.75	142.92	nr	**179.67**
8 crates; 2.00 x 1.00 x 800; 1520 litres	216.00	3.20	48.00	16.39	259.68	nr	**324.08**
12 crates; 6.00 x 500 x 800; 2280 litres	324.00	4.80	72.00	34.00	410.04	nr	**516.05**
16 crates; 4.00 x 1.00 x 800; 3040 litres	432.00	6.40	96.00	30.36	509.28	nr	**635.64**
20 crates; 5.00 x 1.00 x 800; 3800 litres	540.00	8.00	120.00	30.06	620.79	nr	**770.85**
30 crates; 15.00 x 1.00 x 400; 5700 litres	810.00	12.00	180.00	81.97	1015.69	nr	**1277.66**
60 crates; 15.00 x 1.00 x 800; 11400 litres	1620.00	20.00	300.00	107.17	1894.79	nr	**2301.96**

A PRELIMINARIES
Excluding overheads and profit

A11 TENDER AND CONTRACT DOCUMENTS

Description	PC £	Labour hours	Labour £	Plant £	Material £	Unit	Total rate £
Health and Safety							
Produce health and safety file including preliminary meeting and subsequent progress meetings with external planning officer in connection with health and safety: Project value:							
£35,000	-	8.00	200.00	-	-	nr	200.00
£75000	-	12.00	300.00	-	-	nr	300.00
£100000	-	16.00	400.00	-	-	nr	400.00
£200,000 to £500,000	-	40.00	1000.00	-	-	nr	1000.00
Maintain health and safety file for project duration							
£35,000	-	4.00	100.00	-	-	week	100.00
£75000	-	4.00	100.00	-	-	week	100.00
£100000	-	8.00	200.00	-	-	week	200.00
£200,000 to £500,000	-	8.00	200.00	-	-	week	200.00
Produce written risk assessments on all areas of operations within the scope of works of the contract. Project value:							
£35,000	-	2.00	50.00	-	-	week	50.00
£75000	-	3.00	75.00	-	-	week	75.00
£100000	-	5.00	125.00	-	-	week	125.00
£200,000 to £500,000	-	8.00	200.00	-	-	week	200.00
Produce Coshh assessments on all substances to be used in connection with the contract. Project value:							
£35,000	-	1.50	37.50	-	-	nr	37.50
£75000	-	2.00	50.00	-	-	nr	50.00
£100000	-	3.00	75.00	-	-	nr	75.00
£200,000 to £500,000	-	3.00	75.00	-	-	nr	75.00
Method Statements							
Provide detailed method statements on all aspects of the works; Project value:							
£ 30,000	-	2.50	62.50	-	-	nr	62.50
£ 50,000	-	3.50	87.50	-	-	nr	87.50
£ 75,000	-	4.00	100.00	-	-	nr	100.00
£ 100,000	-	5.00	125.00	-	-	nr	125.00
£ 200,000	-	6.00	150.00	-	-	nr	150.00

A PRELIMINARIES Excluding overheads and profit	PC £	Labour hours	Labour £	Plant £	Material £	Unit	Total rate £
A32 EMPLOYERS REQUIREMENTS: MANAGEMENT OF WORKS							
Programmes							
Allow for production of works programmes prior to the start of the works; Project value:							
£ 30,000	-	3.00	75.00	-	-	nr	75.00
£ 50,000	-	6.00	150.00	-	-	nr	150.00
£ 75,000	-	8.00	200.00	-	-	nr	200.00
£ 100,000	-	10.00	250.00	-	-	nr	250.00
£ 200,000	-	14.00	350.00	-	-	nr	350.00
Allow for updating the works programme during the course of the works; Project value							
£ 30,000	-	1.00	25.00	-	-	nr	25.00
£ 50,000	-	1.50	37.50	-	-	nr	37.50
£ 75,000	-	2.00	50.00	-	-	nr	50.00
£ 100,000	-	3.00	75.00	-	-	nr	75.00
£ 200,000	-	5.00	125.00	-	-	nr	125.00
Setting out							
Setting out for external works operations comprising hard and soft works elements. Placing of pegs and string lines to Landscape Architects drawings; surveying levels and placing level pegs; Obtaining approval from the Landscape Architect to commence works; areas of entire site							
1,000 m²	-	5.00	75.00	-	5.60	nr	80.60
2,500 m²	-	8.00	120.00	-	11.20	nr	131.20
5,000 m²	-	8.00	120.00	-	11.20	nr	131.20
10,000 m²	-	32.00	480.00	-	28.00	nr	508.00
A34 SECURITY/SAFETY/PROTECTION							
Jacksons Fencing; Express fence; Framed mesh unclimbable fencing; including precast concrete supports and couplings							
weekly hire							
2.0 m high; weekly hire rate	-	-	-	0.50	-	m	0.50
erection of fencing; labour only	-	0.10	1.50	-	-	m	1.50
delivery charge	-	-	-	0.80	-	m	0.80
return haulage charge	-	-	-	0.60	-	m	0.60
A41 CONTRACTORS GENERAL COST ITEMS: SITE ACCOMMODATION							
General							
The following items are instances of the commonly found preliminary costs associated with external works contracts. The assumption is made that the external works contractor is sub-contracted to a main contractor.							

A PRELIMINARIES Excluding overheads and profit	PC £	Labour hours	Labour £	Plant £	Material £	Unit	Total rate £
Erect temporary office on concrete base measured separately. Elliot Workspace; Prefabricated office hire; jackleg; open plan							
3.0 x 2.4 m	25.00	-	-	-	25.00	week	25.00
4.8 x 2.4 m	27.00	-	-	-	27.00	week	27.00
Delivery and collection charges on site offices							
Delivery charge	-	-	-	-	98.00	load	98.00
Collection charge	-	-	-	-	98.00	load	98.00
Erect temporary secure storage container for tools and equipment Armoured store							
3.0 x 2.4 m	-	-	-	-	9.50	week	9.50
2.4 x 2.4 m	-	-	-	-	9.50	week	9.50
Toilet facilities John Anderson Ltd. Serviced self contained toilet delivered to and collected from site. Maintained by toilet supply company.							
single chemical toilet	-	-	-	-	20.00	week	20.00
delivery and collection; each way	-	-	-	-	17.00	nr	17.00
A44 TEMPORARY ROADS							
Eve Trackway; portable roadway systems Corrugated temporary road system laid directly onto existing surface or onto compacted removable granular base to protect existing surface. Most systems are based on a weekly hire charge and laid and removed by the supplier							
Heavy Duty Trakpanel - per panel (3.05m x 2.59m) per week	-	-	-	-	-	m²	4.15
Delivery charge for Trakpanel	-	-	-	-	-	m²	2.23
Outrigger Mats for use in conjunction with Heavy Duty Trakpanels - per set of 4 mats per week	-	-	-	-	-	set	82.40
Type 6 - Single Trak Roadway - per section (3.9m x 1.22m) based on a weekly hire charge	-	-	-	-	-	m²	3.04
LD20 Eveolution - Light Duty Trakway - roll out system minimum delivery 50 m	-	-	-	-	-	m²	5.02
Terraplas Walkways - turf protection system - per section (1m x 1m) per week	-	-	-	-	-	m²	5.15

B COMPLETE BUILDINGS/STRUCTURES/UNITS Excluding overheads and profit	PC £	Labour hours	Labour £	Plant £	Material £	Unit	Total rate £
B10 PREFABRICATED BUILDINGS/STRUCTURES							
Cast stone buildings; Haddonstone Ltd; **Ornamental garden buildings in Portland** **Bath or Terracotta finished cast stone;** **Prices for stonework and facades only;** **excavations, foundations, reinforcement,** **concrete infill, roofing and floors all** **priced separately**							
Pavilion Venetian Folly L9400; Tuscan columns, pedimented arch, quoins and optional balustrading							
4184 high x 4728 mm wide x 3147 deep	7603.00	175.00	2625.00	196.00	7706.96	nr	**10527.96**
Pavilion L9300;Tuscan columns							
3496 mm high x 3634 wide	4350.00	144.00	2160.00	168.00	4453.96	nr	**6781.96**
Small Classical Temple L9250; 6 column with fibre glass lead effect finish dome roof.							
Overall height 3610 mm, diameter 2540 mm	5020.00	130.00	1950.00	140.00	5123.96	nr	**7213.96**
Large Classical Temple L9100; 8 column with fibre glass lead effect finish dome roof.							
overall height 4664 mm, diameter 3190 mm	8270.00	165.00	2475.00	140.00	8373.96	nr	**10988.96**
Stepped floors to temples							
Single step; Large Classical Temple	955.00	26.00	390.00	-	1020 51	nr	**1410.51**
Single step; Small Classical Temple	675.00	24.00	360.00	-	729.60	nr	**1089.60**

D GROUNDWORK Excluding overheads and profit	PC £	Labour hours	Labour £	Plant £	Material £	Unit	Total rate £
D11 SOIL STABILIZATION							
Soil Stabilization - General Preamble - Earth-retaining and stabilizing materials are often specified as part of the earth-forming work in landscape contracts, and therefore this section lists a number of products specially designed for large-scale earth control. There are two types: rigid units for structural retention of earth on steep slopes; and flexible meshes and sheets for control of soil erosion where structural strength is not required. Prices for these items depend on quantity, difficulty of access to the site and availability of suitable filling material: estimates should be obtained from the manufacturer when the site conditions have been determined.							
Crib Walls; Phi Ltd "Andacrib" dowelless system; precast concrete crib units; dry joints; machine filled with crushed rock (excavation and foundations not included)							
"Mini"; to 1.75m high	-	-	-	-	-	m²	109.72
"Maxi"; to 4.15 m high	-	-	-	-	-	m²	139.65
"Super Maxi"; to 5.35 m high	-	-	-	-	-	m²	159.60
Crib Walls; Keller Comtec "Timbercrib" timber crib walling system; machine filled with crushed rock inclusive of reinforced concrete footing cribfill stone, rear wall land drain and rear wall drainage/ separation membrane; excluding excavation							
ref 450/38; for retaining walls up to 1.20 m high	-	-	-	-	-	m²	105.00
ref 600/38; for retaining walls up to 2.20 m high	-	-	-	-	-	m²	110.00
ref 750/38; for retaining walls up to 3.10 m high	-	-	-	-	-	m²	115.00
ref 900/48; for retaining walls up to 3.60 m high	-	-	-	-	-	m²	140.00
ref 1050/48; for retaining walls up to 4.40 m high	-	-	-	-	-	m²	150.00
ref 1200/48; for retaining walls up to 5.20 m high	-	-	-	-	-	m²	160.00
ref 1500/48; for retaining walls up to 6.50 m high	-	-	-	-	-	m²	180.00
ref 1800/48; for retaining walls up to 8.20 m high	-	-	-	-	-	m²	205.00
Retaining Walls; Keller Comtec "Textomur" reinforced soil system embankments reinforced soil slopes at angles of 60 - 70 degrees to the horizontal; as an alternative to re-inforced concrete or gabion solutions; finishing with excavated material/grass or shrubs; "Textomur"	-	-	-	-	-	m²	90.00

D GROUNDWORK Excluding overheads and profit	PC £	Labour hours	Labour £	Plant £	Material £	Unit	Total rate £
D11 SOIL STABILIZATION - cont'd							
Retaining Walls; RCC Ltd							
Retaining walls of units with plain concrete finish; prices based on 24 tonne loads but other quantities available (excavation, temporary shoring, foundations and backfilling not included)							
1000 wide x 1250 mm high	83.65	1.27	19.06	12.32	86.23	m	**117.61**
1000 wide x 1750 mm high	100.35	1.27	19.06	12.32	103.91	m	**135.29**
1000 wide x 2400 mm high	146.90	1.27	19.06	12.32	152.07	m	**183.44**
1000 wide x 2690 mm high	184.30	1.27	19.06	31.22	190.94	m	**241.21**
1000 wide x 3000 mm high	194.30	1.47	22.05	39.02	202.77	m	**263.84**
1000 wide x 3750 mm high	290.35	1.57	23.63	46.82	303.81	m	**374.26**
Retaining Walls; Maccaferri Ltd							
Wire mesh gabions; galvanized mesh 80 mm x 100 mm; filling with broken stones 125 mm - 200 mm size; wire down securely to manufacturer's instructions; filling front face by hand							
2 x 1 x 0.50 m	22.79	2.00	30.00	8.21	86.63	nr	**124.84**
2 x 1 x 1.00 m	31.93	4.00	60.00	16.43	159.61	nr	**236.04**
3 x 1 x 0.50 m	32.10	3.00	45.00	6.16	127.86	nr	**179.02**
3 x 1 x 1.00 m	45.00	6.00	90.00	12.32	236.52	nr	**338.84**
"Reno" mattress gabions							
6 x 2 x 0.17 m	69.39	3.00	45.00	12.32	199.62	nr	**256.94**
6 x 2 x 0.23 m	76.31	4.50	67.50	16.43	252.51	nr	**336.44**
6 x 2 x 0.30 m	62.96	6.00	90.00	18.48	292.78	nr	**401.26**
Retaining Walls; Tensar International							
"Tensar" retaining wall system; modular dry laid concrete blocks; 220 mm x 400 mm long x 150 mm high connected to "Tensar SR" geogrid with proprietary connectors; geogrid laid horizontally within the fill at 300 mm centres; on 150 x 450 concrete foundation; filling with imported granular material.							
1.00 m high	65.00	3.00	45.00	7.59	73.31	m²	**125.90**
2.00 m high	65.00	4.00	60.00	7.59	73.31	m²	**140.90**
3.00 m high	65.00	4.50	67.50	7.59	73.31	m²	**148.40**
Retaining Walls; Grass Concrete International Ltd							
"Betoflor" precast concrete landscape retaining walls including soil filling to pockets (excavation, concrete foundations, backfilling stones to rear of walls and planting not included)							
"Betoflor" interlocking units; 250 mm long x 250 mm x 200 mm modular deep; in walls 250 mm wide	-	-	-	-	-	m²	**70.16**
extra over "Betoflor" interlocking units for colours	-	-	-	-	-	m²	**1.67**
"Betoatlas" interlocking units; 250 mm long x 500 mm wide x 200 mm modular deep; in walls 500 mm wide	-	-	-	-	-	m²	**90.65**
extra over "Betoatlas" interlocking units for colours	-	-	-	-	-	m²	**3.44**

D GROUNDWORK Excluding overheads and profit	PC £	Labour hours	Labour £	Plant £	Material £	Unit	Total rate £
Excavating existing banks							
1000 x 1000 mm deep	-	0.67	10.00	3.45	-	m	13.45
2000 x 2000 mm deep	-	1.50	22.50	7.76	-	m	30.26
3000 x 2000 mm deep	-	1.50	22.50	18.48	-	m	40.98
Setting out; grading and levelling; compacting bottoms of excavations	-	0.13	2.00	0.38	-	m	2.38
Extra over "Betoflor" retaining walls for C7P concrete foundations							
700 mm x 300 mm deep	14.82	0.27	4.00	-	14.82	m	18.83
Retaining Walls; Forticrete Engineering Systems Ltd							
"Keystone" pc concrete block retaining wall; geogrid included for walls over 1.00 m high; excavation, concrete foundation, stone backfill to rear of wall all measured separately							
1.0m high	52.00	2.40	36.00	-	53.30	m²	89.30
2.0m high	52.00	2.40	36.00	7.00	66.28	m²	109.28
3.0m high	52.00	2.40	36.00	9.33	75.30	m²	120.63
4.0m high	52.00	2.40	36.00	11.20	82.28	m²	129.49
Retaining Walls; Forticrete Architectural Masonry;							
Stepoc Blocks; interlocking blocks; 10 mm reinforcing laid loose horizontally to preformed notches and vertical reinforcing nominal size 10 mm fixed to starter bars; infilling with concrete; foundations and starter bars measured separately							
Type 256 400 x 225 x 256 mm	33.50	1.20	18.00	-	51.38	m²	69.38
Type 190 400 x 225 x 190 mm	25.00	1.00	15.00	-	39.43	m²	54.43
Embankments; Tensar International							
Embankments; reinforced with "Tensar Uniaxial Geogrid"; ref 40 RE; 40 kN/m width; Geogrid laid horizontally within fill to 100% of vertical height of slope at 1.00 m centres; filling with excavated material							
slopes less than 45 degrees	2.15	0.19	2.78	2.69	2.26	m³	7.73
slopes exceeding 45 degrees	4.62	0.19	2.78	4.26	4.85	m³	11.90
Embankments; reinforced with "Tensar Uniaxial Geogrid", ref 55 RE; 55 kN/m width; Geogrid laid horizontally within fill to 100% of vertical height of slope at 1.00 m centres; filling with excavated material							
slopes less than 45 degrees	2.75	0.19	2.78	2.69	2.89	m³	8.36
slopes exceeding 45 degrees	5.91	0.19	2.78	4.26	6.21	m³	13.25
Extra for "Tensar Mat"; erosion control mats; to faces of slopes of 45 degrees or less; filling with 20 mm fine topsoil; seeding	3.00	0.02	0.30	0.15	3.87	m²	4.33
Embankments; reinforced with "Tensar Uniaxial Geogrid" ref "55 RE"; 55 kN/m width; Geogrid laid horizontally within fill to 50% of horizontal length of slopes at specified centres; filling with imported fill PC £16.64/m³							
300 mm centres; slopes up to 45 degrees	4.57	0.23	3.50	3.14	24.76	m³	31.41
600 mm centres; slopes up to 45 degrees	2.28	0.19	2.78	4.26	22.36	m³	29.41

D GROUNDWORK Excluding overheads and profit	PC £	Labour hours	Labour £	Plant £	Material £	Unit	Total rate £
D11 SOIL STABILIZATION - cont'd							
Embankments; Tensar International - cont'd							
Extra for "Tensar Mat"; erosion control mats; to faces of slopes of 45 degrees or less; filling with 20 mm fine topsoil; seeding	3.00	0.02	0.30	0.15	3.87	m²	4.33
Embankments; reinforced with "Tensar Uniaxial Geogrid" ref "55 RE"; 55 kN/m width; Geogrid laid horizontally within fill; filling with excavated material; wrapping around at faces							
300 mm centres; slopes exceeding 45 degrees	9 16	0.19	2.78	2.69	9.62	m³	15.09
600 mm centres; slopes exceeding 45 degrees	4.57	0.19	2.78	2.69	4.79	m³	10.27
Extra over embankments for bagwork face supports for slopes exceeding 45 degrees	-	-	-	-	12.50	m²	12.50
Extra over embankments for seeding of bags	-	-	-	-	1.00	m²	1.00
Extra over embankments for "Bodkin" joints	-	-	-	-	1.00	m	1.00
Extra over embankments for temporary shuttering to slopes exceeding 45 degrees	-	-	-	-	10.00	m²	10.00
Anchoring Systems; Surface stabilization; Platipus Engineering Systems; Centres shown should be verified with a design engineer. They may vary either way dependent on circumstances							
Concrete Revetments; Hard anodised stealth anchors with stainless steel accessories installed to a depth of 1 m							
"SO4" anchors to a depth of 1.00 -1.50 m at 1 00 m ccs	15.00	0.33	5.00	0.89	15.00	m²	20.89
Loadlocking and crimping tool; for stressing and crimping the anchors							
Purchase price	350.00	-	-	-	350.00	nr	350.00
Hire rate	60.00	-	-	-	60.00	nr	60.00
Surface erosion; geotextile anchoring on slopes; Aluminium alloy Stealth anchors							
"SO4", "SO6" to a depth of 1.0 - 2.0 m at 2.00 m ccs	10.00	0.20	3.00	0.89	10.00	m²	13.89
Brick block or in-situ concrete, distressed retaining walls; anchors installed in two rows at varying tensions between 18 kN - 50 kN along the length of the retaining wall. Core drilling of wall not included							
"SO8", "BO6", "BO8" bronze Stealth and Bat anchors installed in combination with stainless steel accessories to a depth of 6 m. Average cost base on 1.50 m ccs; long term solution	300.00	0.50	7.50	0.89	300.00	m²	308.39
"SO8", "BO6", "BO8" bronze Stealth and Bat anchors installed in combination with stainless steel accessories to a depth of 6 m. Average cost base on 1.50 m ccs; short term solution	150.00	0.50	7.50	0.89	112.50	m²	120.89

D GROUNDWORK Excluding overheads and profit	PC £	Labour hours	Labour £	Plant £	Material £	Unit	Total rate £
Timber retaining wall support; anchors installed 19 kN along the length of the retaining wall. "SO6", "SO8"; bronze Stealth anchors installed in combination with stainless steel accessories to a depth of 3.0 m. 1.50 centres; Average cost;	45.00	0.50	7.50	0.71	45.00	m²	53.21
Gabions; anchors for support and stability to moving, overturning or rotating gabion retaining walls "SO8", "BO6" anchors including base plate at 1 00 m centres to a depth of 4.00 m	100.00	0.33	5.00	0.89	100.00	m²	105.89
Soil nailing of geofabrics; anchors fixed through surface of fabric or erosion control surface treatment (not included) "SO8", "BO6" anchors including base plate at 1.50 m centres to a depth of 3.00 m	100.00	0.17	2.50	0.59	100.00	m²	103.09
Embankments; Grass Concrete International Ltd "Grasscrete"; in situ reinforced concrete surfacing; to 20 mm thick sand blinding layer (not included); including soiling and seeding							
ref GC1; 100 mm thick	-	-	-	-	-	m²	24.98
ref GC2; 150 mm thick	-	-	-	-	-	m²	29.43
"Grassblock 103"; solid matrix precast concrete blocks; to 20 mm thick sand blinding layer;excluding edge restraint; including soiling and seeding							
406 mm x 406 mm x 103 mm; fully interlocking	-	-	-	-	-	m²	25.75
Embankments; Cooper Clarke Group "Ecoblock" polyethylene; 925 mm x 310 mm x 50 mm; heavy duty for car parking and fire paths							
to firm sub-soil (not included)	11.40	0.04	0.60	-	11.74	m²	12.34
to 100 mm granular fill and "Geotextile"	11.40	0.08	1.20	0.15	13.74	m²	15.09
to 250 mm granular fill and "Geotextile"	11.40	0.20	3.00	0.25	15.92	m²	19.17
Extra for filling "Ecoblock" with topsoil; seeding with rye grass at 50 g/m²	0.60	0.02	0.30	0.21	0.72	m²	1.22
"Geoweb" polyethylene soil-stabilizing panels; to soil surfaces brought to grade (not included); filling with excavated material							
panels; 2.40 x 6.10 m x 100 mm deep	5.86	0.10	1.50	2.95	5.98	m²	10.42
panels; 2.40 x 6.10 m x 200 mm deep	11.79	0.13	2.00	3.52	12.03	m²	17.55
"Geoweb" polyethylene soil-stabilizing panels; to soil surfaces brought to grade (not included); filling with ballast							
panels; 2.40 x 6.10 m x 100 mm deep	5.86	0.11	1.67	2.95	7.53	m²	12.14
panels; 2.40 x 6.10 m x 200 mm deep	11.79	0.15	2.31	3.52	15.13	m²	20.96
"Geoweb" polyethylene soil-stabilizing panels; to soil surfaces brought to grade (not included); filling with ST2 10 N/mm² concrete							
panels; 2.40 x 6.10 m x 100 mm deep	5.86	0.16	2.40	-	12.73	m²	15.13
panels; 2.40 x 6.10 m x 200 mm deep	11.79	0.20	3.00	-	25.54	m²	28.54
Extra over for filling "Geoweb" with imported topsoil; seeding with rye grass at 50 g/m²	1.27	0.07	1.00	0.34	1.51	m²	2.86

D GROUNDWORK Excluding overheads and profit	PC £	Labour hours	Labour £	Plant £	Material £	Unit	Total rate £
D11 SOIL STABILIZATION - cont'd							
Timber log retaining walls; Western Log Company							
Machine rounded softwood logs to trenches priced separately; disposal of excavated material priced separately; inclusive of 75 mm hardcore blinding to trench and backfilling trench with site mixed concrete 1:3:6; geofabric pinned to rear of logs; Heights of logs above ground							
500 mm (constructed from 1.80 m lengths)	26.40	1.50	22.50	-	36.39	m	58.89
1 20 m (constructed from 1.80 m lengths)	52.80	1.30	19.50	-	74.90	m	94.40
1 60 m (constructed from 2.40 m lengths)	70.40	2.50	37.50	-	98.35	m	135.85
2 00 m (constructed from 3.00 m lengths)	88.50	3.50	52.50	-	122.29	m	174.79
As above but with 150 mm machine rounded timbers							
500 mm (constructed from 1.80 m lengths)	-	2.50	37.50	-	76.04	m	113.54
1 20 m (constructed from 1.80 m lengths)	88.07	1.75	26.25	-	110.17	m	136.42
1 60 m (constructed from 2.40 m lengths)	117.47	3.00	45.00	-	145.42	m	190.42
2 00 m (constructed from 3.00 m lengths)	147.67	4.00	60.00	-	181.46	m	241.46
2 40 m (constructed from 3.60 m lengths)	-	4.50	67.50	-	210.32	m	277.82
Grass Reinforcement; Farmura Environmental Ltd							
"Matrix" Grass Paver; recycled polyethylene and polypropylene mixed interlocking erosion control and grass reinforcement system laid to rootzone prepared separately and filled with screened topsoil and seeded with grass seed; green							
640 x 330 x 38 mm	10.00	0.13	1.88	-	12.10	m²	13.97
330 x 330 x 38 mm	10.00	0.17	2.50	-	12.10	m²	14.60
extra over for coloured material	4.00	-	-	-	4.00	m²	4.00
Willow Walling to River banks; L.D.C. Ltd							
Woven willow walling as retention to river banks; driving or concreting posts in at 2 m centres; intermediate posts at 500 mm centres							
1.20 m high	-	-	-	-	-	m	61.00
1.50 m high	-	-	-	-	-	m	83.00
Flexible sheet materials; Tensar International							
"Tensar Mat"; erosion mats; 3 m - 4.50 m wide; securing with "Tensar" pegs; lap rolls 100 mm; anchors at top and bottom of slopes; in trenches	3.13	0.02	0.25	-	3.44	m²	3.69
Topsoil filling to "Tensar Mat"; including brushing and raking	0.29	0.02	0.25	0.11	0.36	m²	0.73
"Tensar Bi-axial Geogrid"; to graded compacted base; filling with 200 mm granular fill; compacting (turf or paving to surfaces not included); 400 mm laps							
ref SS20; 39 mm x 39 mm mesh	1.40	0.01	0.20	0.25	5.39	m²	5.84
ref SS30; 39 mm x 39 mm mesh	2.20	0.01	0.20	0.25	6.19	m²	6.64

D GROUNDWORK Excluding overheads and profit	PC £	Labour hours	Labour £	Plant £	Material £	Unit	Total rate £
Flexible sheet materials; Malcolm Ogilvie							
"Wiretex" polypropylene/wire-reinforced fabric; to graded base (not included); fixing with steel pegs at 500 mm centres; 25 mm imported topsoil (seeding not included)							
ref Nr.8 (roll 150 m²)	2.57	0.08	1.20	0.06	3.76	m²	5.01
ref Nr.9 (roll 150 m²)	1.72	0.08	1.20	0.18	2.91	m²	4.28
Flexible sheet materials; Terram Ltd							
"Terram" synthetic fibre filter fabric; to graded base (not included)							
"Terram 1000", 0.70 mm thick; mean water flow 50 litre/m²/s	0.43	0.02	0.25	-	0.43	m²	0.68
"Terram 2000"; 1.00 mm thick; mean water flow 33 litre/m²/s	0.87	0.02	0.25	-	1.04	m²	1.29
"Terram Minipack"	1.32	-	0.05	-	1.32	m²	1.37
Flexible sheet materials; Greenfix Ltd							
"Greenfix"; erosion control mats; 10 mm -15 mm thick; fixing with 4 nr crimped pins in accordance with manufacturer's instructions, to graded surface (not included)							
unseeded "Eromat Light"; 2.40 m wide	135.00	2.00	30.00	-	200.50	100m²	230.50
unseeded "Eromat Standard"; 2.40 m wide	145.00	2.00	30.00	-	211.50	100m²	241.50
unseeded "Eromat Coco"; 2.40 m wide	165.00	2.00	30.00	-	233.50	100m²	263.50
seeded "Covamat Standard"; 2.40 m wide	200.00	2.00	30.00	-	262.00	100m²	292.00
seeded "Covamat Special"; 2.40 m wide	216.67	2.00	30.00	-	279.50	100m²	309.50
seeded "Covamat Coco"; 2.40 m wide	216.67	2.00	30.00	-	279.50	100m²	309.50
"Bio Roll" 300 mm diameter to river banks and revetments	10.00	0.13	1.88	-	11.50	m	13.38
Extra over "Greenfix" erosion control mats for fertilizer applied at 70 g/m²	3.36	0.20	3.00	-	3.36	100m²	6.36
Extra over "Greenfix" erosion control mats for polymer mesh; fixing with steel pins	238.00	1.33	20.00	-	238.00	100m²	258.00
Extra over "Greenfix" erosion control mats for laying to slopes exceeding 30 degrees	-	-	-	-	-	25%	-
Extra for the following operations							
Spreading 25 mm approved topsoil							
by machine	0.29	0.01	0.09	0.09	0.35	m²	0.54
by hand	0.29	0.04	0.53	-	0.35	m²	0.87
Grass seed; PC £2.46/kg; spreading in two operations; by hand							
35 g/m²	8.61	0.17	2.50	-	8.61	100m²	11.11
50 g/m²	12.30	0.17	2.50	-	12.30	100m²	14.80
70 g/m²	17.22	0.17	2.50	-	17.22	100m²	19.72
100 g/m²	24.60	0.20	3.00	-	24.60	100m²	27.60
125 g/m²	30.75	0.20	3.00	-	30.75	100m²	33.75
Extra over seeding by hand for slopes over 30 degrees (allowing for the actual area but measured in plan)							
35 g/m²	1.28	-	0.06	-	1.28	100 m²	1.33
50 g/m²	1.84	-	0.06	-	1.84	100 m²	1.90
70 g/m²	2.58	-	0.06	-	2.58	100 m²	2.64
100 g/m²	3.69	-	0.06	-	3.69	100 m²	3.75
125 g/m²	4.60	-	0.06	-	4.60	100 m²	4.66

D GROUNDWORK Excluding overheads and profit	PC £	Labour hours	Labour £	Plant £	Material £	Unit	Total rate £
D11 SOIL STABILIZATION - cont'd							
Extra for the following operations - cont'd Grass seed; PC £2.46/kg; spreading in two operations; by machine							
35 g/m²	8.61	-	-	0.49	8.61	100m²	9.10
50 g/m²	12.30	-	-	0.49	12.30	100m²	12.79
70 g/m²	17.22	-	-	0.49	17.22	100m²	17.71
100 g/m²	24.60	-	-	0.49	24.60	100m²	25.09
125 kg/ha	307.50	-	-	49.03	307.50	ha	356.53
150 kg/ha	369.00	-	-	49.03	369.00	ha	418.03
200 kg/ha	492.00	-	-	49.03	492.00	ha	541.03
250 kg/ha	615.00	-	-	49.03	615.00	ha	664.03
300 kg/ha	738.00	-	-	49.03	738.00	ha	787.03
350 kg/ha	861.00	-	-	49.03	861.00	ha	910.03
400 kg/ha	984.00	-	-	49.03	984.00	ha	1033.03
500 kg/ha	1230.00	-	-	49.03	1230.00	ha	1279.03
700 kg/ha	1722.00	-	-	49.03	1722.00	ha	1771.03
1400 kg/ha	3444.00	-	-	49.03	3444.00	ha	3493.03
Extra over seeding by machine for slopes over 30 degrees (allowing for the actual area but measured in plan)							
35 g/m²	8.61	-	-	0.07	1.29	100m²	1.36
50 g/m²	12.30	-	-	0.07	1.84	100m²	1.92
70 g/m²	17.22	-	-	0.07	2.58	100m²	2.66
100 g/m²	24.60	-	-	0.07	3.69	100m²	3.76
125 kg/ha	307.50	-	-	7.36	46.13	ha	53.48
150 kg/ha	369.00	-	-	7.36	55.35	ha	62.71
200 kg/ha	492.00	-	-	7.36	73.80	ha	81.16
250 kg/ha	615.00	-	-	7.36	92.25	ha	99.61
300 kg/ha	738.00	-	-	7.36	110.70	ha	118.06
350 kg/ha	861.00	-	-	7.36	129.15	ha	136.51
400 kg/ha	984.00	-	-	7.36	147.60	ha	154.96
500 kg/ha	1230.00	-	-	7.36	184.50	ha	191.86
700 kg/ha	1722.00	-	-	7.36	258.30	ha	265.66
1400 kg/ha	3444.00	-	-	7.36	516.60	ha	523.96
D20 EXCAVATION AND FILLING							
Market Prices of Topsoil; Charles Morris Fertilizers Ltd							
Multiple source screened topsoil	-	-	-	-	8.33	m³	8.33
Single source screened topsoil	-	-	-	-	11.66	m³	11.66
"P30" high grade topsoil	-	-	-	-	22.33	m³	22.33
Site preparation Removing trees							
girth 600 mm - 1.50 m	-	5.00	75.00	16.05	-	nr	91.05
girth 1.50 m - 3.00 m	-	17.00	255.00	64.20	-	nr	319.20
girth over 3.00 m girth	-	48.53	727.92	102.72	-	nr	830.64
Removing tree stumps							
girth 600 mm - 1.50 m	-	2.00	30.00	60.72	-	nr	90.72
girth 1.50 m - 3.00 m	-	7.00	105.00	172.48	-	nr	277.48
girth over 3.00 m	-	12.00	180.00	295.68	-	nr	475.68

D GROUNDWORK Excluding overheads and profit	PC £	Labour hours	Labour £	Plant £	Material £	Unit	Total rate £
Stump grinding; disposing to spoil heaps							
girth 600 mm - 1.50 m	-	2.00	30.00	25.22	-	nr	55.22
girth 1.50 m - 3.00 m	-	2.50	37.50	50.44	-	nr	87.94
girth over 3.00 m	-	4.00	60.00	44.13	-	nr	104.13
Clearing site vegetation							
mechanical clearance	-	0.25	3.75	9.11	-	100m²	12.86
hand clearance	-	2.00	30.00	-	-	100m²	30.00
Lifting turf for preservation							
machine lift and stack	-	0.75	11.25	9.25	-	100m²	20.50
hand lift and stack	-	8.33	125.00	-	-	100m²	125.00
Excavating; mechanical							
Topsoil for preservation; JCB Sitemaster 3CX							
average depth 100 mm	-	0.70	10.50	21.25	-	100m²	31.75
average depth 150 mm	-	1.00	15.00	30.36	-	100m²	45.36
average depth 200 mm	-	1.40	21.00	42.50	-	100m²	63.50
average depth 250 mm	-	1.68	25.20	51.00	-	100m²	76.20
average depth 300 mm	-	2.00	30.00	60.72	-	100m²	90.72
To reduce levels							
maximum depth not exceeding 0.25 m; JCB sitemaster	-	0.07	1.05	2.13	-	m³	3.18
maximum depth not exceeding 1.00 m; JCB sitemaster	-	0.05	0.75	1.52	-	m³	2.27
maximum depth not exceeding 1.00 m; 360 Tracked excavator	-	0.04	0.60	1.19	-	m³	1.79
maximum depth not exceeding 2.00 m; 360 Tracked excavator	-	0.06	0.90	1.78	-	m³	2.68
Pits; 3 ton tracked excavator							
maximum depth not exceeding 0.25 m	-	0.58	8.70	6.23	-	m³	14.93
maximum depth not exceeding 1.00 m	-	0.50	7.50	5.37	-	m³	12.87
maximum depth not exceeding 2.00 m	-	0.60	9.00	6.44	-	m³	15.44
Trenches; width not exceeding 0.30 m							
maximum depth not exceeding 0.25 m	-	0.64	9.60	3.31	-	m³	12.91
maximum depth not exceeding 1.00 m	-	0.54	8.10	2.79	-	m³	10.89
maximum depth not exceeding 2.00 m	-	0.60	9.00	3.10	-	m³	12.10
Trenches; width exceeding 0.30 m							
maximum depth not exceeding 0.25 m	-	0.60	9.00	3.10	-	m³	12.10
maximum depth not exceeding 1.00 m	-	0.50	7.50	2.59	-	m³	10.09
maximum depth not exceeding 2.00 m	-	0.56	8.40	2.90	-	m³	11.30
Extra over any types of excavating irrespective of depth for breaking out existing materials; JCB with breaker attachment							
hard rock	-	0.50	7.50	55.20	-	m³	62.70
concrete	-	0.50	7.50	20.70	-	m³	28.20
reinforced concrete	-	1.00	15.00	27.60	-	m³	42.60
brickwork, blockwork or stonework	-	0.25	3.75	20.70	-	m³	24.45

D GROUNDWORK Excluding overheads and profit	PC £	Labour hours	Labour £	Plant £	Material £	Unit	Total rate £
D20 EXCAVATION AND FILLING - cont'd							
Excavating; mechanical - cont'd							
Extra over any types of excavating irrespective of depth for breaking out existing hard pavings; JCB with breaker attachment							
concrete, 100 mm thick	-	-	-	1.52	-	m²	1.52
concrete; 150 mm thick	-	-	-	2.53	-	m²	2.53
concrete; 200 mm thick	-	-	-	3.04	-	m²	3.04
concrete; 300 mm thick	-	-	-	4.55	-	m²	4.55
reinforced concrete; 100 mm thick	-	0.08	1.25	2.10	-	m²	3.35
reinforced concrete; 150 mm thick	-	0.08	1.13	2.72	-	m²	3.84
reinforced concrete; 200 mm thick	-	0.10	1.50	3.62	-	m²	5.12
reinforced concrete; 300 mm thick	-	0.15	2.25	5.43	-	m²	7.68
tarmacadam, 75 mm thick	-	-	-	1 52	-	m²	1.52
tarmacadam and hardcore; 150 mm thick	-	-	-	2.43	-	m²	2.43
Extra over any types of excavating irrespective of depth for taking up							
precast concrete paving slabs	-	0.07	1.00	0.36	-	m²	1.36
natural stone paving	-	0.10	1.50	0.53	-	m²	2.03
cobbles	-	0.13	1.88	0.67	-	m²	2.54
brick paviors	-	0.13	1.88	0.67	-	m²	2.54
Filling to make up levels; mechanical							
Arising from the excavations							
average thickness 100 mm maximum thickness	-	-	0.06	0.12	-	m²	0.18
average thickness 150 mm maximum thickness	-	0.01	0.09	0.18	-	m²	0.27
average thickness 200 mm maximum thickness	-	0.01	0.12	0.24	-	m²	0.36
average thickness 250 mm maximum thickness	-	0.01	0.15	0.30	-	m²	0.45
average thickness 300 mm maximum thickness	-	0.01	0.20	0.40	-	m²	0.60
average thickness 400 mm maximum thickness	-	0.02	0.24	0.49	-	m²	0.73
average thickness 500 mm maximum thickness	-	0.02	0.30	0.61	-	m²	0.91
average thickness 750 mm maximum thickness	-	0.03	0.45	0.91	-	m²	1.36
Obtained from on site spoil heaps; average 25 m distance; multiple handling							
average thickness 250 mm	-	0.02	0.30	0.74	-	m²	1.04
average thickness 300 mm	-	0.02	0.36	0.89	-	m²	1.25
average thickness 400 mm	-	0.03	0.48	1.19	-	m²	1.67
average thickness 500 mm	-	0.04	0.63	1.55	-	m²	2.17
average thickness 750 mm	-	0.06	0.94	2.32	-	m²	3.26
Obtained off site; soil PC £11.66 / m³							
average thickness 100 mm	1.17	0.01	0.10	0.20	1.40	m²	1.70
average thickness 200 mm	2.33	0.01	0.20	0.40	2.80	m²	3.40
average thickness 250 mm	2.92	0.02	0.25	0.51	3.50	m²	4.26
average thickness 300 mm	3.50	0.02	0.23	0.46	4.20	m²	4.88
average thickness 400 mm	4.66	0.02	0.30	0.61	5.60	m²	6.50
average thickness 500 mm	5.83	0.03	0.38	0.76	7.00	m²	8.13
average thickness 750 mm	8.74	0.04	0.56	1.14	10.49	m²	12.20
Obtained off site; hardcore; PC £16.00 / m³							
average thickness 100 mm	1.60	0.01	0.10	0.20	1.76	m²	2.06
average thickness 200 mm	3.20	0.01	0.20	0.40	3.52	m²	4.12
average thickness 250 mm	4.00	0.02	0.25	0.51	4.40	m²	5.16
average thickness 300 mm	4.80	0.02	0.23	0.46	5.28	m²	5.96
average thickness 400 mm	6.40	0.02	0.30	0.61	7.04	m²	7.95
average thickness 500 mm	8.00	0.03	0.38	0.76	8.80	m²	9.93
average thickness 750 mm	12.00	0.04	0.56	1.14	13.20	m²	14.90

D GROUNDWORK Excluding overheads and profit	PC £	Labour hours	Labour £	Plant £	Material £	Unit	Total rate £
Excavating; hand							
Topsoil for preservation							
average depth 100 mm	-	18.00	270.00	-	-	100 m²	270.00
average depth 150 mm	-	25.20	378.00	-	-	100 m²	378.00
average depth 200 mm	-	32.40	486.00	-	-	100 m²	486.00
average depth 250 mm	-	43.20	648.00	-	-	100 m²	648.00
average depth 300 mm	-	46.80	702.00	-	-	100 m²	702.00
Excavating; hand							
Topsoil to reduce levels							
average depth 100 mm	-	0.16	2.40	-	-	m²	2.40
average depth 150 mm	-	0.25	3.79	-	-	m²	3.79
average depth 200 mm	-	0.35	5.29	-	-	m²	5.29
average depth 250 mm	-	0.48	7.20	-	-	m²	7.20
average depth 300 mm	-	0.63	9.38	-	-	m²	9.38
average depth 400 mm	-	0.90	13.46	-	-	m²	13.46
average depth 600 mm	-	1.44	21.63	-	-	m²	21.63
average depth 750 mm	-	1.93	28.92	-	-	m²	28.92
average depth 1.00 mm	-	3.23	48.39	-	-	m²	48.39
Pits							
maximum depth not exceeding 0.25 m	-	2.15	32.25	-	-	m³	32.25
maximum depth not exceeding 1.00 m	-	2.25	33.75	-	-	m³	33.75
maximum depth not exceeding 2.00 m	-	2.60	39.00	-	-	m³	39.00
Trenches; width not exceeding 0.30 m							
maximum depth not exceeding 0.25 m	-	2.10	31.50	-	-	m³	31.50
maximum depth not exceeding 1.00 m	-	2.20	33.00	-	-	m³	33.00
maximum depth not exceeding 2.00 m	-	2.50	37.50	-	-	m³	37.50
Trenches; width exceeding 0.30 m wide							
maximum depth not exceeding 0.25 m	-	1.90	28.50	-	-	m³	28.50
maximum depth not exceeding 1.00 m	-	2.00	30.00	-	-	m³	30.00
maximum depth not exceeding 2.00 m	-	2.30	34.50	-	-	m³	34.50
Extra over any types of excavating irrespective of depth for breaking out existing materials; hand held pneumatic breaker							
rock	-	5.00	75.00	17.80	-	m³	92.80
concrete	-	2.50	37.50	8.90	-	m³	46.40
reinforced concrete	-	4.00	60.00	19.99	-	m³	79.99
brickwork, blockwork or stonework	-	1.50	22.50	5.34	-	m³	27.84
Disposal; mechanical; Recycled Materials Ltd							
Excavated material; off site; to tip not exceeding 13 km; mechanically loaded.							
inert material	-	0.03	0.50	1.01	10.00	m³	11.51
slightly contaminated	-	0.05	0.75	1.52	16.00	m³	18.27
rubbish	-	0.05	0.75	1.52	16.00	m³	18.27
Excavated material; off site; to tip; mechanically loaded by grab; capacity of load 13m³							
inert material	-	-	-	-	-	m³	15.38

D GROUNDWORK Excluding overheads and profit	PC £	Labour hours	Labour £	Plant £	Material £	Unit	Total rate £
D20 EXCAVATION AND FILLING - cont'd							
Disposal by skip; 7 yd³ (5.35 m³)							
Excavated material; off site; to tip							
by machine	-	0.13	1.88	1.29	21.80	m³	24.97
by hand	-	1.00	15.00	-	21.80	m³	36.80
Excavated material; on site; in spoil heaps							
average 25 m distance	-	0.02	0.34	0.87	-	m³	1.20
average 50 m distance	-	0.04	0.58	1.50	-	m³	2.08
average 100 m distance	-	0.06	0.93	2.38	-	m³	3.31
average 200 m distance	-	0.12	1.80	4.61	-	m³	6.41
Excavated material; spreading on site							
average 25 m distance	-	0.02	0.34	3.79	-	m³	4.13
average 50 m distance	-	0.04	0.58	4.31	-	m³	4.90
average 100 m distance	-	0.06	0.93	5.32	-	m³	6.25
average 200 m distance	-	0.12	1.80	5.22	-	m³	7.02
Disposal; hand							
Excavated material; on site; in spoil heaps							
average 25 m distance	-	1.56	23.40	-	-	m³	23.40
average 50 m distance	-	1.92	28.80	-	-	m³	28.80
average 100 m distance	-	2.71	40.68	-	-	m³	40.68
average 200 m distance	-	3.30	49.50	-	-	m³	49.50
Excavated material; spreading on site							
average 25 m distance	-	1.96	29.34	-	-	m³	29.34
average 50 m distance	-	2.32	34.80	-	-	m³	34.80
average 100 m distance	-	3.11	46.68	-	-	m³	46.68
average 200 m distance	-	3.40	51.00	-	-	m³	51.00
Cultivating							
Ripping up subsoil; using approved subsoiling machine; minimum depth 250 mm below topsoil; at 1 20 m centres; in							
gravel or sandy clay	-	-	-	2.31	-	100m²	2.31
soil compacted by machines	-	-	-	2.70	-	100m²	2.70
clay	-	-	-	2.89	-	100m²	2.89
chalk or other soft rock	-	-	-	5.78	-	100m²	5.78
Extra for subsoiling at 1 m centres	-	-	-	0.58	-	100m²	0.58
Breaking up existing ground; using pedestrian operated tine cultivator or rotavator; loam or sandy soil							
100 mm deep	-	0.22	3.30	2.13	-	100m²	5.43
150 mm deep	-	0.28	4.13	2.67	-	100m²	6.79
200 mm deep	-	0.37	5.50	3.56	-	100m²	9.05
As above but in heavy clay or wet soils							
100 mm deep	-	0.44	6.60	4.27	-	100m²	10.87
150 mm deep	-	0.66	9.90	6.40	-	100m²	16.30
200 mm deep	-	0.82	12.38	8.00	-	100m²	20.38
Breaking up existing ground; using tractor drawn tine cultivator or rotavator							
100 mm deep	-	-	-	0.52	-	100m²	0.52
150 mm deep	-	-	-	0.65	-	100m²	0.65
200 mm deep	-	-	-	0.86	-	100m²	0.86
400 mm deep	-	-	-	2.58	-	100m²	2.58

D GROUNDWORK Excluding overheads and profit	PC £	Labour hours	Labour £	Plant £	Material £	Unit	Total rate £
Cultivating ploughed ground; using disc, drag, or chain harrow							
4 passes	-	-	-	3.10	-	100 m²	3.10
Rolling cultivated ground lightly; using self-propelled agricultural roller	-	0.06	0.83	0.58	-	100 m²	1.41
Importing and storing selected and approved topsoil; to BS 3882; from source not exceeding 13 km from site							
1 - 14 m³	11.66	-	-	-	34.98	m³	34.98
over 15 m³	11.66	-	-	-	11.66	m³	11.66
Filling to make up levels; mechanical							
Arising from the excavations							
average thickness not exceeding 0.25 m; depositing in layers 150 mm maximum thickness	-	0.04	0.60	1.21	-	m³	1.81
average thickness exceeding 0.25 m; depositing in layers 150 mm maximum thickness	-	0.03	0.50	1.01	-	m³	1.51
Obtained from on site spoil heaps; average 25 m distance; multiple handling							
average thickness exceeding 0.25 m; depositing in layers 150 mm maximum thickness	-	0.08	1.25	3.09	-	m³	4.34
Obtained off site; soil							
average thickness not exceeding 0.25 m; depositing in layers 150 maximum thickness	11.66	0.07	1.00	2.03	13.99	m³	17.02
average thickness exceeding 0.25 m; depositing in layers 150 maximum thickness	11.66	0.05	0.75	1.52	13.99	m³	16.26
Filling to make up levels; hand							
Arising from the excavations							
average thickness not exceeding 0.25 m; depositing in layers 150 mm maximum thickness	-	0.67	10.01	-	-	m³	10.01
average thickness exceeding 0.25 m; depositing in layers 150 mm maximum thickness	-	0.60	9.00	-	-	m³	9.00
Obtained from on site spoil heaps; average 25 m distance, multiple handling							
average thickness exceeding 0.25 m thick; depositing in layers 150 mm maximum thickness	-	1.00	15.00	-	-	m³	15.00
Obtained off site; soil							
average thickness not exceeding 0.25 m; depositing in layers 150 maximum thickness	11.66	0.50	7.50	-	13.99	m³	21.49
average thickness exceeding 0.25 m; depositing in layers 150 maximum thickness	11.66	0.75	11.25	-	13.99	m³	25.24
Surface treatments							
Compacting							
bottoms of excavations	-	0.01	0.08	0.03	-	m²	0.10
Grading							
Trimming surfaces of cultivated ground to final levels, removing roots stones and debris exceeding 50 mm in any direction to tip offsite; slopes less than 15 degrees							
clean ground with minimal stone content	-	0.25	3.75	-	-	100 m²	3.75
slightly stony - 0.5 kg stones per m²	-	0.33	5.00	-	-	100 m²	5.00
very stony - 1.0 - 3.00 kg stones per m²	-	0.50	7.50	-	0.01	100 m²	7.51
clearing mixed slightly contaminated rubble inclusive of roots and vegetation	-	0.50	7.50	-	0.05	100 m²	7.54
clearing brick-bats stones and clean rubble	-	0.60	9.00	-	0.03	100 m²	9.03

E IN SITU CONCRETE/LARGE PC CONCRETE Excluding overheads and profit	PC £	Labour hours	Labour £	Plant £	Material £	Unit	Total rate £
E10 IN SITU CONCRETE							
General The concrete mixes used here are referred to as "Design" "Standard" and "Designated" mixes. The BS references on these are used to denote the concrete strength and mix volumes. Please refer to the definitions and the tables in the Memoranda for each Trade, Concrete Work, at the back of this book							
Designed mix User specified performance of the concrete. Producer responsible for selecting appropriate mix. Strength testing is essential							
Prescribed mix User specified mix constituents and is responsible for ensuring that the concrete meets performance requirements. The mix proportion is essential							
Standard mix Specified from the list in BS 5328 Pt 2 1991 s.4. Made with a restricted range of materials. Specification to include the proposed use of the material as well as; the standard mix reference, the type of cement, type and size of aggregate, slump (workability). Quality assurance required							
Designated mix Mix specified in BS 5328 Pt 2: 1991 s.5. Producer to hold current product conformity certification and quality approval to BS 5750 Pt 1 (EN 29001). Quality assurance essential. The mix may not be modified.							
Concrete mixes; mixed on site; costs for producing concrete; prices for commonly used mixes for various types of work; based on bulk load 20 tonne rates for aggregates; Roughest type mass concrete such as footings, road haunchings 300 thick							
1:3:6	47.78	1.00	15.00	-	47.78	m³	62.78
1:3:6 sulphate resisting	64.12	0.75	11.25	-	64.12	m³	75.37
As above but aggregates delivered in 1 tonne bags							
1:3:6	72.98	1.00	15.00	-	72.98	m³	87.98
1:3:6 sulphate resisting	89.32	0.75	11.25	-	89.32	m³	100.57
Most ordinary use of concrete such as mass walls above ground, road slabs etc. and general reinforced concrete work							
1:2:4	56.32	0.75	11.25	-	56.32	m³	67.57
1:2:4 sulphate resisting	79.43	0.75	11.25	-	79.43	m³	90.68

E IN SITU CONCRETE/LARGE PC CONCRETE Excluding overheads and profit	PC £	Labour hours	Labour £	Plant £	Material £	Unit	Total rate £
As above but aggregates delivered in 1 tonne bags							
1 2:4	81.53	0.75	11.25	-	81.53	m³	92.78
1 2:4 sulphate resisting	104.63	0.75	11.25	-	104.63	m³	115.88
Watertight floors, pavements and walls, tanks pits steps paths surface of two course roads, reinforced concrete where extra strength is required.							
1.1.5 3	64.48	1.00	15.00	-	64.48	m³	79.48
As above but aggregates delivered in 1 tonne bags							
1.1.5:3	89.69	1.00	15.00	-	89.69	m³	104.69
Plain in situ concrete; site mixed; 10 N/mm² - 40 aggregate (1:3:6); aggregates delivered in 1 tonne bags							
Foundations							
ordinary portland cement	89 02	1.50	22.50	-	91.25	m³	113.75
sulphate resistant cement	89 32	2.25	33.75	-	96.02	m³	129.77
Foundations; poured on or against earth or unblinded hardcore							
1 3·6	-	1.65	24.75	-	91.25	m³	116.00
1 3:6 sulphate resisting	89.32	2.40	36.00	-	98.25	m³	134.25
Isolated foundations							
1·3:6	-	2.15	32.25	-	91.25	m³	123.50
1·3:6 sulphate resisting	89.32	2.75	41.25	-	96.02	m³	137.27
Plain in situ concrete; site mixed; 21 N/mm² - 20 aggregate (1:2:4)							
Foundations							
ordinary portland cement	94.24	1.61	24.19	-	96.60	m³	120.78
sulphate resistant cement	104.63	2.25	33.75	-	112.48	m³	146.23
Foundations; poured on or against earth or unblinded hardcore							
ordinary portland cement	94 24	1.50	22.50	-	96.60	m³	119.10
sulphate resistant cement	104.63	2.25	33.75	-	115.09	m³	148.84
Isolated foundations							
ordinary portland cement	94.24	2.00	30.00	-	96.60	m³	126.60
sulphate resistant cement	104.63	2.75	41.25	-	104.63	m³	145.88
Reinforced insitu concrete; site mixed; 21 N/mm² - 20 aggregate (1:2:4)							
Foundations							
ordinary portland cement	94.24	2.00	30.00	-	96.60	m³	126.60
sulphate resistant cement	104.63	2.75	41.25	-	112.48	m³	153.73
Foundations; poured on or against earth or unblinded hardcore							
ordinary portland cement	94.24	2.00	30.00	-	96.60	m³	126.60
sulphate resistant cement	104.63	2.75	41.25	-	115.09	m³	156.34
Isolated foundations							
ordinary portland cement	94.24	2.50	37.50	-	96.60	m³	134.10
sulphate resistant cement	104.63	3.25	48.75	-	112.48	m³	161.23

E IN SITU CONCRETE/LARGE PC CONCRETE Excluding overheads and profit	PC £	Labour hours	Labour £	Plant £	Material £	Unit	Total rate £
E10 IN SITU CONCRETE - cont'd							
Plain in situ concrete; ready mixed; **Tarmac Topmix 10 N/mm mixes; suitable** **for mass concrete fill and blinding** Foundations							
GEN1; Designated mix	62.50	1.50	22.50	-	62.50	m³	85.00
ST2; Standard mix	67.55	1.50	22.50	-	70.93	m³	93.43
Foundations; poured on or against earth or unblinded hardcore							
GEN1; Designated mix	62.50	1.57	23.63	-	62.50	m³	86.13
ST2; Standard mix	67.55	1.57	23.63	-	72.62	m³	96.24
Isolated foundations							
GEN1; Designated mix	-	2.00	30.00	-	62.50	m³	92.50
ST2; Standard mix	67.55	2.00	30.00	-	70.93	m³	100.93
Plain in situ concrete; ready mixed; **Tarmac Topmix 15 N/mm mixes; suitable** **for oversite below suspended slabs and** **strip footings in non aggressive soils** Foundations							
GEN 2; Designated mix	66.42	1.50	22.50	-	69.74	m³	92.24
ST 3; Standard mix	69.07	1.50	22.50	-	72.52	m³	95.02
Foundations; poured on or against earth or unblinded hardcore							
GEN2; Designated mix	66.42	1.57	23.63	-	66.42	m³	90.05
ST3; Standard mix	69.07	1.57	23.63	-	69.07	m³	92.69
Isolated foundations							
GEN2; Designated mix	66.42	2.00	30.00	-	66.42	m³	96.42
ST3; Standard mix	69.07	2.00	30.00	-	69.07	m³	99.07
Plain in situ concrete; ready mixed; **Tarmac Topmix; Air entrained mixes** **suitable for paving** Beds or slabs; House drives parking and external paving							
PAV 1; 35 N/mm²; Designated mix	71.86	1.50	22.50	-	75.45	m³	97.95
Beds or slabs; Heavy duty external paving							
PAV 2; 40 N/mm²; Designated mix	73.64	1.50	22.50	-	77.32	m³	99.82
Reinforced in situ concrete; ready mixed; **Tarmac Topmix; 35N/mm² mix;** **suitable for foundations in class 2** **sulphate conditions** Foundations							
RC 35; Designated mix	72.11	2.00	30.00	-	75.72	m³	105.72
Foundations; poured on or against earth or unblinded hardcore							
RC 35; Designated mix	72.11	2.10	31.50	-	77.52	m³	109.02
Isolated foundations							
RC 35; Designated mix	72.11	2.00	30.00	-	72.11	m³	102.11
Ready mix concrete; Extra for loads of **less than 6 m³; Ace Minimix**							
Ready mix concrete small load surcharge	-	-	-	-	15.00	m³	15.00

E IN SITU CONCRETE/LARGE PC CONCRETE Excluding overheads and profit	PC £	Labour hours	Labour £	Plant £	Material £	Unit	Total rate £
E20 FORMWORK FOR IN SITU CONCRETE							
Plain vertical formwork; basic finish							
Sides of foundations							
height exceeding 1.00 m	-	2.00	30.00	-	8.46	m²	38.46
height not exceeding 250 mm	-	1.00	15.00	-	2.26	m	17.26
height 250 - 500 mm	-	1.00	15.00	-	4.23	m	19.23
height 500 mm - 1.00 m	-	1.50	22.50	-	8.46	m	30.96
Sides of foundations; left in							
height over 1.00 m	-	2.00	30.00	-	29.45	m²	59.45
height not exceeding 250 mm	-	1.00	15.00	-	8.64	m	23.64
height 250 - 500 mm	-	1.00	15.00	-	16.98	m	31.98
height 500 mm - 1.00 m	-	1.50	22.50	-	33.95	m	56.45
E30 REINFORCEMENT FOR IN SITU CONCRETE							
The rates for reinforcement shown for steel bar below are based on prices which would be supplied on a typical landscape contract. The steel prices shown have been priced on a selection of steel delivered to site where the total order quantity is in the region of 2 tonnes. The assumption is that should larger quantities be required, the work would fall outside the scope of the typical landscape contract defined in the front of this book. Keener rates can be obtained for larger orders.							
Reinforcement bars; BS 4449; hot rolled plain round mild steel; straight							
Bars							
8 mm nominal size	451.00	27.00	405.00	-	451.00	tonne	856.00
10 mm nominal size	466.38	26.00	390.00	-	466.38	tonne	856.38
12 mm nominal size	440.75	25.00	375.00	-	440.75	tonne	815.75
16 mm nominal size	420.25	24.00	360.00	-	420.25	tonne	780.25
20 mm nominal size	379.25	23.00	345.00	-	379.25	tonne	724.25
Reinforcement bars; BS 4449; hot rolled plain round mild steel; bent							
Bars							
8 mm nominal size	492.00	27.00	405.00	-	492.00	tonne	897.00
10 mm nominal size	481.75	26.00	390.00	-	481.75	tonne	871.75
12 mm nominal size	461.25	25.00	375.00	-	461.25	tonne	836.25
16 mm nominal size	451.00	24.00	360.00	-	451.00	tonne	811.00
20 mm nominal size	399.75	23.00	345.00	-	399.75	tonne	744.75
Reinforcement fabric; BS 4483; lapped; in beds or suspended slabs							
Fabric							
ref A98 (1.54 kg/m²)	0.74	0.22	3.30	-	0.81	m²	4.11
ref A142 (2.22 kg/m²)	0.74	0.22	3.30	-	0.82	m²	4.12
ref A193 (3.02 kg/m²)	3.99	0.22	3.30	-	4.39	m²	7.69
ref A252 (3.95 kg/m²)	5.21	0.24	3.60	-	5.73	m²	9.33
ref A393 (6.16 kg/m²)	8.16	0.28	4.20	-	8.98	m²	13.18

F MASONRY Excluding overheads and profit	PC £	Labour hours	Labour £	Plant £	Material £	Unit	Total rate £
Market Prices of Materials							
Cement; Builder Centre							
Portland cement	-	-	-	-	2.40	25kg	2.40
Sulphate resistant cement	-	-	-	-	4.30	25kg	4.30
White cement	-	-	-	-	6.37	25kg	6.37
Sand; Builder Centre							
Building sand							
loose	-	-	-	-	49.24	m³	49.24
1 tonne bags	-	-	-	-	27.36	nr	27.36
Sharp sand							
loose	-	-	-	-	49.24	m³	49.24
1 tonne bags	-	-	-	-	27.36	nr	27.36
Sand; Yeoman Aggregates							
Sharp sand	-	-	-	-	14.56	tonne	14.56
Bricks; E.T. Clay Products							
Ibstock; facing bricks; 215 x 102.5 x 65 mm							
Leicester Red Stock	-	-	-	-	385.00	1000	385.00
Leicester Yellow Stock	-	-	-	-	325.00	1000	325.00
Himley Mixed Russet	-	-	-	-	335.50	1000	335.50
Himley Worcs. Mixture	-	-	-	-	325.00	1000	325.00
Roughdales Red Multi Rustic	-	-	-	-	357.50	1000	357.50
Ashdown Cottage Mixture	-	-	-	-	396.00	1000	396.00
Ashdown Crowborough Multi	-	-	-	-	459.80	1000	459.80
Ashdown Pevensey Multi	-	-	-	-	412.50	1000	412.50
Chailey Stock	-	-	-	-	449.00	1000	449.00
Dorking Multi Coloured	-	-	-	-	341.00	1000	341.00
Holbrook Smooth Red	-	-	-	-	423.50	1000	423.50
Stourbridge Kenilworth Multi	-	-	-	-	324.50	1000	324.50
Stourbridge Pennine Pastone	-	-	-	-	295.00	1000	295.00
Stratford Red Rustic	-	-	-	-	308.00	1000	308.00
Swanage Restoration Red	-	-	-	-	643.50	1000	643.50
Laybrook Sevenoaks Yellow	-	-	-	-	320.00	1000	320.00
Laybrook Arundel Yellow	-	-	-	-	340.00	1000	340.00
Laybrook Thakeham Red	-	-	-	-	290.00	1000	290.00
Funton Second Hard Stock	-	-	-	-	390.00	1000	390.00
Hanson Brick Ltd, London Brand; facing bricks; 215 x 102.5 x 65 mm							
Capel Multi Stock	-	-	-	-	290.00	1000	290.00
Rusper Stock	-	-	-	-	295.00	1000	295.00
Delph Autumn	-	-	-	-	235.00	1000	235.00
Regency	-	-	-	-	266.00	1000	266.00
Sandfaced	-	-	-	-	266.00	1000	266.00
Saxon Gold	-	-	-	-	222.00	1000	222.00
Tudor	-	-	-	-	245.00	1000	245.00
Windsor	-	-	-	-	239.00	1000	239.00
Autumn Leaf	-	-	-	-	185.00	1000	185.00
Claydon Red Multi	-	-	-	-	205.00	1000	205.00
Other facing bricks; 215 x 102.5 x 65 mm							
Soft Reds - Milton Hall	-	-	-	-	375.00	1000	375.00
Soft Reds - Blenheim Reds	-	-	-	-	530.00	1000	530.00
Staffs Blues - Blue Smooth	-	-	-	-	465.00	1000	465.00
Staffs Blues - Blue Brindle	-	-	-	-	425.00	1000	425.00

F MASONRY Excluding overheads and profit	PC £	Labour hours	Labour £	Plant £	Material £	Unit	Total rate £
Reclaimed (second hand) bricks							
Yellows	-	-	-	-	850.00	1000	850.00
Yellow Multi	-	-	-	-	850.00	1000	850.00
Mixed London Stock	-	-	-	-	650.00	1000	650.00
Red Multi	-	-	-	-	530.00	1000	530.00
Gaults	-	-	-	-	460.00	1000	460.00
Red Rubbers	-	-	-	-	740.00	1000	740.00
Common bricks	-	-	-	-	185.00	1000	185.00
Engineering bricks	-	-	-	-	180.00	1000	180.00
Kempston facing bricks; 215 x 102.5 x 65 mm							
Melford Yellow	-	-	-	-	269.00	1000	269.00
Anglian Blend	-	-	-	-	280.00	1000	280.00

F10 BRICK/BLOCK WALLING

Mortar mixes; Common mixes for various types of work; Mortar mixed on site; Prices based on builders merchant rates for cement; aggregates delivered in 1 tonne bags; mechanically mixed; Batching quantities for these mortar mixes may be found in the memorandum section of this book

	PC £	Labour hours	Labour £	Plant £	Material £	Unit	Total rate £
1:3	-	0.75	11.25	-	106.01	m³	117.26
1:4	-	0.75	11.25	-	97.50	m³	108.75
1:1:6	-	0.75	11.25	-	93.09	m³	104.34
1:1:6 Sulphate resisting	-	0.75	11.25	-	113.61	m³	124.86

Mortar mixes; Common mixes for various types of work; Mortar mixed on site; Prices based on builders merchant rates for cement; aggregates delivered in 1 tonne bags; hand mixed

1:3	-	1.00	15.00	-	106.01	m³	121.01
1:4	-	1.00	15.00	-	97.50	m³	112.50
1:1:6	-	1.00	15.00	-	93.09	m³	108.09
1:1:6 Sulphate resisting	-	1.00	15.00	-	113.61	m³	128.61

Common bricks; PC £180.00 /1000; English garden wall bond; in gauged mortar (1:1:6)
Mechanically offloading; maximum 25 m distance; loading to wheel barrows and transporting to location; per 215 mm thick walls

	-	0.42	6.25	-	-	m²	6.25
Walls							
half brick thick	-	1.80	27.00	-	12.90	m²	39.90
one brick thick	-	2.40	36.00	-	25.37	m²	61.37
one and a half brick thick	-	3.60	54.00	-	38.05	m²	92.05
two brick thick	-	4.80	71.99	-	50.73	m²	122.73
Walls; curved; mean radius 6 m							
half brick thick	-	1.67	25.00	-	13.96	m²	38.96
one brick thick	-	3.36	50.40	-	26.88	m²	77.28

F MASONRY Excluding overheads and profit	PC £	Labour hours	Labour £	Plant £	Material £	Unit	Total rate £
F10 BRICK/BLOCK WALLING - cont'd							
Common bricks; PC £180.00 /1000; **English garden wall bond; in gauged** **mortar (1:1:6)** - cont'd							
Walls, curved; mean radius 1.50 m							
half brick thick	-	1.89	28.30	-	14.50	m²	**42.80**
one brick thick	-	4.00	60.00	-	27.96	m²	**87.96**
Extra for cement mortar (1:3) in lieu of gauged mortar							
half brick thick	-	-	-	-	0.22	m²	**0.22**
one brick thick	-	-	-	-	0.44	m²	**0.44**
one and a half brick thick	-	-	-	-	0.65	m²	**0.65**
two brick thick	-	-	-	-	0.87	m²	**0.87**
Walls; stretcher bond; wall ties at 450 centres vertically and horizontally							
one brick thick	-	1.71	25.72	-	25.73	m²	**51.45**
one and a half brick thick	-	2.56	38.40	-	38.78	m²	**77.18**
two brick thick	-	3.43	51.43	-	51.83	m²	**103.26**
Facing bricks; PC £300.00 /1000; **English garden wall bond; in gauged** **mortar (1:1:6); facework one side**							
Mechanically offloading; maximum 25m distance; loading to wheel barrows and transporting to location; per 215 mm thick walls	-	0.42	6.25	-	-	m²	**6.25**
Walls							
half brick thick	-	1.40	21.00	-	20.46	m²	**41.46**
one brick thick	-	2.82	42.30	-	40.92	m²	**83.22**
one and a half brick thick	-	4.24	63.60	-	61.38	m²	**124.98**
two brick thick	-	5.65	84.75	-	81.84	m²	**166.59**
Walls; curved; mean radius 6 m							
half brick thick	-	2.10	31.50	-	21.88	m²	**53.38**
one brick thick	-	4.21	63.16	-	42.72	m²	**105.88**
Walls, curved; mean radius 1.50 m							
half brick thick	-	2.40	36.00	-	21.43	m²	**57.43**
one brick thick	-	4.80	72.01	-	41.82	m²	**113.83**
Walls; tapering; one face battering; average							
one and a half brick thick	-	5.60	84.01	-	64.08	m²	**148.09**
two brick thick	-	7.47	111.98	-	85.44	m²	**197.42**
Walls; battering (retaining)							
one and a half brick thick	-	5.60	84.01	-	64.08	m²	**148.09**
two brick thick	-	7.47	111.98	-	85.44	m²	**197.42**
Isolated piers; English bond; facework all round							
one brick thick	-	7.00	105.00	-	44.91	m²	**149.91**
one and a half brick thick	-	9.00	135.00	-	67.37	m²	**202.37**
two brick thick	-	10.00	150.00	-	91.00	m²	**241.00**
three brick thick	-	12.40	186.00	-	134.73	m²	**320.73**
Projections; vertical							
one brick x half brick	-	0.70	10.50	-	4.56	m	**15.06**
one brick x one brick	-	1.40	21.00	-	9.13	m	**30.13**
one and a half brick x one brick	-	2.10	31.50	-	13.69	m	**45.19**
two brick x one brick	-	2.30	34.50	-	18.25	m	**52.75**
Brickwork fair faced both sides; facing bricks in gauged mortar (1:1:6)							
Extra for fair face both sides; flush, struck, weathered, or bucket-handle pointing	-	0.67	10.00	-	-	m²	**10.00**

F MASONRY Excluding overheads and profit	PC £	Labour hours	Labour £	Plant £	Material £	Unit	Total rate £
Extra for cement mortar (1:3) in lieu of gauged mortar							
half brick thick	-	-	-	-	0.22	m²	0.22
one brick thick	-	-	-	-	0.44	m²	0.44
one and a half brick thick	-	-	-	-	0.65	m²	0.65
two brick thick	-	-	-	-	0.87	m²	0.87
Class B engineering bricks; PC £260.00/1000; double Flemish bond in cement mortar (1:3) Mechanically offloading; maximum 25m distance; loading to wheel barrows; transporting to location; per 215 mm thick walls	-	0.42	6.25	-	-	m²	6.25
Walls							
half brick thick	-	1.41	21.13	-	18.16	m²	39.28
one brick thick	-	2.82	42.35	-	36.31	m²	78.67
one and a half brick thick	-	4.24	63.53	-	62.73	m²	126.26
two brick thick	-	5.65	84.70	-	72.63	m²	157.33
Walls; curved; mean radius 6 m							
half brick thick	-	2.10	31.50	-	18.82	m²	50.33
one brick thick	-	4.21	63.16	-	36.31	m²	99.47
Walls; curved; mean radius 1.50 m							
half brick thick	-	2.40	36.00	-	18.75	m²	54.75
one brick thick	-	4.80	72.01	-	36.31	m²	108.33
Walls; tapering; one face battering; average							
one and a half brick thick	-	5.60	84.01	-	54.47	m²	138.48
two brick thick	-	7.47	111.98	-	72.63	m²	184.61
Walls; battering (retaining)							
one and a half brick thick	-	5.60	84.01	-	54.47	m²	138.48
two brick thick	-	7.47	111.98	-	72.63	m²	184.61
Isolated piers							
one brick thick	-	7.00	105.00	-	39.87	m²	144.87
one and a half brick thick	-	9.00	135.00	-	59.80	m²	194.81
two brick thick	-	10.00	150.00	-	80.92	m²	230.93
three brick thick	-	12.40	186.00	-	119.61	m²	305.61
Projections, vertical							
one brick x half brick	-	0.70	10.50	-	4.05	m	14.55
one brick x one brick	-	1.40	21.00	-	8.11	m	29.11
one and a half brick x one brick	-	2.10	31.50	-	12.16	m	43.66
two brick by one brick	-	2.30	34.50	-	16.22	m	50.72
Walls; half brick thick							
in honeycomb bond	-	1.80	27.00	-	13.03	m²	40.03
in quarter bond	-	1.67	25.00	-	17.77	m²	42.77
Brickwork fair faced both sides; facing bricks in gauged mortar (1:1:6)							
Extra for fair face both sides; flush, struck, weathered, or bucket-handle pointing	-	0.67	10.00	-	-	m²	10.00
Brick copings Copings; all brick headers-on-edge; to BS 4729; two angles rounded 53 mm radius; flush pointing top and both sides as work proceeds; one brick wide; horizontal							
machine-made specials	19.33	0.16	2.34	-	20.62	m	22.96
hand-made specials	19.33	0.16	2.34	-	20.62	m	22.97

F MASONRY Excluding overheads and profit	PC £	Labour hours	Labour £	Plant £	Material £	Unit	Total rate £
F10 BRICK/BLOCK WALLING - cont'd							
Brick copings - cont'd							
Extra over copings for two courses machine-made tile creasings, projecting 25 mm each side; 260 mm wide copings; horizontal Copings, all brick headers-on-edge; flush pointing top and both sides as work proceeds; one brick wide; horizontal	4 46	0.50	7.50	-	5.04	m	12.54
facing bricks PC £300.00/1000	4.00	0.31	4.69	-	4.30	m	8.99
engineering bricks PC £260.00/1000	3.47	0.31	4.69	-	3.64	m	8.33
Variation in brick prices							
Add or subtract the following amounts for every £1 00/1000 difference in the PC price							
half brick thick	-	-	-	-	0.06	m²	0.06
one brick thick	-	-	-	-	0.13	m²	0.13
one and a half brick thick	-	-	-	-	0.19	m²	0.19
two brick thick	-	-	-	-	0.25	m²	0.25
Dense aggregate concrete blocks; "Tarmac Topblock" or other equal and approved; in gauged mortar (1:2:9)							
Walls							
Solid Blocks 10 n/mm²							
440 x 215 x 100 mm thick	6 30	1.20	18.00	-	7.25	m²	25.25
440 x 215 x 140 mm thick	9.09	1.30	19.50	-	10.16	m²	29.66
440 x 215 x 215 mm thick	10.24	1.50	22.50	-	11.90	m²	34.40
Solid Blocks 10N/mm² laid flat							
440 x 100 x 215 mm thick	12 85	3.22	48.30	-	15.40	m²	63.70
Hollow Concrete blocks							
440 x 215 x 215 mm thick	10.25	1.30	19.50	-	11.32	m²	30.82
Filling of Hollow concrete blocks with concrete as work proceeds; tamping and compacting							
440 x 215 x 215 mm thick	6.70	0.20	3.00	-	6.70	m²	9.70
F20 NATURAL STONE RUBBLE WALLING							
Granite walls							
Granite random rubble walls; laid dry							
200 mm thick; single faced	27.33	6.66	99.90	-	27.33	m²	127.23
Granite walls; one face battering to 50 degrees; pointing faces							
450 mm (average) thick	61.50	10.00	150.00	-	66.86	m²	216.86
Dry stone walling - General							
Preamble: In rural areas where natural stone is a traditional material, it may be possible to use dry stone walling or dyking as an alternative to fences or brick walls. Many local authorities are willing to meet the extra cost of stone walling in areas of high landscape value, and they may hold lists of available craftsmen. Dry Stone Association, YFC Centre, National Agricultural Centre, Stoneleigh Park, Kenilworth, Warwickshire. CV8 2LG. Tel: (01203) 696 544.							

F MASONRY Excluding overheads and profit	PC £	Labour hours	Labour £	Plant £	Material £	Unit	Total rate £
Rockery stone - General Preamble: Rockery stone prices vary considerably with source, carriage, distance and load. Typical garden centre prices for small quantities are in the range of £30 - £40 per tonne collected.							
Rockery stone; Breedon Lumping limestone 9 - 12 inches; minimum 15/20 tonne loads							
boulders; maximum diameter 400 mm; to positions maximum 25 m distance from offload	55.00	2.60	39.00	-	55.00	tonne	94.00
Rockery stone; Civil Engineering Developments Glacial boulders; maximum distance 25 m; by machine							
750 mm diameter	110.00	0.90	13.50	12.32	110.00	nr	135.82
1 m diameter	230.00	2.00	30.00	24.64	230.00	nr	284.64
1.5 m diameter	765.00	2.00	30.00	32.00	765.00	nr	827.00
2 m diameter	1735.00	2.00	30.00	32.00	1735.00	nr	1797.00
Glacial boulders; maximum distance 25 m; by hand							
not exceeding 750 mm diameter	48.50	0.75	11.25	-	48.50	nr	59.75
not exceeding 1 m diameter	110.00	1.89	28.28	-	110.00	nr	138.28
F22 CAST STONE WALLING							
Atlas Stone Products; dwarf walls; in Cotswold stone; grey, brown or red "Campden Pitched" stone walls							
uniform size precast units	12.07	2.50	37.50	-	19.18	m²	56.68
random size precast units	12.07	4.00	60.00	-	19.18	m²	79.18
Atlas Stone Products; cast stone characteristic walls with irregular size stone facing "Weathered Cotswold" drystone walls							
600 x 100 x 125 mm	39.48	0.67	10.00	-	43.04	m²	53.04
600 x 100 x 125 mm	59.99	1.50	22.50	-	63.55	m²	86.05
Copings 400 x 150 x 150	3.45	0.40	6.00	-	7.01	m	13.00
Copings							
flat	3.45	0.20	3.00	-	3.53	m	6.53
Haddonstone Ltd; Cast stone piers; Ornamental masonry in Portland Bath or Terracotta finished cast stone Gate Pier S120; to foundations and underground work measured separately; concrete infill							
S 120G base unit to pier; 699 x 699 x 172 mm	104.00	0.75	11.25	-	108.60	nr	119.85
S 120F shaft base unit; 533 x 533 x 280 mm	83.00	1.00	15.00	-	87.72	nr	102.72
S 120E main shaft unit; 533 x 533 x 280 mm; nr of units required dependent on height of pier	83.00	1.00	15.00	-	87.72	nr	102.72

F MASONRY Excluding overheads and profit	PC £	Labour hours	Labour £	Plant £	Material £	Unit	Total rate £
F22 CAST STONE WALLING - cont'd							
Haddonstone Ltd; Cast stone piers; **Ornamental masonry in Portland Bath or** **Terracotta finished cast stone** - cont'd							
S 120D top shaft unit; 33 x 533 x 280 mm	83.00	1.00	15.00	-	87.72	nr	102.72
S 120C pier cap unit; 737 x 737 x 114 mm	99.00	0.50	7.50	-	103.72	nr	111.22
S 120B pier block unit; base for finial							
533 x 533 x 64 mm	43.00	0.33	4.95	-	43.12	nr	48.07
Pier blocks; flat to receive gate finial							
S100B, 440 x 440 x 63 mm	28.00	0.33	5.00	-	28.29	nr	33.29
S120B, 546 x 546 x 64 mm	43.00	0.33	5.00	-	43.29	nr	48.29
S150B, 330 x 330 x 51 mm	14.00	0.33	5.00	-	14.29	nr	19.29
Pier caps; part weathered							
S100C; 915 x 915 x 150 mm	234.00	0.50	7.50	-	234.58	nr	242.08
S120C; 737 x 737 x 114 mm	99.00	0 50	7.50	-	99.29	nr	106.79
S150C; 584 x 584 x 120 mm	71.00	0.50	7.50	-	71.29	nr	78.79
Pier caps; weathered							
S230C; 1029 x 1029 x 175 mm	308.00	0.50	7.50	-	308.29	nr	315.79
S215C; 687 x 687 x 175 mm	134.00	0.50	7.50	-	134.29	nr	141.79
S210C; 584 x 584 x 175 mm	85.00	0.50	7.50	-	85.29	nr	92.79
Pier strings							
S100S, 800 x 800 x 55 mm	80.00	0.50	7.50	-	80.18	nr	87.68
S120S, 555 x 555 x 44 mm	36.00	0.50	7.50	-	36.18	nr	43.68
S150S; 457 x 457 x 48 mm	23.00	0.50	7.50	-	23.18	nr	30.68
Balls and Bases							
E150A Ball 535 mm and E150C collared base	206.81	0.50	7.50	-	206.99	nr	214.49
E120A Ball 330 mm and E120C collared base	72.34	0.50	7.50	-	72.52	nr	80.02
E110A Ball 230 mm and E110C collared base	54.47	0.50	7.50	-	54.65	nr	62.15
E100A Ball 170 mm and E100B plain base	33.19	0.50	7.50	-	33.37	nr	40.87
Haddonstone Ltd; Cast stone copings; **Ornamental masonry in Portland Bath or** **Terracotta finished cast stone**							
Copings for walls; bedded, jointed and pointed in approved coloured cement-lime mortar 1:2:9							
T100; weathered coping 102 mm high 178 mm wide x 914 mm	26.16	0.33	5.00	-	26.41	m	31.40
T140; weathered coping 102 mm high 337 mm wide x 914 mm	42.51	0.33	5.00	-	42.76	m	47.76
T200; weathered coping 127 mm high 508 mm wide x 750 mm	65.40	0.33	5.00	-	65.65	m	70.65
T170; weathered coping 108 mm high 483 mm wide x 914	69.76	0.33	5.00	-	70.01	m	75.01
T340 raked coping 100 -75 mm high 290 wide x 900 mm	39.24	0.33	5.00	-	39.49	m	44.49
T310 raked coping 89 -76 mm high 381 wide x 914 mm	46.87	0.33	5.00	-	47.12	m	52.12

F MASONRY Excluding overheads and profit	PC £	Labour hours	Labour £	Plant £	Material £	Unit	Total rate £
F30 ACCESSORIES/SUNDRY ITEMS FOR BRICK/BLOCK/STONE WALLING							
Damp proof courses; bitumen hessian based; 200 mm laps							
Horizontal							
width not exceeding 225 mm	3.51	0.58	8.70	-	3.77	m²	12.47
width exceeding 225 mm	3.51	0.60	9.00	-	4.39	m²	13.39
Vertical							
not exceeding 225 mm	3.51	1.72	25.80	-	3.77	m²	29.57
Damp proof courses; "Ledkore" grade A; bitumen based; lead cored; 200 mm laps							
Horizontal							
width not exceeding 225 mm	18.00	1.00	15.00	-	19.35	m²	34.35
width exceeding 225 mm	18.00	0.76	11.40	-	22.50	m²	33.90
Vertical							
width not exceeding 225 mm wide	18.00	1.72	25.80	-	19.35	m²	45.15
Damp proof courses; pitch polymer; 150 mm laps							
Horizontal							
width not exceeding 225 mm	3.90	1.14	17.10	-	5.34	m²	22.44
width exceeding 225 mm	3.90	0.58	8.70	-	5.82	m²	14.52
Vertical							
width not exceeding 225 mm	3.90	1.72	25.80	-	5.34	m²	31.14
Two courses slates in cement mortar (1:3)							
Horizontal							
width exceeding 225 mm	8.43	3.46	51.90	-	11.99	m²	63.89
Vertical							
width exceeding 225 mm	8.43	5.18	77.70	-	11.99	m²	89.69
F31 PRECAST CONCRETE SILLS/LINTELS/COPINGS/FEATURES							
Mix 21.00 N/mm² - 20 aggregate (1:2:4)							
Copings; once weathered; twice grooved							
152 x 75 mm	4.68	0.40	6.00	-	4.99	m	10.99
178 x 65 mm	6.85	0.40	6.00	-	7.15	m	13.15
305 x 75 mm	9.78	0.50	7.50	-	10.25	m	17.75
Pier caps; four sides weathered							
305 x 305 mm	4.75	1.00	15.00	-	4.89	nr	19.89
381 x 381 mm	4.97	1.00	15.00	-	5.11	nr	20.11
533 x 533 mm	5.95	1.20	18.00	-	6.09	nr	24.09
Copings; Milner Delvaux; "Blanc de Bierges"							
200 x 50 mm	9.65	0.50	7.50	-	10.29	m	17.80
300 x 50 mm	13.85	0.50	7.50	-	14.62	m	22.12
400 x 50 mm	15.19	0.50	7.50	-	16.00	m	23.50
300 x 80 mm	20.91	0.50	7.50	-	21.89	m	29.39
400 x 80 mm	24.55	0.50	7.50	-	25.82	m	33.32

G STRUCTURAL/ CARCASSING METAL/TIMBER Excluding overheads and profit	PC £	Labour hours	Labour £	Plant £	Material £	Unit	Total rate £
G31 PREFABRICATED TIMBER UNIT DECKING							
Timber Decking; Board Walk Timber Decking Co							
Supports for timber decking; treated softwood joists to receive decking boards; joists at 400 mm centres							
50 x 150 mm	5.75	1.00	15.00	-	9.79	m²	24.79
50 x 200 mm	8.75	1.00	15.00	-	13.10	m²	28.09
50 x 250 mm;	11.60	1.00	15.00	-	16.23	m²	31.23
Decking planks; Top Deck 150							
Yellow Cedar; planks 141 mm wide x 42 mm thick	25.76	1.00	15.00	-	31.98	m²	46.98
Red Cedar ; planks 131 mm wide x 42 mm thick	-	1.00	15.00	-	34.36	m²	49.36
Decking planks; Top Deck 100							
Yellow Cedar; planks 91 mm wide x 42 mm thick	23.87	1.00	15.00	-	29.82	m²	44.82
Red Cedar; planks 91 mm wide x 40 mm thick	-	1.00	15.00	-	32.12	m²	47.12
Decking planks; Maxicover 150							
Yellow Cedar; planks 141 mm wide x 26 mm thick	20.51	0.75	11.25	-	26.20	m²	37.45
Red Cedar; planks 141 mm wide x 26 mm thick	23.66	0.75	11.25	-	29.67	m²	40.92
Decking planks; Maxicover 100							
Yellow Cedar; planks 90 mm wide x 26 mm thick	14.63	1.00	15.00	-	19.73	m²	34.73
Red Cedar; planks 90 mm wide x 26 mm thick	17.16	1.00	15.00	-	22.52	m²	37.52
Decking planks; Econocover; Pressure treated Western Hemlock							
planks 88 mm wide x 38 mm thick	21.45	1.00	15.00	-	27.16	m²	42.16
Handrails and base rail; fixed to posts at 2.00 m centres							
Posts;100 x 100 x 1370 high	48.31	1.00	15.00	-	51.48	m	66.48
Posts turned 1220 high	40.67	1.00	15.00	-	43.83	m	58.83
Hand rails; balusters							
Square balusters at 100 mm centres	33.60	0.50	7.50	-	34.29	m	41.79
Square balusters at 300 mm centres	11.19	0.33	5.00	-	11.74	m	16.74
Turned balusters at 100 mm centres	52.80	0.50	7.50	-	53.49	m	60.99
Turned balusters at 300 mm centres	17.58	0.33	4.95	-	18.14	m	23.09

H CLADDING/COVERING Excluding overheads and profit	PC £	Labour hours	Labour £	Plant £	Material £	Unit	**Total rate £**
H51 NATURAL STONE/SLAB CLADDING FEATURES							
Sawn Yorkstone cladding; Johnsons Stone Ltd; Six sides sawn stone; rubbed face ; sawn and jointed edges; fixed to blockwork (not included) with stainless steel fixings "Ancon Ltd" grade 304 stainless steel frame cramp and dowel 7mm; cladding units drilled 4 x to receive dowels 440 x 200 x 50 mm thick	74.95	1.87	28.05	-	77.11	m²	**105.16**
H52 CAST STONE SLAB CLADDING FEATURES							
Cast Stone cladding; Haddonstone Ltd; Reconstituted stone in Portland Bath or Terracotta; fixed to blockwork or concrete (not included) with stainless steel fixings "Ancon Ltd" grade 304 stainless steel frame cramp and dowel M6 mm; cladding units drilled 4 x to receive dowels 440 x 200 x 50 mm thick	85.73	1.87	28.05	-	87.89	m²	**115.94**

J WATERPROOFING Excluding overheads and profit	PC £	Labour hours	Labour £	Plant £	Material £	Unit	Total rate £
J10 SPECIALIST WATERPROOF RENDERING							
"Sika" waterproof rendering; steel trowelled							
Walls; 20 thick; three coats; to concrete base							
width exceeding 300 mm	-	-	-	-	-	m²	46.51
width not exceeding 300 mm	-	-	-	-	-	m²	72.89
Walls, 25 thick; three coats; to concrete base							
width exceeding 300 mm	-	-	-	-	-	m²	52.74
width not exceeding 300 mm	-	-	-	-	-	m²	83.74
J20 MASTIC ASPHALT TANKING/DAMP PROOFING							
Tanking and damp proofing; mastic asphalt; to BS 6925; type T 1097; Coverite Ltd							
13 mm thick; one coat covering; to concrete base; flat; work subsequently covered							
width exceeding 300 mm	-	-	-	-	-	m²	9.36
20 mm thick; two coat coverings; to concrete base, flat; work subsequently covered							
width exceeding 300 mm	-	-	-	-	-	m²	11.72
30 mm thick; three coat coverings; to concrete base; flat; work subsequently covered							
width exceeding 300 mm	-	-	-	-	-	m²	15.92
13 mm thick; two coat coverings; to brickwork base; vertical; work subsequently covered							
width exceeding 300 mm	-	-	-	-	-	m²	31.08
20 mm thick; three coat coverings; to brickwork base; vertical; work subsequently covered							
width exceeding 300 mm	-	-	-	-	-	m²	42.44
Internal angle fillets; work subsequently covered	-	-	-	-	-	m	3.30
Turning asphalt nibs into grooves; 20 mm deep	-	-	-	-	-	m	2.10
J30 LIQUID APPLIED TANKING/DAMP PROOFING							
Tanking and damp proofing; Ruberoid Building Products, "Synthaprufe" cold applied bituminous emulsion waterproof coating							
"Synthaprufe"; to smooth finished concrete or screeded slabs; flat; blinding with sand							
two coats	2.26	0.22	3.33	-	2.60	m²	5.93
three coats	3.39	0.31	4.65	-	3.84	m²	8.49
"Synthaprufe"; to fair faced brickwork with flush joints, rendered brickwork, or smooth finished concrete walls; vertical							
two coats	2.54	0.29	4.29	-	2.91	m²	7.20
three coats	3.73	0.40	6.00	-	4.21	m²	10.21

J WATERPROOFING Excluding overheads and profit	PC £	Labour hours	Labour £	Plant £	Material £	Unit	Total rate £
Tanking and damp proofing; RIW Ltd Liquid asphaltic composition; to smooth finished concrete screeded slabs or screeded slabs; flat							
two coats	6.93	0.33	5.00	-	6.93	m²	11.93
Liquid asphaltic composition; fair faced brickwork with flush joints, rendered brickwork, or smooth finished concrete walls; vertical							
two coats	6.93	0.50	7.50	-	6.93	m²	14.43
"Heviseal"; to smooth finished concrete or screeded slabs; to surfaces of ponds, tanks, planters; flat							
two coats	3.21	0.33	5.00	-	3.53	m²	8.53
"Heviseal"; to fair faced brickwork with flush joints, rendered brickwork, or smooth finished concrete walls; to surfaces of retaining walls, ponds, tanks, planters; vertical							
two coats	3.21	0.50	7.50	-	3.53	m²	11.03
J40 FLEXIBLE SHEET TANKING/DAMP PROOFING							
Tanking and damp proofing; Servicised Ltd "Bitu-thene 500x"; 1.50 mm thick; overlapping and bonding; including sealing all edges							
to concrete slabs; flat	3.69	0.25	3.75	-	4.06	m²	7.81
to brick/concrete walls; vertical	3.69	0.40	6.00	-	4.32	m²	10.32
J50 GREEN ROOF SYSTEMS							
Preamble - A variety of systems are available which address all the varied requirements for a successful Green Roof. For installation by approved contractors only the prices shown are for budgeting purposes only as each installation is site specific and may incorporate some or all of the resources shown. Specifiers should verify that the systems specified include for design liability and inspections by the suppliers. The systems below assume commercial insulation levels are required to the space below the proposed Green Roof. Intensive Green roofs are those of generally lightweight construction with low maintenance planting and shallow soil designed for aesthetics only; Extensive Green roofs are designed to allow use for recreation and trafficking. They require more maintenance and allow a greater variety of surfaces and plant types. **Intensive Green Roof; Erisco Bauder Ltd;** soil based systems able to provide a variety of hard and soft landscaping; laid to the surface of an unprepared roof deck.							

J WATERPROOFING Excluding overheads and profit	PC £	Labour hours	Labour £	Plant £	Material £	Unit	Total rate £
J50 GREEN ROOF SYSTEMS - cont'd							
Vapour barrier laid to prevent intersticial condensation from spaces below the roof applied by torching to the roof deck							
VB4-Expal aluminium lined	-	-	-	-	-	m²	7.91
Insulation laid and hot bitumen bonded to vapour barrier							
PUR-V Polyurethane	-	-	-	-	-	m²	15.51
Underlayers to receive rootbarriers partially bonded to insulation by torching							
G4E	-	-	-	-	-	m²	8.82
K5E	-	-	-	-	-	m²	10.99
Root barriers							
"Plant E"; chemically treated root resistant capping sheet fully bonded to G4E underlayer by torching	-	-	-	-	-	m²	10.10
"Rootbar"; copper lined mechanical root resistant capping sheet fully bonded to K5E underlayer by torching	-	-	-	-	-	m²	18.61
Slip layers to absorb differential movement							
PE Foil 2 layers laid to root barriers	-	-	-	-	-	m²	2.29
Optional protection layer to prevent mechanical damage							
"Protection mat" 6 mm thick rubber matting loose laid	-	-	-	-	-	m²	8.70
Drainage medium laid to root barrier							
"Drainage board" free draining EPS 50 mm thick	-	-	-	-	-	m²	5.42
"Reservoir board" up to 35 litre water storage capacity EPS 62 mm thick	-	-	-	-	-	m²	13.61
Filtration to prevent soil migration to drainage system							
"Filter fleece" 3 mm thick polyester geotextile loose laid over drainage/reservoir layer	-	-	-	-	-	m²	2.49
Extensive Green Roof System; Erisco Bauder Ltd; low maintenance soil free system incorporating single layer growing and planting medium.							
Vapour barrier laid to prevent intersticial condensation applied by torching to the roof deck							
VB4-Expal aluminium lined	-	-	-	-	-	m²	7.91
Insulation laid and hot bitumen bonded to vapour barrier							
PUR-V Polyurethane	-	-	-	-	-	m²	15.51
Underlayer to receive rootbarrier partially bonded to insulation by torching							
G4E	-	-	-	-	-	m²	8.82
Root barrier							
"Plant E"; chemically treated root resistant capping sheet fully bonded to G4E underlayer by torching	-	-	-	-	-	m²	10.10

J WATERPROOFING Excluding overheads and profit	PC £	Labour hours	Labour £	Plant £	Material £	Unit	Total rate £
Inverted waterproofing systems; Alumasc Exterior Building Products Ltd (Euroroof); to roof surfaces to receive Green Roof systems.							
"Hydrotech 6125"; monolithic hot melt rubberised bitumen; Applied in two 3 mm layers incorporating a polyester reinforcing sheet with 4 mm thick protection sheet and chemically impregnated root barrier; fully bonded into the Hydrotech; applied to plywood or suitably prepared wood float finish and primed concrete deck or screeds.							
10 mm thick	-	-	-	-	-	m²	21.00
"Topfoam 350"; extruded polystyrene insulation; Optional system; thickness to suit required U value; calculated at design stage; indicative thicknesses; laid to Hydrotech 6125							
0 25 U value; average requirement 125 mm	-	-	-	-	-	m²	15.00
0 35 U value; average requirement 80 mm thick	-	-	-	-	-	m²	10.00
0 45 U value average requirement 60 mm	-	-	-	-	-	m²	7.00
Warm Roof waterproofing systems; Alumasc Exterior Building Products Ltd (Euroroof); to roof surfaces to receive Green Roof systems.							
"Derbigum" system							
"Nilperm" aluminium lined vapour barrier; 2 mm thick bonded in hot bitumen to the roof deck	-	-	-	-	-	m²	6.00
"Korklite" insulation bonded to the vapour barrier in hot bitumen; U value dependent; 80 mm thick	-	-	-	-	-	m²	13.00
"Hi-Ten Bond" 2 mm thick underlayer; fully bonded to the insulation	-	-	-	-	-	m²	4.00
"Derbigum SP" cap sheet impregnated with root resisting chemical bonded to the underlayer	-	-	-	-	-	m²	14.00
Intensive Green Roof Systems; Alumasc Exterior Building Products Ltd (Euroroof); components laid to the insulation over the Hydrotech or Derbigum waterproofing above.							
Optional inclusion; moisture retention layer; SSM-45; moisture mat	-	-	-	-	-	m²	4.00
Drainage layer; "Floradrain" recycled polypropylene; providing water reservoir, multi-directional drainage and mechanical damage protection;							
FD25; 25 mm deep; rolls inclusive of filter sheet SF	-	-	-	-	-	m²	10.00
FD40; 40 mm deep; rolls inclusive of filter sheet SF	-	-	-	-	-	m²	14.00
FD25; 25 mm deep; sheets excluding filter sheet	-	-	-	-	-	m²	9.00
FD40; 40 mm deep; sheets excluding filter sheet	-	-	-	-	-	m²	13.00
FD60; 60 mm deep; sheets excluding filter sheet	-	-	-	-	-	m²	18.00
Drainage layer; "Elastodrain" recycled rubber mat; providing multi-directional drainage and mechanical damage protection;							
EL200; 20 mm deep	-	-	-	-	-	m²	10.00

J WATERPROOFING Excluding overheads and profit	PC £	Labour hours	Labour £	Plant £	Material £	Unit	Total rate £
J50 GREEN ROOF SYSTEMS - cont'd							
Intensive Green Roof Systems; Alumasc Exterior Building Products Ltd (Euroroof); components laid to the insulation over the Hydrotech or Derbigum waterproofing above. - cont'd							
"Zincolit" recycled crushed brick; Optional drainage infil to Floradrain layers;							
FD25; 10 litres per m²	-	0.02	0.30	-	40.00	m²	40.30
FD40; 17 litres per m²	-	0.03	0.51	-	40.00	m²	40.51
FD60; 27 litres per m²	-	0.05	0.81	-	40.00	m²	40.81
Filter sheet; rolled out onto drainage layer							
"Filter sheet TG" for Elastodrain EL200	-	-	-	-	-	m²	2.00
"Filter sheet SF" for Floradrain range	-	-	-	-	-	m²	1.50
Intensive soil; lightweight growing medium laid to filter sheet	-	-	-	-	-	m³	100.00
Extensive Green Roof Systems; Alumasc Exterior Building Products Ltd (Euroroof); components laid to the Hydrotech or insulation layers above.							
Moisture retention layer;							
SSM-45; moisture mat	-	-	-	-	-	m²	4.00
Drainage layer for flat roofs; "Floradrain"; recycled polypropylene; providing water reservoir, multi-directional drainage and mechanical damage protection; supplied in sheet or roll form							
FD25; 25 mm deep; rolls inclusive of filter sheet SF	-	-	-	-	-	m²	10.00
FD40; 40 mm deep; rolls inclusive of filter sheet SF	-	-	-	-	-	m²	14.00
FD25; 25 mm deep; sheets excluding filter sheet	-	-	-	-	-	m²	9.00
FD40; 40 mm deep; sheets excluding filter sheet	-	-	-	-	-	m²	13.00
FD60; 60 mm deep; sheets excluding filter sheet	-	-	-	-	-	m²	18.00
Drainage layer for pitched roofs; "Floratec"; recycled polystyrene; providing water reservoir and multi-directional drainage; laid to moisture mat or to the waterproofing.							
FS50	-	-	-	-	-	m²	10.00
FS75	-	-	-	-	-	m²	13.00
FS100	-	-	-	-	-	m²	14.00
Filter sheet; only required at perimeters and protrusions	-	-	-	-	-	m	1.50
Extensive soil; lightweight growing medium laid to filter sheet; choice of soil types is dependent on planting design requirements							
"E110" 10% humus 90% Zincolit	-	-	-	-	-	m²	100.00
"E120" 20% humus 80% Zincolit	-	-	-	-	-	m²	100.00
"E130" 30% humus 70% Zincolit	-	-	-	-	-	m²	100.00

J WATERPROOFING

J WATERPROOFING Excluding overheads and profit	PC £	Labour hours	Labour £	Plant £	Material £	Unit	Total rate £
Landscape Options							
Sedum Mat vegetation layer; Alumasc Exterior Building Products Ltd (Euroroof); to Extensive Green Roof as detailed above	-	-	-	-	-	m²	22.00
Hydroplanting system; Erisco Bauder Ltd; to Extensive Green Roof as detailed above							
Ecomat 3 mm thick geotextile loose laid	-	-	-	-	-	m²	2.19
Hydroplanting with Sedum and succulent coagulant directly onto plant substrate to 20 mm deep	-	-	-	-	-	m²	10.08
Xeroflor vegetation blanket; Erisco Bauder Ltd; to Extensive Green Roof as detailed above							
Ecomat 6 mm thick geotextile loose laid	-	-	-	-	-	m²	2.19
Sedum or sedum plus pre cultivated vegetation blanket laid loose	-	-	-	-	-	m²	26.69
Green Roof Components; Alumasc Exterior Building Products Ltd (Euroroof)							
Outlet inspection chambers							
"KS 10" 100 mm deep	-	-	-	-	-	nr	50.00
"KS 15" 150 mm deep	-	-	-	-	-	nr	60.00
Height extension piece 100 mm	-	-	-	-	-	nr	20.00
Height extension piece 200 mm	-	-	-	-	-	nr	30.00
Outlet and Irrigation control chambers							
"B32" 300 x 300 x 300 mm high	-	-	-	-	-	nr	250.00
"B52" 400 x 500 x 500 mm high	-	-	-	-	-	nr	300.00
Outlet damming piece for water retention	-	-	-	-	-	nr	30.00
Linear drainage channel; collects surface water from adjacent hard surfaces or down pipes for distribution to the drainage layer.	-	-	-	-	-	m	50.00

M SURFACE FINISHES Excluding overheads and profit	PC £	Labour hours	Labour £	Plant £	Material £	Unit	Total rate £
M10 CEMENT: SAND/CONCRETE SCREEDS/TOPPINGS **Granolithic paving; cement and granite chippings 5 mm down (1:2:5); steel trowelled; Jonathan James Ltd** Floors; one coat; laid on concrete while green; width exceeding 300 mm							
20 mm thick	-	-	-	-	-	m²	14.84
25 mm thick	-	-	-	-	-	m²	16.25
Floors; two coats; laid on hacked concrete with slurry; width exceeding 300 mm							
38 mm thick	-	-	-	-	-	m²	19.91
50 mm thick	-	-	-	-	-	m²	23.30
75 mm thick	-	-	-	-	-	m²	30.34
M20 PLASTERED/RENDERED/ROUGHCAST COATING							
Cement:lime:sand (1:1:6); 19 mm thick; two coats; wood floated finish Walls width exceeding 300 mm; to brickwork or blockwork base	-	-	-	-	-	m²	14.43
Extra over cement:sand:lime (1:1:6) coatings for decorative texture finish combed or floated finish	-	-	-	-	-	m²	1.90
M40 STONE/CONCRETE/QUARRY/CERAMIC TILING /MOSAIC							
Ceramic tiles; unglazed slip resistant; various colours and textures; jointing Floors							
level or to falls only not exceeding 15 degrees from horizontal; 150 x 150 x 8 mm thick	16.00	1.00	15.00	-	19.17	m²	34.17
level or to falls only not exceeding 15 degrees from horizontal; 150 x 150 x 12 mm thick	19.50	1.00	15.00	-	22.85	m²	37.84
Clay tiles - General Preamble: Typical Specification - Clay tiles should be to BS 6431 and shall be reasonably true to shape, flat, free from flaws, frost resistant and true to sample approved by the Landscape Architect prior to laying. Quarry Tiles (or semi-vitrified tiles) shall be of external quality, either heather brown or blue, to size specified, laid on 1.2:4 concrete, with 20 maximum aggregate 100 thick, on 100 hardcore. The hardened concrete should be well wetted and the surplus water taken off. Clay tiles shall be thoroughly wetted immediately before laying and then drained and shall be bedded to 19 thick cement: sand (1:3) screed. Joints should be approximately 4mm (or 3mm for vitrified tiles) grouted in cement: sand (1:2) and cleaned off immediately.							

M SURFACE FINISHES Excluding overheads and profit	PC £	Labour hours	Labour £	Plant £	Material £	Unit	Total rate £
Quarry tiles; external quality; including bedding; jointing							
Floors							
level or to falls only not exceeding 15 degrees from horizontal; 150 x 150 x 12.5 mm thick; heather brown	20.00	0.80	12.00	-	23.37	m²	**35.37**
level or to falls only not exceeding 15 degrees from horizontal; 225 x 225 x 29 mm thick; heather brown	20.00	0.67	10.00	-	23.37	m²	**33.37**
level or to falls only not exceeding 15 degrees from horizontal; 150 x 150 x 12.5 mm thick; blue/black	12.70	1.00	15.00	-	15.71	m²	**30.71**
level or to falls only not exceeding 15 degrees from horizontal; 194 x 194 x 12.5 mm thick; heather brown	11.70	0.80	12.00	-	14.66	m²	**26.66**
M60 PAINTING/CLEAR FINISHING - EXTERNALLY							
Prepare; touch up primer; two undercoats and one finishing coat of gloss oil paint; on metal surfaces							
General surfaces							
girth exceeding 300 mm	0.75	0.33	5.00	-	0.75	m²	**5.75**
isolated surfaces; girth not exceeding 300 mm	0.21	0.13	2.00	-	0.21	m	**2.21**
isolated areas not exceeding 0.50 m² irrespective of girth	0.40	0.13	2.00	-	0.40	nr	**2.40**
Ornamental railings; each side measured overall							
girth exceeding 300 mm	0.75	0.75	11.25	-	0.75	m²	**12.00**
Prepare; one coat primer; two undercoats and one finishing coat of gloss oil paint; on wood surfaces							
General Surfaces							
girth exceeding 300 mm	1.01	0.40	6.00	-	1.01	m²	**7.01**
isolated areas not exceeding 0.50 m² irrespective of girth	0.39	0.36	5.45	-	0.39	nr	**5.84**
isolated surfaces; girth not exceeding 300 mm	0.33	0.20	3.00	-	0.33	m	**3.33**
Prepare; two coats of creosote; on wood surfaces							
General surfaces							
girth exceeding 300 mm	0.17	0.21	3.11	-	0.17	m²	**3.28**
isolated surfaces; girth not exceeding 300 mm	0.05	0.12	1.73	-	0.05	m	**1.78**
Prepare, proprietary solution primer; two coats of dark stain; on wood surfaces							
General surfaces							
girth exceeding 300 mm	1.72	0.10	1.50	-	1.77	m²	**3.27**
isolated surfaces; girth not exceeding 300 mm	0.22	0.05	0.75	-	0.23	m	**0.98**

M SURFACE FINISHES Excluding overheads and profit	PC £	Labour hours	Labour £	Plant £	Material £	Unit	Total rate £
M60 PAINTING/CLEAR FINISHING - EXTERNALLY - cont'd							
Three coats "Dimex Shield"; to clean, dry surfaces; in accordance with manufacturer's instructions							
Brick or block walls							
girth exceeding 300 mm	0.92	0.28	4.20	-	0.92	m²	**5.12**
Cement render or concrete walls							
girth exceeding 300 mm	0.72	0.25	3.75	-	0.72	m²	**4.47**
Two coats resin based paint; "Sandtex Matt"; in accordance with manufacturer's instructions							
Brick or block walls							
girth exceeding 300 mm	6.16	0.20	3.00	-	6.16	m²	**9.16**
Cement render or concrete walls							
girth exceeding 300 mm	4.19	0.17	2.50	-	4.19	m²	**6.69**

P BUILDING FABRIC SUNDRIES Excluding overheads and profit	PC £	Labour hours	Labour £	Plant £	Material £	Unit	Total rate £
P30 TRENCHES/PIPEWAYS/PITS FOR BURIED ENGINEERING SERVICES							
Excavating trenches; using 3 tonne tracked excavator; to receive pipes; grading bottoms; earthwork support; filling with excavated material to within 150 mm of finished surfaces and compacting; completing fill with topsoil; disposal of surplus soil							
Services not exceeding 200 mm nominal size							
average depth of run not exceeding 0.50 m	0.52	0.12	1.80	1.15	0.63	m	3.58
average depth of run not exceeding 0.75 m	0.52	0.16	2.44	1.59	0.63	m	4.67
average depth of run not exceeding 1.00 m	0.52	0.28	4.25	2.78	0.63	m	7.66
average depth of run not exceeding 1.25 m	0.52	0.38	5.75	3.74	0.52	m	10.01
Excavating trenches; using 3 tonne tracked excavator; to receive pipes; grading bottoms; earthwork support; filling with imported granular material and compacting; disposal of surplus soil							
Services not exceeding 200 mm nominal size							
average depth of run not exceeding 0.50 m	4.37	0.09	1.30	0.80	6.17	m	8.28
average depth of run not exceeding 0.75 m	6.56	0.11	1.62	1.01	9.26	m	11.89
average depth of run not exceeding 1.00 m	8.75	0.14	2.10	1.31	12.35	m	15.76
average depth of run not exceeding 1.25 m	10.94	0.23	3.43	2.23	15.44	m	21.10
Excavating trenches, using 3 tonne tracked excavator, to receive pipes; grading bottoms; earthwork support; filling with lean mix concrete; disposal of surplus soil							
Services not exceeding 200 mm nominal size							
average depth of run not exceeding 0.50 m	10.13	0.11	1.60	0.41	11.93	m	13.95
average depth of run not exceeding 0.75 m	15.20	0.13	1.95	0.52	17.90	m	20.37
average depth of run not exceeding 1.00 m	20.27	0.17	2.50	0.69	23.87	m	27.06
average depth of run not exceeding 1.25 m	25.33	0.23	3.38	1.03	29.83	m	34.24
Earthwork Support; providing support to opposing faces of excavation; moving along as work proceeds; A Plant Acrow							
Maximum depth not exceeding 2.00 m							
distance between opposing faces not exceeding 2.00 m	-	0.80	12.00	15.39	-	m	27.39

Q PAVING/PLANTING/FENCING/SITE FURNITURE Excluding overheads and profit	PC £	Labour hours	Labour £	Plant £	Material £	Unit	Total rate £
Q10 STONE/CONCRETE/BRICK KERBS/EDGINGS/CHANNELS							
Foundations to kerbs							
Excavating trenches; 450 x 150 mm; for concrete foundations to kerbs, edgings or channels							
by machine	-	0.14	2.10	0.72	-	m	2.82
by hand	-	0.66	9.90	-	-	m	9.90
Foundations; to kerbs, edgings, or channels; site mixed reinforced in situ concrete; 21 N/mm² - 20 aggregate (1:2:4) site mixed); one side against earth face, other against formwork (not included)							
150 wide x 100 mm deep	-	0.13	2.00	-	1.41	m	3.41
150 wide x 150 mm deep	-	0.17	2.50	-	2.12	m	4.62
200 wide x 150 mm deep	-	0.20	3.00	-	2.83	m	5.83
300 wide x 150 mm deep	-	0.23	3.50	-	4.24	m	7.74
600 wide x 200 mm deep	-	0.29	4.29	-	11.31	m	15.59
Formwork; sides of foundations (this will usually be required to one side of each kerb foundation adjacent to road sub-bases)							
100 mm deep	-	0.04	0.62	-	0.17	m	0.80
150 mm deep	-	0.04	0.62	-	0.26	m	0.88
Precast concrete kerbs, channels, edgings etc.; to BS 340; Marshalls Mono; bedding, jointing and pointing in cement mortar (1:3); on 150 mm deep concrete foundation; including haunching with in situ concrete; 11.50 N/mm² - 40 aggregate one side							
Kerbs; straight							
150 x 305 mm; ref HB1	5.65	0.50	7.50	-	8.32	m	15.82
125 x 255 mm; ref HB2; SP	2.93	0.44	6.67	-	5.60	m	12.26
125 x 150 mm; ref BN	2.38	0.40	6.00	-	5.05	m	11.05
Dropper kerbs; left and right handed							
125 x 255 - 150 mm; ref DL1 or DR1	4.15	0.50	7.50	-	6.82	m	14.32
125 x 255 - 150 mm; ref DL2 or DR2	4.15	0.50	7.50	-	6.82	m	14.32
Quadrant kerbs							
305 mm radius	7.85	0.50	7.50	-	9.63	nr	17.13
455 mm radius	8.40	0.50	7.50	-	11.07	nr	18.57
Internal or external angles							
127 x 254 mm section	12.29	0.50	7.50	-	14.08	m	21.58
Straight kerbs or channels; to radius; 125 x 255 mm							
0.90 m radius (2 units per quarter circle)	5.65	0.80	12.00	-	8.39	m	20.39
1.80 m radius (4 units per quarter circle)	5.65	0.73	10.91	-	8.39	m	19.30
2.40 m radius (5 units per quarter circle)	5.65	0.68	10.17	-	8.39	m	18.56
3.00 m radius (5 units per quarter circle)	5.65	0.67	10.00	-	8.39	m	18.39
4.50 m radius (2 units per quarter circle)	5.65	0.60	9.01	-	8.39	m	17.40
6.10 m radius (11 units per quarter circle)	5.65	0.58	8.70	-	8.39	m	17.09
7.60 m radius (14 units per quarter circle)	5.65	0.57	8.57	-	8.39	m	16.96
9.15 m radius (17 units per quarter circle)	5.65	0.56	8.33	-	8.39	m	16.73
10.70 m radius (20 units per quarter circle)	5.65	0.56	8.33	-	8.39	m	16.73
12.20 m radius (22 units per quarter circle)	5.65	0.53	7.90	-	8.39	m	16.29

Q PAVING/PLANTING/FENCING/SITE FURNITURE Excluding overheads and profit	PC £	Labour hours	Labour £	Plant £	Material £	Unit	Total rate £
Kerbs; "Conservation Kerb" units; to simulate natural granite kerbs							
255 x 150 x 914 mm; laid flat	14.11	0.57	8.57	-	18.90	m	27.47
150 x 255 x 914 mm; laid vertical	14.11	0.57	8.57	-	18.12	m	26.69
145 x 255 mm; radius internal 3.25 m	17.95	0.83	12.50	-	22.41	m	34.91
150 x 255 mm; radius external 3.40 m	17.15	0.83	12.50	-	21.58	m	34.08
145 x 255 mm; radius internal 6.50 m	16.44	0.67	10.00	-	20.45	m	30.45
145 x 255 mm; radius external 6.70 m	15.95	0.67	10.00	-	20.35	m	30.35
150 x 255 mm; radius internal 9.80 m	15.86	0.67	10.00	-	20.26	m	30.26
150 x 255 mm; radius external 10.00 m	15.41	0.67	10.00	-	19.42	m	29.42
305 x 305 x 255 mm; solid quadrants	20.32	0.57	8.57	-	24.33	nr	32.90
Channels; dished							
125 x 225 x 915 mm long; square; ref CS	3.10	0.40	6.00	-	11.17	m	17.17
125 x 150 x 915 mm long; square; ref CS2	2.15	0.40	6.00	-	8.73	m	14.73
305 x 150 x 915 mm	7.01	0.40	6.00	-	15.78	m	21.78
150 x 125 x 915 mm	5.56	0.40	6.00	-	13.31	m	19.31
152 x 102 x 915 mm	5.56	0.40	6.00	-	10.50	m	16.50
Precast concrete edging units; including haunching with in situ concrete; 11.50 N/mm² - 40 aggregate both sides							
Edgings; rectangular, bullnosed, or chamfered							
50 x 150 mm	1.60	0.33	5.00	-	6.05	m	11.05
125 x 150 mm bullnosed	1.93	0.33	5.00	-	6.38	m	11.38
50 x 200 mm	1.96	0.33	5.00	-	6.41	m	11.41
50 x 250 mm	2.46	0.33	5.00	-	6.91	m	11.91
50 x 250 mm flat top	2.13	0.33	5.00	-	6.58	m	11.58
Precast concrete kerbs; to BS 340; BDC Concrete Products; on 150 mm deep concrete foundation; including haunching with in situ concrete; 11.50 N/mm² - 40 aggregate one side							
Kerbs; "Kerb-sett" units (prices shown are for grey units, prices for other colours vary)							
standard blocks; 100 wide x 190 x 160 mm (set as high or low rise)	3.62	0.80	12.00	-	13.84	m	25.85
external angles; high or low profiles	4.59	0.80	12.00	-	14.81	m	26.82
Dropper kerbs (handed pairs); ref H-L	5.40	1.00	15.00	-	8.23	set	23.23
Crossovers; low; ref L-X	5.40	0.20	3.00	-	6.03	set	9.03
Crossovers; high; ref H-X	5.40	0.20	3.00	-	6.58	set	9.58
Curves; internal; low; ref IR/L	22.50	0.80	12.00	-	25.33	m	37.33
Curves; external; high; ref ER/H	22.50	0.80	12.00	-	25.33	m	37.33
Dressed natural stone kerbs - General							
Preamble: BS 435 includes the following conditions for dressed natural stone kerbs. The kerbs are to be good, sound, uniform in texture and free from defects; worked straight or to radius, square and out of wind, with the top front and back edges parallel or concentric to the dimensions specified. All drill and pick holes shall be removed from dressed faces. Standard dressings shall be in accordance with one of three illustrations in BS 435; and designated as either fine picked, single axed or nidged, or rough punched.							

Q PAVING/PLANTING/FENCING/SITE FURNITURE Excluding overheads and profit	PC £	Labour hours	Labour £	Plant £	Material £	Unit	Total rate £
Q10 STONE/CONCRETE/BRICK **KERBS/EDGINGS/CHANNELS** - cont'd							
Dressed natural stone kerbs; to BS435; **CED Ltd; on concrete foundations (not** **included); including haunching with in situ** **concrete; 11.50 N/mm² - 40 aggregate one side**							
Granite kerbs; 125 x 250 mm							
rough, straight, random lengths	12.00	0.80	12.00	-	14.95	m	26.95
special quality, straight, random lengths	17.00	0.80	12.00	-	19.95	m	31.95
Granite kerbs; 125 x 250 mm; curved to mean radius 3 m							
special quality, random lengths	23.00	0.91	13.65	-	25.95	m	39.60
rough, random lengths	15.00	0.91	13.65	-	17.95	m	31.60
York stone kerbs; on 150 mm deep **concrete foundations; including haunching** **with in situ concrete; 11.50 N/mm² - 40** **aggregate one side**							
Kerbs							
150 x 254 mm; sawn	41.00	0.80	12.00	-	43.95	m	55.95
Kerbs; curved to mean radius 5 m							
150 x 254 mm; fine picked	192.00	0.80	12.00	-	194.95	m	206.95
Second-hand granite setts; 100 x 100 mm; **bedding in cement mortar (1:4); on** **150 mm deep concrete foundations;** **including haunching with in situ** **concrete; 11.50 N/mm² - 40 aggregate** **one side**							
Edgings							
300 mm wide	10.08	1.33	20.00	-	14.02	m	34.02
Concrete setts; Blanc de Bierges; **bedding in cement mortar (1:4); on 150 mm** **deep concrete foundations, including** **haunching with in situ concrete;** **11.50 N/mm² - 40 aggregate one side**							
Edgings							
70 x 70 x 70 mm; in strips 230 mm wide	6.26	1.33	20.00	-	10.21	m	30.21
140 x 140 x 80 mm; in strips 310 mm wide	8.04	1.33	20.00	-	12.39	m	32.39
210 x 140 x 80 mm; in strips 290 mm wide	7.53	1.33	20.00	-	11.85	m	31.85
"Kidney" flint cobbles; bedding in 50 mm **concrete (1:2:4); on concrete** **foundations (not included)**							
Edgings							
200 mm wide	13.00	1.00	15.00	-	13.94	m	28.94

Q PAVING/PLANTING/FENCING/SITE FURNITURE Excluding overheads and profit	PC £	Labour hours	Labour £	Plant £	Material £	Unit	Total rate £
Brick or block stretchers; bedding in cement mortar (1:4); on 150 mm deep concrete foundations, including haunching with in situ concrete; 11.50 N/mm² - 40 a aggregate one side							
Single course							
concrete paving blocks; PC £8.85 /m²; 200 x 100 x 60 mm	1.77	0.31	4.61	-	4.27	m	8.88
engineering bricks; PC £260.00/1000; 215 x 102.5 x 65 mm	1.16	0.40	6.00	-	3.71	m	9.71
paving bricks; PC £450.00/1000; 215 x 102.5 x 65 mm	2.00	0.40	6.00	-	4.50	m	10.50
Two courses							
concrete paving blocks; PC £8.85 /m²; 200 x 100 x 60 mm	3.54	0.40	6.00	-	6.53	m	12.53
engineering bricks; PC £260.00/1000; 215 x 102.5 x 65 mm	2.31	0.57	8.57	-	5.41	m	13.99
paving bricks; PC £450.00/1000; 215 x 102.5 x 65 mm	4.00	0.57	8.57	-	6.99	m	15.56
Three courses							
concrete paving blocks; PC £8.85/m²; 200 x 100 x 60 mm	5.31	0.44	6.67	-	8.30	m	14.96
engineering bricks; PC £260.00/1000; 215 x 102.5 x 65 mm	3.47	0.67	10.00	-	6.63	m	16.63
paving bricks; PC £450.00/1000; 215 x 102.5 x 65 mm	6.00	0.67	10.00	-	8.99	m	18.99
Bricks on edge; bedding in cement mortar (1:4); on 150 mm deep concrete foundations; including haunching with in situ concrete; 11.50 N/mm² - 40 aggregate one side							
One brick wide							
engineering bricks; 215 x 102.5 x 65 mm	3.47	0.57	8.57	-	6.75	m	15.32
paving bricks; 215 x 102.5 x 65 mm	6.00	0.57	8.57	-	9.11	m	17.68
Two courses, stretchers laid on edge; cobble infill; 200 mm wide							
engineering bricks; 215 x 102.5 x 65 mm	6.93	1.33	20.00	-	24.48	m	44.48
paving bricks; 215 x 102.5 x 65 mm	12.00	1.33	20.00	-	29.20	m	49.20
Extra over bricks on edge for standard kerbs to one side; haunching in concrete							
125 x 255 mm; ref HB2; SP	2.93	0.44	6.67	-	5.60	m	12.26
Channels; bedding in cement mortar (1:3); joints pointed flush; on concrete foundations (not included)							
Three courses stretchers; 350 mm wide; quarter bond to form dished channels							
engineering bricks; PC £260.00/1000 ; 215 x 102.5 x 65 mm	3.47	1.00	15.00	-	4.74	m	19.74
paving bricks; PC £450.00/1000; 215 x 102.5 x 65 mm	6.00	1.00	15.00	-	7.10	m	22.10
Three courses granite setts; 340 mm wide; to form dished channels							
340 mm wide	10.08	2.00	30.00	-	12.61	m	42.61
Precast concrete dished channels; sides haunching up							
255 x 125 mm	2.92	0.50	7.50	-	5.45	m	12.95

Q PAVING/PLANTING/FENCING/SITE FURNITURE Excluding overheads and profit	PC £	Labour hours	Labour £	Plant £	Material £	Unit	Total rate £
Q10 STONE/CONCRETE/BRICK KERBS/EDGINGS/CHANNELS - cont'd							
Precast concrete channels; Charcon Hard Landscaping; on 150 mm deep concrete foundations; including haunching with in situ concrete; 21.00 N/mm² - 20 aggregate; both sides							
"Charcon Safeticurb"; slotted safety channels; for pedestrians and light vehicles							
ref DBJ; 305 x 305 mm	37.30	0.67	10.00	-	53.32	m	63.32
ref DBA; 250 x 250 mm	22.65	0.67	10.00	-	38.67	m	48.67
"Charcon Safeticurb"; slotted safety channels; for heavy vehicles							
ref DBM; 248 x 248 mm	39.35	0.80	12.00	-	55.37	m	67.37
ref Clearway; 305 x 305 mm	45.75	0.80	12.00	-	61.77	m	73.77
Inspection units; cast iron lids; including jointing to drainage channels							
248 x 248 x 914 mm	59.75	1.50	22.50	-	60.69	nr	83.19
Silt box tops; concrete frame; cast iron grid lids; type 1; set over gully							
457 x 610 mm	267.60	2.00	30.00	-	268.54	nr	298.54
Manhole covers; type K; cast iron; providing inspection to blocks and back gullies	289.00	2.00	30.00	-	291.23	nr	321.23
Kerb drainage systems; Marshalls Mono "Beany Block"; on foundations (not included); including haunching with class "E" concrete one side							
base block with top block	39.36	1.00	15.00	-	42.19	m	57.19
base block and cover plate	68.62	2.00	30.00	-	71.45	m	101.45
Channels; Ensor Building Products; on 100 mm deep concrete bed; 100 mm concrete fill both sides							
Polymer concrete drain units; tapered to falls							
channel units; Grade 'A', ref Stora-drain 100; 100 mm wide including galvanised grate and locks	19.18	0.80	12.00	-	25.31	m	37.31
channel units; Grade 'B', ref Stora-drain 100; 100 mm wide including galvanised grate and locks	24.52	0.80	12.00	-	30.65	m	42.65
channel units; Grade 'C', ref Stora-drain 100; 100 mm wide including galvanised grate and locks	26.80	0.80	12.00	-	32.93	m	44.93
channel units; Grade 'C', ref Stora-drain 100; 100 mm wide including anti-heel grate and locks	25.44	0.80	12.00	-	31.57	m	43.57
Stora-drain sump unit inclusive of steel bucket, 500 mm long	34.48	2.00	30.00	-	40.61	nr	70.61

Q PAVING/PLANTING/FENCING/SITE FURNITURE Excluding overheads and profit	PC £	Labour hours	Labour £	Plant £	Material £	Unit	Total rate £
Channels; Polybau; on 100 mm deep concrete bed; 100 mm concrete fill both sides							
"Polychannel"; polyester concrete channel units; tapered to falls							
galvanized mild steel slotted grating; 100 mm wide internally	25.93	1.00	15.00	-	29.94	m	44.94
stainless steel slotted grating; 100 mm wide internally	94.39	1.00	15.00	-	98.40	m	113.40
Channels; Hodkin & Jones; on well rammed subsoil base; set to gradient; backfilling with compacted excavated material							
"Decathlon"; precast concrete drainage channels							
150 mm wide channels	42.00	1.00	15.00	-	42.00	m	57.00
250 mm wide channels	53.00	1.00	15.00	-	53.00	m	68.00
375 mm wide channels	71.40	1.10	16.50	-	71.40	m	87.90
525 mm wide channels	84.00	1.20	18.00	0.29	84.00	m	102.29
600 mm wide channels	96.60	1.60	24.00	0.36	96.60	m	120.96
Q20 BASES AND SUB-BASES TO ROADS/PAVINGS							
Filling to make up levels							
Obtained off site; hardcore; PC £16.64 /m³							
100 mm thick	1.66	0.05	0.75	0.90	1.66	m²	3.32
100 mm thick	16.64	0.50	7.50	9.03	16.64	m³	33.17
150 mm thick	2.50	0.07	1.00	1.15	2.50	m²	4.65
200 mm thick	3.33	0.08	1.20	1.24	3.33	m²	5.77
300 mm thick	4.99	0.07	1.00	1.28	4.99	m²	7.27
exceeding 300 mm thick	16.64	0.17	2.50	3.01	19.97	m³	25.48
Obtained off site; granular fill type 1; PC £14.45 /tonne (£31.80/m³ compacted)							
100 mm thick	3.18	0.03	0.42	0.30	3.18	m²	3.90
150 mm thick	4.77	0.03	0.38	0.45	4.77	m²	5.60
250 mm thick	7.95	0.02	0.31	0.75	7.95	m²	9.01
Difference for each £1.00 increase/decrease in PC price per m³; price will vary with type and source of hardcore							
average 75 mm thick	-	-	-	-	0.08	m²	0.08
average 100 mm thick	-	-	-	-	0.10	m²	0.10
average 150 mm thick	-	-	-	-	0.16	m²	0.16
average 200 mm thick	-	-	-	-	0.21	m²	0.21
average 250 mm thick	-	-	-	-	0.26	m²	0.26
average 300 mm thick	-	-	-	-	0.31	m²	0.31
exceeding 300 mm thick	-	-	-	-	1.14	m³	1.14
Surface treatments							
Sand blinding; to hardcore base (not included); 25 mm thick	0.66	0.03	0.50	-	0.66	m²	1.15
Sand blinding; to hardcore base (not included); 50 mm thick	1.31	0.05	0.75	-	1.31	m²	2.06
Filter fabrics; to hardcore base (not included)	0.43	0.01	0.15	-	0.46	m²	0.60

Q PAVING/PLANTING/FENCING/SITE FURNITURE Excluding overheads and profit	PC £	Labour hours	Labour £	Plant £	Material £	Unit	Total rate £
Q20 BASES AND SUB-BASES TO ROADS/PAVINGS - cont'd							
Herbicides; ICI							
"Casoron G" (residual) herbicide; treating substrate before laying base							
at 1kg / 125m²	3.04	0.05	0.75	-	3.34	100 m²	4.09
Q21 IN SITU CONCRETE ROADS/PAVINGS/BASES							
Unreinforced concrete; on prepared sub-base (not included)							
Roads; 21.00 N/mm² - 20 aggregate (1:2:4) mechanically mixed on site							
100 mm thick	9.42	0.13	1.88	-	9.90	m²	11.77
150 mm thick	14.14	0.17	2.50	-	14.84	m²	17.34
Reinforced in situ concrete; mechanically mixed on site; normal Portland cement; on hardcore base (not included); reinforcement (not included)							
Roads; 11.50 N/mm² - 40 aggregate (1:3:6)							
100 mm thick	8.90	0.40	6.00	-	9.12	m²	15.12
150 mm thick	13.35	0.60	9.00	-	13.69	m²	22.69
200 mm thick	17.80	0.80	12.00	-	18.69	m²	30.69
250 mm thick	22.25	1.00	15.00	0.40	22.81	m²	38.21
300 mm thick	26.71	1.20	18.00	0.40	27.37	m²	45.78
Roads; 21.00 N/mm² - 20 aggregate (1:2:4)							
100 mm thick	9.42	0.40	6.00	-	9.66	m²	15.66
150 mm thick	14.14	0.60	9.00	-	14.49	m²	23.49
200 mm thick	18.85	0.80	12.00	-	19.32	m²	31.32
250 mm thick	23.56	1.00	15.00	0.40	24.14	m²	39.55
300 mm thick	28.27	1.20	18.00	0.40	28.98	m²	47.38
Roads; 25.00 N/mm² - 20 aggregate GEN 4 ready mixed							
100 mm thick	6.86	0.40	6.00	-	7.03	m²	13.03
150 mm thick	10.28	0.60	9.00	-	10.54	m²	19.54
200 mm thick	13.71	0.80	12.00	-	14.05	m²	26.05
250 mm thick	17.14	1.00	15.00	0.40	17.57	m²	32.97
300 mm thick	20.57	1.20	18.00	0.40	21.08	m²	39.49
Reinforced in situ concrete; ready mixed; discharged directly into location from supply lorry; normal Portland cement; on hardcore base (not included); reinforcement (not included)							
Roads; 11.50 N/mm² - 40 aggregate (1:3:6)							
100 mm thick	6.75	0.16	2.40	-	6.75	m²	9.15
150 mm thick	10.13	0.24	3.60	-	10.13	m²	13.73
200 mm thick	13.51	0.36	5.40	-	13.51	m²	18.91
250 mm thick	16.89	0.54	8.10	0.40	16.89	m²	25.39
300 mm thick	20.27	0.66	9.90	0.40	20.27	m²	30.57

Q PAVING/PLANTING/FENCING/SITE FURNITURE Excluding overheads and profit	PC £	Labour hours	Labour £	Plant £	Material £	Unit	Total rate £
Roads, 21.00 N/mm² - 20 aggregate (1:2:4)							
100 mm thick	7.06	0.16	2.40	-	7.06	m²	9.46
150 mm thick	10.59	0.24	3.60	-	10.59	m²	14.19
200 mm thick	14.12	0.36	5.40	-	14.12	m²	19.52
250 mm thick	17.65	0.54	8.10	0.40	17.65	m²	26.15
300 mm thick	21.18	0.66	9.90	0.40	21.18	m²	31.48
Roads, 26.00 N/mm² - 20 aggregate (1:1:5:3)							
100 mm thick	7.26	0.16	2.40	-	7.26	m²	9.66
150 mm thick	10.89	0.24	3.60	-	10.89	m²	14.49
200 mm thick	14.12	0.36	5.40	-	14.12	m²	19.52
250 mm thick	18.16	0.54	8.10	0.40	18.16	m²	26.66
300 mm thick	21.79	0.66	9.90	0.40	21 79	m²	32.09
Roads; PAV1 Concrete - 35 N/mm²; Designated mix							
100 mm thick	7.19	0.16	2.40	-	7.37	m²	9.77
150 mm thick	10.78	0.24	3.60	-	11.05	m²	14.65
200 mm thick	14.37	0.36	5.40	-	14.73	m²	20.13
250 mm thick	17.96	0.54	8.10	0.40	18.41	m²	26.91
300 mm thick	21.56	0.66	9.90	0.40	22.10	m²	32.40
Concrete sundries							
Treating surfaces of unset concrete; grading to cambers, tamping with 75 mm thick steel shod tamper or similar	-	0.13	2.00	-	-	m²	2.00
Expansion joints							
13 mm thick joint filler; formwork							
width or depth not exceeding 150 mm	0.95	0.20	3.00	-	2.14	m	5.14
width or depth 150 - 300 mm	0.90	0.25	3.75	-	3.28	m	7.03
width or depth 300 - 450 mm	1.76	0.30	4.50	-	5.34	m	9.84
25 mm thick joint filler; formwork							
width or depth not exceeding 150 mm	1.18	0.20	3.00	-	2.37	m	5.37
width or depth 150 - 300 mm	2.35	0.25	3.75	-	4.73	m	8.48
width or depth 300 - 450 mm	3.53	0.30	4.50	-	7.10	m	11.60
Sealants; sealing top 25 mm of joint with rubberized bituminous compound	1.01	0.25	3.75	-	1.01	m	4.76
Formwork for in situ concrete							
Sides of foundations							
height exceeding 1.00 m	6.94	2.00	30.00	-	8.46	m²	38.46
height not exceeding 250 mm	1.73	1.00	15.00	-	2.26	m²	17.26
height 250 - 500 mm	3.47	1.00	15.00	-	4.23	m²	19.23
height 500 mm - 1.00 m	6.94	1.50	22.50	-	8.46	m²	30.96
Extra over formwork for curved work 6 m radius	-	0.25	3.75	-	-	m²	3.75
Steel road forms; to edges of beds or faces of foundations							
150 mm wide	-	0.20	3.00	-	0.53	m	3.53
Reinforcement; fabric; BS 4483; side laps 150 mm; head laps 300 mm; mesh 200 x 200 mm; in roads, footpaths or pavings							
Fabric							
ref A142 (2.22 kg/m²)	0.74	0.08	1.25	-	0.82	m²	2.06
ref A193 (3.02 kg/m²)	3.99	0.08	1.25	-	4.39	m²	5.64

Q PAVING/PLANTING/FENCING/SITE FURNITURE Excluding overheads and profit	PC £	Labour hours	Labour £	Plant £	Material £	Unit	Total rate £
Q22 COATED MACADAM/ASPHALT ROADS/PAVINGS							
Coated macadam/asphalt roads/pavings - General Preamble: The prices for all in situ finishings to roads and footpaths include for work to falls, crossfalls or slopes not exceeding 15 degrees from horizontal; for laying on prepared bases (not included) and for rolling with an appropriate roller.							
Preamble: Users should note the new terminology for the surfaces described below which is to European standard descriptions. The now redundant descriptions for each course are shown in brackets.							
Macadam Surfacing; **Spadeoak Construction Co Ltd;** **Surface (Wearing) Course; 20 mm of 6 mm dense bitumen macadam to BS4987-1 2001 ref 7.5** Machine lay; areas 1000 m² and over							
limestone aggregate	-	-	-	-	-	m²	3.92
granite aggregate	-	-	-	-	-	m²	3.98
red	-	-	-	-	-	m²	7.33
Hand Lay; areas 400 m² and over							
limestone aggregate	-	-	-	-	-	m²	5.02
granite aggregate	-	-	-	-	-	m²	5.08
red	-	-	-	-	-	m²	9.10
Macadam Surfacing; **Spadeoak Construction Co Ltd;** **Surface (Wearing) Course; 30 mm of 10 mm dense bitumen macadam to BS4987-1 2001 ref 7.4** Machine lay; areas 1000 m² and over							
limestone aggregate	-	-	-	-	-	m²	4.69
granite aggregate	-	-	-	-	-	m²	4.77
red	-	-	-	-	-	m²	10.25
Hand Lay; areas 400 m² and over							
limestone aggregate	-	-	-	-	-	m²	5.83
granite aggregate	-	-	-	-	-	m²	5.92
red	-	-	-	-	-	m²	12.21

Q PAVING/PLANTING/FENCING/SITE FURNITURE Excluding overheads and profit	PC £	Labour hours	Labour £	Plant £	Material £	Unit	Total rate £
Macadam Surfacing; **Spadeoak Construction Co Ltd;** **Surface (Wearing) Course; 40 mm of 10 mm** **dense bitumen macadam to BS4987-1 2001** **ref 7.4**							
Machine lay; areas 1000 m² and over							
limestone aggregate	-	-	-	-	-	m²	5.94
granite aggregate	-	-	-	-	-	m²	6.05
red	-	-	-	-	-	m²	11.62
Hand Lay; areas 400 m² and over							
limestone aggregate	-	-	-	-	-	m²	7.17
granite aggregate	-	-	-	-	-	m²	7.29
red	-	-	-	-	-	m²	13.68
Macadam Surfacing; **Spadeoak Construction Co Ltd;** **Binder (Base) Course; 50 mm of 20 mm** **dense bitumen macadam to BS4987-1 2001** **ref 6.5**							
Machine lay; areas 1000 m² and over							
limestone aggregate	-	-	-	-	-	m²	6.32
granite aggregate	-	-	-	-	-	m²	6.45
Hand Lay; areas 400 m² and over							
limestone aggregate	-	-	-	-	-	m²	7.57
granite aggregate	-	-	-	-	-	m²	7.71
Macadam Surfacing; **Spadeoak Construction Co Ltd;** **Binder (Base) Course; 60 mm of 20 mm** **dense bitumen macadam to BS4987-1 2001** **ref 6.5**							
Machine lay; areas 1000 m² and over							
limestone aggregate	-	-	-	-	-	m²	6.95
granite aggregate	-	-	-	-	-	m²	7.10
Hand Lay; areas 400 m² and over							
limestone aggregate	-	-	-	-	-	m²	8.25
granite aggregate	-	-	-	-	-	m²	8.41
Macadam Surfacing; **Spadeoak Construction Co Ltd;** **Base (Roadbase) Course; 75 mm of 28 mm** **dense bitumen macadam to BS4987-1 2001** **ref 5.2**							
Machine lay; areas 1000 m² and over							
limestone aggregate	-	-	-	-	-	m²	8.34
granite aggregate	-	-	-	-	-	m²	8.54
Hand Lay; areas 400 m² and over							
limestone aggregate	-	-	-	-	-	m²	9.73
granite aggregate	-	-	-	-	-	m²	9.94

Q PAVING/PLANTING/FENCING/SITE FURNITURE Excluding overheads and profit	PC £	Labour hours	Labour £	Plant £	Material £	Unit	Total rate £
Q22 COATED MACADAM/ASPHALT ROADS/PAVINGS - cont'd							
Macadam Surfacing; Spadeoak Construction Co Ltd; Base (Roadbase) Course; 100 mm of 28 mm dense bitumen macadam to BS4987-1 2001 ref 5.2							
Machine lay; areas 1000 m² and over							
limestone aggregate	-	-	-	-	-	m²	**10.52**
granite aggregate	-	-	-	-	-	m²	**10.78**
Hand Lay; areas 400 m² and over							
limestone aggregate	-	-	-	-	-	m²	**12.51**
granite aggregate	-	-	-	-	-	m²	**12.78**
Macadam Surfacing; Spadeoak Construction Co Ltd; Base (Roadbase) Course; 150 mm of 28 mm dense bitumen macadam in two layers to BS4987-1 2001 ref 5.2							
Machine lay; areas 1000 m² and over							
limestone aggregate	-	-	-	-	-	m²	**16.59**
granite aggregate	-	-	-	-	-	m²	**16.97**
Hand Lay; areas 400 m² and over							
limestone aggregate	-	-	-	-	-	m²	**20.26**
granite aggregate	-	-	-	-	-	m²	**20.67**
Macadam Surfacing; Spadeoak Construction Co Ltd; Base (Roadbase) Course; 200 mm of 28 mm dense bitumen macadam in two layers to BS4987-1 2001 ref 5.2							
Machine lay; areas 1000 m² and over							
limestone aggregate	-	-	-	-	-	m²	**21.08**
granite aggregate	-	-	-	-	-	m²	**21.60**
Hand Lay; areas 400 m² and over							
limestone aggregate	-	-	-	-	-	m²	**25.06**
granite aggregate	-	-	-	-	-	m²	**25.61**
Resin Bound Macadam Pavings; Machine laid; Colourpave Ltd; "Naturatex" Clear resin bound macadam to pedestrian or vehicular hard landscape areas; laid to base course (not included)							
20 mm thick	-	-	-	-	-	m²	**18.00**
"Buff Chiptex" Buff asphalt with pre-coated gravel chippings rolled in							
20 mm thick to pedestrian areas	-	-	-	-	-	m²	**20.00**
30 mm thick to vehicular areas	-	-	-	-	-	m²	**23.00**
"Colourtex" Coloured resin bound macadam to pedestrian or vehicular hard landscape areas							
20 mm thick	-	-	-	-	-	m²	**17.00**

Q PAVING/PLANTING/FENCING/SITE FURNITURE Excluding overheads and profit	PC £	Labour hours	Labour £	Plant £	Material £	Unit	Total rate £
Road repairs; Colas Ltd "Colas Pre-Formed Surface Dressing." preformed membrane; non-slip surface dressing; on clean smooth surface; concrete, wood, steel etc (not included); treated with "Cariphalte CP Primer" "Calcinated Bauxite" finish	-	0.20	3.00	-	12.67	m²	15.67
Marking car parks Car parking space division strips; in accordance with BS 3262; laid hot at 115 degrees C; on bitumen macadam surfacing							
100 mm wide thermoplastic strips	-	-	-	-	-	m	0.95
Stainless metal road studs							
100 x 100 mm	4.90	0.25	3.75	-	4.90	nr	8.65
Q23 GRAVEL/HOGGIN ROADS/PAVINGS							
Excavation and path preparation Excavating; 300 mm deep; to width of path; depositing excavated material at sides of excavation							
width 1.00 m	-	-	-	2.03	-	m²	2.03
width 1.50 m	-	-	-	1.69	-	m²	1.69
width 2.00 m	-	-	-	1.45	-	m²	1.45
width 3.00 m	-	-	-	1.21	-	m²	1.21
Excavating trenches; in centre of pathways; 100 flexible drain pipes; filling with clean broken stone or gravel rejects							
300 x 450 mm deep	3.51	0.10	1.50	1.01	3.58	m	6.09
Hand trimming and compacting reduced surface of pathway; by machine							
width 1.00 m	-	0.05	0.75	0.14	-	m	0.89
width 1.50 m	-	0.04	0.67	0.13	-	m	0.79
width 2.00 m	-	0.04	0.60	0.12	-	m	0.72
width 3.00 m	-	0.04	0.60	0.12	-	m	0.72
Permeable membranes; to trimmed and compacted surface of pathway							
"Terram 1000"	0.43	0.02	0.30	-	0.46	m²	0.76

Q PAVING/PLANTING/FENCING/SITE FURNITURE Excluding overheads and profit	PC £	Labour hours	Labour £	Plant £	Material £	Unit	Total rate £
Q23 GRAVEL/HOGGIN ROADS/PAVINGS - cont'd							
Edging boards Boards; 50 x 50 x 750 mm timber pegs at 1000 mm centres (excavations and hardcore under edgings not included)							
38 x 150 mm treated softwood edge boards	2.19	0.10	1.50	-	2.19	m	3.69
50 x 150 mm treated softwood edge boards	2.46	0.10	1.50	-	2.46	m	3.96
38 x 150 mm hardwood (iroko) edge boards	7.45	0.10	1.50	-	7.45	m	8.95
50 x 150 mm hardwood (iroko) edge boards	9.41	0.10	1.50	-	9.41	m	10.91
Filling to make up levels Obtained off site; hardcore; PC £16.64 /m³							
150 mm thick	2.66	0.04	0.60	0.42	2.66	m²	3.68
Obtained off site; granular fill type 1; PC £14.45 /tonne (£31.80/m³ compacted)							
100 mm thick	3.18	0.03	0.42	0.30	3.18	m²	3.90
150 mm thick	4.77	0.03	0.38	0.45	4.77	m²	5.60
Surface treatments Sand blinding; to hardcore (not included)							
50 mm thick	1.31	0.04	0.60	-	1.44	m²	2.04
Filter fabric; to hardcore (not included)	0.43	0.01	0.15	-	0.46	m²	0.60
Herbicides; ICI "Casoron G" (residual) herbicide; treating substrate before laying base							
at 1kg / 125m²	3.04	0.05	0.75	-	3.34	100m²	4.09
Granular pavings Approved washed river or pit gravel; on base layer of coarse aggregate; wearing layer 10 mm gravel graded; watering; rolling; on hardcore base (not included); for pavements; to falls and crossfalls and to slopes not exceeding 15 degrees from horizontal; over 300 mm wide							
50 mm thick	1.53	0.09	1.31	-	1.84	m²	3.15
63 mm thick	2.08	0.13	1.89	-	2.50	m²	4.39
Breedon PLC; "Golden Gravel" or equivalent; rolling wet; on hardcore base (not included); for pavements; to falls and crossfalls and to slopes not exceeding 15 degrees from horizontal; over 300 mm wide							
50 mm thick	7.20	0.10	1.50	0.47	7.20	m²	9.17
75 mm thick	10.80	0.10	1.50	0.56	10.80	m²	12.86
Breedon PLC; 'Wayfarer' specially formulated fine gravel for use on golf course pathways							
50 mm thick	5.63	0.10	1.50	0.47	5.63	m²	7.59
75 mm thick	8.43	0.11	1.67	0.56	8.43	m²	10.66
Breedon PLC; Breedon Buff decorative limestone chippings							
50 mm thick	-	0.01	0.15	0.10	4.72	m²	4.98
75 mm thick	9.00	0.01	0.19	0.13	9.46	m²	9.77

Q PAVING/PLANTING/FENCING/SITE FURNITURE Excluding overheads and profit	PC £	Labour hours	Labour £	Plant £	Material £	Unit	Total rate £
Hoggin (stabilized); PC £14.50/m³ on hardcore base (not included); to falls and crossfalls and to slopes not exceeding 15 degrees from horizontal; over 300 mm wide							
100 mm thick	1.45	0.13	2.00	1.07	2.17	m²	5.25
150 mm thick	2.17	0.13	1.89	1.01	3.26	m²	6.17
Washed shingle; on prepared base (not included)							
25 - 50 size, 25 mm thick	0.65	0.02	0.26	0.09	0.69	m²	1.04
25 - 50 size, 75 mm thick	1.96	0.05	0.79	0.27	2.06	m²	3.12
50 - 75 size, 25 mm thick	0.65	0.02	0.30	0.10	0.69	m²	1.09
50 - 75 size, 75 mm thick	1.96	0.07	1.00	0.34	2.06	m²	3.40
Pea shingle; on prepared base (not included)							
10 - 15 size, 25 mm thick	0.68	0.02	0.26	0.09	0.71	m²	1.07
5 - 10 size, 75 mm thick	2.04	0.05	0.79	0.27	2.14	m²	3.20
Ballast ; as dug; watering; rolling; on hardcore base (not included)							
100 mm thick	2.71	0.13	2.00	1.07	3.39	m²	6.47
150 mm thick	4.07	0.13	1.89	1.01	5.09	m²	7.99
CED Ltd; Cedec golden gravel; self binding; laid to base measured separately; compacting							
50 mm thick	6.70	0.03	0.43	0.06	6.70	m²	7.19
Grundon Ltd; Coxwell self binding path gravels laid and compacted to excavation or base measured separately							
50 mm thick	1.56	0.03	0.43	0.06	1.56	m²	2.05
Bark surfaces; ICI							
"Casoron G" (residual) herbicide; treating substrate before laying base							
at 1kg / 125m²	3.04	0.05	0.75	-	3.34	100m²	4.09
Bark surfaces; Melcourt Industries							
Bark chips; to surface of pathways by machine; material delivered in 70 m³ loads; levelling and spreading by hand (excavation and preparation not included)							
conifer walk chips; 100 mm thick	1.65	0.03	0.50	0.25	1.73	m²	2.48
wood fibre; 100 mm thick	1.59	0.03	0.50	0.25	1.67	m²	2.43
hardwood chips; 100 mm thick	1.72	0.03	0.50	0.25	1.80	m²	2.55
Bound aggregates; Addagrip Surface Treatments UK Ltd; Natural decorative Resin bonded surface dressing laid to concrete, macadam or to plywood panels priced separately.							
Primer coat to macadam or concrete base	-	-	-	-	-	m²	4.00
Golden pea Gravel 1-3 mm							
Buff adhesive	-	-	-	-	-	m²	19.36
Red adhesive	-	-	-	-	-	m²	19.36
Green adhesive	-	-	-	-	-	m²	19.36
Golden pea Gravel 2-5 mm							
Buff adhesive	-	-	-	-	-	m²	22.99
Chinese bauxite 1-3 mm							
Buff adhesive	-	-	-	-	-	m²	19.36

Q PAVING/PLANTING/FENCING/SITE FURNITURE Excluding overheads and profit	PC £	Labour hours	Labour £	Plant £	Material £	Unit	Total rate £
Q24 INTERLOCKING BRICK/BLOCK ROADS/PAVINGS							
Precast concrete block edgings; PC £8.18/m²; 200 x 100 x 60 mm; on prepared base (not included); haunching one side							
Edgings; butt joints							
stretcher course	1.64	0.17	2.50	-	4.08	m	6.58
header course	0.82	0.27	4.00	-	3.40	m	7.40
Precast concrete vehicular paving blocks; Marshalls Mono; on prepared base (not included); on 50 mm compacted sharp sand bed; blocks laid in 7 mm loose sand and vibrated; joints filled with sharp sand and vibrated; level and to falls only							
"Trafica" paving blocks; 450 x 450 x 70 mm							
"Perfecta" finish; colour natural	20.56	0.50	7.50	0.09	22.51	m²	30.10
"Perfecta" finish; colour buff	24.06	0.50	7.50	0.09	26.10	m²	33.69
"Saxon" finish; colour natural	17.35	0.50	7.50	0.09	19.23	m²	26.82
"Saxon" finish; colour buff	20.46	0.50	7.50	0.09	22.41	m²	30.00
Precast concrete vehicular paving blocks; Marshalls Mono; on prepared base (not included); on 50 mm compacted sharp sand bed; blocks laid in 7 mm loose sand and vibrated; joints filled with sharp sand and vibrated; level and to falls only							
"Keyblock" concrete blocks							
200 x 100 x 60 mm; natural grey	8.18	0.40	6.00	0.09	10.19	m²	16.27
200 x 100 x 60 mm; colours	8.85	0.40	6.00	0.09	10.89	m²	16.98
200 x 100 x 80 mm; natural grey	9.06	0.40	6.00	0.09	11.11	m²	17.20
200 x 100 x 80 mm; colours	10.50	0.40	6.00	0.09	12.62	m²	18.71
Precast concrete vehicular paving blocks; Charcon Hard Landscaping; on prepared base (not included); on 50 mm compacted sharp sand bed; blocks laid in 7 mm loose sand and vibrated; joints filled with sharp sand and vibrated; level and to falls only							
"Europa" concrete blocks							
200 x 100 x 65 mm; natural grey	6.97	0.40	6.00	0.09	8.92	m²	15.00
200 x 100 x 65 mm; colours	7.31	0.40	6.00	0.09	9.27	m²	15.36
"Parliament" concrete blocks							
200 x 100 x 65 mm; natural grey	17.80	0.40	6.00	0.09	19.84	m²	25.93
200 x 100 x 65 mm; colours	17.80	0.40	6.00	0.09	19.84	m²	25.93

Q PAVING/PLANTING/FENCING/SITE FURNITURE Excluding overheads and profit	PC £	Labour hours	Labour £	Plant £	Material £	Unit	Total rate £
Recycled polyethylene grassblocks; **Netlon Ltd; interlocking units laid to** **prepared base or rootzone (not** **included);** ""Netpave 50"; load bearing 50 tonnes per m²; 500 x 500 x 50 mm deep							
minimum area 40 m²	15.00	0.20	3.00	0.21	16.68	m²	19.89
40 -1039 m²	10.40	0.20	3.00	0.21	12.08	m²	15.29
over 1039 m²	9.50	0.20	3.00	0.21	11.18	m²	14.39
"Netpave 25"; load bearing :light vehicles and pedestrians; 500 x 500 x 25 mm deep; laid onto established grass surface							
minimum area 40 m²	14.00	0.10	1.50	-	14.00	m²	15.50
40 -1039 m³	9.40	0.10	1.50	-	9.40	m²	10.90
over 1039 m²	8.50	0.10	1.50	-	8 50	m²	10.00
"Grassroad"; Cooper Clarke Group; **heavy duty for car parking and fire paths** **verge hardening and shallow** **embankments** Honeycomb cellular polypropylene interconnecting paviors with integral downstead anti-shear cleats including topsoil but excluding edge restraints; to granular sub-base (not included)							
635 x 330 x 42 overall laid to a module of 622 x 311 x 32	-	-	-	-	-	m²	21.23
Extra over for green colour	-	-	-	-	-	m²	0.26
Q25 SLAB/BRICK/SETT/COBBLE PAVINGS							
Bricks - General Preamble: BS 3921 includes the following specification for bricks for paving: bricks shall be hard, well burnt, non-dusting, resistant to frost and sulphate attack and true to shape, size and sample.							
Movement of materials Mechanically offloading bricks; loading wheel barrows; transporting maximum 25 m distance	-	0.20	3.00	-	-	m²	3.00
Edge restraints; to brick paving; on **prepared base (not included); 65 mm** **thick bricks; PC £300.00 /1000;** **haunching one side** Header Course							
200 x 100 mm; butt joints	3.00	0.27	4.00	-	5.58	m	9.58
210 x 105 mm; mortar joints	2.67	0.50	7.50	-	5.57	m	13.07
Stretcher course							
200 x 100 mm; butt joints	1.50	0.17	2.50	-	3.95	m	6.44
210 x 105 mm; mortar joints	1.36	0.33	5.00	-	3.96	m	8.96

Q PAVING/PLANTING/FENCING/SITE FURNITURE Excluding overheads and profit	PC £	Labour hours	Labour £	Plant £	Material £	Unit	Total rate £
Q25 SLAB/BRICK/SETT/COBBLE PAVINGS - cont'd							
Variation in brick prices; Add or subtract the following amounts for every £1.00/1000 difference in the PC price							
Edgings							
100 wide	-	-	-	-	0.05	10 m	0.05
200 wide	-	-	-	-	0.11	10 m	0.11
102.5 wide	-	-	-	-	0.05	10 m	0.05
215 wide	-	-	-	-	0.09	10 m	0.09
Clay brick pavings; on prepared base (not included); bedding on 50 mm sharp sand; kiln dried sand joints							
Pavings; Marshalls Clay Products; "Nori"; 200 x 100 x 65 mm wirecut chamfered paviors							
"Sherbourne Red"; PC £16.05/m²	16.05	1.06	15.88	0.26	18.20	m²	34.33
"Allandale Gold"; PC £19.88/m²	19.88	1.06	15.88	0.26	21.63	m²	37.76
Paving bricks; PC £300.00 per 1000							
215 x 102.5 x 65 mm	12.40	1.00	15.00	-	15.74	m²	30.74
200 x 100.5 x 50 mm	15.00	1.10	16.50	-	18.34	m²	34.84
Reduction for bedding on 50 mm compacted sharp sand; setting; vibrating; sand joints							
Clay brick pavings; on prepared base (not included); bedding on cement:sand (1:4) pointing mortar as work proceeds							
Pavings; Hanson Brick							
Buff Multi Chamfered Paver; 200 x 100 x 50 mm; PC £314.80/1000	13 00	2.20	33.00	-	17.11	m²	50.11
Antique Chamfered Paver; 200 x 100 x 50 mm; PC £272.80/1000	11 27	2.20	33.00	-	15.33	m²	48.33
Blue Brown Chamfered Cobble Paver; 200 x 100 x 50 mm; PC £367.67/1000	13.00	2.20	33.00	-	17.11	m²	50.11
Paving bricks; PC £300.00 per 1000							
215 x 102.5 x 65 mm	11.85	2.22	33.33	-	15.19	m²	48.52
200 x 100 x 50 mm	12.99	2.50	37.50	-	16.33	m²	53.83
Extra over clay brick pavings for labours							
basketweave bond	-	0.20	3.00	-	-	m²	3.00
herringbone, laid on edge	-	2.00	30.00	-	-	m²	30.00
Cutting							
raking cutting	-	0.33	5.00	4.20	-	m	9.20
curved cutting	-	0.44	6.67	5.27	-	m	11.93
Add or subtract the following amounts for every £10.00 /1000 difference in the prime cost of bricks							
Butt joints							
200 x 100	-	-	-	-	0.50	m²	0.50
215 x 102.5	-	-	-	-	0.45	m²	0.45
10 mm mortar joints							
200 x 100	-	-	-	-	0.43	m²	0.43
215 x 102.5	-	-	-	-	0.40	m²	0.40

Q PAVING/PLANTING/FENCING/SITE FURNITURE Excluding overheads and profit	PC £	Labour hours	Labour £	Plant £	Material £	Unit	Total rate £
Precast concrete pavings; Charcon Hard Landscaping; to BS 7263; on prepared sub-base (not included); bedding on 25 mm thick cement:sand mortar (1:4); butt joints; straight both ways; jointing in cement:sand (1:3) brushed in; on 50 mm thick sharp sand base							
Pavings; natural grey							
450 x 450 x 70 mm chamfered	10.62	0.44	6.67	-	14.66	m²	**21.32**
450 x 450 x 50 mm chamfered	9.63	0.44	6.67	-	13.67	m²	**20.34**
600 x 300 x 50 mm	7.61	0.44	6.67	-	11.65	m²	**18.32**
400 x 400 x 65 mm chamfered	17.44	0.40	6.00	-	21.48	m²	**27.48**
450 x 600 x 50 mm	7.59	0.44	6.67	-	11.63	m²	**18.30**
600 x 600 x 50 mm	5.92	0.40	6.00	-	9.96	m²	**15.96**
750 x 600 x 50 mm	5.60	0.40	6.00	-	9.64	m²	**15.64**
900 x 600 x 50 mm	5.15	0.40	6.00	-	9.19	m²	**15.19**
Pavings; coloured							
450 x 450 x 70 mm chamfered	17.63	0.44	6.67	-	21.67	m²	**28.34**
450 x 600 x 50 mm	11.78	0.44	6.67	-	15.82	m²	**22.48**
400 x 400 x 65 mm chamfered	17.44	0.40	6.00	-	21.48	m²	**27.48**
600 x 600 x 50 mm	9.86	0.40	6.00	-	13.90	m²	**19.90**
750 x 600 x 50 mm	9.11	0.40	6.00	-	13.15	m²	**19.15**
900 x 600 x 50 mm	8.14	0.40	6.00	-	12.18	m²	**18.18**
Precast concrete pavings; Charcon Hard Landscaping; to BS 7263; on prepared sub-base (not included); bedding on 25 mm thick cement:sand mortar (1:4); butt joints; straight both ways; jointing in cement:sand (1:3) brushed in; on 50 mm thick sharp sand base							
"Appalacian" rough textured exposed aggregate pebble paving							
600 mm x 600 mm x 65 mm	17.00	0.50	7.50	-	20.15	m²	**27.65**
Pavings; Atlas Stone Products; on prepared base (not included);							
"French Limestone" precast slab in random sizes laid to full mortar bed; butt jointed							
340 x 340 x 25 mm	16.61	0.44	6.67	-	19.34	m²	**26.00**
685 x 340 x 25 mm	16.82	0.44	6.67	-	19.97	m²	**26.63**
455 x 340 x 25 mm	16.86	0.57	8.57	-	20.01	m²	**28.58**
"Broadway paving" in Cotswold; spot bedded on 5 pads of sand:cement mortar (1:4) on sharp sand							
600 x 600 mm	7.64	0.40	6.00	-	9.19	m²	**15.19**
450 x 450 mm	6.57	0.50	7.50	-	8.10	m²	**15.60**
600 x 300 mm	9.44	0.50	7.50	-	11.05	m²	**18.55**

Q PAVING/PLANTING/FENCING/SITE FURNITURE Excluding overheads and profit	PC £	Labour hours	Labour £	Plant £	Material £	Unit	Total rate £
Q25 SLAB/BRICK/SETT/COBBLE PAVINGS - cont'd							
Pavings; Marshalls Mono; spot bedding on 5 nr pads of cement:sand mortar (1:4); on sharp sand							
"Blister Tactile" pavings; specially textured slabs for guidance of blind pedestrians; red or buff							
400 x 400 x 50 mm	19.06	0.50	7.50	-	20.90	m²	28.40
450 x 450 x 50 mm	16.84	0.50	7.50	-	18.63	m²	26.13
"Metric Four Square" pavings							
496 x 496 x 50 mm; exposed river gravel aggregate	48.92	0.50	7.50	-	50.29	m²	57.79
"Metric Four Square" cycle blocks							
496 x 496 x 50 mm; exposed aggregate	48.92	0.25	3.75	-	50.92	m²	54.67
Precast concrete pavings; Marshalls; "Heritage" imitation riven yorkstone paving; on prepared sub-base measured separately; bedding on 25 mm thick cement:sand mortar (1:4); pointed straight both ways cement:sand (1:3)							
Square and rectangular paving							
300 x 300 x 38 mm	22.05	1.00	15.00	-	25.41	m²	40.41
600 x 300 x 38 mm	17.38	0.80	12.00	-	20.75	m²	32.75
600 x 450 x 38 mm	17.17	0.75	11.25	-	20.53	m²	31.78
450 x 450 x 38 mm	15.48	0.75	11.25	-	18.85	m²	30.10
600 x 600 x 38 mm	17.11	0.50	7.50	-	20.53	m²	28.03
Extra labours for laying the a selection of the above sizes to random rectangular pattern	-	0.33	5.00	-	-	m²	5.00
Radial paving for circles							
circle with centre stone and first ring (8 slabs), 450 x 230/560 x 38 mm; diameter 1.54 m (total area 1.86 m²)	47.93	1.50	22.50	-	52.97	nr	75.47
circle with second ring (16 slabs), 450 x 300/460 x 38 mm; diameter 2.48 m (total area 4.83 m²)	47.93	4.00	60.00	-	134.97	nr	194.97
circle with third ring (16 slabs), 450 x 470/625 x 38 mm; diameter 3.42 m (total area 9.18 m²)	47.93	8.00	120.00	-	245.78	nr	365.78
Stepping stones							
600 dia x 38 mm	8.41	0.20	3.00	-	13.87	nr	16.87
380 dia x 38 mm	3.34	0.20	3.00	-	8.80	nr	11.80
Asymmetrical 560 x 420 x 38 mm	5.00	0.20	3.00	-	10.46	nr	13.46
Precast concrete pavings; Marshalls; "Chancery" imitation reclaimed riven yorkstone paving; on prepared sub-base measured separately; bedding on 25 mm thick cement:sand mortar (1:4); pointed straight both ways cement:sand (1:3)							
Square and rectangular paving							
300 x 300 x 45 mm	21.11	1.00	15.00	-	24.48	m²	39.48
450 x 300 x 45 mm	19.94	0.90	13.50	-	23.31	m²	36.81
600 x 300 x 45 mm	19.45	0.80	12.00	-	22.82	m²	34.82
600 x 450 x 45 mm	19.46	0.75	11.25	-	22.83	m²	34.08
450 x 450 x 45 mm	17.59	0.75	11.25	-	20.96	m²	32.21
600 x 600 x 45 mm	19.94	0.50	7.50	-	23.30	m²	30.80

Q PAVING/PLANTING/FENCING/SITE FURNITURE Excluding overheads and profit	PC £	Labour hours	Labour £	Plant £	Material £	Unit	Total rate £
Extra labours for laying the a selection of the above sizes to random rectangular pattern	-	0.33	5.00	-	-	m²	5.00
Radial paving for circles							
circle with centre stone and first ring (8 slabs), 450 x 230/560 x 38 mm; diameter 1.54 m (total area 1.86 m²)	63.49	1.50	22.50	-	68.53	nr	91.03
circle with second ring (16 slabs), 450 x 300/460 x 38 mm; diameter 2.48 m (total area 4.83 m²)	163.49	4.00	60.00	-	176.45	nr	236.45
circle with third ring (16 slabs), 450 x 470/625 x 38 mm; diameter 3.42 m (total area 9.18 m²)	297.25	8.00	120.00	-	321.66	nr	441.66
Squaring off set for 2 ring circle							
16 slabs; 2.72 m²	91.74	1.00	15.00	-	91.74	nr	106.74
Pavings; Milner Delvaux Ltd; bedding on 50 mm cement:sand mortar (1:4)							
"Blanc de Bierges" precast concrete pavings							
200 x 200 mm	28.12	0.80	12.00	-	34.28	m²	46.28
400 x 200 mm	26.74	0.74	11.11	-	32.87	m²	43.98
400 x 300 mm	26.96	0.71	10.71	-	33.09	m²	43.81
400 x 400 mm	26.74	0.71	10.71	-	32.87	m²	43.58
600 x 300 mm	25.27	0.61	9.09	-	31.36	m²	40.45
400 x 600 mm	25.27	0.50	7.54	-	31.36	m²	38.90
600 x 600 mm	25.27	0.44	6.64	-	31.36	m²	38.00
300 mm diameter	3.40	0.40	6.00	-	3.78	nr	9.78
Pedestrian deterrent pavings; Marshalls Mono; on prepared base (not included); bedding on 25 mm cement:sand (1:3); cement:sand (1:3) joints							
"Lambeth" pyramidal pavings							
600 x 600 x 75 mm	33.16	0.25	3.75	-	36.53	m²	40.28
"Thaxted" pavings; granite sett appearance							
600 x 600 x 75 mm	38.47	0.33	5.00	-	41.84	m²	46.83
"Contour" undulating plain pavings; grey							
498 x 498 x 125 mm	63.40	0.29	4.29	-	66.77	m²	71.05
Pedestrian deterrent pavings; Townscape Products; on prepared base (not included); bedding on 25 mm cement:sand (1:3); cement:sand (1:3) joints							
"Strata" striated textured slab pavings, giving bonded appearance; grey							
600 x 600 x 60 mm	22.75	0.50	7.50	-	26.12	m²	33.62
"Geoset" raised chamfered studs pavings; grey							
600 x 600 x 60 mm	22.75	0.50	7.50	-	26.12	m²	33.62
"Abbey" square cobble pattern pavings; reinforced							
600 x 600 x 65 mm	21.50	0.80	12.00	-	24.87	m²	36.87
Surface treatments; ICI							
Treating substrate before laying base							
"Casoron G" (residual); at 1kg / 125m²	3.04	0.05	0.75	-	3.34	100m²	4.09
Slip-resistant surfaces							
Resin-base coat 1 - 2 kg/m² with applied grit or aggregate finish; 0.50 - 0.60 kg/m²; on screeded sub-base	-	-	-	-	-	m²	10.34

Q PAVING/PLANTING/FENCING/SITE FURNITURE Excluding overheads and profit	PC £	Labour hours	Labour £	Plant £	Material £	Unit	Total rate £
Q25 SLAB/BRICK/SETT/COBBLE PAVINGS - cont'd							
Edge restraints; to block paving; on prepared base (not included); 200 x 100 x 80 mm; PC £181.00/1000); haunching one side							
Header course							
200 x 100 mm; butt joints	1.81	0.27	4.00	-	4.39	m	8.39
Stretcher course							
200 x 100 mm; butt joints	0.91	0.17	2.50	-	3.35	m	5.85
Concrete paviors; Marshalls Mono; on prepared base (not included); bedding on 50 mm sand; kiln dried sand joints swept in							
"Keyblock" paviors							
200 x 100 x 60 mm; grey	8.18	0.40	6.00	0.09	9.78	m²	15.87
200 x 100 x 60 mm; colours	8.85	0.40	6.00	0.09	10.45	m²	16.54
200 x 100 x 80 mm; grey	9.06	0.44	6.67	0.09	10.66	m²	17.41
200 x 100 x 80 mm; colours	10.50	0.44	6.67	0.09	12.10	m²	18.85
Concrete cobble paviors; Charcon Hard Landscaping; Concrete Products; on prepared base (not included); bedding on 50 mm sand; kiln dried sand joints swept in							
Paviors							
"Woburn" blocks; 100 - 201 x 134 x 80 mm; random sizes	18.75	0.67	10.00	0.09	20.82	m²	30.90
"Woburn" blocks; 100 - 201 x 134 x 80 mm; single size	18.75	0.50	7.50	0.09	20.82	m²	28.40
"Woburn" blocks; 100 - 201 x 134 x 60 mm; random sizes	15.15	0.67	10.00	0.09	17.13	m²	27.21
"Woburn" blocks; 100 - 201 x 134 x 60 mm; single size	15.15	0.50	7.50	0.09	17.11	m²	24.70
Concrete setts; on 25 mm sand; compacted; vibrated; joints filled with sand; natural or coloured; well rammed hardcore base (not included)							
Marshalls Mono; Tegula Paving							
60 mm thick; random sizes	16.85	0.57	8.57	0.09	19.42	m²	28.08
60 mm thick; single size	16.85	0.45	6.82	0.09	19.42	m²	26.33
80 mm thick; random sizes	18.96	0.57	8.57	0.09	21.64	m²	30.30
80 mm thick; single size	18.96	0.45	6.82	0.09	21.64	m²	28.54
Charcon Hard Landscaping; Country setts							
100 mm thick; random sizes	23.10	1.00	15.00	0.09	24.83	m²	39.92
100 mm thick; single size	23.10	0.67	10.00	0.09	25.41	m²	35.49
Concrete setts; Milner Delvaux Ltd; on prepared base (not included); bedding on 30 mm cement:sand (1:3); cement:sand (1:3) joints							
"Blanc de Bierges" concrete setts							
140 x 140 x 80 mm	25.95	2.50	37.50	-	30.98	m²	68.48
210 x 140 x 80 mm	25.95	2.30	34.50	-	30.98	m²	65.48
70 x 70 x 70 mm	27.22	2.75	41.25	-	32.31	m²	73.56

Q PAVING/PLANTING/FENCING/SITE FURNITURE Excluding overheads and profit	PC £	Labour hours	Labour £	Plant £	Material £	Unit	Total rate £
Natural stone, slab or granite paving - General Preamble: provide paving slabs of the specified thickness in random sizes but not less than 25 slabs per 10 m² of surface area, to be laid in parallel courses with joints alternately broken and laid to falls.							
Reconstituted Yorkstone aggregate pavings; Marshalls; "Saxon" on prepared sub-base measured separately; bedding on 25 mm thick cement:sand mortar (1:4); on 50 mm thick sharp sand base Square and rectangular paving in various colours; butt joints straight both ways							
300 x 300 x 35 mm	23.64	0.88	13.20	-	29.60	m²	42.80
600 x 300 x 35 mm	15.51	0.71	10.72	-	21.47	m²	32.19
450 x 450 x 35 mm	10.68	0.71	10.72	-	16.64	m²	27.36
600 x 600 x 35 mm	11.62	0.55	8.25	-	17.58	m²	25.83
450 x 450 x 50 mm	17.59	0.77	11.55	-	23.55	m²	35.10
600 x 600 x 50 mm	15.43	0.60	9.07	-	21.39	m²	30.46
Square and rectangular paving in natural; butt joints straight both ways							
300 x 300 x 35 mm	19.65	0.88	13.20	-	25.60	m²	38.80
600 x 300 x 35 mm	13.66	0.82	12.38	-	19.62	m²	32.00
450 x 450 x 35 mm	9.36	0.71	10.72	-	15.32	m²	26.04
600 x 600 x 35 mm	9.97	0.55	8.25	-	15.93	m²	24.18
450 x 450 x 50 mm	15.34	0.77	11.55	-	21.29	m²	32.84
600 x 600 x 50 mm	13.08	0.66	9.90	-	19.03	m²	28.93
Radial paving for circles; 20 mm joints							
circle with centre stone and first ring (8 slabs), 450 x 230/560 x 35 mm; diameter 1.54 m (total area 1.86 m²)	47.12	1.50	22.50	-	52.45	nr	74.95
circle with second ring (16 slabs), 450 x 300/460 x 35 mm; diameter 2.48 m (total area 4.83 m²)	113.52	4.00	60.00	-	127.28	nr	187.28
circle with third ring (24 slabs), 450 x 310/430 x 35 mm; diameter 3.42 m (total area 9.18 m²)	213.12	8.00	120.00	-	239.43	nr	359.43
stepping stones							
600 dia x 38 mm	7.44	0.05	0.75	-	8.98	ea	9.73
Granite setts; bedding on 25 mm cement:sand (1:3) Natural granite setts; 100 x 100 mm to 125 x 150 mm; x 150 to 250 mm length; riven surface; silver grey							
new	23.50	2.50	37.50	-	30.84	m²	68.33
reclaimed; cleaned	33.50	2.50	37.50	-	41.51	m²	79.00
Natural stone, slate or granite flag pavings; CED Limited; on prepared base (not included); bedding on 25 mm cement:sand (1:3); cement:sand (1:3) joints York stone; riven laid random rectangular							
new slabs; 40 - 60 mm thick	38.00	1.71	25.71	-	44.92	m²	70.63
reclaimed slabs, Cathedral grade; 50 - 75 mm thick	52.00	2.80	42.00	-	60.32	m²	102.32
Donegal quartzite slabs; standard tiles							
595 x 595 x 7 - 13 mm	51.25	2.33	35.00	-	56.93	m²	91.93

Q PAVING/PLANTING/FENCING/SITE FURNITURE Excluding overheads and profit	PC £	Labour hours	Labour £	Plant £	Material £	Unit	Total rate £
Q25 SLAB/BRICK/SETT/COBBLE PAVINGS - cont'd							
Natural Yorkstone, pavings or edgings; Johnsons Wellfield Quarries; sawn 6 sides; 50 mm thick; on prepared base measured separately; bedding on 25 mm cement:sand (1:3); cement:sand (1:3) joints Paving							
laid to random rectangular pattern	45.00	1.71	25.71	-	50.66	m²	76.37
laid to coursed laying pattern; 3 sizes	49.00	1.72	25.86	-	54.41	m²	80.27
Paving, single size							
600 x 600 mm	54.50	0.85	12.75	-	60.19	m²	72.94
600 x 400 mm	54.50	1.00	15.00	-	60.19	m²	75.19
300 x 200 mm	62.50	2.00	30.00	-	68.59	m²	98.59
215 x 102.5 mm	65.50	2.50	37.50	-	71.74	m²	109.24
Paving; cut to template off site; 600 x 600; radius:							
1 00 m	160.00	3.33	50.00	-	162.96	m²	212.96
2.50 m	155.00	2.00	30.00	-	157.96	m²	187.96
5.00 m	150.00	2.00	30.00	-	152.96	m²	182.96
Edgings							
100 mm wide x random lengths	6.25	0.50	7.50	-	6.86	m	14.36
100 mm x 100 mm	6.25	0.50	7.50	-	9.53	m	17.02
100 mm x 200 mm	12.50	0.50	7.50	-	16.09	m	23.59
250 mm wide x random lengths	12.25	0.40	6.00	-	15.83	m	21.83
500 mm wide x random lengths	24.50	0.33	5.00	-	28.69	m	33.69
York stone edgings; 600 mm long x 250 mm wide; cut to radius							
1 00 m to 3.00	57.50	0.50	7.50	-	60.67	m	68.17
3 00 m to 5.00 m	56.25	0.44	6.67	-	59.36	m	66.02
exceeding 5.00 m	55.00	0.40	6.00	-	58.05	m	64.05
Natural Yorkstone, pavings or edgings; Johnsons Wellfield Quarries; sawn 6 sides; 75 mm thick; on prepared base measured separately; bedding on 25 mm cement:sand (1:3); cement:sand (1:3) joints Paving							
laid to random rectangular pattern	52.50	0.95	14.25	-	58.09	m²	72.34
laid to coursed laying pattern; 3 sizes	59.00	0.95	14.25	-	64.91	m²	79.16
Paving; single size							
600 x 600 mm	65.75	0.95	14.25	-	72.00	m²	86.25
600 x 400 mm	65.75	0.95	14.25	-	72.00	m²	86.25
300 x 200 mm	75.50	0.75	11.25	-	82.24	m²	93.49
215 x 102.5 mm	79.25	2.50	37.50	-	86.18	m²	123.68
Paving; cut to template off site; 600 x 600; radius:							
1.00 m	190.00	4.00	60.00	-	192.96	m²	252.96
2 50 m	185.00	2.50	37.50	-	187.96	m²	225.46
5.00 m	180.00	2.50	37.50	-	182.96	m²	220.46

Q PAVING/PLANTING/FENCING/SITE FURNITURE Excluding overheads and profit	PC £	Labour hours	Labour £	Plant £	Material £	Unit	Total rate £
Edgings							
100 mm wide x random lengths	8.00	0.60	9.00	-	8.70	m	17.70
100 mm x 100 mm	7.92	0.60	9.00	-	11.28	m	20.28
100 mm x 200 mm	15.85	0.50	7.50	-	19.60	m	27.10
250 mm wide x random lengths	14.75	0.50	7.50	-	18.45	m	25.95
500 mm wide x random lengths	29.50	0.40	6.00	-	33.94	m	39.94
Edgings; 600 mm long x 250 mm wide; cut to radius							
1.00 m to 3.00	70.00	0.60	9.00	-	73.80	m	82.80
3.00 m to 5.00 m	68.75	0.50	7.50	-	72.48	m	79.98
exceeding 5.00 m	67.50	0.44	6.67	-	71.17	m	77.84

Cobble pavings - General
Preamble: Excavate for and supply and lay hard coarse clinker, consolidated to a thickness of 75 mm. Lay a 50 mm bed of concrete (1:2:4) with small aggregate. Cobbles should be embedded by hand, tight-butted, endwise to a depth of 60% of their length. The tops of cobbles generally shall be about 25 mm above the level of the adjoining paving, except where cobbled areas abut flag paving when the last three rows of cobbles shall be graded down so that the tops of the cobbles shall be level with the flags. Each flint shall be consolidated, using a wooden mallet.
Preamble: Upon completion of each area, a dry grout of rapid-hardening cement:sand (1:2) shall be brushed over the cobbles until the interstices are filled to the level of the adjoining paving. Surplus grout shall then be brushed off and a light, fine spray of water applied over the area. The area shall again be brushed after 24 hours to remove any free grout adhering to exposed faces of cobbles. Cement for grouting shall be rapid-hardening Portland Cement conforming to BS12:1978, and shall be stored on site in a proper manner to avoid deterioration (manufacturer's instructions to be carefully followed).

Cobble pavings
Cobbles; to present a uniform colour in panels; or varied in colour as required

Scottish Beach Cobbles; 200 - 100 mm	24.50	2.00	30.00	-	30.91	m²	60.91
Scottish Beach Cobbles; 100 - 75 mm	22.25	2.50	37.50	-	28.66	m²	66.16
Scottish Beach Cobbles; 75 - 50 mm	21.25	3.33	50.00	-	27.66	m²	77.66

Single layer of kidney flint cobbles; on prepared base (not included); bedding in sand; on 50 mm compacted sand

150 - 100 mm average size	21.00	3.00	45.00	-	25.71	m²	70.71
100 - 75 mm average size	21.00	3.00	45.00	-	25.71	m²	70.71
75 - 50 mm average size	21.75	2.50	37.50	-	26.46	m²	63.96

Q PAVING/PLANTING/FENCING/SITE FURNITURE Excluding overheads and profit	PC £	Labour hours	Labour £	Plant £	Material £	Unit	Total rate £
Q25 SLAB/BRICK/SETT/COBBLE PAVINGS - cont'd							
Concrete cycle blocks; Marshalls Mono; **on 100 mm concrete (1:2:4); on 150 mm** **hardcore; bedding in cement:sand (1:4)** "Metric 4 Square" cycle stand blocks, smooth grey concrete							
496 x 496 x 100 mm	20.72	0.25	3.75	-	23.98	nr	27.73
Concrete cycle blocks; Townscape **Products; on 100 mm concrete (1:2:4);** **on 150 mm hardcore; bedding in** **cement:sand (1:4)** Cycle Blocks							
"Cycle Bloc"; in white concrete	16.94	0.25	3.75	-	22.61	nr	26.36
"Mountain Cycle Bloc"; in white concrete	23.44	0.25	3.75	-	29.11	nr	32.86
Grass concrete - General Preamble: Grass seed should be a perennial ryegrass mixture, with the proportion depending on expected traffic. Hardwearing winter sportsground mixtures are suitable for public areas. Loose gravel, shingle or sand is liable to be kicked out of the blocks; rammed hoggin or other stabilized material should be specified.							
Grass concrete; Grass Concrete Ltd; on **blinded Granular Type 1 sub-base (not** **included)** "Grasscrete" in situ concrete continuously reinforced surfacing; including expansion joints at 10 m centres; soiling; seeding							
ref GC2; 150 mm thick; traffic up to 40.00 tonnes	-	-	-	-	-	m²	29.43
ref GC1; 100 mm thick; traffic up to 13.30 tonnes	-	-	-	-	-	m²	24.98
ref GC3; 76 mm thick; traffic up to 4.30 tonnes	-	-	-	-	-	m²	21.95
Grass concrete; Grass Concrete Ltd; **406 x 406 mm blocks; on 20 mm sharp** **sand; on blinded MOT type 1 sub-base** **(not included); level and to falls only;** **including filling with topsoil; seeding with** **dwarf rye grass at £2.40/kg** Pavings							
ref GB103; 103 mm thick	11.67	0.40	6.00	0.21	13.05	m²	19.26
ref GB83; 83 mm thick	10.64	0.36	5.45	0.21	12.02	m²	17.68
Grass concrete; Charcon Hard **Landscaping; on 25 mm sharp sand;** **including filling with topsoil; seeding with** **dwarf rye grass at £2.40/kg** "Grassgrid" grass/concrete paving blocks							
366 x 274 x 100 mm thick	14.20	0.38	5.63	0.21	15.58	m²	21.42

Q PAVING/PLANTING/FENCING/SITE FURNITURE Excluding overheads and profit	PC £	Labour hours	Labour £	Plant £	Material £	Unit	Total rate £
Grass concrete; Marshalls Mono; on 25 mm sharp sand; including filling with topsoil; seeding with dwarf rye grass at £2.40/kg Concrete grass pavings							
"Grassguard 130"; for light duty applications; (80 mm prepared base not included)	12.01	0.38	5.63	0.21	13.83	m²	19.67
"Grassguard 160"; for medium duty applications; (80 - 150 mm prepared base not included)	14.18	0.46	6.92	0.21	16.06	m²	23.19
"Grassguard 180"; for heavy duty applications; (150 mm prepared base not included)	15.89	0.60	9.00	0.21	17.81	m²	27.02
Precast concrete units; Marshalls Mono; including filling with topsoil; seeding with dwarf rye grass at £2.40/kg "Mono Hexpot" paving; 100 mm thick							
standard; 292 x 339 mm	23.60	0.33	5.00	0.21	27.46	m²	32.67
large; 360 x 415 mm	20.68	0.33	5.00	0.21	24.50	m²	29.71
Full mortar bedding Extra over pavings for bedding on 25 mm cement:sand (1:4); in lieu of spot bedding on sharp sand	-	0.03	0.38	-	1.53	m²	1.90
Q26 SPECIAL SURFACINGS FOR SPORT							
Market Prices of Surfacing Materials Surfacings; Melcourt Industries Ltd							
Playbark 10/50; per 20m³ load	-	-	-	-	46.00	m³	46.00
Playbark 10/50; per 70m³ load	-	-	-	-	33.60	m³	33.60
Playbark 8/25; per 20m³ load	-	-	-	-	43.50	m³	43.50
Playbark 8/25; per 70m³ load	-	-	-	-	31.10	m³	31.10
Adventure Bark 30; per 20m³ load	-	-	-	-	44.90	m³	44.90
Adventure Bark 30; per 70m³ load	-	-	-	-	32.50	m³	32.50
Playchips; per 20m³ load	-	-	-	-	31.40	m³	31.40
Playchips; per 70m³ load	-	-	-	-	19.00	m³	19.00
Kushyfall; per 20m³ load	-	-	-	-	30.90	m³	30.90
Kushyfall; per 70m³ load	-	-	-	-	18.50	m³	18.50
Softfall; per 20m³ load	-	-	-	-	24.60	m³	24.60
Softfall; per 70m³ load	-	-	-	-	12.20	m³	12.20
Playsand; per 10t load	-	-	-	-	78.30	m³	78.30
Playsand; per 20t load	-	-	-	-	66.06	m³	66.06
Hardwood walk chips; per 20m³ load	-	-	-	-	29.85	m³	29.85
Hardwood walk chips; per 70m³ load	-	-	-	-	17.15	m³	17.15
Conifer walk chips; per 20m³ load	-	-	-	-	28.85	m³	28.85
Conifer walk chips; per 70m³ load	-	-	-	-	16.45	m³	16.45
Woodfibre; per 20m³ load	-	-	-	-	28.35	m³	28.35
Woodfibre; per 70m³ load	-	-	-	-	15.95	m³	15.95

Q PAVING/PLANTING/FENCING/SITE FURNITURE Excluding overheads and profit	PC £	Labour hours	Labour £	Plant £	Material £	Unit	Total rate £
Q26 SPECIAL SURFACINGS FOR SPORT - cont'd							
Artificial surfaces and finishes - General Preamble: Manufacturer's should be consulted before specifying artificial playing surfaces as usage, local site conditions and maintenance will affect the choice of material. Advice should also be sought from the Technical Unit for Sport, the appropriate regional office of the Sports Council or the National Playing Fields Association as many surfaces have not yet been fully tested in use so that final opinions on durability, value for money or permanence cannot always be formed. Note that finishes which are inexpensive to lay are often expensive to maintain. Prices for the construction of average facilities are given in the Approximate Estimates section (book only). Some of the following prices include base work whereas others are for a specialist surface only on to a base prepared and costed separately							
Sports areas; Baylis Landscapes Sports tracks, polyurethane rubber surfacing, on bitumen-macadam (not included); (prices for 5500 m² minimum)							
"International"	-	-	-	-	-	m²	42.00
"Club Grade"	-	-	-	-	-	m²	25.00
Sports areas; multi-component polyurethane-rubber surfacing; on bitumen-macadam; finished with polyurethane or acrylic coat; green or red							
"Permaprene"	-	-	-	-	-	m²	58.80
Sports areas; Exclusive Leisure "Tom Graveney Cricketweave" artificial wicket system; including proprietary sub-base (all by specialist sub-contractor)							
28 x 2.74 m	-	-	-	-	-	nr	5500.00
"Tom Graveney Cricketweave" practice batting end							
14 x 2 m	-	-	-	-	-	each	2800.00
"Tom Graveney Cricketweave" practice bowling end							
7 x 2 m	-	-	-	-	-	each	1700.00
Supply and erection of single bay Cricket cage 18.3 x 3.65 x 3.2 m high	-	-	-	-	-	each	1700.00
Tennis courts; Duracourt Hard playing surfaces to SAPCA Code of Practice minimum requirements; laid on 65 mm thick macadam base on 150 mm thick stone foundation, to include lines, nets, posts, brick edging, 2.75 m high fence and perimeter drainage (excluding excavation, levelling or additional foundation). Based on court size 36.6 m x 18.3 m - 670 m²							

Q PAVING/PLANTING/FENCING/SITE FURNITURE Excluding overheads and profit	PC £	Labour hours	Labour £	Plant £	Material £	Unit	Total rate £
"Durapore"; 'All Weather' porous macadam and acrylic colour coating	-	-	-	-	-	m²	29.75
"Tiger Turf TT"; sand filled artificial grass	-	-	-	-	-	m²	43.66
"Tiger Turf Grand Prix"; short pile sand filled artificial grass	-	-	-	-	-	m²	46.87
"DecoColour"; impervious acrylic hardcourt (200 mm foundation)	-	-	-	-	-	m²	42.59
"DecoTurf"; cushioned impervious acrylic tournament surface (200 mm foundation)	-	-	-	-	-	m²	50.93
"Porous Kushion Kourt"; porous cushioned acrylic surface	-	-	-	-	-	m²	52.11
"Porous Klay Kourt"; synthetic clay system	-	-	-	-	-	m²	53.39
"Canada Tenn"; American green clay, fast dry surface (no macadam but including irrigation)	-	-	-	-	-	m²	53.39
Playgrounds; Baylis Landscapes; **"Rubaflex" in-situ playground surfacing,** **porous; on prepared stone/granular base** **Type 1 (not included) and macadam** **base course (not included)** Black							
15 mm thick (0.50 m critical fall height)	-	-	-	-	-	m²	24.00
35 mm thick (1.00 m critical fall height)	-	-	-	-	-	m²	37.00
60 mm thick (1.50 m critical fall height)	-	-	-	-	-	m²	49.00
Playgrounds; Baylis Landscapes; **"Rubaflex" in-situ playground surfacing,** **porous; on prepared stone/granular base** **Type 1 (not included) and macadam** **base course (not included)** Coloured							
15 mm thick (0.50 m critical fall height)	-	-	-	-	-	m²	57.00
35 mm thick (1.00 m critical fall height)	-	-	-	-	-	m²	62.00
60 mm thick (1.50 m critical fall height)	-	-	-	-	-	m²	70.00
Playgrounds; Wicksteed Leisure Safety tiles; on prepared base (not included)							
1000 x 1000 x 60 mm; red or green	-	-	-	-	43.00	m²	43.00
1000 x 1000 x 60 mm; black	-	-	-	-	43.00	m²	43.00
1000 x 1000 x 43 mm; red or green	-	-	-	-	43.00	m²	43.00
1000 x 1000 x 43 mm; black	-	-	-	-	43.00	m²	43.00
Playgrounds; SMP Tiles; on prepared base (not included)							
ref FX22; 500 x 500 x 22 mm tiles; for general use around playground equipment, black	22.00	0.20	3.00	-	27.00	m²	30.00
ref FX100; 500 x 500 x 100 mm tiles; for use around high equipment; black	88.00	0.29	4.29	-	93.00	m²	97.29
Playgrounds; Melcourt Industries "Playbark"; on drainage layer (not included); to BS5696; minimum 300 mm settled depth							
"Playbark 8/25", 8 - 25 mm particles; red/brown	9.33	0.35	5.26	-	10.26	m²	15.52
"Playbark 10/50"; 10 - 50 mm particles; red/brown	11.20	0.35	5.26	-	12.32	m²	17.58

Prices for Measured Works

Q PAVING/PLANTING/FENCING/SITE FURNITURE Excluding overheads and profit	PC £	Labour hours	Labour £	Plant £	Material £	Unit	Total rate £
Q26 SPECIAL SURFACINGS FOR SPORT - cont'd							
Playgrounds; timber edgings Timber edging boards; 50 x 50 x 750 mm timber pegs at 1000 mm centres; excavations and hardcore under edgings (not included)							
50 x 150 mm; hardwood (iroko) edge boards	9.41	0.10	1.50	-	9.41	m	10.91
38 x 150 mm; hardwood (iroko) edge boards	7.45	0.10	1.50	-	7.45	m	8.95
50 x 150 mm; treated softwood edge boards	2.46	0.10	1.50	-	2.46	m	3.96
38 x 150 mm; treated softwood edge boards	2.19	0.10	1.50	-	2.19	m	3.69
Q30 SEEDING/TURFING							
Seeding/turfing - General Preamble: The following market prices generally reflect the manufacturer's recommended retail prices. Trade and bulk discounts are often available on the prices shown. The manufacturer's of these products generally recommend application rates. Note the following rates reflect the average rate for each product.							
Market Prices of pre-seeding materials Scotts UK Professional							
herbicide; "Casoron G"; minimum application rate	-	-	-	-	2.13	100m²	2.13
"Casoron G" as a selective herbicide	-	-	-	-	4.75	100m²	4.75
herbicide; "Casoron G"; maximum application rate	-	-	-	-	8.55	100m²	8.55
"Dextrone X"	-	-	-	-	0.43	100m²	0.43
"Intrepid"	-	-	-	-	0.52	100m²	0.52
"Speedway"	-	-	-	-	0.86	100m²	0.86
Rigby Taylor Ltd							
outfield fertilizer; "Taylor's Pre-Seeding"	-	-	-	-	1.39	100m²	1.39
Boughton Loam							
Screened topsoil 100 mm	-	-	-	-	44.10	m³	44.10
Screened Kettering loam 3 mm	-	-	-	-	62.10	m³	62.10
Top dressing; sand soil mixtures; 90/10 to 50/50	-	-	-	-	49.50	m³	49.50
Market Prices of grass maintenance materials; recommended application rates for most products vary. The rates below reflect an application rate for "normal" application Scotts UK Professional; Moss killers							
mosskiller; "Enforcer"	-	-	-	-	5.30	100m²	5.30
Scotts UK Professional; Fungicides							
fungicide; "Daconil"; turf	-	-	-	-	5.95	100m²	5.95
fungicide; "Greenshield"	-	-	-	-	6.65	100m²	6.65
fungicide; "Nimrod-T"	-	-	-	-	18.11	litre	18.11
Scotts UK Professional; Rodent control							
rodenticide; "Ratak"; per bait station	-	-	-	-	0.32	nr	0.32

Q PAVING/PLANTING/FENCING/SITE FURNITURE Excluding overheads and profit	PC £	Labour hours	Labour £	Plant £	Material £	Unit	Total rate £
Scotts UK Professional; Turf fertilizers							
grass fertilizer; "Longlife Fine Turf - Spring & Summer"	-	-	-	-	3.99	100m²	3.99
grass fertilizer; "Longlife Fine Turf - Autumn Feed"	-	-	-	-	2.29	100m²	2.29
grass fertilizer; "Longlife Fine Turf - Finegreen NK"	-	-	-	-	3.86	100m²	3.86
grass fertilizer; "Longlife Fine Turf - Nitrogen with Iron"	-	-	-	-	2.51	100m²	2.51
grass fertilizer; "Longlife Sportsfield - Spring & Summer"	-	-	-	-	2.78	100m²	2.78
grass fertilizer; "Longlife Sportsfield - Autumn Feed"	-	-	-	-	2.55	100m²	2.55
grass fertilizer; "Longlife Sportsfield - Nitrogen"	-	-	-	-	3.54	100m²	3.54
grass fertilizer; "Longlife Standard - Spring & Summer"	-	-	-	-	1.34	100m²	1.34
grass fertilizer; "Longlife Standard - Autumn Feed"	-	-	-	-	1.63	100m²	1.63
grass fertilizer; "Longlife Standard - Cleanrun"	-	-	-	-	2.77	100m²	2.77
grass fertilizer; "Longlife - Invigorator"	-	-	-	-	2.14	100m²	2.14
grass fertilizer; "Longlife - Renovator"	-	-	-	-	4.88	100m²	4.88
grass fertilizer; slow release; "Sierraform", 18+24+5	-	-	-	-	-	100m²	-
grass fertilizer; slow release; "Sierraform", 18+9+18+Fe+Mn	-	-	-	-	3.36	100m²	3.36
grass fertilizer; slow release; "Sierraform", 15+0+26	-	-	-	-	4.03	100m²	4.03
grass fertilizer; slow release; "Sierraform", 22+5+10	-	-	-	-	2.69	100m²	2.69
grass fertilizer; slow release; "Sierraform", 16+0+15+Fe+Mn	-	-	-	-	3.36	100m²	3.36
grass fertilizer; water soluble; "Sierrasol", 28+5+18+TE	-	-	-	-	5.29	100m²	5.29
grass fertilizer; water soluble; "Sierrasol", 20+5+30+TE	-	-	-	-	4.11	100m²	4.11
grass fertilizer; water soluble; "Sierrasol Extra", 28+5+18+TE+SW	-	-	-	-	8.51	100m²	8.51
grass fertilizer; water soluble; "Sierrasol Extra", 34+0+11+TE+SW	-	-	-	-	8.51	100m²	8.51
grass fertilizer; controlled release; "Sierrablen", 28+5+5+Fe	-	-	-	-	3.80	100m²	3.80
grass fertilizer; controlled release; "Sierrablen", 27+5+5+Fe	-	-	-	-	3.98	100m²	3.98
grass fertilizer; controlled release; "Sierrablen", 15+0+22+Fe	-	-	-	-	3.80	100m²	3.80
grass fertilizer; controlled release; "Sierrablen Fine", 38+0+0	-	-	-	-	1.49	100m²	1.49
grass fertilizer; controlled release; "Sierrablen Fine", 25+5+12	-	-	-	-	2.40	100m²	2.40
grass fertilizer; controlled release; "Sierrablen Fine", 21+0+20	-	-	-	-	2.48	100m²	2.48
grass fertilizer; controlled release; "Sierrablen Fine", 15+0+29	-	-	-	-	2.48	100m²	2.48
grass fertilizer; controlled release; "Sierrablen Mini", 22+5+10	-	-	-	-	7.57	100m²	7.57
grass fertilizer; controlled release; "Sierrablen Mini", 0+0+45	-	-	-	-	3.77	100m²	3.77

Q PAVING/PLANTING/FENCING/SITE FURNITURE Excluding overheads and profit	PC £	Labour hours	Labour £	Plant £	Material £	Unit	Total rate £
Q30 SEEDING/TURFING - cont'd							
Market Prices of grass maintenance materials; recommended application rates for most products vary. The rates below reflect an application rate for "normal" application - cont'd							
Scotts UK Professional; Turf chemicals							
growth regulator; "Shortcut"	-	-	-	-	1.43	100m²	1.43
spray indicator; "Turf Mark"; per litre of water	-	-	-	-	0.33	litre	0.33
grass fertilizer; "Greenmaster Turf Tonic"	-	-	-	-	1.93	100m²	1.93
grass fertilizer; "Greenmaster Spring & Summer"	-	-	-	-	2.76	100m²	2.76
grass fertilizer; "Greenmaster Zero Phosphate"	-	-	-	-	2.76	100m²	2.76
grass fertilizer; "Greenmaster Super N"	-	-	-	-	4.76	100m²	4.76
grass fertilizer; "Greenmaster Mosskiller"	-	-	-	-	3.07	100m²	3.07
grass fertilizer; "Greenmaster Extra"	-	-	-	-	3.49	100m²	3.49
grass fertilizer; "Greenmaster Autumn"	-	-	-	-	2.97	100m²	2.97
grass fertilizer; "Greenmaster Fine Turf"	-	-	-	-	3.33	100m²	3.33
grass fertilizer; "Greenmaster Double K"	-	-	-	-	2.97	100m²	2.97
grass fertilizer; "Greenmaster NK"	-	-	-	-	2.97	100m²	2.97
grass fertilizer; "Greenmaster Liquid N Plus"	-	-	-	-	2.23	100m²	2.23
grass & soil fertilizer; "Greenmaster Liquid Seafeed"	-	-	-	-	0.61	100m²	0.61
outfield turf fertilizer; "Sportsmaster PS3"	-	-	-	-	1.74	100m²	1.74
outfield turf fertilizer; "Sportsmaster PS4"	-	-	-	-	1.74	100m²	1.74
outfield turf fertilizer; "Sportsmaster PS5"	-	-	-	-	1.74	100m²	1.74
outfield turf fertilizer; "Sportsmaster Fairway"	-	-	-	-	2.79	100m²	2.79
outfield turf fertilizer; "Sportsmaster Municipality"	-	-	-	-	3.64	100m²	3.64
wetting and penetrating agent; "Aquamaster"; applied quarterly	-	-	-	-	2.14	100m²	2.14
turf herbicide; "Tritox", 5 litre	-	-	-	-	2.43	100m²	2.43
turf herbicide; "Tritox", 20 litre	-	-	-	-	2.27	100m²	2.27
turf fungicide and wormcast control; "Turfclear"; 5 litre	-	-	-	-	1.00	100m²	1.00
turf fungicide and wormcast control; "Turfclear"; 800 ml	-	-	-	-	23.52	800 ml	23.52
turf fungicide and wormcast control; "Turfclear WDG"; sachet	-	-	-	-	31.47	100m²	31.47
fine turf top dressing; "Topmaster TD4"	-	-	-	-	5.74	80 lt	5.74
"TPMC"	-	-	-	-	3.84	80 lt	3.84
Rigby Taylor Ltd; 35 g/m²							
grass fertilizer; "Mascot Microfine 20-0-15" + 2% Mg	-	-	-	-	6.80	100m²	6.80
grass fertilizer; "Mascot Microfine 15-5-15" +2% Mg	-	-	-	-	4.87	100m²	4.87
grass fertilizer; "Mascot Microfine 18-0-0" + 6% Fe	-	-	-	-	5.94	100m²	5.94
grass fertilizer; "Mascot Microfine 14-4-7" + 2% Mg	-	-	-	-	4.34	100m²	4.34
grass fertilizer; "Mascot Microfine 8-0-0" + 4% Fe + 2% Mg	-	-	-	-	3.12	100m²	3.12
grass fertilizer; "Mascot Microfine 4-0-8" + 6% Fe + 2% Mg	-	-	-	-	4.82	100m²	4.82

Q PAVING/PLANTING/FENCING/SITE FURNITURE Excluding overheads and profit	PC £	Labour hours	Labour £	Plant £	Material £	Unit	Total rate £
grass fertilizer; "Taylor's Organic SS2"	-	-	-	-	2.79	100m²	2.79
grass fertilizer; "Taylor's Mini-Granular 11-2-7"	-	-	-	-	2.96	100m²	2.96
grass fertilizer/weedkiller; "Taylor's Weed & Feed"	-	-	-	-	2.56	100m²	2.56
grass fertilizer/mosskiller; "Taylor's Lawn Sand"	-	-	-	-	1.39	100m²	1.39
outfield fertilizer; "Taylor's Spring/Summer Outfield 9-7-7"	-	-	-	-	2.03	100m²	2.03
outfield fertilizer; "Taylor's Spring/Summer Outfield 7-7-7"	-	-	-	-	1.68	100m²	1.68
outfield fertilizer; "Taylor's Spring/Summer Outfield 20-10-10"	-	-	-	-	2.37	100m²	2.37
outfield fertilizer; "Taylor's Autumn/Winter Outfield"	-	-	-	-	1.97	100m²	1.97
outfield fertilizer; "Taylor's Spring/Summer Sportsfield"	-	-	-	-	2.37	100m²	2.37
liquid fertilizer; "Vitax 50/50 Fine Turf"; 56 ml/100m²	-	-	-	-	13.58	100m²	13.58
liquid fertilizer; "Vitax 50/50 Fine Turf Special"; 56 ml/100m²	-	-	-	-	11.62	100m²	11.62
liquid fertilizer; "Vitax 50/50 Autumn & Winter"; 56 ml/100m²	-	-	-	-	12.46	100m²	12.46
Cultivation							
Treating soil with "Paraquat-Diquat" weedkiller at rate of 5 litre/ha; PC £22.20 per litre; in accordance with manufacturer's instructions; including all safety precautions							
by machine	-	-	-	0.24	1.11	100 m²	1.35
by hand	-	0.13	2.00	-	1.11	100 m²	3.11
Ripping up subsoil; using approved subsoiling machine; minimum depth 250 mm below topsoil; at 1.20 m centres; in							
gravel or sandy clay	-	-	-	2.31	-	100m²	2.31
soil compacted by machines	-	-	-	2.70	-	100m²	2.70
clay	-	-	-	2.89	-	100m²	2.89
chalk or other soft rock	-	-	-	5.78	-	100m²	5.78
Extra for subsoiling at 1 m centres	-	-	-	0.58	-	100m²	0.58
Breaking up existing ground; using pedestrian operated tine cultivator or rotavator							
100 mm deep	-	0.22	3.30	2.13	-	100m²	5.43
150 mm deep	-	0.28	4.13	2.67	-	100m²	6.79
200 mm deep	-	0.37	5.50	3.56	-	100m²	9.05
As above but in heavy clay or wet soils							
100 mm deep	-	0.44	6.60	4.27	-	100m²	10.87
150 mm deep	-	0.66	9.90	6.40	-	100m²	16.30
200 mm deep	-	0.82	12.38	8.00	-	100m²	20.38
Breaking up existing ground; using tractor drawn tine cultivator or rotavator							
single pass							
100 mm deep	-	-	-	0.52	-	100m²	0.52
150 mm deep	-	-	-	0.65	-	100m²	0.65
200 mm deep	-	-	-	0.86	-	100m²	0.86
600 mm deep	-	-	-	2.58	-	100m²	2.58
Cultivating ploughed ground; using disc, drag, or chain harrow							
4 passes	-	-	-	3.10	-	100m²	3.10

Q PAVING/PLANTING/FENCING/SITE FURNITURE Excluding overheads and profit	PC £	Labour hours	Labour £	Plant £	Material £	Unit	Total rate £
Q30 SEEDING/TURFING - cont'd							
Breaking up existing ground; using tractor drawn tine cultivator or rotavator - cont'd							
Rolling cultivated ground lightly; using self-propelled agricultural roller	-	0.06	0.83	0.58	-	100m²	1.41
Importing and storing selected and approved topsoil; to BS 3882; from source not exceeding 13 km from site; inclusive of settlement							
small quantities, less than 15 m³	26.00	-	-	-	31.20	m³	31.20
over 15 m³	11.66	-	-	-	13.99	m³	13.99
Spreading and lightly consolidating approved topsoil (imported or from spoil heaps); in layers not exceeding 150 mm; travel distance from spoil heaps not exceeding 100 m; by machine (imported topsoil not included)							
minimum depth 100 mm	-	1.55	23.25	35.53	-	100m²	58.78
minimum depth 150 mm	-	2.33	35.00	53.47	-	100m²	88.47
minimum depth 300 mm	-	4.67	70.00	106.94	-	100m²	176.94
minimum depth 450 mm	-	6.99	104.85	160.24	-	100m²	265.09
Spreading and lightly consolidating approved topsoil (imported or from spoil heaps); in layers not exceeding 150 mm; travel distance from spoil heaps not exceeding 100 m; by hand (imported topsoil not included)							
minimum depth 100 mm	-	20.00	300.06	-	-	100m²	300.06
minimum depth 150 mm	-	30.01	450.09	-	-	100m²	450.09
minimum depth 300 mm	-	60.01	900.18	-	-	100m²	900.18
minimum depth 450 mm	-	90.02	1350.27	-	-	100m²	1350.27
Extra over for spreading topsoil to slopes 15 - 30 degrees by machine or hand	-	-	-	-	-	10%	-
Extra over for spreading topsoil to slopes over 30 degrees by machine or hand	-	-	-	-	-	25%	-
Extra over for spreading topsoil from spoil heaps travel exceeding 100 m; by machine							
100 - 150 m	-	0.01	0.18	0.10	-	m³	0.28
150 - 200 m	-	0.02	0.28	0.15	-	m³	0.43
200 - 300 m	-	0.03	0.42	0.22	-	m³	0.64
Extra over spreading topsoil for travel exceeding 100 m; by hand							
100 m	-	0.83	12.50	-	-	m³	12.50
200 m	-	1.67	25.00	-	-	m³	25.00
300 m	-	2.50	37.50	-	-	m³	37.50
Evenly grading; to general surfaces to bring to finished levels							
by machine (tractor mounted rotavator)	-	-	-	0.03	-	m²	0.03
by pedestrian operated rotavator	-	-	0.06	0.04	-	m²	0.10
by hand	-	0.01	0.15	-	-	m²	0.15
Extra over grading for slopes 15 - 30 degrees by machine or hand	-	-	-	-	-	10%	-
Extra over grading for slopes over 30 degrees by machine or hand	-	-	-	-	-	25%	-

Q PAVING/PLANTING/FENCING/SITE FURNITURE Excluding overheads and profit	PC £	Labour hours	Labour £	Plant £	Material £	Unit	Total rate £
Apply screened topdressing to grass surfaces. Spread using Tru-Lute							
Sand soil mixes 90/10 to 50/50	0.07	-	0.03	0.03	0.07	m²	0.13
Spread only existing cultivated soil to final levels using Tru-Lute							
Cultivated soil	-	-	0.03	0.03	-	m²	0.06
Clearing stones; disposing off site; to distance not exceeding 13 km							
by hand; stones not exceeding 50 mm in any direction; loading to skip 5.35 m³	-	0.01	0.15	0.02	-	m²	0.17
by mechanical stone rake; stones not exceeding 50 mm in any direction; loading to 15 m³ truck							
by mechanical loader	-	-	0.03	0.07	-	m²	0.10
Lightly cultivating; weeding; to fallow areas; disposing debris off site; to distance not exceeding 13 km							
by hand	-	0.01	0.21	-	0.08	m²	0.29
Surface applications; soil additives; pre-seeding; material delivered to a maximum of 25 m from area of application; applied; by machine							
Soil conditioners; to cultivated ground; ground limestone; PC £9.30/tonne; including turning in							
0.25 kg/m² = 2.50 tonnes/ha	0.23	-	-	4.03	0.23	100m²	4.26
0.50 kg/m² = 5.00 tonnes/ha	0.47	-	-	4.03	0.47	100m²	4.49
0.75 kg/m² = 7.50 tonnes/ha	0.70	-	-	4.03	0.70	100m²	4.73
1.00 kg/m² = 10.00 tonnes/ha	0.93	-	-	4.03	0.93	100m²	4.96
Soil conditioners; to cultivated ground; medium bark; based on deliveries of 15 m³ loads; PC £29.95/m³; including turning in							
1 m³ per 40 m² = 25 mm thick	0.75	-	-	0.09	0.75	m²	0.84
1 m³ per 20 m² = 50 mm thick	-	-	-	0.14	1.50	m²	1.64
1 m³ per 13.33 m² = 75 mm thick	-	-	-	0.19	2.25	m²	2.44
1 m³ per 10 m² = 100 mm thick	3.00	-	-	0.24	3.00	m²	3.24
Soil conditioners; to cultivated ground; peat loose 55 m³ loads; PC £27.00/m³; including turning in							
1 m³ per 40 m² = 25 mm thick	0.68	-	-	0.09	0.68	m²	0.77
1 m³ per 20 m² = 50 mm thick	1.35	-	-	0.14	1.35	m²	1.49
1 m³ per 13.33 m² = 75 mm thick	2.02	-	-	0.19	2.02	m²	2.22
1 m³ per 10 m² = 100 mm thick	2.70	-	-	0.24	2.70	m²	2.94
Soil conditioners; to cultivated ground; mushroom compost; delivered in 20 m³ loads; PC £8.00/m³; including turning in							
1 m³ per 40 m² = 25 mm thick	0.20	0.02	0.25	-	0.20	m²	0.45
1 m³ per 40 m² = 25 mm thick	0.20	0.02	0.25	-	0.20	m²	0.45
1 m³ per 20 m² = 50 mm thick	0.40	0.03	0.46	-	0.40	m²	0.86
1 m³ per 13.33 m² = 75 mm thick	0.60	0.04	0.60	-	0.60	m²	1.20
1 m³ per 10 m² = 100 mm thick	0.80	0.05	0.75	-	0.80	m²	1.55
Soil conditioners; to cultivated ground; mushroom compost; delivered in 55 m³ loads; PC £4.00/m³; including turning in							
1 m³ per 40 m² = 25 mm thick	0.10	0.02	0.25	-	0.10	m²	0.35
1 m³ per 20 m² = 50 mm thick	0.20	0.03	0.46	-	0.20	m²	0.66
1 m³ per 13.33 m² = 75 mm thick	0.30	0.04	0.60	-	0.30	m²	0.90
1 m³ per 10 m² = 100 mm thick	0.40	0.05	0.75	-	0.40	m²	1.15

Q PAVING/PLANTING/FENCING/SITE FURNITURE Excluding overheads and profit	PC £	Labour hours	Labour £	Plant £	Material £	Unit	Total rate £
Q30 SEEDING/TURFING - cont'd							
Surface applications and soil additives; pre-seeding; material delivered to a maximum of 25 m from area of application; applied; by hand							
Soil conditioners; to cultivated ground; ground limestone; PC £9.30/tonne; including turning in							
0.25 kg/m² = 2.50 tonnes/ha	0.23	1.20	18.00	-	0.23	100m²	18.23
0.50 kg/m² = 5.00 tonnes/ha	0.47	1.33	20.00	-	0.47	100m²	20.46
0.75 kg/m² = 7.50 tonnes/ha	0.70	1.50	22.50	-	0.70	100m²	23.20
1.00 kg/m² = 10.00 tonnes/ha	0.93	1.71	25.71	-	0.93	100m²	26.64
Soil conditioners; to cultivated ground; medium bark; based on deliveries of 15 m³ loads; PC £29.95 / m³; including turning in							
1 m³ per 40 m² = 25 mm thick	0.75	0.02	0.33	-	0.75	m²	1.08
1 m³ per 20 m² = 50 mm thick	1.50	0.04	0.67	-	1.50	m²	2.16
1 m³ per 13.33 m² = 75 mm thick	2.25	0.07	1.00	-	2.25	m²	3.25
1 m³ per 10 m² = 100 mm thick	3.00	0.08	1.20	-	3.00	m²	4.20
Soil conditioners; to cultivated ground; peat loose 55 m³ loads; PC £27.00 /m³; including turning in							
1 m³ per 40 m² = 25 mm thick	0.68	0.02	0.33	-	0.68	m²	1.01
1 m³ per 20 m² = 50 mm thick	1.35	0.04	0.67	-	1.35	m²	2.02
1 m³ per 13.33 m² = 75 mm thick	2.02	0.07	1.00	-	2.02	m²	3.02
1 m³ per 10 m² = 100 mm thick	2.70	0.08	1.20	-	2.70	m²	3.90
Soil conditioners; to cultivated ground; mushroom compost; delivered in 20 m³ loads; PC £10.50/m³; including turning in							
1 m³ per 40 m² = 25 mm thick	0.20	0.02	0.33	-	0.20	m²	0.53
1 m³ per 20 m² = 50 mm thick	0.40	0.04	0.67	-	0.40	m²	1.07
1 m³ per 13.33 m² = 75 mm thick	0.60	0.07	1.00	-	0.60	m²	1.60
1 m³ per 10 m² = 100 mm thick	0.80	0.08	1.20	-	0.80	m²	2.00
Soil conditioners; to cultivated ground; mushroom compost; delivered in 55 m³ loads; PC £4.00/m³; including turning in							
1 m³ per 40 m² = 25 mm thick	0.10	0.02	0.33	-	0.10	m²	0.43
1 m³ per 20 m² = 50 mm thick	0.20	0.04	0.67	-	0.20	m²	0.87
1 m³ per 13.33 m² = 75 mm thick	0.30	0.07	1.00	-	0.30	m²	1.30
1 m³ per 10 m² = 100 mm thick	0.40	0.08	1.20	-	0.40	m²	1.60
Preparation of seedbeds - General Preamble: For preliminary operations see "Cultivation" section.							
Preparation of seedbeds; soil preparation Lifting selected and approved topsoil from spoil heaps; passing through 6 mm screen; removing debris	-	0.08	1.25	2.94	0.01	m³	4.21
Topsoil; supply only; PC £11.66 /m³; allowing for 20% settlement							
25 mm	0.29	-	-	-	0.35	m²	0.35
50 mm	0.58	-	-	-	0.70	m²	0.70
100 mm	1.17	-	-	-	1.40	m²	1.40
150 mm	1.75	-	-	-	2.10	m²	2.10
200 mm	2.33	-	-	-	2.80	m²	2.80

Q PAVING/PLANTING/FENCING/SITE FURNITURE Excluding overheads and profit	PC £	Labour hours	Labour £	Plant £	Material £	Unit	Total rate £
250 mm	2 92	-	-	-	3.50	m²	3.50
300 mm	3.50	-	-	-	4.20	m²	4.20
400 mm	4.66	-	-	-	5.60	m²	5.60
450 mm	5.25	-	-	-	6.30	m²	6.30
Spreading topsoil to form seedbeds (topsoil not included); by machine							
25 mm deep	-	-	0.04	0.10	-	m²	0.13
50 mm deep	-	-	0.05	0.13	-	m²	0.18
75 mm deep	-	-	0.06	0.15	-	m²	0.20
100 mm deep	-	0.01	0.08	0.19	-	m²	0.27
150 mm deep	-	0.01	0.11	0.29	-	m²	0.40
Spreading only topsoil to form seedbeds (topsoil not included); by hand							
25 mm deep	-	0.03	0.38	-	-	m²	0.38
50 mm deep	-	0.03	0.50	-	-	m²	0.50
75 mm deep	-	0.04	0.64	-	-	m²	0.64
100 mm deep	-	0.05	0.75	-	-	m²	0.75
150 mm deep	-	0.08	1.13	-	-	m²	1.13
Bringing existing topsoil to a fine tilth for seeding; by raking or harrowing; stones not to exceed 6 mm; by machine	-	-	0.06	0.04	-	m²	0.10
Bringing existing topsoil to a fine tilth for seeding; by raking or harrowing; stones not to exceed 6 mm; by hand	-	0.01	0.13	-	-	m²	0.13
Preparation of seedbeds; soil treatments							
For the following topsoil improvement and seeding operations add or subtract the following amounts for every £0.10 difference in the material cost price							
35 g/m²	-	-	-	-	0.35	100m²	0.35
50 g/m²	-	-	-	-	0.50	100m²	0.50
70 g/m²	-	-	-	-	0.70	100m²	0.70
100 g/m²	-	-	-	-	1.00	100m²	1.00
125 kg/ ha	-	-	-	-	12.50	ha	12.50
150 kg/ ha	-	-	-	-	15.00	ha	15.00
175 kg/ ha	-	-	-	-	17.50	ha	17.50
200 kg/ ha	-	-	-	-	20.00	ha	20.00
225 kg/ ha	-	-	-	-	22.50	ha	22.50
250 kg/ ha	-	-	-	-	25.00	ha	25.00
300 kg/ ha	-	-	-	-	30.00	ha	30.00
350 kg/ ha	-	-	-	-	35.00	ha	35.00
400 kg/ ha	-	-	-	-	40.00	ha	40.00
500 kg/ ha	-	-	-	-	50.00	ha	50.00
700 kg/ ha	-	-	-	-	70.00	ha	70.00
1000 kg/ ha	-	-	-	-	100.00	ha	100.00
1250 kg/ ha	-	-	-	-	125.00	ha	125.00
Pre-seeding fertilizers (6:9:6); PC £0.33/kg; to seedbeds; by machine							
35 g/m²	1.16	-	-	0.18	1.16	100m²	1.34
50 g/m²	1.65	-	-	0.18	1.65	100m²	1.84
70 g/m²	2 32	-	-	0.18	2.32	100m²	2.50
100 g/m²	3.31	-	-	0.18	3.31	100m²	3.49
125 g/m²	4.13	-	-	0.18	4.13	100m²	4.32

Q PAVING/PLANTING/FENCING/SITE FURNITURE Excluding overheads and profit	PC £	Labour hours	Labour £	Plant £	Material £	Unit	Total rate £
Q30 SEEDING/TURFING - cont'd							
Preparation of seedbeds; soil treatments - cont'd							
125 kg/ha	41.35	-	-	18.05	41.35	ha	59.40
250 kg/ha	82.70	-	-	18.05	82.70	ha	100.75
300 kg/ha	99.24	-	-	18.05	99.24	ha	117.29
350 kg/ha	115.78	-	-	18.05	115.78	ha	133.83
400 kg/ha	132.32	-	-	18.05	132.32	ha	150.37
500 kg/ha	165.40	-	-	28.89	165.40	ha	194.29
700 kg/ha	231.56	-	-	28.89	231.56	ha	260.45
1250 kg/ha	413.50	-	-	48.14	413.50	ha	461.64
Pre-seeding fertilizers (6:9:6); PC £0.33/kg; to seedbeds; by hand							
35 g/m²	1.16	0.17	2.50	-	1.16	100m²	3.66
50 g/m²	1.65	0.17	2.50	-	1.65	100m²	4.15
70 g/m²	2.32	0.17	2.50	-	2.32	100m²	4.82
100 g/m²	3.31	0.20	3.00	-	3.31	100m²	6.31
125 g/m²	4.13	0.20	3.00	-	4.13	100m²	7.13
Pre-seeding pesticides; to seedbeds; in accordance with manufacturer's instructions; by machine							
35 g/m²	16.45	-	-	0.18	16.45	100m²	16.63
50 g/m²	23.50	-	-	0.18	23.50	100m²	23.68
70 g/m²	32.90	-	-	0.18	32.90	100m²	33.08
100 g/m²	47.00	-	-	0.18	47.00	100m²	47.18
125 g/m²	58.75	-	-	0.18	58.75	100m²	58.93
125 kg/ha	587.50	-	-	18.05	587.50	ha	605.55
250 kg/ha	1175.00	-	-	18.05	1175.00	ha	1193.05
300 kg/ha	1410.00	-	-	18.05	1410.00	ha	1428.05
350 kg/ha	1645.00	-	-	18.05	1645.00	ha	1663.05
400 kg/ha	1880.00	-	-	18.05	1880.00	ha	1898.05
500 kg/ha	2350.00	-	-	28.89	2350.00	ha	2378.89
700 kg/ha	3290.00	-	-	28.89	3290.00	ha	3318.89
1250 kg/ha	5875.00	-	-	18.05	5875.00	ha	5893.05
Pre-seeding pesticides; to seedbeds; in accordance with manufacturer's instructions; by hand							
35 g/m²	16.45	0.17	2.50	-	16.45	100m²	18.95
50 g/m²	23.50	0.17	2.50	-	23.50	100m²	26.00
70 g/m²	32.90	0.17	2.50	-	32.90	100m²	35.40
100 g/m²	47.00	0.20	3.00	-	47.00	100m²	50.00
125 g/m²	58.75	0.20	3.00	-	58.75	100m²	61.75
Pre-emergent weedkillers; in accordance with manufacturer's instructions; including all safety precautions; by machine (coverage 0.80 ha /hr)							
15 ml	1.26	-	-	0.18	1.26	100m²	1.45
20 ml	1.69	-	-	0.18	1.69	100m²	1.87
25 ml	2.11	-	-	0.18	2.11	100m²	2.29
50 ml	4.21	-	-	0.18	4.21	100m²	4.40

Q PAVING/PLANTING/FENCING/SITE FURNITURE Excluding overheads and profit	PC £	Labour hours	Labour £	Plant £	Material £	Unit	Total rate £
Pre-emergent liquid applied weedkillers; in accordance with manufacturer's instructions; including all safety precautions; by hand							
10 ml	0.04	0.50	7.50	-	0.04	100m²	7.54
12 ml	0.05	0.50	7.50	-	0.05	100m²	7.55
15 ml	0.06	0.50	7.50	-	0.06	100m²	7.56
20 ml	0.08	0.50	7.50	-	0.08	100m²	7.58
25 ml	0.10	0.50	7.50	-	0.10	100m²	7.60
30 ml	0.12	0.50	7.50	-	0.12	100m²	7.62
35 ml	0.14	0.50	7.50	-	0.14	100m²	7.64
40 ml	0.16	0.50	7.50	-	0.16	100m²	7.66
45 ml	0.18	0.50	7.50	-	0.18	100m²	7.68
50 ml	0.20	0.50	7.50	-	0.20	100m²	7.70
Lime granules; on seedbeds; by machine							
35 g/m²	0.73	-	-	0.18	0.73	100m²	0.92
50 g/m²	1.05	-	-	0.18	1.05	100m²	1.23
70 g/m²	1.47	-	-	0.18	1.47	100m²	1.65
100 g/m²	2.10	-	-	0.18	2.10	100m²	2.28
125 g/m²	2.63	-	-	0.18	2.63	100m²	2.81
125 kg/ha	26.25	-	-	18.05	26.25	ha	44.30
250 kg/ha	52.50	-	-	18.05	52.50	ha	70.55
300 kg/ha	63.00	-	-	18.05	63.00	ha	81.05
350 kg/ha	73.50	-	-	18.05	73.50	ha	91.55
400 kg/ha	84.00	-	-	18.05	84.00	ha	102.05
500 kg/ha	105.00	-	-	28.89	105.00	ha	133.89
700 kg/ha	147.00	-	-	72.22	147.00	ha	219.22
1250 kg/ha	262.50	-	-	72.22	262.50	ha	334.72
Lime granules, on seedbeds; by hand							
35 g/m²	0.73	0.17	2.50	-	0.73	100m²	3.24
50 g/m²	1.05	0.17	2.50	-	1.05	100m²	3.55
70 g/m²	1.47	0.17	2.50	-	1.47	100m²	3.97
100 g/m²	2.10	0.17	2.50	-	2.10	100m²	4.60
125 g/m²	2.63	0.17	2.50	-	2.63	100m²	5.13

Seeding grass areas - General
Preamble: The British Standard recommendations for seed and seeding of grass areas are contained in BS 4428: 1989. The prices given in this section are based on compliance with the standard.

Market Prices of grass seed
Preamble: The prices shown are for supply only at one number 25 kg bag purchase price unless otherwise stated. Rates shown are based on the manufacturer's maximum recommendation for each seed type. Trade and bulk discounts are often available on the prices shown for quantities of more than one bag.
Bowling greens; fine lawns; ornamental turf; croquet lawns

	PC £	Labour hours	Labour £	Plant £	Material £	Unit	Total rate £
"British Seed Houses"; ref A1Greens; 35g/m² (New Formulation)	-	-	-	-	20.58	100m²	20.58
"Johnsons Seeds"; ref JL 21; 34 - 50g/m²	-	-	-	-	12.80	100m²	12.80
"Johnsons Seeds"; ref J1; 34 - 50g/m²	-	-	-	-	20.70	100m²	20.70
"Johnsons Seeds"; ref J2; 34 - 50g/m²	-	-	-	-	19.70	100m²	19.70

Q PAVING/PLANTING/FENCING/SITE FURNITURE Excluding overheads and profit	PC £	Labour hours	Labour £	Plant £	Material £	Unit	Total rate £
Q30 SEEDING/TURFING - cont'd							
Market Prices of grass seed - cont'd							
"Perryfields"; ref Pro 20; 35 - 50g/m²	-	-	-	-	14.00	100m²	14.00
"Perryfields"; ref Pro 10; 35 - 50g/m²	-	-	-	-	18.40	100m²	18.40
Tennis courts; cricket squares							
"British Seed Houses"; ref A2 Lawn Tennis; 35g/m²	-	-	-	-	12.39	100m²	12.39
"British Seed Houses"; ref A5 Cricket Square; 35g/m²	-	-	-	-	14.70	100m²	14.70
"Johnsons Seeds"; ref JL 21; 34 - 50g/m²	-	-	-	-	12.80	100m²	12.80
"Johnsons Seeds"; ref J2; 34 - 50g/m²	-	-	-	-	19.70	100m²	19.70
"Johnsons Seeds"; ref Taskmaster; 18 - 25g/m²	-	-	-	-	7.15	100m²	7.15
"Perryfields"; ref Pro 35; 35g/m²	-	-	-	-	7.91	100m²	7.91
Amenity grassed areas; general purpose lawns							
"British Seed Houses"; ref A3 Landscape; 25 - 50g/m²	-	-	-	-	15.25	100m²	15.25
"Johnsons Seeds"; ref J5; 18 - 25g/m²	-	-	-	-	6.95	100m²	6.95
"Johnsons Seeds"; ref Taskmaster; 18 - 25gm²	-	-	-	-	7.15	100m²	7.15
"Perryfields"; ref Pro 50; 25 - 35g/m²	-	-	-	-	10.36	100m²	10.36
"Perryfields"; ref Pro 120; 25 - 35g/m²	-	-	-	-	11.55	100m²	11.55
Conservation; country parks; slopes and banks							
"British Seed Houses"; ref A4 Parkland; 17 - 35g/m²	-	-	-	-	11.47	100m²	11.47
"British Seed Houses"; ref A16 Country Park; 8 - 19g/m²	-	-	-	-	7.82	100m²	7.82
"British Seed Houses"; ref A17 (legume and clover); 2g/m²	-	-	-	-	13.00	100m²	13.00
"Johnsons Seeds"; ref J8 Slowgrow; 8 - 30g/m²	-	-	-	-	8.94	100m²	8.94
"Perryfields"; ref Pro 30; 25 - 35g/m²; up to 50g/m² for banks	-	-	-	-	8.61	100m²	8.61
"Perryfields"; ref Pro 30; up to 50g/m² for banks	-	-	-	-	12.30	100m²	12.30
Shaded areas							
"British Seed Houses"; ref A6 Supra Shade; 50g/m²	-	-	-	-	22.99	100m²	22.99
"Johnsons Seeds"; ref JL 21; 34 - 50g/m²	-	-	-	-	12.80	100m²	12.80
"Johnsons Seeds"; ref J1; 34 - 50g/m²	-	-	-	-	20.70	100m²	20.70
"Perryfields"; ref Pro 60; 35 - 50g/m²	-	-	-	-	14.80	100m²	14.80
Sports pitches; rugby; soccer pitches							
"British Seed Houses"; ref A7 Sports Ground; 20g/m²	-	-	-	-	6.30	100m²	6.30
"Johnson Seeds"; ref JL 24; 18 - 30g/m²	-	-	-	-	6.36	100m²	6.36
"Johnson Seeds"; ref J7; 18 - 25g/m²	-	-	-	-	5.75	100m²	5.75
"Johnson Seeds"; ref Sportsmaster (winter wearing); 18 - 30g/m²	-	-	-	-	7.02	100m²	7.02
"Perryfields"; ref Pro 70; 15 - 35g/m²	-	-	-	-	8.12	100m²	8.12
"Perryfields"; ref Pro 75; 15 - 35g/m²	-	-	-	-	9.59	100m²	9.59
"Perryfields"; ref Pro 80; 17 - 35g/m²	-	-	-	-	6.30	100m²	6.30
Outfields							
"British Seed Houses"; ref A7 Sports Ground; 20g/m²	-	-	-	-	6.30	100m²	6.30
"British Seed Houses"; ref A9 Universal; 17 - 35g/m²	-	-	-	-	9.05	100m²	9.05
"Johnsons Seeds"; ref J4; 12 - 25g/m²	-	-	-	-	6.55	100m²	6.55
"Perryfields"; ref Pro 40; 35g/m²	-	-	-	-	11.62	100m²	11.62
"Perryfields"; ref Pro 70; 15 - 35g/m²	-	-	-	-	8.12	100m²	8.12

Q PAVING/PLANTING/FENCING/SITE FURNITURE Excluding overheads and profit	PC £	Labour hours	Labour £	Plant £	Material £	Unit	Total rate £
Hockey pitches							
"Johnsons Seeds"; ref J4; 12 - 25g/m²	-	-	-	-	6.55	100m²	6.55
"Perryfields"; ref Pro 70; 15 - 35g/m²	-	-	-	-	8.12	100m²	8.12
Parks							
"British Seed Houses"; ref A7; Sportsground 20g/m²	-	-	-	-	6.30	100m²	6.30
"British Seed Houses"; ref A9 Universal; 17 - 35g/m²	-	-	-	-	9.05	100m²	9.05
"Johnson Seeds"; ref JL 24; 18 - 30g/m²	-	-	-	-	6.36	100m²	6.36
"Perryfields"; ref Pro 120; 25 - 35g/m²	-	-	-	-	11.55	100m²	11.55
Informal playing fields							
"Johnsons seeds"; ref J5; 18 - 25g/m²	-	-	-	-	6.95	100m²	6.95
"Perryfields"; ref Pro 45; 35g/m²	-	-	-	-	10.78	100m²	10.78
Caravan sites							
"British Seed Houses"; ref A9 Universal; 17 - 35g/m²	-	-	-	-	9.05	100m²	9.05
"Johnson Seeds"; ref JL 22; 18 - 30g/m²	-	-	-	-	7.62	100m²	7.62
Sports pitch re-seeding and repair							
"British Seed Houses"; ref A8 Supra Sport; 20 - 35g/m²	-	-	-	-	13.34	100m²	13.34
"British Seed Houses"; ref A20 Ryesport; 20 - 35g/m²	-	-	-	-	8.71	100m²	8.71
"Johnson Seeds"; ref JL 24; 18 - 30g/m²	-	-	-	-	6.36	100m²	6.36
"Johnson Seeds"; ref Sportsmaster (winter wearing); 18 - 30g/m²	-	-	-	-	7.02	100m²	7.02
"Perryfields"; ref Pro 80; 17 - 35g/m²	-	-	-	-	6.30	100m²	6.30
"Perryfields"; ref Pro 81; 17 - 35g/m²	-	-	-	-	7.28	100m²	7.28
Racecourses; gallops; polo grounds; horse rides							
"British Seed Houses"; ref A14; 25 - 30g/m²	-	-	-	-	8.05	100m²	8.05
"Johnsons Seeds"; ref Taskmaster; 18 - 25g/m²	-	-	-	-	7.15	100m²	7.15
"Perryfields"; ref Pro 65; 17 - 35g/m²	-	-	-	-	11.06	100m²	11.06
Motorway and road verges							
"British Seed Houses"; ref A18; 6 - 15g/m²	-	-	-	-	4.39	100m²	4.39
"Johnsons Seeds"; ref J8; 18 - 30g/m²	-	-	-	-	8.94	100m²	8.94
"Perryfields"; ref Pro 120; 25 - 35g/m²	-	-	-	-	11 55	100m²	11.55
"Perryfields"; ref Pro 85 (DOT Official Mix); 10g/m²	-	-	-	-	2.68	100m²	2.68
Golf courses; tees							
"British Seed Houses"; ref A10, 35 - 50g/m²	-	-	-	-	19.80	100m²	19.80
"Johnsons Seeds"; ref J3; 18 - 30g/m²	-	-	-	-	9.30	100m²	9.30
"Johnsons Seeds"; ref Taskmaster; 18 - 25g/m²	-	-	-	-	7.15	100m²	7.15
"Perryfields"; ref Pro 40; 35g/m²	-	-	-	-	11.62	100m²	11.62
"Perryfields"; ref Pro 45; 35g/m²	-	-	-	-	10.78	100m²	10.78
Golf courses; greens							
"British Seed Houses"; ref A11; 35g/m²	-	-	-	-	21.84	100m²	21.84
"British Seed Houses"; ref A13 (for overseeding); 8g/m²	-	-	-	-	11.80	100m²	11.80
"Johnsons Seeds"; ref J1; 34 - 50g/m²	-	-	-	-	20.70	100m²	20.70
"Johnsons Seeds"; ref Greensmaster (for overseeding); 8g/m²	-	-	-	-	8.00	100m²	8.00
"Perryfields"; ref Pro 5; 35 - 50g/m²	-	-	-	-	12.70	100m²	12.70
"Perryfields"; ref Pro 55; 20g/m²	-	-	-	-	11.60	100m²	11.60
Golf courses; fairways							
"British Seed Houses"; ref A12; 15 - 25g/m²	-	-	-	-	9.49	100m²	9.49
"Johnsons Seeds"; ref J3; 18 - 30g/m²	-	-	-	-	9.30	100m²	9.30
"Johnsons Seeds"; ref J5; 18 - 30g/m²	-	-	-	-	8.34	100m²	8.34

Q PAVING/PLANTING/FENCING/SITE FURNITURE Excluding overheads and profit	PC £	Labour hours	Labour £	Plant £	Material £	Unit	Total rate £
Q30 SEEDING/TURFING - cont'd							
Market Prices of grass seed - cont'd							
"Perryfields"; ref Pro 40; 35g/m²	-	-	-	-	11.62	100m²	11.62
"Perryfields"; ref Pro 45; 35g/m²	-	-	-	-	10.78	100m²	10.78
Golf courses; roughs							
"Johnsons Seeds"; ref J8; 18 - 30g/m²	-	-	-	-	8.94	100m²	8.94
"Perryfields"; ref Pro 25; 17 - 35g/m²	-	-	-	-	10.15	100m²	10.15
Waste land; spoil heaps; quarries							
"British Seed Houses"; ref A15 (ley type; can be grazed); 15 - 20g/m²	-	-	-	-	7.25	100m²	7.25
"Johnsons Seeds"; ref JR33 (acid to alkaline); 5 - 10g/m²	-	-	-	-	3.40	100m²	3.40
"Johnsons Seeds"; ref JR35 (acid to neutral); 5 - 10g/m²	-	-	-	-	8.80	100m²	8.80
"Johnsons Seeds"; ref JR39 (fertility builders); 5 - 10g/m²	-	-	-	-	4.40	100m²	4.40
"Perryfields"; ref Pro 95; 12 - 35g/m²	-	-	-	-	16.38	100m²	16.38
"Perryfields"; ref Pro 105 Pioneer Mix (fertility builders); 5g/m²	-	-	-	-	2.27	100m²	2.27
Low maintenance; housing estates; amenity grassed areas							
"British Seed Houses"; ref A19; 25 - 35g/m²	-	-	-	-	9.70	100m²	9.70
"British Seed Houses"; ref A22; 25 - 35g/m²	-	-	-	-	14.70	100m²	14.70
"Perryfields"; ref Pro 120; 25 - 35g/m²	-	-	-	-	11.55	100m²	11.55
Saline coastal; road side areas							
"British Seed Houses"; ref A21; 15 - 20g/m²	-	-	-	-	7.92	100m²	7.92
"Johnsons Seeds"; ref JR37; 10 - 20g/m²	-	-	-	-	7.20	100m²	7.20
"Perryfields"; ref Pro 90; 15 - 35g/m²	-	-	-	-	11.90	100m²	11.90
Turf production							
"British Seed Houses"; ref A23 (fine turf); 160 kg/ha	-	-	-	-	432.00	ha	432.00
"British Seed Houses"; ref A24 (hardwearing turf); 185 kg/ha	-	-	-	-	452.21	ha	452.21
Market Prices of wild flora seed mixtures							
Acid soils							
"British Seed Houses"; ref WF1 (annual flowering); 1.00 - 2.00 g/m²	-	-	-	-	16.43	100m²	16.43
Neutral soils							
"British Seed Houses"; ref WF3; 0.50 - 1.00 g/m²	-	-	-	-	8.21	100m²	8.21
Market Prices of wild flora and grass seed mixtures							
General purpose							
"Johnsons Seeds"; ref JF40 Simplicity; 3 g/m²	-	-	-	-	7.35	100m²	7.35
"Johnsons Seeds"; ref JF41 General Purpose; 3 g/m²	-	-	-	-	7.35	100m²	7.35
"Perryfields"; ref Pro Flora 8 Old English Country Meadow Mix; 5 g/m²	-	-	-	-	15.00	100m²	15.00
"Perryfields"; ref Pro Flora 9; 5 g/m²	-	-	-	-	15.00	100m²	15.00

Q PAVING/PLANTING/FENCING/SITE FURNITURE Excluding overheads and profit	PC £	Labour hours	Labour £	Plant £	Material £	Unit	Total rate £
Acid soils							
"British Seed Houses"; ref WFG2 (meadow mixture), 5 g/m²	-	-	-	-	22.25	100m²	22.25
"Johnsons Seeds"; ref JF47W (Wet); 3 g/m²	-	-	-	-	14.85	100m²	14.85
"Johnsons Seeds"; ref JF47D (Dry); 3 g/m²	-	-	-	-	14.85	100m²	14.85
"Perryfields"; ref Pro Flora 2; 5 g/m²	-	-	-	-	11.25	100m²	11.25
Neutral soils							
"British Seed Houses"; ref WFG4; 5 g/m²	-	-	-	-	24.45	100m²	24.45
"British Seed Houses"; ref WFG13 (Scotland); 5 g/m²	-	-	-	-	21.55	100m²	21.55
"Perryfields"; ref Pro Flora 3; 5 g/m²	-	-	-	-	15.75	100m²	15.75
Calcareous soils							
"British Seed Houses"; ref WFG5; 5 g/m²	-	-	-	-	22.80	100m²	22.80
"Johnsons Seeds"; ref JF44 (calcareous loam); 3 g/m²	-	-	-	-	10.80	100m²	10.80
"Johnsons Seeds"; ref JF45; 3 g/m²	-	-	-	-	10.80	100m²	10.80
"Perryfields"; ref Pro Flora 4; 5 g/m²	-	-	-	-	13.50	100m²	13.50
Heavy clay soils							
"British Seed Houses"; ref WFG6; 5 g/m²	-	-	-	-	27.50	100m²	27.50
"British Seed Houses"; ref WFG12 (Ireland); 5 g/m²	-	-	-	-	21.63	100m²	21.63
"Perryfields"; ref Pro Flora 5; 5 g/m²	-	-	-	-	24.75	100m²	24.75
Sandy soils							
"British Seed Houses"; ref WFG7; 5 g/m²	-	-	-	-	32.25	100m²	32.25
"British Seed Houses"; ref WFG11 (Ireland); 5 g/m²	-	-	-	-	22.75	100m²	22.75
"British Seed Houses"; ref WFG14 (Scotland); 5 g/m²	-	-	-	-	25.00	100m²	25.00
"Perryfields"; ref Pro Flora 6; 5 g/m²	-	-	-	-	29.25	100m²	29.25
Loam and alluvial soils							
"Johnsons Seeds"; ref JF42; 3 g/m²	-	-	-	-	10.80	100m²	10.80
Shaded areas							
"British Seed Houses"; ref WFG8; 5 g/m²	-	-	-	-	24.00	100m²	24.00
"Johnsons Seeds"; ref JF46; 3 g/m²	-	-	-	-	10.80	100m²	10.80
"Perryfields"; ref Pro Flora 7; 5 g/m²	-	-	-	-	33.75	100m²	33.75
Educational							
"British Seed Houses"; ref WFG15; 5 g/m²	-	-	-	-	33.75	100m²	33.75
Wetlands							
"British Seed Houses"; ref WFG9; 5 g/m²	-	-	-	-	29.75	100m²	29.75
"Johnsons Seeds"; ref JF43; 3 g/m²	-	-	-	-	14.85	100m²	14.85
"Perryfields"; ref Pro Flora 5; 5 g/m²	-	-	-	-	24.75	100m²	24.75
Scrub and moorland							
"British Seed Houses"; ref WFG10; 5 g/m²	-	-	-	-	32.80	100m²	32.80
"Johnsons Seeds"; ref JF45 (chalk downland or quarries); 3 g/m²	-	-	-	-	10.80	100m²	10.80
Hedgerow							
"Johnsons Seeds"; ref JF46; 3 g/m²	-	-	-	-	10.80	100m²	10.80
"Perryfields"; ref Pro Flora 7; 5 g/m²	-	-	-	-	33.75	100m²	33.75
Vacant sites							
"Johnsons seeds"; ref JF49S (short); 30 g/m²	-	-	-	-	108.00	100m²	108.00
"Johnsons Seeds"; ref JF49T (tall); 30 g/m²	-	-	-	-	54.00	100m²	54.00
"Perryfields"; ref Pro Flora 1; 5 g/m²	-	-	-	-	30.00	100m²	30.00

Q PAVING/PLANTING/FENCING/SITE FURNITURE Excluding overheads and profit	PC £	Labour hours	Labour £	Plant £	Material £	Unit	Total rate £
Q30 SEEDING/TURFING - cont'd							
Seeding							
Seeding labours only in two operations; by machine (for seed prices see above)							
35 g/m²	-	-	-	0.49	-	100m²	0.49
Grass seed; spreading in two operations; PC £2.46/kg; (for changes in material prices please refer to table above); by machine							
35 g/m²	-	-	-	0.49	8.61	100m²	9.10
50 g/m²	-	-	-	0.49	12.30	100m²	12.79
70 g/m²	-	-	-	0.49	17.22	100m²	17.71
100 g/m²	-	-	-	0.49	24.60	100m²	25.09
125 kg/ha	-	-	-	49.03	307.50	ha	356.53
150 kg/ha	-	-	-	49.03	369.00	ha	418.03
200 kg/ha	-	-	-	49.03	492.00	ha	541.03
250 kg/ha	-	-	-	49.03	615.00	ha	664.03
300 kg/ha	-	-	-	49.03	738.00	ha	787.03
350 kg/ha	-	-	-	49.03	861.00	ha	910.03
400 kg/ha	-	-	-	49.03	984.00	ha	1033.03
500 kg/ha	-	-	-	49.03	1230.00	ha	1279.03
700 kg/ha	-	-	-	49.03	1722.00	ha	1771.03
1400 kg/ha	-	-	-	49.03	3444.00	ha	3493.03
Extra over seeding by machine for slopes over 30 degrees (allowing for the actual area but measured in plan)							
35 g/m²	-	-	-	0.07	1.29	100m²	1.36
50 g/m²	-	-	-	0.07	1.84	100m²	1.92
70 g/m²	-	-	-	0.07	2.58	100m²	2.66
100 g/m²	-	-	-	0.07	3.69	100m²	3.76
125 kg/ha	-	-	-	7.36	46.13	ha	53.48
150 kg/ha	-	-	-	7.36	55.35	ha	62.71
200 kg/ha	-	-	-	7.36	73.80	ha	81.16
250 kg/ha	-	-	-	7.36	92.25	ha	99.61
300 kg/ha	-	-	-	7.36	110.70	ha	118.06
350 kg/ha	-	-	-	7.36	129.15	ha	136.51
400 kg/ha	-	-	-	7.36	147.60	ha	154.96
500 kg/ha	-	-	-	7.36	184.50	ha	191.86
700 kg/ha	-	-	-	7.36	258.30	ha	265.66
1400 kg/ha	-	-	-	7.36	516.60	ha	523.96
Seeding labours only in two operations; by machine (for seed prices see above)							
35 g/m²	-	0.17	2.50	-	-	100m²	2.50
Grass seed; spreading in two operations; PC £2.46/kg; (for changes in material prices please refer to table above); by hand							
35 g/m²	-	0.17	2.50	-	8.61	100m²	11.11
50 g/m²	-	0.17	2.50	-	12.30	100m²	14.80
70 g/m²	-	0.17	2.50	-	17.22	100m²	19.72
100 g/m²	-	0.20	3.00	-	24.60	100m²	27.60
125 g/m²	-	0.20	3.00	-	30.75	100m²	33.75

Q PAVING/PLANTING/FENCING/SITE FURNITURE Excluding overheads and profit	PC £	Labour hours	Labour £	Plant £	Material £	Unit	Total rate £
Extra over seeding by hand for slopes over 30 degrees (allowing for the actual area but measured in plan)							
35 g/m²	1.28	-	0.06	-	1.28	100m²	1.33
50 g/m²	1.84	-	0.06	-	1.84	100m²	1.90
70 g/m²	2.58	-	0.06	-	2.58	100m²	2.64
100 g/m²	3.69	-	0.06	-	3.69	100m²	3.75
125 g/m²	4.60	-	0.06	-	4.60	100m²	4.66
Harrowing seeded areas; light chain harrow	-	-	-	0.09	-	100m²	0.09
Raking over seeded areas							
by mechanical stone rake	-	-	-	2.10	-	100m²	2.10
by hand	-	0.80	12.00	-	-	100m²	12.00
Rolling seeded areas; light roller							
by tractor drawn roller	-	-	-	0.52	-	100m²	0.52
by pedestrian operated mechanical roller	-	0.08	1.25	0.48	-	100m²	1.73
by hand drawn roller	-	0.17	2.50	-	-	100m²	2.50
Extra over harrowing, raking or rolling seeded areas for slopes over 30 degrees; by machine or hand	-	-	-	-	-	25%	-
Turf edging; to seeded areas; 300 mm wide	1.25	0.05	0.71	-	1.25	m²	1.96

Hydroseeding - General
Preamble: Hydroseeding may be specified where conventional seeding is difficult or impossible, ie. where cultivation is hard on certain subsoils, on steep banks, in dangerous situations, where soil erosion by rain or wind is likely. Hydroseeding consists of high pressure spraying a mixture of water, seed, fertilizer, mulches and binders from a special machine which can spray up to 60 m horizontally and 25 m vertically. Hose extensions allow seeding to be carried out up to 200 m depending on how thick a mix is used.
Preamble: Seed can be grass, wild flowers, certain shrubs and trees. Seed mixes should be site specific as all ecological requirements for selected species must be present on site if hydroseeding is to be successful. Special soil conditioners, mulches, fertilizers and binders can be used in the hydromix and each contractor has a preference to use his own. Specification and prices vary widely according to the location, hydromix required, and size of contract. The following are price ranges.

Hydroseeding; Keller Comtec
"Terra-soil" process (used where grass is to be established on topsoil or easy grow sites); typical solids application rate 250 g/m²

from:	-	-	-	-	-	m²	0.15
to:	-	-	-	-	-	m²	0.25

Q PAVING/PLANTING/FENCING/SITE FURNITURE Excluding overheads and profit	PC £	Labour hours	Labour £	Plant £	Material £	Unit	Total rate £
Q30 SEEDING/TURFING - cont'd							
Hydroseeding; Keller Comtec - cont'd							
"Terraseal" process (used on most subsoils including sand, clays, silts, shale and rock; the hydramulch includes materials which will kick start the generation of topsoil); typical solids application rate 400 - 500 g/m²							
from	-	-	-	-	-	m²	0.30
to:	-	-	-	-	-	m²	0.70
"Terrablanket" process (used on steep slopes to assist in soil stability and erosion control; can be used in place of straw or erosion geotextiles and is more effective in establishing grass cover); typical solids application rate 600 - 700 g/m²							
from:	-	-	-	-	-	m²	0.90
to:	-	-	-	-	-	m²	1.50
"Terraseal Runway" process (used on airport runway aprons to stabilise soil against jet blast); typical solids application rate 300 g/m²							
from:	-	-	-	-	-	m²	0.30
to:	-	-	-	-	-	m²	0.40
Liquid Sod; Turf Management							
Spray on grass system of grass plantlets fertilizer, bio-degradable mulch carrier, root enhancer and water							
to prepared ground	-	-	-	-	-	m²	1.00
Preparation of turf beds							
Rolling turf to be lifted; lifting by hand or mechanical turf stripper; stacks to be not more than 1 m high							
cutting only preparing to lift; pedestrian turf cutter	-	0.75	11.25	9.25	-	100m²	20.50
lifting and stacking; by hand	-	8.33	125.00	-	-	100m²	125.00
Rolling up; moving to stacks							
distance not exceeding 100 m	-	2.50	37.50	-	-	100m²	37.50
extra over rolling and moving turf to stacks to transport per additional 100 m	-	0.83	12.50	-	-	100m²	12.50
Lifting selected and approved topsoil from spoil heaps							
passing through 6 mm screen; removing debris	-	0.17	2.50	5.89	-	m³	8.39
Extra over lifting topsoil and passing through screen for imported topsoil; plus 20% allowance for settlement	11.66	-	-	-	13.99	m³	13.99
Topsoil; PC £11.66 /m³; plus 20% allowance for settlement							
25 mm deep	-	-	-	-	0.35	m²	0.35
50 mm deep	-	-	-	-	0.70	m²	0.70
100 mm deep	-	-	-	-	1.40	m²	1.40
150 mm deep	-	-	-	-	2.10	m²	2.10
200 mm deep	-	-	-	-	2.80	m²	2.80
250 mm deep	-	-	-	-	3.50	m²	3.50
300 mm deep	-	-	-	-	4.20	m²	4.20
400 mm deep	-	-	-	-	5.60	m²	5.60
450 mm deep	-	-	-	-	6.30	m²	6.30

Q PAVING/PLANTING/FENCING/SITE FURNITURE Excluding overheads and profit	PC £	Labour hours	Labour £	Plant £	Material £	Unit	Total rate £
Spreading topsoil to form turfbeds (topsoil not included); by machine							
25 mm deep	-	-	0.04	0.10	-	m²	0.13
50 mm deep	-	-	0.05	0.13	-	m²	0.18
75 mm deep	-	-	0.06	0.15	-	m²	0.20
100 mm deep	-	0.01	0.08	0.19	-	m²	0.27
150 mm deep	-	0.01	0.11	0.29	-	m²	0.40
Spreading topsoil to form turfbeds (topsoil not included); by hand							
25 mm deep	-	0.03	0.38	-	-	m²	0.38
50 mm deep	-	0.03	0.50	-	-	m²	0.50
75 mm deep	-	0.04	0.64	-	-	m²	0.64
100 mm deep	-	0.05	0.75	-	-	m²	0.75
150 mm deep	-	0.08	1.13	-	-	m²	1.13
Bringing existing topsoil to a fine tilth for turfing by raking or harrowing; stones not to exceed 6 mm; by machine	-	-	0.06	0.04	-	m²	0.10
Bringing existing topsoil to a fine tilth for turfing by raking or harrowing; stones not to exceed 6 mm; by hand	-	0.01	0.14	-	-	m²	0.14
Preparation of cricket squares							
Excavating cricket square size 20 x 22 m to a depth of 150 mm; screening and returning topsoil; mixing with imported marl or clay loam at the rate of 1 kg/m²; bringing to accurate levels; hand	-	-	-	-	-	nr	4514.37
Turfing							
Turfing; laying only; to stretcher bond; butt joints; including providing and working from barrow plank runs where necessary to surfaces not exceeding 30 degrees from horizontal; average size of turves 0.50 yd2 (0.418 m²); 2.39 turves per m²							
specially selected lawn turves from previously lifted stockpile	-	0.08	1.13	-	-	m²	1.13
cultivated lawn turves; to large open areas	-	0.06	0.87	-	-	m²	0.87
cultivated lawn turves; to domestic or garden areas	-	0.08	1.17	-	-	m²	1.17
meadow turf; to amenity areas	-	0.05	0.75	-	-	m²	0.75
road verge quality turf	-	0.04	0.60	-	-	m²	0.60
Industrially grown turf; PC prices listed represent the general range of industrial turf prices for sportsfields and amenity purposes. Prices will vary with quantity and site location "Rolawn"							
ref Standard; golf tees and tennis courts, fine domestic lawns, ornamental landscaping	2.83	0.08	1.13	-	2.83	m²	3.96
ref RB Medallion; sports fields, domestic lawns, general landscape	2.03	0.07	1.05	-	2.03	m²	3.08

Q PAVING/PLANTING/FENCING/SITE FURNITURE Excluding overheads and profit	PC £	Labour hours	Labour £	Plant £	Material £	Unit	Total rate £
Q30 SEEDING/TURFING - cont'd							
Industrially grown turf; PC prices listed represent the general range of industrial turf prices for sportsfields and amenity purposes. Prices will vary with quantity and site location - cont'd							
Greenkeeper Ltd;							
"Turfliner" light reinforced turf for low maintenance, low fertility areas to steep slopes	2.30	0.10	1.43	-	2.30	m²	3.73
"Turf Carpet" grown on sterile soil on polythene	2.25	0.10	1.43	-	2.25	m²	3.68
"Tensar Turf Mat" reinforced turf for embankments; laid to embankments 3.3 m² turves	5.95	0.10	1.50	-	5.95	m²	7.45
"Tensar Turf Mat" reinforced turf for embankments; laid to embankments 16.5 m² rolls	5.95	0.04	0.53	1.07	5.95	m²	7.54
"Inturf"							
ref SS5; fine texture, special golf and bowling greens	1.75	0.10	1.43	-	1.75	m²	3.18
ref SS1; fine lawns, golf greens, bowling greens	2.15	0.10	1.43	-	2.15	m²	3.58
ref SS3; hockey grounds, polo, medium wearing areas	1.67	0.05	0.81	-	1.67	m²	2.48
ref SS4; hardwearing fine turf, low maintenance areas	1.76	0.04	0.63	-	1.76	m²	2.39
ref SS2; football grounds, parks, hardwearing areas	1.67	0.05	0.79	-	1.67	m²	2.46
Millennium Turf; for difficult site conditions	1.67	0.08	1.20	-	1.67	m²	2.87
Custom Grown Turf; specific seed mixtures to suit soil or site conditions	4.75	0.08	1.20	-	4.75	m²	5.95
Turf Tiles; instant repairs to goal mouths and playing fields providing instant stable surface	-	-	-	-	-	m²	26.50
Turf Modules; for trade shows exhibitions and events	-	-	-	-	-	m²	120.00
Reinforced turf; Netlon Advanced Turf; Netlon Ltd; Blended mesh fibre elements incorporated into root zone; Rootzone spread and levelled over cultivated, prepared and reduced and levelled ground (not included); compacted with light roller.							
"ATS 300" with "R300" topping, seed and fertilizer; 100 thick							
100 -199 m²	16.15	0.03	0.38	0.60	16.15	m²	17.13
200 - 299 m²	14.65	0.03	0.38	0.60	14.65	m²	15.63
300 - 499 m²	14.10	0.03	0.38	0.60	14.10	m²	15.08
over 500 m²	13.70	0.03	0.38	0.60	13.70	m²	14.68
"ATS 300" with "R300" topping, seed and fertilizer; 150 thick							
100 -199 m²	21.10	0.03	0.50	0.80	21.10	m²	22.40
200 - 299 m²	19.60	0.03	0.50	0.80	19.60	m²	20.90
300 - 499 m²	18.95	0.03	0.50	0.80	18.95	m²	20.25
over 500 m²	18.50	0.03	0.50	0.80	18.50	m²	19.80

Q PAVING/PLANTING/FENCING/SITE FURNITURE Excluding overheads and profit	PC £	Labour hours	Labour £	Plant £	Material £	Unit	Total rate £
"ATS 300" with washed turf and fertilizer ("R300" topping not included) 100 thick							
100 -199 m²	19.70	0.03	0.50	0.80	19.70	m²	21.00
200 - 499 m²	18.10	0.03	0.50	0.80	18.10	m²	19.40
over 500 m²	17.15	0.03	0.50	0.80	17.15	m²	18.45
"ATS 300" with washed turf and fertilizer ("R300" topping not included) 150 thick							
100 -199 m²	24.55	0.04	0.60	0.90	24.55	m²	26.05
200 - 499 m²	22.95	0.04	0.60	0.90	22.95	m²	24.45
over 500 m²	21.85	0.04	0.60	0.90	21.85	m²	23.35
Firming turves with wooden beater	-	0.01	0.15	-	-	m²	0.15
Rolling turfed areas; light roller							
by tractor with turf tyres and roller	-	-	-	0.52	-	100m²	0.52
by pedestrian operated mechanical roller	-	0.08	1.25	0.48	-	100m²	1.73
by hand drawn roller	-	0.17	2.50	-	-	100m²	2.50
Dressing with finely sifted topsoil; brushing into joints	0.01	0.05	0.75	-	0.01	m²	0.76
Turfing; laying only to slopes over 30 degrees; to diagonal bond (measured as plan area - add 15% to these rates for the incline area of 30 degree slopes)	-	0.12	1.80	-	-	m²	1.80
Extra over laying turfing for pegging down turves wooden or galvanized wire pegs; 200 mm long; 2 pegs per 0.50 yd2	0.19	0.01	0.20	-	0.19	m²	0.39
Initial cutting; to turfed areas 20 mm high; using pedestrian guided power driven cylinder mower; including boxing off cuttings (stone picking and rolling not included)	-	0.18	2.70	0.25	-	100m²	2.95
Maintenance operations (Note: the following rates apply to aftercare maintenance executed as part of a landscaping contract only)							
Repairing damaged grass areas scraping out; removing slurry; from ruts and holes							
average 100 mm deep	-	0.13	2.00	-	-	m²	2.00
100 mm topsoil	-	0.13	2.00	-	1.40	m²	3.40
Repairing damaged grass areas; sowing grass seed to match existing or as specified; to individually prepared worn patches							
35 g/m²	0.09	0.01	0.15	-	0.10	m²	0.25
50 g/m²	0.12	0.01	0.15	-	0.14	m²	0.29
Sweeping leaves; disposing off site; motorized vacuum sweeper or rotary brush sweeper areas of maximum 2500 m² with occasional large tree and established boundary planting; 5.35 m³ (1 skip of material to be removed)	-	0.40	6.00	2.54	-	100m²	8.54

Prices for Measured Works

Q PAVING/PLANTING/FENCING/SITE FURNITURE Excluding overheads and profit	PC £	Labour hours	Labour £	Plant £	Material £	Unit	Total rate £
Q30 SEEDING/TURFING - cont'd							
Maintenance operations (Note: the following rates apply to aftercare maintenance executed as part of a landscaping contract only) - cont'd							
Leaf Clearance; clearing grassed area of leaves and other extraneous debris							
using equipment towed by tractor							
large grassed areas with perimeters of mature trees such as sports fields and amenity areas	-	0.01	0.19	0.27	-	100m²	0.46
large grassed areas containing ornamental trees and shrub beds	-	0.03	0.38	0.38	-	100m²	0.76
using pedestrian operated mechanical equipment and blowers							
grassed areas with perimeters of mature trees such as sports fields and amenity areas	-	0.04	0.60	0.09	-	100m²	0.69
grassed areas containing ornamental trees and shrub beds	-	0.10	1.50	0.24	-	100m²	1.74
verges	-	0.07	1.00	0.16	-	100m²	1.16
by hand							
grassed areas with perimeters of mature trees such as sports fields and amenity areas	-	0.05	0.75	0.09	-	100m²	0.84
grassed areas containing ornamental trees and shrub beds	-	0.08	1.25	0.15	-	100m²	1.40
verges	-	1.00	15.00	1.84	-	100m²	16.84
removal of arisings							
areas with perimeters of mature trees	-	0.01	0.08	0.08	0.20	100m²	0.36
areas containing ornamental trees and shrub beds	-	0.02	0.24	0.31	0.50	100m²	1.05
Cutting grass to specified height; per cut							
multi unit gang mower	-	0.59	8.82	15.89	-	ha	24.71
ride-on triple cylinder mower	-	0.01	0.21	0.12	-	100m²	0.33
ride-on triple rotary mower	-	0.01	0.21	-	-	100m²	0.21
pedestrian mower	-	0.18	2.70	0.59	-	100m²	3.29
Cutting grass to banks; per cut							
side arm cutter bar mower	-	0.02	0.35	0.29	-	100m²	0.64
Cutting rough grass; per cut power flail or scythe cutter	-	0.04	0.53	-	-	100m²	0.53
Extra over cutting grass for slopes not exceeding 30 degrees	-	-	-	-	-	10%	-
Extra over cutting grass for slopes exceeding 30 degrees	-	-	-	-	-	40%	-
Cutting fine sward							
pedestrian operated seven-blade cylinder lawn mower	-	0.14	2.10	0.19	-	100m²	2.29
Extra over cutting fine sward for boxing off cuttings							
pedestrian mower	-	0.03	0.42	0.04	-	100m²	0.46
Cutting areas of rough grass							
scythe	-	1.00	15.00	-	-	100m²	15.00
sickle	-	2.00	30.00	-	-	100m²	30.00
petrol operated strimmer	-	0.30	4.50	0.41	-	100m²	4.91
Cutting areas of rough grass which contain trees or whips							
petrol operated strimmer	-	0.40	6.00	0.54	-	100m²	6.54
Extra over cutting rough grass for on site raking up and dumping	-	0.33	5.00	-	-	100m²	5.00

Q PAVING/PLANTING/FENCING/SITE FURNITURE Excluding overheads and profit	PC £	Labour hours	Labour £	Plant £	Material £	Unit	Total rate £
Trimming edge of grass areas; edging tool							
with petrol powered strimmer	-	0.13	2.00	0.18	-	100m	2.18
by hand	-	0.67	10.00	-	-	100m	10.00
Marking out pitches using approved line marking compound; including initial setting out and marking							
discus, hammer, javelin or shot putt area	2.79	2.00	30.00	-	2.79	nr	32.79
cricket square	1.86	2.00	30.00	-	1.86	nr	31.86
cricket boundary	6.51	8.00	120.00	-	6.51	nr	126.51
grass tennis court	2.79	4.00	60.00	-	2.79	nr	62.79
hockey pitch	9.30	8.00	120.00	-	9.30	nr	129.30
football pitch	9.30	8.00	120.00	-	9.30	nr	129.30
rugby pitch	9.30	8.00	120.00	-	9.30	nr	129.30
eight lane running track; 400 m	18.60	16.00	240.00	-	18.60	nr	258.60
Re-marking out pitches using approved line marking compound							
discus, hammer, javelin or shot putt area	1.86	0.50	7.50	-	1.86	nr	9.36
cricket square	1.86	0.50	7.50	-	1.86	nr	9.36
grass tennis court	6.51	1.00	15.00	-	6.51	nr	21.51
hockey pitch	6.51	1.00	15.00	-	6.51	nr	21.51
football pitch	6.51	1.00	15.00	-	6.51	nr	21.51
rugby pitch	6.51	1.00	15.00	-	6.51	nr	21.51
eight lane running track; 400 m	18.60	2.50	37.50	-	18.60	nr	56.10
Rolling grass areas, light roller							
by tractor drawn roller	-	-	-	0.52	-	100m²	0.52
by pedestrian operated mechanical roller	-	0.08	1.25	0.48	-	100m²	1.73
by hand drawn roller	-	0.17	2.50	-	-	100m²	2.50
Aerating grass areas; to a depth of 100 mm							
using tractor-drawn aerator	-	0.06	0.87	2.08	-	100m²	2.95
using pedestrian-guided motor powered solid or slitting tine turf aerator	-	0.18	2.63	2.75	-	100m²	5.38
using hollow tine aerator; including sweeping up and dumping corings	-	0.50	7.50	5.50	-	100m²	13.00
using hand aerator or fork	-	1.67	25.00	-	-	100m²	25.00
Extra over aerating grass areas for on site sweeping up and dumping corings	-	0.17	2.50	-	-	100m²	2.50
Switching off dew; from fine turf areas	-	0.20	3.00	-	-	100m²	3.00
Scarifying grass areas to break up thatch; removing dead grass							
using tractor-drawn scarifier	-	0.07	1.05	1.45	-	100m²	2.50
using self-propelled scarifier; including removing and disposing of grass on site	-	0.33	5.00	0.44	-	100m²	5.44
Harrowing grass areas							
using drag mat	-	0.03	0.42	0.29	-	100m²	0.71
using chain harrow	-	0.04	0.53	0.36	-	100m²	0.89
using drag mat	-	2.80	42.00	29.11	-	ha	71.11
using chain harrow	-	3.50	52.50	36.38	-	ha	88.89
Extra for scarifying and harrowing grass areas for disposing excavated material off site; to tip not exceeding 13 km; loading by machine							
slightly contaminated	-	-	-	1.52	16.00	m³	17.52
rubbish	-	-	-	1.52	16.00	m³	17.52
inert material	-	-	-	1.01	10.00	m³	11.01

Q PAVING/PLANTING/FENCING/SITE FURNITURE Excluding overheads and profit	PC £	Labour hours	Labour £	Plant £	Material £	Unit	Total rate £
Q30 SEEDING/TURFING - cont'd							
Maintenance operations (Note: the following rates apply to aftercare maintenance executed as part of a landscaping contract only) - cont'd							
For the following topsoil improvement and seeding operations add or subtract the following amounts for every £0.10 difference in the material cost price							
35 g/m²	-	-	-	-	0.35	100m²	0.35
50 g/m²	-	-	-	-	0.50	100m²	0.50
70 g/m²	-	-	-	-	0.70	100m²	0.70
100 g/m²	-	-	-	-	1.00	100m²	1.00
125 kg/ha	-	-	-	-	12.50	ha	12.50
150 kg/ha	-	-	-	-	15.00	ha	15.00
175 kg/ha	-	-	-	-	17.50	ha	17.50
200 kg/ha	-	-	-	-	20.00	ha	20.00
225 kg/ha	-	-	-	-	22.50	ha	22.50
250 kg/ha	-	-	-	-	25.00	ha	25.00
300 kg/ha	-	-	-	-	30.00	ha	30.00
350 kg/ha	-	-	-	-	35.00	ha	35.00
400 kg/ha	-	-	-	-	40.00	ha	40.00
500 kg/ha	-	-	-	-	50.00	ha	50.00
700 kg/ha	-	-	-	-	70.00	ha	70.00
1000 kg/ha	-	-	-	-	100.00	ha	100.00
1250 kg/ha	-	-	-	-	125 00	ha	125.00
Selective residual pre-emergent weedkiller; Rigby Taylor Ltd; "Flexidor 125"; in accordance with manufacturer's instructions; PC £84.30 per litre; application rate 1.20 - 2 litre/hectare							
1.20 l/ha	-	0.28	4.20	-	1.01	100m²	5.21
2 00 l/ha	-	0.28	4.20	-	1.69	100m²	5.89
Top dressing fertilizers (7:7:7); PC £0.48/kg; to seedbeds; by machine							
35 g/m²	1.68	-	-	0.18	1.68	100m²	1.86
50 g/m²	2.40	-	-	0.18	2.40	100m²	2.58
300 kg/ha	144.00	-	-	18.05	144.00	ha	162.05
350 kg/ha	168.00	-	-	18.05	168.00	ha	186.05
400 kg/ha	192.00	-	-	18.05	192.00	ha	210.05
500 kg/ha	240.00	-	-	28.89	240.00	ha	268.89
Top dressing fertilizers (7:7:7); PC £0.48/kg; to seedbeds; by hand							
35 g/m²	1.68	0.17	2.50	-	1.68	100m²	4.18
50 g/m²	2.40	0.17	2.50	-	2.40	100m²	4.90
70 g/m²	3.36	0.17	2.50	-	3.36	100m²	5.86
Watering turf; evenly; at a rate of 5 litre/m²							
using movable spray lines powering 3 nr sprinkler heads with a radius of 15 m and allowing for 60% overlap (irrigation machinery costs not included)	-	0.02	0.24	-	-	100m²	0.24
using sprinkler equipment and with sufficient water pressure to run 1 nr 15 m radius sprinkler	-	0.02	0.30	-	-	100m²	0.30
using hand-held watering equipment	-	0.25	3.75	-	-	100m²	3.75

Q PAVING/PLANTING/FENCING/SITE FURNITURE Excluding overheads and profit	PC £	Labour hours	Labour £	Plant £	Material £	Unit	Total rate £
Q31 PLANTING							
Market prices of mulching materials							
Melcourt Industries Ltd; 20m³ loads							
"Ornamental Bark Mulch"	-	-	-	-	42.50	m³	42.50
"Bark Nuggets"	-	-	-	-	41.40	m³	41.40
"Graded Bark Flakes"	-	-	-	-	44.00	m³	44.00
"Amenity Bark Mulch"	-	-	-	-	30.15	m³	30.15
"Contract Bark Mulch"	-	-	-	-	27.80	m³	27.80
"Spruce Ornamental"	-	-	-	-	31.55	m³	31.55
"Spruce Flakes"	-	-	-	-	33.85	m³	33.85
"Decor Biomulch"	-	-	-	-	28.80	m³	28.80
"Rustic Biomulch"	-	-	-	-	33.55	m³	33.55
"Mulchip"	-	-	-	-	29.30	m³	29.30
"Mulch 2000"	-	-	-	-	24.80	m³	24.80
"Mulch 3000"	-	-	-	-	21.80	m³	21.80
"Forest Bio Mulch"	-	-	-	-	26.30	m³	26.30
Melcourt Industries Ltd; 70m³ loads							
"Ornamental Bark Mulch"	-	-	-	-	30.10	m³	30.10
"Bark Nuggets"	-	-	-	-	29.00	m³	29.00
"Graded Bark Flakes"	-	-	-	-	31.60	m³	31.60
"Amenity Bark Mulch"	-	-	-	-	17.75	m³	17.75
"Contract Bark Mulch"	-	-	-	-	16.10	m³	16.10
"Spruce Ornamental"	-	-	-	-	19.15	m³	19.15
"Spruce Flakes"	-	-	-	-	21.45	m³	21.45
"Decor Biomulch"	-	-	-	-	16.40	m³	16.40
"Rustic Biomulch"	-	-	-	-	21.15	m³	21.15
"Mulchip"	-	-	-	-	16.90	m³	16.90
"Mulch 2000"	-	-	-	-	14.30	m³	14.30
"Mulch 3000"	-	-	-	-	11.30	m³	11.30
"Forest Bio Mulch"	-	-	-	-	13.90	m³	13.90
Market prices of planting materials (Note: The rates shown generally reflect the manufacturer's recommended retail prices. Trade and bulk discounts are often available on the prices shown)							
Topsoil; Boughton Loam Ltd; 50/50; as dug / 10 mm screened	-	-	-	-	35.10	m³	35.10
Mulch; Melcourt Industries Ltd; 20 m³ loads							
"Landscape Bark"	-	-	-	-	31.00	m³	31.00
"Humus 2000"	-	-	-	-	23.05	m³	23.05
"Spent Mushroom Compost"	-	-	-	-	17.90	m³	17.90
"Topgrow"	-	-	-	-	27.00	m³	27.00
Mulch; Melcourt Industries Ltd; 70 m³ loads							
"Super Humus"	-	-	-	-	14.30	m³	14.30
"Landscape Bark"	-	-	-	-	19.30	m³	19.30
"Humus 2000"	-	-	-	-	12.55	m³	12.55
"Spent Mushroom Compost"	-	-	-	-	6.20	m³	6.20
"Topgrow"	-	-	-	-	15.30	m³	15.30
Fertilizers; Scotts UK Professional							
"Enmag"; 70 g/m²	-	-	-	-	7.80	100m²	7.80
fertilizer; controlled release; "Osmocote Plus"; 15+10+12+2MgO+TE, tablet; 5gr each	-	-	-	-	2.53	100 nr	2.53

Q PAVING/PLANTING/FENCING/SITE FURNITURE Excluding overheads and profit	PC £	Labour hours	Labour £	Plant £	Material £	Unit	Total rate £
Q31 PLANTING - cont'd							
Market prices of planting materials (Note: The rates shown generally reflect the manufacturer's recommended retail prices. Trade and bulk discounts are often available on the prices shown) - cont'd							
Fertilizers; Scotts UK Professional; granular "Sierrablen Flora", 15+9+9+3 MgO; controlled release fertilizer							
transplant	-	-	-	-	0.08	nr	0.08
whip	-	-	-	-	0.12	nr	0.12
feathered	-	-	-	-	0.16	nr	0.16
light standard	-	-	-	-	0.16	nr	0.16
standard	-	-	-	-	0.19	nr	0.19
selected standard	-	-	-	-	0.27	nr	0.27
heavy standard	-	-	-	-	0.31	nr	0.31
extra heavy standard	-	-	-	-	0.39	nr	0.39
16 - 18 cm girth	-	-	-	-	0.43	nr	0.43
18 - 20 cm girth	-	-	-	-	0.47	nr	0.47
20 - 22 cm girth	-	-	-	-	0.54	nr	0.54
22 - 24 cm girth	-	-	-	-	0.58	nr	0.58
24 - 26 cm girth	-	-	-	-	0.62	nr	0.62
Fertilizers; Farmura Products Ltd; Seanure Root Dip							
to transplants	-	-	-	-	0.22	10 nr	0.22
medium whips	-	-	-	-	0.06	each	0.06
standard trees	-	-	-	-	0.37	each	0.37
Fertilizers; Farmura Products Ltd; Seanure Soilbuilder							
soil amelioration; 70g/m²	-	-	-	-	7.83	100m²	7.83
to plant pits; 300 x 300 x 300	-	-	-	-	0.03	each	0.03
to plant pits; 600 x 600 x600	-	-	-	-	0.36	each	0.36
to tree pits; 1.00 x 1.00 x 1.00	-	-	-	-	1.68	each	1.68
Fertilizers; Levington Horticulture Ltd							
grass and soil fertilizer; "Greenmaster Liquid Seafeed"	-	-	-	-	3.05	litre	3.05
"TPMC" tree and shrub planting compost	-	-	-	-	3.84	bag	3.84
general fertilizer; "Ficote 70" (14:14:14) top dressing	-	-	-	-	2.81	kg	2.81
general fertilizer; "Ficote 140" (14:14:14) top dressing	-	-	-	-	2.81	kg	2.81
general fertilizer; "Ficote 180" (14:14:14) top dressing	-	-	-	-	2.81	kg	2.81
Fertilizers and anti-desiccants; Rigby Taylor Ltd; fertilizer application rates 35 g/m² unless otherwise shown							
straight fertilizer; "Bone Meal"; at 70 g/m²	-	-	-	-	6.02	100m²	6.02
straight fertilizer; "Sulphate of Ammonia"	-	-	-	-	1.73	100m²	1.73
straight fertilizer; "Sulphate of Iron"	-	-	-	-	1.61	100m²	1.61
straight fertilizer; "Sulphate of Potash"	-	-	-	-	2.13	100m²	2.13
straight fertilizer; "Super Phosphate Powder"	-	-	-	-	0.45	100m²	0.45
liquid fertilizer; "Vitax 50/50 Soluble Iron"; 56 ml/100m²	-	-	-	-	6.59	100m²	6.59

Q PAVING/PLANTING/FENCING/SITE FURNITURE Excluding overheads and profit	PC £	Labour hours	Labour £	Plant £	Material £	Unit	Total rate £
liquid fertilizer; "Vitax 50/50 Standard"; 56 ml/100m²	-	-	-	-	10.78	100m²	10.78
liquid fertilizer; "Vitax 50/50 Extra"; 56 ml/100m²	-	-	-	-	13.02	100m²	13.02
wetting agent; "Breaker liquid"; 10 lt	-	-	-	-	1.21	100m²	1.21
wetting agent; "Breaker granules"; 25 Kg	-	-	-	-	7.57	100m²	7.57
Herbicides; Scotts UK Professional; application rates used to produce these rates are the maximum rate recommended by the manufacturer in each case (Note: lower rates may often be appropriate in many cases)							
"Dextrone X"	-	-	-	-	0.43	100m²	0.43
"Intrepid"	-	-	-	-	0.52	100m²	0.52
"Speedway"	-	-	-	-	0.86	100m²	0.86
Herbicides; Rigby Taylor Ltd; application rates used to produce these are the maximum rate recommended by the manufacturer in each case (Note: these application rates will vary dependent on season)							
"Roundup Pro Biactive"; 5 lt/ha	-	-	-	-	1.03	100m²	1.03
"Flexidor 125 (Isoxaben)"; 2lt/ha	-	-	-	-	1.69	100m²	1.69
"Kerb Flowable (Propyzamide)"	-	-	-	-	2.16	100m²	2.16
"Kerb Granules" (15 x 120 tree pack)	-	-	-	-	0.31	100m²	0.31
"Timbrel (Triclopyr)"	-	-	-	-	4.72	100m²	4.72
"Casoron G" as a selective herbicide	-	-	-	-	4.75	100m²	4.75
"Casoron G" as a residual herbicide	-	-	-	-	8.55	100m²	8.55
Market prices of trees, shrubs and plants: Notcutts Nurseries Ltd							
Shrubs							
Acer palmatum Dissectum Atropurpureum; 30-45 3 L	-	-	-	-	12.50	nr	12.50
Arbutus unedo rubra; 30-45 3 L	-	-	-	-	5.95	nr	5.95
Aucuba japonica Variegata; 30-45 3 L	-	-	-	-	2.95	nr	2.95
Berberis candidula; 20-30 2 L	-	-	-	-	1.85	nr	1.85
Berberis thunbergii; 30-45 3 L	-	-	-	-	1.40	nr	1.40
Berberis thunbergii Rose Glow; 30-45 2 L	-	-	-	-	2.80	nr	2.80
Buddleja davidii Black Knight; 45-60 3 L	-	-	-	-	1.80	nr	1.80
Buddleja globosa; 45-60 3 L	-	-	-	-	2.90	nr	2.90
Buxus sempervirens; 30-45 3 L	-	-	-	-	2.90	nr	2.90
Caryopteris clandonensis; 30-45 3 L	-	-	-	-	2.45	nr	2.45
Ceanothus dentatus; 60-90 3 L	-	-	-	-	3.20	nr	3.20
Ceanothus thyrsiflorus Repens; 30-45 2 L	-	-	-	-	2.20	nr	2.20
Chaenomeles superba Jet Trail; 45-60 3 L	-	-	-	-	2.20	nr	2.20
Choisya ternata; 30-45 3 L	-	-	-	-	2.45	nr	2.45
Cornus alba; 60-90 3 L	-	-	-	-	1.40	nr	1.40
Cornus alba Elegantissima; 60-90 3 L	-	-	-	-	2.40	nr	2.40
Cornus stolonifera Flayiramea; 60-90 3 L	-	-	-	-	1.80	nr	1.80
Corylus avellana; 45-60 3 L	-	-	-	-	1.40	nr	1.40
Cotinus coggygria Royal Purple; 30-45 3 L	-	-	-	-	3.30	nr	3.30
Cotoneaster salicifolius Exburyensis; 45-50 3 L	-	-	-	-	2.40	nr	2.40
Cotoneaster suecicus watereri; 60-90 3 L	-	-	-	-	1.80	nr	1.80
Elaeagnus pungens maculata; 45-60 3 L	-	-	-	-	3.20	nr	3.20
Elaeagnus x ebbingei; 45-60 3 L	-	-	-	-	2.20	nr	2.20
Elaeagnus x ebbingei Limelight; 45-60 3 L	-	-	-	-	3.20	nr	3.20
Escallonia Edinensis; 45-60 3 L	-	-	-	-	2.10	nr	2.10

Q PAVING/PLANTING/FENCING/SITE FURNITURE Excluding overheads and profit	PC £	Labour hours	Labour £	Plant £	Material £	Unit	Total rate £
Q31 PLANTING - cont'd							
Market prices of trees, shrubs and plants: Notcutts Nurseries Ltd - cont'd							
Euonymus fortunei Blondy; 20-30 2 L	-	-	-	-	2.80	nr	2.80
Euonymus fortunei Darts Blanket; 30-45 2 L	-	-	-	-	1.40	nr	1.40
Forsythia intermedia Lynwood; 60-90 3 L	-	-	-	-	1.80	nr	1.80
Genista hispanica; 20-30 2 L	-	-	-	-	1.80	nr	1.80
Griselinia littoralis; 45-60 3 L	-	-	-	-	2.95	nr	2.95
Hebe albicans; 20-30 2 L	-	-	-	-	1.80	nr	1.80
Hebe Great Orme; 20-30 3 L	-	-	-	-	2.60	nr	2.60
Hebe pinguifolia Pagei; 15-20 2 L	-	-	-	-	1.40	nr	1.40
Hebe rakaiensis; 20-30 2 L	-	-	-	-	1.40	nr	1.40
Hedera helix Glacier; 45-60 1.5 L	-	-	-	-	2.10	nr	2.10
Hippophae rhamnoides; 45-60 3 L	-	-	-	-	1.60	nr	1.60
Hydrangea macrophylla Blue Wave; 45-60 4 L	-	-	-	-	4.20	nr	4.20
Hypericum Hidcote; 30-45 3 L	-	-	-	-	1.80	nr	1.80
Ilex aquifolium Pyramidalis; 60-90 3 L	-	-	-	-	3.90	nr	3.90
Lavandula angustifolia Hidcote; 15-20 1.5 L	-	-	-	-	1.40	nr	1.40
Lavandula angustifolia Hidcote; 20-30 3 L	-	-	-	-	2.20	nr	2.20
Lavatera Barnsley; 45-60 3 L	-	-	-	-	2.45	nr	2.45
Leucothoe walteri Rainbow; 30-45 3 L	-	-	-	-	4.50	nr	4.50
Leycesteria formosa; 30-45 3 L	-	-	-	-	2.80	nr	2.80
Ligustrum ovalifolium; 45-60 2 L	-	-	-	-	1.40	nr	1.40
Ligustrum ovalifolium Aureum; 30-45 2 L	-	-	-	-	1.80	nr	1.80
Ligustrum nitida Baggesens Gold; 30-45 3 L	-	-	-	-	1.90	nr	1.90
Mahonia japonica; 45-60 3 L	-	-	-	-	2.60	nr	2.60
Olearia haastii; 30-45 3 L	-	-	-	-	2.20	nr	2.20
Osmanthus delavayi; 30-45 3 L	-	-	-	-	3.20	nr	3.20
Pachysandra terminalis; 15-20 1.5 L	-	-	-	-	1.60	nr	1.60
Pernettya mucronata Male; 20-30 3 L	-	-	-	-	2.60	nr	2.60
Philadelphus Belle Etoile; 45-60 3 L	-	-	-	-	1.80	nr	1.80
Philadelphus coronarius Aureus; 30-45 3 L	-	-	-	-	2.80	nr	2.80
Philadelphus Virginal; 45-60 3 L	-	-	-	-	2.40	nr	2.40
Phormium tenax; 30-45 3 L	-	-	-	-	3.50	nr	3.50
Photinia x fraseri Red Robin; 30-45 3 L	-	-	-	-	3.60	nr	3.60
Potentilla arbuscula; 20-30 3 L	-	-	-	-	1.60	nr	1.60
Potentilla fruticosa Pretty Polly; 20-30 3 L	-	-	-	-	2.45	nr	2.45
Prunus lusitanica; 45-60 3 L	-	-	-	-	1.80	nr	1.80
Pyracantha Orange Glow; 60-90 3 L	-	-	-	-	1.80	nr	1.80
Pyracantha Teton; 30-45 3 L	-	-	-	-	2.20	nr	2.20
Ribes sanguineum King Edward VII; 45-60 3 L	-	-	-	-	2.10	nr	2.10
Rosmarinus officinalis; 20-30 3 L	-	-	-	-	2.20	nr	2.20
Rubus Betty Ashburner; 45-60 2 L	-	-	-	-	1.30	nr	1.30
Rubus Tridel Benenden; 45-60 3 L	-	-	-	-	2.40	nr	2.40
Salix lanata; 45-60 3 L	-	-	-	-	2.60	nr	2.60
Salvia officinalis Icterina; 20-30 3 L	-	-	-	-	2.60	nr	2.60
Sambucus nigra Aurea; 45-60 3 L	-	-	-	-	1.40	nr	1.40
Sambucus racemosa Plumosa Aurea; 45-60 3 L	-	-	-	-	3.95	nr	3.95
Santolina chamaecyparissus; 20-30 3 L	-	-	-	-	1.60	nr	1.60
Sarcococca confusa; 20-30 2 L	-	-	-	-	2.80	nr	2.80
Senecio monroi; 20-25 2 L	-	-	-	-	2.60	nr	2.60
Senecio Sunshine; 30-45 3 L	-	-	-	-	1.60	nr	1.60
Skimmia japonica; 30-45 3 L	-	-	-	-	2.60	nr	2.60

Q PAVING/PLANTING/FENCING/SITE FURNITURE Excluding overheads and profit	PC £	Labour hours	Labour £	Plant £	Material £	Unit	Total rate £
Spiraea x arguta; 45-60 3 L	-	-	-	-	1.80	nr	1.80
Spiraea bumalda x Goldflame; 20-30 2 L	-	-	-	-	1.60	nr	1.60
Stephanandra incisa Crispa; 30-45 2 L	-	-	-	-	1.60	nr	1.60
Symphoricarpos albus; 45-60 3 L	-	-	-	-	1.60	nr	1.60
Symphoricarpos x chenaultii Hancock; 45-60 1.5 L	-	-	-	-	1.20	nr	1.20
Tamarix tetrandra; 45-60 3 L	-	-	-	-	2.40	nr	2.40
Viburnum x bodnantense Dawn; 45-60 3 L	-	-	-	-	3.40	nr	3.40
Viburnum opulus; 60-90 3 L	-	-	-	-	1.60	nr	1.60
Viburnum opulus Sterile; 45-60 3 L	-	-	-	-	2.40	nr	2.40
Viburnum plicatum Pink Beauty; 45-60 3 L	-	-	-	-	4.30	nr	4.30
Viburnum tinus; 30-45 3 L	-	-	-	-	1.95	nr	1.95
Viburnum tinus Variegatum; 20-30 3 L	-	-	-	-	2.90	nr	2.90
Container grown climbers							
Clematis Jackmanii; 60-90 2 L	-	-	-	-	3.25	nr	3.25
Clematis montana Elizabeth; 60-90 2 L	-	-	-	-	3.25	nr	3.25
Hydrangea petiolaris; 45-60 3 L	-	-	-	-	3.55	nr	3.55
Jasminum nudiflorum; 45-60 2 L	-	-	-	-	2.60	nr	2.60
Lonicera japonica Halliana; 60-90 3 L	-	-	-	-	2.60	nr	2.60
Parthenocissus quinquefolia; 60-90 2 L	-	-	-	-	2.40	nr	2.40
Passiflora caerulea; 60-90 2 L	-	-	-	-	3.50	nr	3.50
Vitis coignetiae; 60-90 2 L	-	-	-	-	4.75	nr	4.75
Wisteria sinensis; 60-90 3 L	-	-	-	-	6.50	nr	6.50
Container grown specimen shrubs							
Amelanchier lamarckii; 90-120 10 L	-	-	-	-	8.50	nr	8.50
Aralia elata; 120-150 10 L	-	-	-	-	15.00	nr	15.00
Aucuba japonica; 45-60 10 L	-	-	-	-	8.50	nr	8.50
Berberis darwinii; 45-60 10 L	-	-	-	-	8.50	nr	8.50
Choisya ternata; 45-60 10 L	-	-	-	-	8.50	nr	8.50
Cotoneaster cornubia; 90-120 10 L	-	-	-	-	8.50	nr	8.50
Cotoneaster Exburiensis; 90-120 10 L	-	-	-	-	8.50	nr	8.50
Cotoneaster watereri; 90-120 10 L	-	-	-	-	8.50	nr	8.50
Elaeagnus pungens Maculata; 45-60 10 L	-	-	-	-	12.50	nr	12.50
Elaeagnus ebbingei; 60-90 10 L	-	-	-	-	8.50	nr	8.50
Elaeagnus ebbingei Limelight; 60-90 10 L	-	-	-	-	12.50	nr	12.50
Escallonia rubra Crimson Spire; 60-90 10 L	-	-	-	-	8.50	nr	8.50
Euonymus fortunei Emerald N Gold; 30-45 7.3 L	-	-	-	-	8.00	nr	8.00
Fatsia japonica; 90-120 15 L	-	-	-	-	12.50	nr	12.50
Ilex aquifolium; 90-120 10 L	-	-	-	-	8.50	nr	8.50
Ilex aquifolium Golden King; 90-120 10 L	-	-	-	-	15.00	nr	15.00
Ilex aquifolium Argentea Marginata; 90-120 10 L	-	-	-	-	15.00	nr	15.00
Mahonia japonica; 60-90 10 L	-	-	-	-	12.50	nr	12.50
Phormium tenax; 60-90 10 L	-	-	-	-	14.00	nr	14.00
Phormium tenax Purpureum; 60-90 10 L	-	-	-	-	14.00	nr	14.00
Phormium Yellow Wave; 60-90 10 L	-	-	-	-	18.00	nr	18.00
Photinia x Fraseri Red Robin; 60-90 10 L	-	-	-	-	12.50	nr	12.50
Prunus x cistena Crimson Dwarf; 45-60 10 L	-	-	-	-	12.50	nr	12.50
Prunus laurocerasus; 90-120 10 L	-	-	-	-	8.50	nr	8.50
Prunus laurocerasus Otto Luyken; 45-60 10 L	-	-	-	-	8.50	nr	8.50
Prunus laurocerasus Zabeliana; 45-60 10 L	-	-	-	-	8.50	nr	8.50
Prunus lusitanica; 45-60 10 L	-	-	-	-	8.50	nr	8.50
Pyracantha Mohave; 45-60 10 L	-	-	-	-	8.50	nr	8.50
Pyracantha Orange Glow; 60-90 10 L	-	-	-	-	8.50	nr	8.50
Pyracantha Red Column; 60-90 10 L	-	-	-	-	8.50	nr	8.50
Pyracantha Soleil d'Or; 60-90 10 L	-	-	-	-	8.50	nr	8.50
Pyracantha Teton; 45-60 10 L	-	-	-	-	8.50	nr	8.50

Q PAVING/PLANTING/FENCING/SITE FURNITURE Excluding overheads and profit	PC £	Labour hours	Labour £	Plant £	Material £	Unit	Total rate £
Q31 PLANTING - cont'd							
Market prices of trees, shrubs and plants: Notcutts Nurseries Ltd - cont'd							
Rhus typhina; 90-120 10 L	-	-	-	-	8.50	nr	8.50
Rosmarinus officinalis; 30-45 10 L	-	-	-	-	10.00	nr	10.00
Rubus cockburnianus Golden Vale; 90-120 10 L	-	-	-	-	8.50	nr	8.50
Skimmia Rubella; 30-45 7.5 L	-	-	-	-	8.50	nr	8.50
Spiraea japonica Little Princess; 30-45 7.5 L	-	-	-	-	8.50	nr	8.50
Spiraea Snowmound; 45-60 10 L	-	-	-	-	8.50	nr	8.50
Spiraea thunbergii; 45-60 10 L	-	-	-	-	8.50	nr	8.50
Bamboos							
Arundinaria auricoma; 40-50 7.5 L	-	-	-	-	12.50	nr	12.50
Container grown conifers							
X Cupressocyparis leylandii; 120-150 10 L	-	-	-	-	7.50	nr	7.50
X Cupressocyparis leylandii Castlewellan; 120-150 10 L	-	-	-	-	7.50	nr	7.50
Juniperus communis Repanda; 30-45 10 L	-	-	-	-	8.50	nr	8.50
Juniperus x media pfitzeriana Aurea; 45-60 10L	-	-	-	-	8.50	nr	8.50
Juniperus sabina Tamariscifolia; 45-60 10 L	-	-	-	-	8.50	nr	8.50
Pinus mugo; 45-60 10 L	-	-	-	-	12.50	nr	12.50
Pinus sylvestris; 90-120 10 L	-	-	-	-	8.50	nr	8.50
Pinus sylvestris; 150-180 50 L	-	-	-	-	60.00	nr	60.00
Pinus sylvestris; 180-210 80 L	-	-	-	-	105.00	nr	105.00
Chamaecyparis lawsoniana Ellwoodii; 30-45 2 L	-	-	-	-	2.20	nr	2.20
Chamaecyparis lawsoniana Ellwoods Gold; 30-45 2 L	-	-	-	-	2.90	nr	2.90
Chamaecyparis lawsoniana Ellwoods Pillar; 30-45 2 L	-	-	-	-	2.90	nr	2.90
X Cupressocyparis leylandii; 60-90 3 L	-	-	-	-	1.80	nr	1.80
X Cupressocyparis leylandii Castlewellan; 60-90 3 L	-	-	-	-	1.80	nr	1.80
Juniperus communis Compressa; 30-45 2 L	-	-	-	-	2.60	nr	2.60
Juniperus communis Repanda; 30-45 3 L	-	-	-	-	2.60	nr	2.60
Juniperus conferta; 30-45 3 L	-	-	-	-	2.80	nr	2.80
Juniperus media Mint Julep; 30-45 3 L	-	-	-	-	2.90	nr	2.90
Juniperus media pfitzeriana; 30-45 3 L	-	-	-	-	2.70	nr	2.70
Juniperus media pfitzeriana Aurea; 30-45 3 L	-	-	-	-	2.90	nr	2.90
Juniperus sabina Tamariscifolia; 30-45 3 L	-	-	-	-	2.70	nr	2.70
Juniperus squamata Blue Star; 30-45 2 L	-	-	-	-	3.20	nr	3.20
Larix decidua; 60-90 3 L	-	-	-	-	1.60	nr	1.60
Picea albertiana Conica; 20-30 3 L	-	-	-	-	2.80	nr	2.80
Pinus mugo; 20-30 2 L	-	-	-	-	2.20	nr	2.20
Pinus nigra; 45-60 3 L	-	-	-	-	1.90	nr	1.90
Pinus nigra austriaca; 45-60 3 L	-	-	-	-	1.60	nr	1.60
Pinus nigra maritima; 45-60 3 L	-	-	-	-	1.60	nr	1.60
Pinus sylvestris; 45-60 3 L	-	-	-	-	1.60	nr	1.60
Transplants 45-60							
Acer campestre	-	-	-	-	0.19	nr	0.19
Acer platanoides	-	-	-	-	0.19	nr	0.19
Acer pseudoplatanus	-	-	-	-	0.21	nr	0.21
Alnus cordata	-	-	-	-	0.21	nr	0.21
Alnus glutinosa	-	-	-	-	0.21	nr	0.21
Alnus incana	-	-	-	-	0.21	nr	0.21
Betula pendula	-	-	-	-	0.21	nr	0.21

Q PAVING/PLANTING/FENCING/SITE FURNITURE Excluding overheads and profit	PC £	Labour hours	Labour £	Plant £	Material £	Unit	Total rate £
Carpinus betulus	-	-	-	-	0.24	nr	0.24
Cornus sanguinea	-	-	-	-	0.24	nr	0.24
Corylus avellana	-	-	-	-	0.24	nr	0.24
Crataegus monogyna	-	-	-	-	0.16	nr	0.16
Fagus sylvatica	-	-	-	-	0.26	nr	0.26
Fraxinus excelsior	-	-	-	-	0.24	nr	0.24
Hippophae rhamnoides	-	-	-	-	0.28	nr	0.28
Prunus avium	-	-	-	-	0.24	nr	0.24
Quercus robur	-	-	-	-	0.28	nr	0.28
Rosa rugosa	-	-	-	-	0.26	nr	0.26
Rosa rugosa Alba	-	-	-	-	0.26	nr	0.26
Sambucus nigra	-	-	-	-	0.24	nr	0.24
Sorbus aria	-	-	-	-	0.26	nr	0.26
Sorbus intermedia	-	-	-	-	0.32	nr	0.32
Viburnum opulus	-	-	-	-	0.32	nr	0.32
Transplants 60-90							
Acer campestre	-	-	-	-	0.21	nr	0.21
Acer platanoides	-	-	-	-	0.24	nr	0.24
Alnus cordata	-	-	-	-	0.24	nr	0.24
Amelanchier canadensis	-	-	-	-	0.28	nr	0.28
Cornus alba	-	-	-	-	0.24	nr	0.24
Crataegus monogyna	-	-	-	-	0.24	nr	0.24
Fagus sylvatica	-	-	-	-	0.32	nr	0.32
Fraxinus excelsior	-	-	-	-	0.26	nr	0.26
Populus alba	-	-	-	-	0.26	nr	0.26
Populus nigra italica	-	-	-	-	0.28	nr	0.28
Populus trichocarpa	-	-	-	-	0.28	nr	0.28
Prunus avium	-	-	-	-	0.28	nr	0.28
Salix alba	-	-	-	-	0.21	nr	0.21
Salix caprea	-	-	-	-	0.26	nr	0.26
Salix fragilis	-	-	-	-	0.24	nr	0.24
Salix rosmarinifolia	-	-	-	-	0.24	nr	0.24
Salix viminalis	-	-	-	-	0.24	nr	0.24
Salix vitellina	-	-	-	-	0.24	nr	0.24
Sorbus aucuparia	-	-	-	-	0.26	nr	0.26
Herbaceous							
Achillea fil. Cloth of Gold; 10-15 2 L	-	-	-	-	1.80	nr	1.80
Alchemilla mollis; 10-15 2 L	-	-	-	-	1.80	nr	1.80
Anemone hybrida Honorine Jobert; 10-15 3 L	-	-	-	-	2.20	nr	2.20
Artemisia absin. Lambrook Silver; 10-15 3 L	-	-	-	-	2.45	nr	2.45
Artemisia Powis Castle; 10-15 3 L	-	-	-	-	2.45	nr	2.45
Bergenia Bressingham Salmon; 20-30 2 L	-	-	-	-	2.25	nr	2.25
Convallaria majalis; 10-15 2 L	-	-	-	-	1.80	nr	1.80
Euphorbia amygdaloides Purpurea; 15-20 2 L	-	-	-	-	2.20	nr	2.20
Euphorbia wulfenii; 10-15 2 L	-	-	-	-	2.60	nr	2.60
Geranium oxonianum Wargrave Pink; 10-15 3 L	-	-	-	-	1.80	nr	1.80
Geranium Johnson Blue; 10-15 3 L	-	-	-	-	1.80	nr	1.80
Hosta August Moon; 10-15 2 L	-	-	-	-	2.50	nr	2.50
Hosta sieboldii; 10-15 2 L	-	-	-	-	2.50	nr	2.50
Iris foetidissima; 10-15 2 L	-	-	-	-	2.20	nr	2.20
Papaver orientale; 10-15 3 L	-	-	-	-	1.80	nr	1.80
Penstemon Garnet; 10-15 3 L	-	-	-	-	2.25	nr	2.25
Pulmonaria angustifolia; 10-15 2 L	-	-	-	-	1.80	nr	1.80
Salvia East Friesland; 10-15 3 L	-	-	-	-	1.80	nr	1.80
Sedum spectabilis Autumn Joy; 10-15 3 L	-	-	-	-	1.80	nr	1.80
Tiarella cordifolia; 10-15 2 L	-	-	-	-	1.80	nr	1.80
Waldsteinia ternata; 10-15 2 L	-	-	-	-	1.80	nr	1.80

Q PAVING/PLANTING/FENCING/SITE FURNITURE Excluding overheads and profit	PC £	Labour hours	Labour £	Plant £	Material £	Unit	Total rate £
Q31 PLANTING - cont'd							
Market prices of trees, shrubs and plants: Notcutts Nurseries Ltd - cont'd							
Bare root trees 1.50 - 1.80 feathered							
Acer campestre	-	-	-	-	2.00	nr	2.00
Acer platanoides	-	-	-	-	1.50	nr	1.50
Acer pseudoplatanus	-	-	-	-	1.50	nr	1.50
Alnus cordata	-	-	-	-	1.50	nr	1.50
Alnus glutinosa	-	-	-	-	1.50	nr	1.50
Alnus incana	-	-	-	-	3.00	nr	3.00
Amelanchier canadensis	-	-	-	-	4.00	nr	4.00
Betula pendula	-	-	-	-	1.50	nr	1.50
Carpinus betulus	-	-	-	-	2.50	nr	2.50
Fagus sylvatica	-	-	-	-	2.50	nr	2.50
Fagus sylvatica purpurea	-	-	-	-	5.50	nr	5.50
Fraxinus excelsior	-	-	-	-	2.50	nr	2.50
Populus alba	-	-	-	-	1.50	nr	1.50
Populus nigra italica	-	-	-	-	1.50	nr	1.50
Populus tremula	-	-	-	-	5.50	nr	5.50
Prunus avium	-	-	-	-	2.50	nr	2.50
Prunus avium 'Plena'	-	-	-	-	6.00	nr	6.00
Quercus robur	-	-	-	-	9.50	nr	9.50
Quercus rubra	-	-	-	-	4.50	nr	4.50
Sorbus lutescens	-	-	-	-	3.00	nr	3.00
Sorbus aucuparia	-	-	-	-	2.00	nr	2.00
Bare root trees 1.80 - 2.10 feathered							
Acer campestre	-	-	-	-	2.00	nr	2.00
Alnus cordata	-	-	-	-	3.00	nr	3.00
Acer platanoides	-	-	-	-	3.00	nr	3.00
Acer pseudoplatanus	-	-	-	-	3.00	nr	3.00
Alnus glutinosa	-	-	-	-	3.00	nr	3.00
Alnus incana	-	-	-	-	2.50	nr	2.50
Amelanchier canadensis	-	-	-	-	9.00	nr	9.00
Betula pendula	-	-	-	-	2.50	nr	2.50
Carpinus betulus	-	-	-	-	4.50	nr	4.50
Fagus sylvatica	-	-	-	-	4.50	nr	4.50
Fagus sylvatica purpurea	-	-	-	-	9.50	nr	9.50
Fraxinus excelsior	-	-	-	-	4.00	nr	4.00
Populus alba	-	-	-	-	3.50	nr	3.50
Populus nigra italica	-	-	-	-	3.50	nr	3.50
Populus tremula	-	-	-	-	3.50	nr	3.50
Prunus avium	-	-	-	-	4.00	nr	4.00
Prunus avium 'Plena'	-	-	-	-	8.00	nr	8.00
Quercus robur	-	-	-	-	9.50	nr	9.50
Quercus rubra	-	-	-	-	10.50	nr	10.50
Sorbus aucuparia	-	-	-	-	4.50	nr	4.50
Sorbus lutescens	-	-	-	-	9.00	nr	9.00
Bare root trees 8-10 cm							
Acer campestre	-	-	-	-	11.00	nr	11.00
Acer platanoides	-	-	-	-	9.00	nr	9.00
Acer pseudoplatanus	-	-	-	-	9.00	nr	9.00
Alnus cordata	-	-	-	-	9.00	nr	9.00
Alnus glutinosa	-	-	-	-	9.00	nr	9.00
Alnus incana	-	-	-	-	9.00	nr	9.00
Amelanchier canadensis	-	-	-	-	22.00	nr	22.00

Q PAVING/PLANTING/FENCING/SITE FURNITURE Excluding overheads and profit	PC £	Labour hours	Labour £	Plant £	Material £	Unit	Total rate £
Betula pendula	-	-	-	-	8.00	nr	8.00
Carpinus betulus	-	-	-	-	12.00	nr	12.00
Fagus sylvatica	-	-	-	-	12.00	nr	12.00
Fraxinus excelsior	-	-	-	-	13.95	nr	13.95
Fagus sylvatica Purpurea	-	-	-	-	18.00	nr	18.00
Populus tremula	-	-	-	-	11.00	nr	11.00
Populus alba	-	-	-	-	9.50	nr	9.50
Populus nigra Italica	-	-	-	-	9.50	nr	9.50
Prunus avium	-	-	-	-	10.00	nr	10.00
Prunus avium Plena	-	-	-	-	13.00	nr	13.00
Quercus robor	-	-	-	-	12.00	nr	12.00
Quercus rubra	-	-	-	-	17.00	nr	17.00
Sorbus aucuparia	-	-	-	-	11.00	nr	11.00
Bare root trees 10-12 cm sel. standard							
Acer campestre	-	-	-	-	14.00	nr	14.00
Alnus cordata	-	-	-	-	10.00	nr	10.00
Acer platanoides	-	-	-	-	10.00	nr	10.00
Acer pseudoplatanus	-	-	-	-	10.00	nr	10.00
Amelanchier canadensis	-	-	-	-	35.00	nr	35.00
Betula pendula	-	-	-	-	10.00	nr	10.00
Carpinus betulus	-	-	-	-	23.50	nr	23.50
Fagus sylvatica	-	-	-	-	18.00	nr	18.00
Fagus sylvatica purpurea	-	-	-	-	25.00	nr	25.00
Fraxinus excelsior	-	-	-	-	12.00	nr	12.00
Populus alba	-	-	-	-	12.50	nr	12.50
Populus nigra italica	-	-	-	-	12.50	nr	12.50
Populus tremula	-	-	-	-	16.00	nr	16.00
Prunus avium	-	-	-	-	12.00	nr	12.00
Prunus avium 'Plena'	-	-	-	-	16.00	nr	16.00
Quercus robur	-	-	-	-	21.00	nr	21.00
Quercus rubra	-	-	-	-	21.00	nr	21.00
Sorbus aucuparia	-	-	-	-	15.00	nr	15.00
Sorbus lutescens	-	-	-	-	18.00	nr	18.00
Bare root trees 12-14 cm heavy standard							
Acer campestre	-	-	-	-	24.00	nr	24.00
Alnus cordata	-	-	-	-	20.00	nr	20.00
Acer platanoides	-	-	-	-	20.00	nr	20.00
Acer pseudoplatanus	-	-	-	-	20.00	nr	20.00
Alnus glutinosa	-	-	-	-	20.00	nr	20.00
Alnus incana	-	-	-	-	20.00	nr	20.00
Amelanchier canadensis	-	-	-	-	60.00	nr	60.00
Betula pendula	-	-	-	-	20.00	nr	20.00
Carpinus betulus	-	-	-	-	28.00	nr	28.00
Fagus sylvatica	-	-	-	-	28.00	nr	28.00
Fagus sylvatica Purpurea	-	-	-	-	37.00	nr	37.00
Fraxinus excelsior	-	-	-	-	24.00	nr	24.00
Populus alba	-	-	-	-	22.00	nr	22.00
Populus nigra italica	-	-	-	-	12.50	nr	12.50
Populus tremula	-	-	-	-	16.00	nr	16.00
Prunus avium	-	-	-	-	24.00	nr	24.00
Prunus avium 'Plena'	-	-	-	-	28.00	nr	28.00
Quercus robur	-	-	-	-	32.00	nr	32.00
Quercus rubra	-	-	-	-	32.00	nr	32.00
Sorbus aucuparia	-	-	-	-	22.00	nr	22.00
Sorbus lutescens	-	-	-	-	30.00	nr	30.00

Q PAVING/PLANTING/FENCING/SITE FURNITURE Excluding overheads and profit	PC £	Labour hours	Labour £	Plant £	Material £	Unit	Total rate £
Q31 PLANTING - cont'd							
Market prices of trees; Coblands Nurseries Ltd							
Bare root trees 14-16 cm extra heavy standard							
Acer campestre	-	-	-	-	42.00	nr	42.00
Alnus cordata	-	-	-	-	35.00	nr	35.00
Acer platanoides	-	-	-	-	34.00	nr	34.00
Acer pseudoplatanus	-	-	-	-	34.00	nr	34.00
Alnus glutinosa	-	-	-	-	35.00	nr	35.00
Alnus incana	-	-	-	-	35.00	nr	35.00
Amelanchier canadensis	-	-	-	-	58.00	nr	58.00
Betula pendula	-	-	-	-	34.00	nr	34.00
Carpinus betulus	-	-	-	-	49.00	nr	49.00
Fagus sylvatica	-	-	-	-	49.00	nr	49.00
Fagus sylvatica purpurea	-	-	-	-	49.00	nr	49.00
Fraxinus excelsior	-	-	-	-	49.00	nr	49.00
Populus alba	-	-	-	-	42.00	nr	42.00
Populus nigra italica	-	-	-	-	42.00	nr	42.00
Populus tremula	-	-	-	-	49.00	nr	49.00
Prunus avium	-	-	-	-	34.00	nr	34.00
Prunus avium 'Plena'	-	-	-	-	49.00	nr	49.00
Quercus robur	-	-	-	-	49.00	nr	49.00
Quercus rubra	-	-	-	-	49.00	nr	49.00
Robinia pseudoacasia Frisia	-	-	-	-	75.00	nr	75.00
Sorbus aucuparia	-	-	-	-	34.00	nr	34.00
Sorbus aria	-	-	-	-	49.00	nr	49.00
Root ball trees; extra over for root balling and wrapping in hessian	-	-	-	-	18.00	nr	18.00
Market prices of aquatic plants in prepared growing medium of approved hessian bag filled with soil and fertilizer; tied, slit and weighted and placed in position in pool to receive three submerged floating leafed marginal or swamp plants; Anglo Aquarium Plant							
Water lilies							
Nymphaea in variety - white	-	-	-	-	5.80	nr	5.80
Nymphaea in variety - deep red	-	-	-	-	7.40	nr	7.40
Deep water aquatics							
Aponogeton distachyos	-	-	-	-	2.10	nr	2.10
Nymphoides peltata	-	-	-	-	1.69	nr	1.69
Marginal plants							
Acorus calamus	-	-	-	-	1.69	nr	1.69
Butomus umbellatus	-	-	-	-	1.69	nr	1.69
Caltha palustris	-	-	-	-	1.69	nr	1.69
Carex spp.	-	-	-	-	1.69	nr	1.69
Iris pseudacorus	-	-	-	-	1.69	nr	1.69
Juncus spp.	-	-	-	-	1.69	nr	1.69
Mentha aquatica	-	-	-	-	1.69	nr	1.69
Mimulus luteus	-	-	-	-	1.69	nr	1.69
Phragmites (australis)	-	-	-	-	1.69	nr	1.69
Ranunculus lingua	-	-	-	-	1.69	nr	1.69
Scirpus lacustris	-	-	-	-	1 69	nr	1.69
Sagittaria sagittifolia var. leucopetala	-	-	-	-	1.69	nr	1.69
Typha spp.	-	-	-	-	1.69	nr	1.69

Q PAVING/PLANTING/FENCING/SITE FURNITURE Excluding overheads and profit	PC £	Labour hours	Labour £	Plant £	Material £	Unit	Total rate £
Submerged oxygenating plants							
Ceratophyllum demersum; bunched and containerised	-	-	-	-	51.00	100 nr	51.00
Lagarosiphon major (Elodea crispa); bunched and containerised	-	-	-	-	33.00	100 nr	33.00
Ranunculus aquatilis; bunched and containerised	-	-	-	-	33.00	100 nr	33.00
Market Prices of Bulbs;							
Dutch Hyacinths (prepared for early flowering); 15 to 16 cm							
"Amsterdam", cerise red	-	-	-	-	150.00	100	150.00
"Anna Marie", clear pink	-	-	-	-	135.00	100	135.00
"Carnegie", white, dense spike	-	-	-	-	160.00	100	160.00
"Ostara", deep blue	-	-	-	-	165.00	100	165.00
"City of Haarlem", primrose yellow	-	-	-	-	168.00	100	168.00
Dutch Hyacinths (not prepared); 15 to 16 cm							
"Amsterdam", cerise red, Medium	-	-	-	-	119.00	100	119.00
"Anna Marie", clear pink, Early	-	-	-	-	113.00	100	113.00
"Carnegie", white, dense spike, Late	-	-	-	-	121.00	100	121.00
"Ostara", deep blue, Early	-	-	-	-	116.00	100	116.00
"City of Haarlem", primrose yellow, Late	-	-	-	-	119.00	100	119.00
Tulips, Single Early,							
"Apricot Beauty", soft salmon rose	-	-	-	-	75.00	1000	75.00
"Best Seller", bright coppery orange	-	-	-	-	86.00	1000	86.00
"Superb Mixed", blended colours	-	-	-	-	70.00	1000	70.00
Tulips, Double Early							
"Abba", glowing tomato red	-	-	-	-	69.00	1000	69.00
"Electra", deep cherry red	-	-	-	-	87.00	1000	87.00
"Murillo Sports Mixed", selected species	-	-	-	-	88.00	1000	88.00
Tulips, Triumph							
"Abu Hassan", deep mahogany, broad yellow edge	-	-	-	-	72.00	1000	72.00
"Arabian Mystery", deep purple violet, edged white	-	-	-	-	75.00	1000	75.00
"Coquette", creamy white	-	-	-	-	86.00	1000	86.00
"Superb Mixed", blended colours	-	-	-	-	73.00	1000	73.00
Tulips, Darwin Hybrid							
"Ad Rem", scarlet	-	-	-	-	75.00	1000	75.00
"Golden Parade", deep yellow	-	-	-	-	69.00	1000	69.00
"Superb Mixed", blended colours	-	-	-	-	61.00	1000	61.00
Tulips, Single Late or Cottage							
"Alabaster", pure white	-	-	-	-	83.00	1000	83.00
"Blushing Beauty", red, feathered rosy white	-	-	-	-	102.00	1000	102.00
"Superb Mixed", blended colours	-	-	-	-	73.00	1000	73.00
Tulips, Lily Flowered							
"Aladdin", scarlet, edged yellow	-	-	-	-	78.00	1000	78.00
"Superb Mixed", blended colours	-	-	-	-	87.00	1000	87.00
Tulips, Parrot							
"Apricot Parrot", pale apricot yellow, tinged creamy white	-	-	-	-	81.00	1000	81.00
"Rococo", carmine edged glowing red	-	-	-	-	69.00	1000	69.00
"Superb Mixed", blended colours	-	-	-	-	100.00	1000	100.00
Tulips, Double Late							
"Lilac Perfection", lilac	-	-	-	-	104.00	1000	104.00
"Upstar", pale pink and white	-	-	-	-	66.00	1000	66.00
"Superb Mixed", blended colours	-	-	-	-	85.00	1000	85.00

Prices for Measured Works

Q PAVING/PLANTING/FENCING/SITE FURNITURE Excluding overheads and profit	PC £	Labour hours	Labour £	Plant £	Material £	Unit	Total rate £
Q31 PLANTING - cont'd							
Market Prices of Bulbs; - cont'd							
Tulips, Greigii							
"Ali Baba", rose	-	-	-	-	75.00	1000	75.00
"Mary Ann", carmine red, edged white	-	-	-	-	92.00	1000	92.00
"Superb Mixed", blended colours	-	-	-	-	82.00	1000	82.00
Daffodils, Yellow Trumpets							
"Dutch Master", large golden yellow	-	-	-	-	65.00	1000	65.00
"King Alfred", medium golden yellow	-	-	-	-	65.00	1000	65.00
Daffodils, Bi-Coloured Trumpets							
"Magnet", perianth creamy white, lemon yellow trumpet	-	-	-	-	75.00	1000	75.00
Daffodils, White Trumpets							
"Mount Hood", pure white perianth, pure white frilled trumpet	-	-	-	-	90.00	1000	90.00
Narcissi, Large Cupped							
"Carbineer", rich yellow perianth, bright orange cup	-	-	-	-	65.00	1000	65.00
"St Keverne", deep golden yellow	-	-	-	-	70.00	1000	70.00
"Salome", white perianth, shaded pink trumpet	-	-	-	-	85.00	1000	85.00
Narcissi, Small Cupped							
"Birma", soft yellow perianth, orange scarlet cup	-	-	-	-	70.00	1000	70.00
"Verger", white perianth, orange scarlet cup	-	-	-	-	75.00	1000	75.00
Narcissi, Double							
"Flower Drift", white perianth, orange yellow cup	-	-	-	-	75.00	1000	75.00
"Texas", yellow outer petals, inter-mixed with orange scarlet	-	-	-	-	85.00	1000	85.00
Narcissi, Triandrus							
"Albus" (Angel Tears), reflexing silvery white flowers in clusters	-	-	-	-	90.00	1000	90.00
"Hawera", 2 or 3 creamy yellow flowers on a stem	-	-	-	-	49.00	1000	49.00
Narcissi, Cyclamineus							
"Jack Snipe", long creamy white reflexing petals, short orange yellow cup	-	-	-	-	85.00	1000	85.00
Narcissi, Jonquilla							
"Jonquilla Single", buttercup yellow flowers, sweet scented	-	-	-	-	69.00	1000	69.00
"Sugar Bush", ivory white petals, orange yellow cup	-	-	-	-	91.00	1000	91.00
Narcissi, Tazetta or Poetaz							
"Cheerfulness", creamy white double flowers, scented	-	-	-	-	65.00	1000	65.00
"Sir Winston Churchill", double geranium, white perianth, white trumpet	-	-	-	-	75.00	1000	75.00
Daffodils, Miniature							
"Lobularis", lemon yellow tapering perianth, darker yellow trumpet, very early	-	-	-	-	60.00	1000	60.00
"Minnow", miniature tazetta hybrid, 2-3 soft flowers on each stem	-	-	-	-	43.00	1000	43.00
"Nanus", a delightful yellow miniature	-	-	-	-	107.00	1000	107.00
Daffodils & Narcissi, Special Mixture							
suitable for planting in parkland, river banks etc.	-	-	-	-	65.00	1000	65.00

Q PAVING/PLANTING/FENCING/SITE FURNITURE Excluding overheads and profit	PC £	Labour hours	Labour £	Plant £	Material £	Unit	Total rate £
Crocus, Spring Flowering Dutch							
Blue	-	-	-	-	32.00	1000	32.00
Yellow	-	-	-	-	32.00	1000	32.00
Striped	-	-	-	-	40.00	1000	40.00
Mixed	-	-	-	-	33.00	1000	33.00
Crocus, Winter Flowering Species							
"Ancyrensis" (Golden Bunch), maize yellow with orange interior	-	-	-	-	24.00	1000	24.00
"Chrysanthus Advance", yellow and violet	-	-	-	-	27.00	1000	27.00
"Sieberi Violet Queen", dark blue	-	-	-	-	30.00	1000	30.00
Species Mixed, blended colours	-	-	-	-	22.00	1000	22.00
Iris, Dutch							
"Golden Harvest", golden yellow	-	-	-	-	35.00	1000	35.00
"Imperator", very deep blue	-	-	-	-	32.00	1000	32.00
"Superb Mixed", blended colours	-	-	-	-	30.00	1000	30.00
Iris, Species							
"Danfordiae", sweet scented, deep lemon yellow with brown markings; dwarf variety	-	-	-	-	35.00	1000	35.00
"Reticulata", deep purple violet with yellow blotch; dwarf variety	-	-	-	-	27.00	1000	27.00
"Reticulata Harmony", royal blue with yellow rimmed white blotch	-	-	-	-	27.00	1000	27.00
"Superb Mixed", blended colours	-	-	-	-	28.00	1000	28.00
Allium							
"Giganteum", deep violet, star shaped flowers, height 80 cm	-	-	-	-	147.50	100	147.50
"Neapolitanum", pure white, height 30 cm	-	-	-	-	19.00	1000	19.00
Amaryllis Hippeastrum							
"Apple Blossom", blushing pink	-	-	-	-	16.00	10	16.00
"Striped", red and white	-	-	-	-	16.00	10	16.00
Anemones							
"Single De Caen", mixed shades	-	-	-	-	25.00	1000	25.00
"Blanda Pink Star", phlox purple	-	-	-	-	59.00	1000	59.00
Freesia							
Single Flowering Mixture	-	-	-	-	44.00	1000	44.00
Double Flowering Mixture	-	-	-	-	46.00	1000	46.00
Galanthus (Snowdrop)							
"Nivalis Double"	-	-	-	-	98.00	1000	98.00
"Nivalis Single"	-	-	-	-	43.00	1000	43.00
"Nivalis Single", supplied 'in the green'	-	-	-	-	57.00	1000	57.00
Muscari (Grape Hyacinth)							
"Armeniacum", bright blue	-	-	-	-	22.00	1000	22.00
"Botryoides Album", white	-	-	-	-	54.00	1000	54.00
Ranunculus							
Paeony Flowering Mixed	-	-	-	-	42.00	1000	42.00
Scilla							
"Campanulata Blue"	-	-	-	-	51.00	1000	51.00
Sparaxis							
"Superb Mixed"	-	-	-	-	20.00	1000	20.00

Q PAVING/PLANTING/FENCING/SITE FURNITURE Excluding overheads and profit	PC £	Labour hours	Labour £	Plant £	Material £	Unit	Total rate £
Q31 PLANTING - cont'd							
Planting - General Preamble: The British Standard recommendations for Nursery Stock are in BS 3936: pts 1-5. Planting of trees and shrubs as well as forestry is covered by the appropriate sections of BS 4428. Transplanting of semi-mature trees is covered by BS 4043. Prices for all planting work are deemed to include carrying out planting in accordance with BS 4428 and good horticultural practice.							
Site protection; temporary protective fencing Rolled chestnut pale fencing; to 100 mm diameter chestnut posts; driving into firm ground at 3 m centres; pales at 50 mm centres							
900 mm high	1.76	0.11	1.60	-	3.79	m	5.39
1200 mm high	2.36	0.11	1.60	-	4.31	m	5.90
1500 mm high	-	0.11	1.60	-	5.53	m	7.13
Chain link fencing; to BS 1722 Pt 1; including timber posts at 3m centres							
1.80 m high	6.30	0.13	2.00	-	6.30	m	8.30
2 13 m high	7.30	0.17	2.50	-	7.30	m	9.80
Extra over temporary protective fencing for removing and making good (no allowance for re-use of material)	-	0.07	1.00	0.19	-	m	1.19
Cultivation Treating soil with "Paraquat-Diquat" weedkiller at rate of 5 litre/ha; PC £22.20 / litre; in accordance with manufacturer's instructions; including all safety precautions							
by machine	-	-	-	0.24	1.11	100m²	1.35
by hand	-	0.13	2.00	-	1.11	100m²	3.11
Ripping up subsoil; using approved subsoiling machine; minimum depth 250 mm below topsoil, at 1.20 m centres; in							
gravel or sandy clay	-	-	-	2.31	-	100m²	2.31
soil compacted by machines	-	-	-	2.70	-	100m²	2.70
clay	-	-	-	2.89	-	100m²	2.89
chalk or other soft rock	-	-	-	5.78	-	100m²	5.78
Extra for subsoiling at 1 m centres	-	-	-	0.58	-	100m²	0.58
Breaking up existing ground; using pedestrian operated tine cultivator or rotavator							
100 mm deep	-	0.22	3.30	2.13	-	100m²	5.43
150 mm deep	-	0.28	4.13	2.67	-	100m²	6.79
200 mm deep	-	0.37	5.50	3.56	-	100m²	9.05
As above but in heavy clay or wet soils							
100 mm deep	-	0.44	6.60	4.27	-	100m²	10.87
150 mm deep	-	0.66	9.90	6.40	-	100m²	16.30
200 mm deep	-	0.82	12.38	8.00	-	100m²	20.38

Q PAVING/PLANTING/FENCING/SITE FURNITURE Excluding overheads and profit	PC £	Labour hours	Labour £	Plant £	Material £	Unit	Total rate £
Breaking up existing ground; using tractor drawn tine cultivator or rotavator							
100 mm deep	-	-	-	0.52	-	100m²	0.52
150 mm deep	-	-	-	0.65	-	100m²	0.65
200 mm deep	-	-	-	0.86	-	100m²	0.86
600 mm deep	-	-	-	2.58	-	100m²	2.58
Cultivating ploughed ground; using disc, drag, or chain harrow							
4 passes	-	-	-	3.10	-	100m²	3.10
Rolling cultivated ground lightly; using self-propelled agricultural roller	-	0.06	0.83	0.58	-	100m²	1.41
Importing only selected and approved topsoil; to BS 3882; from source not exceeding 13 km from site							
1 - 14 m³	11.66	-	-	-	34.98	m³	34.98
over 15 m³	11.66	-	-	-	11.66	m³	11.66
Spreading and lightly consolidating approved topsoil (imported or from spoil heaps); in layers not exceeding 150 mm; travel distance from spoil heaps not exceeding 100 m; by machine (imported topsoil not included)							
minimum depth 100 mm	-	1.55	23.25	35.53	-	100m²	58.78
minimum depth 150 mm	-	2.33	35.00	53.47	-	100m²	88.47
minimum depth 300 mm	-	4.67	70.00	106.94	-	100m²	176.94
minimum depth 450 mm	-	6.99	104.85	160.24	-	100m²	265.09
Spreading and lightly consolidating approved topsoil (imported or from spoil heaps); in layers not exceeding 150 mm; travel distance from spoil heaps not exceeding 100 m; by hand (imported topsoil not included)							
minimum depth 100 mm	-	20.00	300.06	-	-	100m²	300.06
minimum depth 150 mm	-	30.01	450.09	-	-	100m²	450.09
minimum depth 300 mm	-	60.01	900.18	-	-	100m²	900.18
minimum depth 450 mm	-	90.02	1350.27	-	-	100m²	1350.27
Extra over for spreading topsoil to slopes 15 - 30 degrees by machine or hand	-	-	-	-	-	10%	-
Extra over for spreading topsoil to slopes over 30 degrees by machine or hand	-	-	-	-	-	25%	-
Extra over spreading topsoil for travel exceeding 100 m; by machine							
100 - 150 m	-	0.02	0.30	0.16	-	m³	0.46
150 - 200 m	-	0.03	0.40	0.21	-	m³	0.62
200 - 300 m	-	0.04	0.53	0.29	-	m³	0.82
Extra over spreading topsoil for travel exceeding 100 m; by hand							
100 m	-	2.50	37.50	-	-	m³	37.50
200 m	-	3.50	52.50	-	-	m³	52.50
300 m	-	4.50	67.50	-	-	m³	67.50
Evenly grading; to general surfaces to bring to finished levels							
by machine (tractor mounted rotavator)	-	-	-	0.03	-	m²	0.03
by pedestrian operated rotavator	-	-	0.06	0.04	-	m²	0.10
by hand	-	0.01	0.15	-	-	m²	0.15
Extra over grading for slopes 15 - 30 degrees by machine or hand	-	-	-	-	-	10%	-
Extra over grading for slopes over 30 degrees by machine or hand	-	-	-	-	-	25%	-

Q PAVING/PLANTING/FENCING/SITE FURNITURE Excluding overheads and profit	PC £	Labour hours	Labour £	Plant £	Material £	Unit	Total rate £
Q31 PLANTING - cont'd							
Cultivation - cont'd							
Clearing stones; disposing off site; to distance not exceeding 13 km							
by hand; stones not exceeding 50 mm in any direction; loading to skip 5.35 m³	-	0.01	0.15	0.02	-	m²	**0.17**
by mechanical stone rake; stones not exceeding 50 mm in any direction; loading to 15 m³ truck							
by mechanical loader	-	-	0.03	0.07	-	m²	**0.10**
Lightly cultivating; weeding; to fallow areas; disposing debris off site; to distance not exceeding 13 km							
by hand	-	0.01	0.21	-	0.08	m²	**0.29**
Preparation of planting operations							
For the following topsoil improvement and planting operations add or subtract the following amounts for every £0.10 difference in the material cost price							
35 g/m²	-	-	-	-	0.35	100m²	**0.35**
50 g/m²	-	-	-	-	0.50	100m²	**0.50**
70 g/m²	-	-	-	-	0.70	100m²	**0.70**
100 g/m²	-	-	-	-	1.00	100m²	**1.00**
150 kg/ ha	-	-	-	-	15.00	ha	**15.00**
200 kg/ ha	-	-	-	-	20.00	ha	**20.00**
250 kg/ ha	-	-	-	-	25.00	ha	**25.00**
300 kg/ ha	-	-	-	-	30.00	ha	**30.00**
400 kg/ ha	-	-	-	-	40.00	ha	**40.00**
500 kg/ ha	-	-	-	-	50.00	ha	**50.00**
700 kg/ ha	-	-	-	-	70.00	ha	**70.00**
1000 kg/ ha	-	-	-	-	100.00	ha	**100.00**
1250 kg/ ha	-	-	-	-	125.00	ha	**125.00**
Selective herbicides; in accordance with manufacturer's instructions; PC £26.47 per litre; by machine; application rate							
30 ml/100m²	-	-	-	0.29	0.79	100m²	**1.08**
35 ml/100m²	-	-	-	0.29	0.93	100m²	**1.22**
40 ml/100m²	-	-	-	0.29	1.06	100m²	**1.35**
50 ml/100m²	-	-	-	0.29	1.32	100m²	**1.61**
3.00 l/ha	-	-	-	28.89	79.42	ha	**108.31**
3.50 l/ ha	-	-	-	28.89	92.66	ha	**121.55**
4.00 l/ha	-	-	-	28.89	105.90	ha	**134.79**
Selective herbicides; in accordance with manufacturer's instructions; PC £26.47 per litre; by hand; application rate							
30 ml/m²	-	0.17	2.50	-	0.79	100m²	**3.29**
35 ml/m²	-	0.17	2.50	-	0.93	100m²	**3.43**
40 ml/m²	-	0.17	2.50	-	1.06	100m²	**3.56**
50 ml/m²	-	0.17	2.50	-	1.32	100m²	**3.82**
3.00 l/ha	-	16.67	250.00	-	79.42	ha	**329.43**
3.50 l/ha	-	16.67	250.00	-	92.66	ha	**342.66**
4.00 l/ha	-	16.67	250.00	-	105.90	ha	**355.90**

Q PAVING/PLANTING/FENCING/SITE FURNITURE Excluding overheads and profit	PC £	Labour hours	Labour £	Plant £	Material £	Unit	Total rate £
General herbicides, in accordance with manufacturer's instructions; "Knapsack" spray application							
"Super Verdone" at 100 ml/100 m²	0.60	0.05	0.75	-	0.60	100m²	1.35
"Roundup Pro" at 40 ml/100 m²	0.16	0.05	0.75	-	0.16	100m²	0.91
"Spasor" at 50 ml/100 m²	0.70	0.05	0.75	-	0.70	100m²	1.45
"Casoron G (residual)" at 1kg/125m²	3 04	0.05	0.75	-	3.34	100m²	4.09
Fertilizers; in top 150 mm of topsoil; at 35g/m²							
fertilizer (18:0:0+Mg +Fe)	5.94	0.12	1.84	-	5.94	100m²	7.77
"Enmag"	3 90	0.12	1.84	-	4.09	100m²	5.93
fertilizer (7:7:7)	1.68	0.12	1.84	-	1.68	100m²	3.52
"Superphosphate"	1.57	0.12	1.84	-	1.65	100m²	3.49
fertilizer (20:10:10)	2 37	0.12	1.84	-	2.37	100m²	4.21
"Hoof and Horn"	2 17	0.12	1.84	-	2.28	100m²	4.12
"Bone meal"	3 01	0.12	1.84	-	3.16	100m²	5.00
Fertilizers; in top 150 mm of topsoil at 70 g/m²							
fertilizer (18:0:0+Mg +Fe)	11 87	0.12	1.84	-	11.87	100m²	13.71
"Enmag"	7 80	0.12	1.84	-	8.19	100m²	10.03
fertilizer (7:7:7)	3.36	0.12	1.84	-	3.36	100m²	5.20
"Superphosphate"	3 15	0.12	1.84	-	3.31	100m²	5.14
fertilizer (20:10:10)	4.75	0.12	1.84	-	4.75	100m²	6.58
"Hoof and Horn"	4.34	0.12	1.84	-	4.56	100m²	6.39
"Bone meal"	6.02	0.12	1.84	-	6.32	100m²	8.16
Spreading topsoil; from dump not exceeding 100 m distance; by machine (topsoil not included)							
at 1 m³ per 13 m²; 75 mm thick	-	0.56	8.33	27.25	-	100m²	35.58
at 1 m³ per 10 m²; 100 mm thick	-	0.89	13.33	28.45	-	100m²	41.78
at 1 m³ per 6.50 m²; 150 mm thick	-	1.11	16.65	49.38	-	100m²	66.03
at 1 m³ per 5 m²; 200 mm thick	-	1.48	22.20	65.65	-	100m²	87.85
Spreading topsoil; from dump not exceeding 100 m distance; by hand (topsoil not included)							
at 1 m³ per 13 m²; 75 mm thick	-	15.00	225.04	-	-	100m²	225.04
at 1 m³ per 10 m²; 100 mm thick	-	20.00	300.06	-	-	100m²	300.06
at 1 m³ per 6.50 m²; 150 mm thick	-	30.01	450.09	-	-	100m²	450.09
at 1 m³ per 5 m²; 200 mm thick	-	36.67	550.11	-	-	100m²	550.11
Imported topsoil; tipped 100 m from area of application; by machine							
at 1 m³ per 13 m²; 75 mm thick	87.45	0.56	8.33	27.25	104.94	100m²	140.52
at 1 m³ per 10 m²; 100 mm thick	116.60	0.89	13.33	28.45	139.92	100m²	181.70
at 1 m³ per 6.50 m²; 150 mm thick	174.90	1.11	16.65	49.38	209.88	100m²	275.91
at 1 m³ per 5 m²; 200 mm thick	233.20	1.48	22.20	65.65	279.84	100m²	367.69
Imported topsoil; tipped 100 m from area of application; by hand							
at 1 m³ per 13 m²; 75 mm thick	87.45	15.00	225.04	-	104.94	100m²	329.99
at 1 m³ per 10 m²; 100 mm thick	116.60	20.00	300.06	-	139.92	100m²	439.98
at 1 m³ per 6.50 m²; 150 mm thick	174.90	30.01	450.09	-	209.88	100m²	659.97
at 1 m³ per 5 m²; 200 mm thick	233 20	36.67	550.11	-	279.84	100m²	829.95
Composted bark and manure soil conditioner (20 m³ loads); from not further than 25m from location; cultivating into topsoil by machine							
50 mm thick	130.00	2.86	42.86	1.78	136.50	100m²	181.14
100 mm thick	260.00	6.05	90.73	1.78	273.00	100m²	365.51
150 mm thick	390.00	8.90	133.57	1.78	409.50	100m²	544.85
200 mm thick	520.00	12.90	193.57	1.78	546.00	100m²	741.35

Q PAVING/PLANTING/FENCING/SITE FURNITURE Excluding overheads and profit	PC £	Labour hours	Labour £	Plant £	Material £	Unit	Total rate £
Q31 PLANTING - cont'd							
Preparation of planting operations - cont'd							
Composted bark soil conditioner (20 m³ loads); placing on beds by mechanical loader; spreading and rotavating into topsoil by machine							
50 mm thick	130.00	-	-	7.38	130.00	100m²	137.38
100 mm thick	260.00	-	-	11.29	273.00	100m²	284.29
150 mm thick	390.00	-	-	15.35	409.50	100m²	424.85
200 mm thick	520.00	-	-	19.34	546.00	100m²	565.34
Mushroom compost (20 m³ loads); from not further than 25m from location; cultivating into topsoil by machine							
50 mm thick	40.00	2.86	42.86	1.78	40.00	100m²	84.64
100 mm thick	80.00	6.05	90.73	1.78	80.00	100m²	172.51
150 mm thick	120.00	8.90	133.57	1.78	120.00	100m²	255.35
200 mm thick	160.00	12.90	193.57	1.78	160.00	100m²	355.35
Mushroom compost (20 m³ loads); placing on beds by mechanical loader; spreading and rotavating into topsoil by machine							
50 mm thick	40.00	-	-	7.38	40.00	100m²	47.38
100 mm thick	80.00	-	-	11.29	80.00	100m²	91.29
150 mm thick	120.00	-	-	15 35	120.00	100m²	135.35
200 mm thick	160.00	-	-	19.34	160.00	100m²	179.34
Manure (60 m³ loads); from not further than 25m from location; cultivating into topsoil by machine							
50 mm thick	55.00	2.86	42.86	1.78	55.00	100 m²	99.64
100 mm thick	110.00	6.05	90.73	1.78	115.50	100 m²	208.01
150 mm thick	165.00	8.90	133.57	1.78	173.25	100 m²	308.60
200 mm thick	220.00	12.90	193.57	1.78	231.00	100 m²	426.35
Surface applications and soil additives; pre-planting; from not further than 25 m from location; by machine							
Ground limestone soil conditioner; including turning in to cultivated ground							
0.25 kg/m² = 2.50 tonnes/ha	0.23	-	-	4.03	0.23	100m²	4.26
0.50 kg/m² = 5.00 tonnes/ha	0.47	-	-	4.03	0.47	100m²	4.49
0.75 kg/m² = 7.50 tonnes/ha	0.70	-	-	4.03	0.70	100m²	4.73
1.00 kg/m² = 10.00 tonnes/ha	0.93	-	-	4.03	0.93	100m²	4.96
Medium bark soil conditioner; A.H.S. Ltd; including turning in to cultivated ground; delivered in 15 m³ loads							
1 m³ per 40 m² = 25 mm thick	0.75	-	-	0.09	0.75	m²	0.84
1 m³ per 20 m² = 50 mm thick	1.50	-	-	0.14	1.50	m²	1.64
1 m³ per 13.33 m² = 75 mm thick	2.25	-	-	0.19	2.25	m²	2.44
1 m³ per 10 m² = 100 mm thick	3.00	-	-	0.24	3.00	m²	3.24
Peat soil conditioner; A.H.S. Ltd; including turning in to cultivated ground; delivered in 55 m³ loads							
1 m³ per 40 m² = 25 mm thick	0.68	-	-	0.09	0.68	m²	0.77
1 m³ per 20 m² = 50 mm thick	1.35	-	-	0.14	1.35	m²	1.49
1 m³ per 13.33 m² = 75 mm thick	2.02	-	-	0.19	2.02	m²	2.22
1 m³ per 10 m² = 100 mm thick	2.70	-	-	0.24	2.70	m²	2.94

Q PAVING/PLANTING/FENCING/SITE FURNITURE Excluding overheads and profit	PC £	Labour hours	Labour £	Plant £	Material £	Unit	Total rate £
Mushroom compost soil conditioner; A.H.S. Ltd; including turning in to cultivated ground; delivered in 20 m³ loads							
1 m³ per 40 m² = 25 mm thick	0.20	0.02	0.25	-	0.20	m²	0.45
1 m³ per 20 m² = 50 mm thick	0.40	0.03	0.46	-	0.40	m²	0.86
1 m³ per 13.33 m² = 75 mm thick	0.60	0.04	0.60	-	0.60	m²	1.20
1 m³ per 10 m² = 100 mm thick	0.80	0.05	0.75	-	0.80	m²	1.55
Mushroom compost soil conditioner; A.H.S. Ltd; including turning in to cultivated ground; delivered in 55 m³ loads							
1 m³ per 40 m² = 25 mm thick	0.10	0.02	0.25	-	0.10	m²	0.35
1 m³ per 20 m² = 50 mm thick	0.20	0.03	0.46	-	0.20	m²	0.66
1 m³ per 13.33 m² = 75 mm thick	0.30	0.04	0.60	-	0.30	m²	0.90
1 m³ per 10 m² = 100 mm thick	0.40	0.05	0.75	-	0.40	m²	1.15
Surface applications and soil additives; pre-planting; from not further than 25 m from location; by hand							
Ground limestone soil conditioner; including turning in to cultivated ground							
0.25 kg/m² = 2.50 tonnes/ha	0.23	1.20	18.00	-	0.23	100m²	18.23
0.50 kg/m² = 5.00 tonnes/ha	0.47	1.33	20.00	-	0.47	100m²	20.46
0.75 kg/m² = 7.50 tonnes/ha	0.70	1.50	22.50	-	0.70	100m²	23.20
1.00 kg/m² = 10.00 tonnes/ha	0.93	1.71	25.71	-	0.93	100m²	26.64
Medium bark soil conditioner; A.H.S. Limited; including turning in to cultivated ground; delivered in 15 m³ loads							
1 m³ per 40 m² = 25 mm thick	0.75	0.02	0.33	-	0.75	m²	1.08
1 m³ per 20 m² = 50 mm thick	1.50	0.04	0.67	-	1.50	m²	2.16
1 m³ per 13.33 m² = 75 mm thick	2.25	0.07	1.00	-	2.25	m²	3.25
1 m³ per 10 m² = 100 mm thick	3.00	0.08	1.20	-	3.00	m²	4.20
Peat soil conditioner; A.H.S. Limited; including turning in to cultivated ground; delivered in 55 m³ loads							
1 m³ per 40 m² = 25 mm thick	0.68	0.02	0.33	-	0.68	m²	1.01
1 m³ per 20 m² = 50 mm thick	1.35	0.04	0.67	-	1.35	m²	2.02
1 m³ per 13.33 m² = 75 mm thick	2.02	0.07	1.00	-	2.02	m²	3.02
1 m³ per 10 m² = 100 mm thick	2.70	0.08	1.20	-	2.70	m²	3.90
Mushroom compost soil conditioner; A.H.S. Limited; including turning in to cultivated ground; delivered in 20 m³ loads							
1 m³ per 40 m² = 25 mm thick	0.20	0.02	0.33	-	0.20	m²	0.53
1 m³ per 20 m² = 50 mm thick	0.40	0.04	0.67	-	0.40	m²	1.07
1 m³ per 13.33 m² = 75 mm thick	0.60	0.07	1.00	-	0.60	m²	1.60
1 m³ per 10 m² = 100 mm thick	0.80	0.08	1.20	-	0.80	m²	2.00
Mushroom compost soil conditioner; A.H.S. Limited; including turning in to cultivated ground; delivered in 55 m³ loads							
1 m³ per 40 m² = 25 mm thick	0.10	0.02	0.33	-	0.10	m²	0.43
1 m³ per 20 m² = 50 mm thick	0.20	0.04	0.67	-	0.20	m²	0.87
1 m³ per 13.33 m² = 75 mm thick	0.30	0.07	1 00	-	0.30	m²	1.30
1 m³ per 10 m² = 100 mm thick	0.40	0.08	1 20	-	0 40	m²	1.60
"Super Humus" mixed bark and manure conditioner; Melcourt Industries Ltd; delivered in 25 m³ loads							
1 m³ per 40 m² = 25 mm thick	0.65	0.02	0 33	-	0.68	m²	1.02
1 m³ per 20 m² = 50 mm thick	1.30	0.04	0.67	-	1 36	m²	2.03
1 m³ per 13.33 m² = 75 mm thick	1.95	0.07	1.00	-	2.05	m²	3.05
1 m³ per 10 m² = 100 mm thick	2.60	0.08	1 20	-	2.73	m²	3.93

Q PAVING/PLANTING/FENCING/SITE FURNITURE Excluding overheads and profit	PC £	Labour hours	Labour £	Plant £	Material £	Unit	Total rate £
Q31 PLANTING - cont'd							
Tree planting; pre-planting operations							
Excavating tree pits; depositing soil alongside pits, by machine							
600 mm x 600 mm x 600 mm deep	-	0.15	2.21	0.76	-	nr	2.97
900 mm x 900 mm x 600 mm deep	-	0.33	4.95	1.71	-	nr	6.66
1.00 m x 1.00 m x 600 mm deep	-	0.61	9.20	2.12	-	nr	11.32
1 25 m x 1.25 m x 600 mm deep	-	0.96	14.41	3.31	-	nr	17.71
1 00 m x 1.00 m x 1.00 m deep	-	1.02	15.34	3.53	-	nr	18.87
1.50 m x 1.50 m x 750 mm deep	-	1.73	25.89	5.95	-	nr	31.84
1.50 m x 1.50 m x 1.00 m deep	-	2.30	34.43	7.91	-	nr	42.34
1.75 m x 1.75 m x 1.00 mm deep	-	3.13	46.98	10.80	-	nr	57.78
2.00 m x 2.00 m x 1.00 mm deep	-	4.09	61.36	14.11	-	nr	75.47
Excavating tree pits; depositing soil alongside pits, by hand							
600 mm x 600 mm x 600 mm deep	-	0.44	6.60	-	-	nr	6.60
900 mm x 900 mm x 600 mm deep	-	1.00	15.00	-	-	nr	15.00
1.00 m x 1.00 m x 600 mm deep	-	1.13	16.88	-	-	nr	16.88
1.25 m x 1.25 m x 600 mm deep	-	1.93	28.95	-	-	nr	28.95
1.00 m x 1.00 m x 1.00 m deep	-	2.06	30.90	-	-	nr	30.90
1.50 m x 1.50 m x 750 mm deep	-	3.47	52.05	-	-	nr	52.05
1.75 m x 1.50 m x 750 mm deep	-	4.05	60 75	-	-	nr	60.75
1 50 m x 1.50 m x 1.00 m deep	-	4.63	69.45	-	-	nr	69.45
2 00 m x 2.00 m x 750 mm deep	-	6.17	92.55	-	-	nr	92.55
2 00 m x 2.00 m x 1.00 m deep	-	8.23	123.46	-	-	nr	123.46
Breaking up subsoil in tree pits; to a depth of 200 mm	-	0.03	0.50	-	-	m^2	0.50
Spreading and lightly consolidating approved topsoil (imported or from spoil heaps); in layers not exceeding 150 mm; distance from spoil heaps not exceeding 100 m (imported topsoil not included); by machine							
minimum depth 100 mm	-	1.55	23.25	35.53	-	$100m^2$	58.78
minimum depth 150 mm	-	2.33	35.00	53.47	-	$100m^2$	88.47
minimum depth 300 mm	-	4.67	70.00	106.94	-	$100m^2$	176.94
minimum depth 450 mm	-	6.99	104.85	160.24	-	$100m^2$	265.09
Spreading and lightly consolidating approved topsoil (imported or from spoil heaps); in layers not exceeding 150 mm; distance from spoil heaps not exceeding 100 m (imported topsoil not included); by hand							
minimum depth 100 mm	-	20.00	300.06	-	-	$100m^2$	300.06
minimum depth 150 mm	-	30.01	450.09	-	-	$100m^2$	450.09
minimum depth 300 mm	-	60.01	900.18	-	-	$100m^2$	900.18
minimum depth 450 mm	-	90.02	1350.27	-	-	$100m^2$	1350.27
Extra for filling tree pits with imported topsoil; PC £11.66 /m^3; plus allowance for 20% settlement							
depth 100 mm	-	-	-	-	1.40	m^2	1.40
depth 150 mm	-	-	-	-	2.10	m^2	2.10
depth 200 mm	-	-	-	-	2.80	m^2	2.80
depth 300 mm	-	-	-	-	4.20	m^2	4.20
depth 400 mm	-	-	-	-	5.60	m^2	5.60
depth 450 mm	-	-	-	-	6.30	m^2	6.30
depth 500 mm	-	-	-	-	7.00	m^2	7.00
depth 600 mm	-	-	-	-	8.40	m^2	8.40

Q PAVING/PLANTING/FENCING/SITE FURNITURE Excluding overheads and profit	PC £	Labour hours	Labour £	Plant £	Material £	Unit	Total rate £
Add or deduct the following amounts for every £0 50 change in the material price of topsoil							
depth 100 mm	-	-	-	-	0.06	m²	0.06
depth 150 mm	-	-	-	-	0.09	m²	0.09
depth 200 mm	-	-	-	-	0.12	m²	0.12
depth 300 mm	-	-	-	-	0.18	m²	0.18
depth 400 mm	-	-	-	-	0.24	m²	0.24
depth 450 mm	-	-	-	-	0.27	m²	0.27
depth 500 mm	-	-	-	-	0.30	m²	0.30
depth 600 mm	-	-	-	-	0.36	m²	0.36
Structural Soils; Amsterdam Tree Sand; Urban Soils Ltd; Non compressive soil mixture for tree planting in areas to receive compressive surface treatments.							
backfilling and light compacting in layers; excavation, disposal, moving of material from delivery position and surface treatments not included; by machine							
individual treepits	45.38	0.50	7.50	2.59	45.38	m³	55.46
in trenches	45.38	0.42	6.25	2.15	45.38	m³	53.78
backfilling and light compacting in layers; excavation, disposal, moving of material from delivery position and surface treatments not included; by hand							
individual treepits	45.38	1.60	24.00	-	45.38	m³	69.38
in trenches	45.38	1.33	19.95	-	45.38	m³	65.33
Tree planting; tree staking and tree pit additives; J Toms Ltd							
Extra over trees for tree stake(s); driving 500 mm into firm ground; trimming to approved height; including two tree ties to approved pattern							
one stake; 100 mm diameter x 3.00 m long	4.18	0.20	3.00	-	4.18	nr	7.18
one stake; 100 mm diameter x 2.40 m long	3.25	0.20	3.00	-	3.25	nr	6.25
two stakes; 100 mm diameter x 3.00 m long	8.36	0.30	4.50	-	8.36	nr	12.86
two stakes; 100 mm diameter x 2.40 m long	6.50	0.30	4.50	-	6.50	nr	11.00
three stakes, 100 mm diameter x 3.00 m long	12.54	0.36	5.40	-	12.54	nr	17.94
three stakes, 100 mm diameter x 2.40 m long	9.75	0.36	5.40	-	9.75	nr	15.15
Platipus Tree Anchoring Systems							
ref RF1 rootball kit; for 75 to 225 mm girth, 2 to 4.5 m high inclusive of "Plati-mat" PM1	25.31	2.50	37.50	-	25.31	nr	62.81
ref RF2; rootball kit; for 225 to 450 mm girth, 4.5 to 7.5 m high inclusive of "Plati-mat" PM²	41.46	2.00	30.00	-	41.46	nr	71.46
ref RF3; rootball kit; for 450 to 750 mm girth, 7.5 to 12 m high "Plati-mat" PM³	87.62	1.33	20.00	-	87.62	nr	107.62
ref CG1; guy fixing kit; 75to 225 mm girth, 2 to 4.5 m high	16.47	1.67	25.00	-	16.47	nr	41.47
ref CG2; guy fixing kit; 225 to 450 mm girth, 4.5 to 7.5 m high	28.99	2.00	30.00	-	28.99	nr	58.99
installation tools; Drive Rod for CG1 Kit	-	-	-	-	52.18	nr	52.18
installation tools; Drive Rod for CG2 Kit	-	-	-	-	78.63	nr	78.63
Extra over trees for land drain to tree pits; 100 mm diameter perforated flexible agricultural drain; including excavating drain trench; laying pipe; backfilling	0.63	1.00	15.00	-	0.69	m	15.69

Q31 PLANTING - cont'd

Tree planting; tree staking and tree pit additives; J Toms Ltd - cont'd

Q PAVING/PLANTING/FENCING/SITE FURNITURE Excluding overheads and profit	PC £	Labour hours	Labour £	Plant £	Material £	Unit	Total rate £
Melcourt Industries Ltd; "Topgrow"; incorporating into topsoil at 1 part "Topgrow" to 3 parts excavated topsoil; supplied in 80 l bags; pit size							
600 mm x 600 mm x 600 mm	0.37	0.02	0.25	-	0.37	nr	0.62
900 mm x 900 mm x 900 mm	3.22	0.05	0.75	-	3.22	nr	3.97
1.00 m x 1.00 m x 1.00 m	5.89	0.20	3.00	-	5.89	nr	8.89
1.25 m x 1.25 m x 1.25 m	11.61	0.33	5.00	-	11.61	nr	16.61
1.50 m x 1.50 m x 1.50 m	20.11	0.75	11.25	-	20.11	nr	31.36
Melcourt Industries Ltd; "Topgrow"; incorporating into topsoil at 1 part "Topgrow" to 3 parts excavated topsoil; supplied in 60 m³ loose loads; pit size							
600 mm x 600 mm x 600 mm	0.83	0.02	0.25	-	0.83	nr	1.08
900 mm x 900 mm x 900 mm	2.79	0.05	0.75	-	2.79	nr	3.54
1.00 m x 1.00 m x 1.00 m	3.83	0.20	3.00	-	3.83	nr	6.83
1.25 m x 1.25 m x 1.25 m	7.47	0.33	5.00	-	7.47	nr	12.47
1.50 m x 1.50 m x 1.50 m	12.91	0.75	11.25	-	12.91	nr	24.16
Alginure Products; "Alginure Root Dip"; to bare rooted plants at 1 part "Alginure Root Dip" to 3 parts water							
transplants; at 3000/15 kg bucket	1.11	0.07	1.00	-	1.11	100 nr	2.11
standard trees; at 112/15 kg bucket of dip	0.30	0.03	0.50	-	0.30	nr	0.80
medium shrubs; at 600/15 kg bucket	0.06	0.02	0.25	-	0.06	nr	0.31
Tree Planting - Root Barriers							
English Woodlands; "Root Director", one-piece root control planters for installation at time of planting, to divert root growth down away from pavements and out for anchorage. Excavation measured separately							
RD 1050; 1050 x 1050 mm to 1300 x 1300 mm at base	68.25	0.25	3.75	-	68.25	nr	72.00
RD 640; 640 x 640 mm to 870 x 870 mm at base	40.95	0.25	3.75	-	40.95	nr	44.70
English Woodlands; linear root deflection barriers; installed to trench measured separately							
Re-Root 2000, 2.0 mm thick	7.46	0.05	0.75	-	7.46	m	8.21
Re-Root 2000, 1.5 mm thick	6.38	0.05	0.75	-	6.38	m	7.13
Re-Root 600, 1.0 mm thick.	5.50	0.05	0.75	-	5.50	m	6.25
Tree Planting - Irrigation Systems							
English Woodlands; "Root Rain" tree pit irrigation systems							
"Precinct"	34.50	0.25	3.75	-	34.50	nr	38.25
"Civic"	26.50	0.25	3.75	-	26.50	nr	30.25
"Metro"	5.40	0.25	3.75	-	5.40	nr	9.15
Mulching of tree pits; Melcourt Industries Ltd							
Spreading mulch; to individual trees; maximum distance 25 m (mulch not included)							
50 mm thick	-	0.05	0.73	-	-	m²	0.73
75 mm thick	-	0.07	1.09	-	-	m²	1.09
100 mm thick	-	0.10	1.46	-	-	m²	1.46

Prices for Measured Works

Q PAVING/PLANTING/FENCING/SITE FURNITURE Excluding overheads and profit	PC £	Labour hours	Labour £	Plant £	Material £	Unit	Total rate £
Mulch; "Bark Nuggets"; to individual trees; delivered in 70 m³ loads; maximum distance 25 m							
50 mm thick	1.45	0.05	0.73	-	1.52	m²	2.25
75 mm thick	2.17	0.07	1.09	-	2.29	m²	3.38
100 mm thick	2.90	0.10	1.46	-	3.04	m²	4.50
Mulch; "Bark Nuggets"; to individual trees; delivered in 20 m³ loads; maximum distance 25 m							
50 mm thick	2.07	0.05	0.73	-	2.17	m²	2.90
75 mm thick	3.10	0.05	0.75	-	3.26	m²	4.01
100 mm thick	4.14	0.07	1.00	-	4.35	m²	5.35
Mulch; "Amenity Bark"; to individual trees; delivered in 70 m³ loads; maximum distance 25 m							
50 mm thick	0.89	0.05	0.73	-	0.93	m²	1.66
75 mm thick	1.33	0.07	1.09	-	1.40	m²	2.49
100 mm thick	1.77	0.10	1.46	-	1.86	m²	3.32
Mulch; "Amenity Bark"; to individual trees; delivered in 20 m³ loads; maximum distance 25 m							
50 mm thick	1.51	0.05	0.73	-	1.58	m²	2.31
75 mm thick	2.26	0.07	1.09	-	2.38	m²	3.47
100 mm thick	3.02	0.07	1.00	-	3.17	m²	4.17
Trees; planting labours only Bare root trees; including backfilling with previously excavated material (all other operations and materials not included)							
light standard; 6 - 8 cm girth	-	0.35	5.25	-	-	nr	5.25
standard; 8 - 10 cm girth	-	0.40	6.00	-	-	nr	6.00
selected standard; 10 - 12 cm girth	-	0.58	8.70	-	-	nr	8.70
heavy standard; 12 - 14 cm girth	-	0.83	12.50	-	-	nr	12.50
extra heavy standard; 14 - 16 cm girth	-	1.00	15.00	-	-	nr	15.00
Root balled trees; including backfilling with previously excavated material (all other operations and materials not included)							
standard; 8 - 10 cm girth	-	0.50	7.50	-	-	nr	7.50
selected standard; 10 - 12 cm girth	-	0.60	9.00	-	-	nr	9.00
heavy standard; 12 - 14 cm girth	-	0.80	12.00	-	-	nr	12.00
extra heavy standard;14 - 16 cm girth	-	1.00	15.00	-	-	nr	15.00
16 - 18 cm girth	-	1.30	19.44	19.67	-	nr	39.11
18 - 20 cm girth	-	1.60	24.00	24.29	-	nr	48.29
20 - 25 cm girth	-	4.50	67.50	68.31	-	nr	135.81
25 - 30 cm girth	-	6.00	90.00	91.08	-	nr	181.08
30 - 35 cm girth	-	11.00	165.00	166.98	-	nr	331.98
Tree planting; root balled trees; **advanced nursery stock and semi-mature** **- General** Preamble: The cost of planting semi-mature trees will depend on the size and species, and on the access to the site for tree handling machines. Prices should be obtained for individual trees and planting.							

Q PAVING/PLANTING/FENCING/SITE FURNITURE Excluding overheads and profit	PC £	Labour hours	Labour £	Plant £	Material £	Unit	Total rate £
Q31 PLANTING - cont'd							
Tree planting; bare root trees; nursery stock							
"Acer platanoides"; including backfillling with excavated material (other operations not included)							
light standard; 6 - 8 cm girth	7.80	0.35	5.25	-	7.80	nr	**13.05**
standard; 8 - 10 cm girth	9.60	0.40	6.00	-	9.60	nr	**15.60**
selected standard; 10 - 12 cm girth	13.95	0.58	8.70	-	13.95	nr	**22.65**
heavy standard; 12 - 14 cm girth	20.00	0.83	12.50	-	20.00	nr	**32.50**
extra heavy standard; 14 - 16 cm girth	34.00	1.00	15.00	-	34.00	nr	**49.00**
"Carpinus betulus"; including backfillling with excavated material (other operations not included)							
light standard; 6 - 8 cm girth	12.25	0.35	5.25	-	12.25	nr	**17.50**
standard; 8 - 10 cm girth	13.95	0.40	6.00	-	13.95	nr	**19.95**
selected standard; 10 - 12 cm girth	23.50	0.58	8.70	-	23.50	nr	**32.20**
heavy standard; 12 - 14 cm girth	39.50	0.83	12.51	-	39.50	nr	**52.01**
extra heavy standard; 14 - 16 cm girth	34.00	1.00	15.00	-	34.00	nr	**49.00**
"Fraxinus excelsior"; including backfillling with excavated material (other operations not included)							
light standard; 6 - 8 cm girth	10.00	0.35	5.25	-	10.00	nr	**15.25**
standard; 8 - 10 cm girth	13.95	0.40	6.00	-	13.95	nr	**19.95**
selected standard; 10 - 12 cm girth	18.80	0.58	8.70	-	18.80	nr	**27.50**
heavy standard; 12 - 14 cm girth	39.50	0.83	12.45	-	39.50	nr	**51.95**
extra heavy standard; 14 - 16 cm girth	49.00	1.00	15.00	-	49.00	nr	**64.00**
"Prunus avium Plena"; including backfillling with excavated material (other operations not included)							
light standard; 6 - 8 cm girth	10.00	0.36	5.46	-	10.00	nr	**15.46**
standard; 8 - 10 cm girth	13.95	0.40	6.00	-	13.95	nr	**19.95**
selected standard; 10 - 12 cm girth	23.50	0.58	8.70	-	23.50	nr	**32.20**
heavy standard; 12 - 14 cm girth	39.50	0.83	12.50	-	39.50	nr	**52.00**
extra heavy standard; 14 - 16 cm girth	49.00	1.00	15.00	-	49.00	nr	**64.00**
"Quercus robur"; including backfillling with excavated material (other operations not included)							
light standard; 6 - 8 cm girth	12.25	0.35	5.25	-	12.25	nr	**17.50**
standard; 8 - 10 cm girth	13.95	0.40	6.00	-	13.95	nr	**19.95**
selected standard; 10 - 12 cm girth	23.50	0.58	8.70	-	23.50	nr	**32.20**
heavy standard; 12 - 14 cm girth	39.50	0.83	12.50	-	39.50	nr	**52.00**
extra heavy standard; 14 - 16 cm girth	49.00	1.00	15.00	-	49.00	nr	**64.00**
"Robinia pseudoacacia Frisia"; including backfillling with excavated material (other operations not included)							
light standard; 6 - 8 cm girth	26.25	0.35	5.25	-	26.25	nr	**31.50**
standard; 8 - 10 cm girth	34.50	0.40	6.00	-	34.50	nr	**40.50**
selected standard; 10 - 12 cm girth	47.00	0.58	8.70	-	47.00	nr	**55.70**
heavy standard; 12 - 14 cm girth	60.00	0.83	12.50	-	60.00	nr	**72.50**
extra heavy standard; 14 - 16 cm girth	75.00	1.00	15.00	-	75.00	nr	**90.00**

Prices for Measured Works

Q PAVING/PLANTING/FENCING/SITE FURNITURE Excluding overheads and profit	PC £	Labour hours	Labour £	Plant £	Material £	Unit	Total rate £
Tree planting; root balled trees; nursery stock							
"Acer platanoides"; including backfillling with excavated material (other operations not included)							
standard; 8 - 10 cm girth	-	0.48	7.20	-	13.60	nr	**20.80**
selected standard; 10 - 12 cm girth	-	0.56	8.41	-	21.95	nr	**30.36**
heavy standard; 12 - 14 cm girth	-	0.76	11.47	-	38.00	nr	**49.47**
extra heavy standard; 14 - 16 cm girth	-	1.20	18.00	-	52.00	nr	**70.00**
"Carpinus betulus"; including backfillling with excavated material (other operations not included)							
standard; 8 - 10 cm girth	-	0.48	7.20	-	17.95	nr	**25.15**
selected standard; 10 - 12 cm girth	-	0.56	8.41	-	31.50	nr	**39.91**
heavy standard; 12 - 14 cm girth	-	0.76	11.47	-	51.50	nr	**62.97**
extra heavy standard; 14 - 16 cm girth	-	1.20	18.00	-	70.00	nr	**88.00**
"Fraxinus excelsior"; including backfillling with excavated material (other operations not included)							
standard; 8 - 10 cm girth	-	0.48	7.20	-	16.00	nr	**23.20**
selected standard, 10 - 12 cm girth	-	0.56	8.41	-	30.00	nr	**38.41**
heavy standard; 12 - 14 cm girth	-	0.76	11.47	-	50.00	nr	**61.47**
extra heavy standard; 14 - 16 cm girth	-	1.20	18.00	-	67.00	nr	**85.00**
"Prunus avium Plena"; including backfillling with excavated material (other operations not included)							
standard; 8 - 10 cm girth	-	0.40	6.00	-	17.95	nr	**23.95**
selected standard, 10 - 12 cm girth	-	0.56	8.41	-	17.95	nr	**26.36**
heavy standard; 12 - 14 cm girth	-	0.76	11.47	-	51.50	nr	**62.97**
extra heavy standard; 14 - 16 cm girth	-	1.20	18.00	-	67.00	nr	**85.00**
"Quercus robur"; including backfillling with excavated material (other operations not included)							
standard; 8 - 10 cm girth	-	0.48	7.20	-	17.95	nr	**25.15**
selected standard; 10 - 12 cm girth	-	0.56	8.41	-	31.50	nr	**39.91**
heavy standard; 12 - 14 cm girth	-	0.76	11.47	-	51.50	nr	**62.97**
extra heavy standard; 14 - 16 cm girth	-	1.20	18.00	-	67.00	nr	**85.00**
"Robinia pseudoacacia Frisia"; including backfillling with excavated material (other operations not included)							
standard; 8 - 10 cm girth	-	0.48	7.20	-	38.50	nr	**45.70**
selected standard; 10 - 12 cm girth	-	0.56	8.41	-	47.00	nr	**55.41**
heavy standard; 12 - 14 cm girth	-	0.76	11.47	-	68.00	nr	**79.47**
extra heavy standard; 14 - 16 cm girth	-	1.20	18.00	-	98.00	nr	**116.00**

Q PAVING/PLANTING/FENCING/SITE FURNITURE Excluding overheads and profit	PC £	Labour hours	Labour £	Plant £	Material £	Unit	Total rate £
Q31 PLANTING - cont'd							
Tree planting; root balled trees; **advanced nursery stock and** **semi-mature; Belwood Nurseries**							
"Acer platanoides Emerald Queen"; including backfilling with excavated material (other operations not included)							
16 - 18 cm girth	69.00	1.30	19.44	19.67	69.00	nr	108.11
18 - 20 cm girth	77.00	1.60	24.00	24.29	77.00	nr	125.29
20 - 25 cm girth	90.00	4.50	67.50	68.31	90.00	nr	225.81
25 - 30 cm girth	140.00	6.00	90.00	91.08	140.00	nr	321.08
30 - 35 cm girth	190.00	11.00	165.00	166.98	190.00	nr	521.98
"Aesculus briotti"; including backfilling with excavated material (other operations not included)							
16 - 18 cm girth	74.00	1.30	19.44	19.67	74.00	nr	113.11
18 - 20 cm girth	89.00	1.60	24.00	24.29	89.00	nr	137.29
20 - 25 cm girth	140.00	4.50	67.50	68.31	140.00	nr	275.81
25 - 30 cm girth	189.00	6.00	90.00	91.08	189.00	nr	370.08
30 - 35 cm girth	240.00	11.00	165.00	166.98	240.00	nr	571.98
"Prunus avium Flora Plena"; including backfilling with excavated material (other operations not included)							
16 - 18 cm girth	60.00	1.30	19.44	19.67	60.00	nr	99.11
18 - 20 cm girth	70.00	1.60	24.00	24.29	70.00	nr	118.29
20 - 25 cm girth	110.00	4.50	67.50	68.31	110.00	nr	245.81
25 - 30 cm girth	150.00	6.00	90.00	91.08	150.00	nr	331.08
30 - 35 cm girth	190.00	11.00	165.00	166.98	190.00	nr	521.98
"Sorbus aucuparia Sheerwater Seedling"; including backfilling with excavated material (other operations not included)							
16 - 18 cm girth	65.00	1.30	19.44	19.67	65.00	nr	104.11
18 - 20 cm girth	70.00	1.60	24.00	24.29	70.00	nr	118.29
20 - 25 cm girth	90.00	4.50	67.50	68.31	90.00	nr	225.81
25 - 30 cm girth	150.00	6.00	90.00	91.08	150.00	nr	331.08
30 - 35 cm girth	190.00	11.00	165.00	166.98	190.00	nr	521.98
"Tilia cordata Green Spire"; including backfilling with excavated material (other operations not included)							
16 - 18 cm girth	67.00	1.30	19.44	19.67	67.00	nr	106.11
18 - 20 cm girth	70.00	1.60	24.00	24.29	70.00	nr	118.29
20 - 25 cm girth	110.00	4.50	67.50	68.31	110.00	nr	245.81
25 - 30 cm girth	160.00	6.00	90.00	91.08	160.00	nr	341.08
30 - 35 cm girth	190.00	11.00	165.00	166.98	190.00	nr	521.98

Prices for Measured Works

Q PAVING/PLANTING/FENCING/SITE FURNITURE Excluding overheads and profit	PC £	Labour hours	Labour £	Plant £	Material £	Unit	Total rate £
Tree planting; root balled trees; semi-mature and mature trees; Belwood Trees; planting and back filling; planted by tele handler or by crane; delivery included; all other operations priced separately							
Semi mature trees							
40 - 45 cm girth	495.00	8.00	120.00	98.36	495.00	nr	713.36
45 - 50 cm girth	540.00	8.00	120.00	138.43	540.00	nr	798.43
55 - 60 cm girth	695.00	10.00	150.00	164.54	695.00	nr	1009.54
67 - 70 cm girth	972.00	15.00	225.00	155.08	972.00	nr	1352.08
75 - 80 cm girth	1500.00	18.00	270.00	136.92	1500.00	nr	1906.92
over 80 cm girth	1700.00	18.00	270.00	136.92	1700.00	nr	2106.92
Tree planting; tree protection - General Preamble: Care must be taken to ensure that tree grids and guards are removed when trees grow beyond the specified diameter of guard.							
Tree planting; tree protection							
Crowders Nurseries, "Crowders Tree Tube"; olive green							
1200 mm high x 80 x 80 mm	0.71	0.07	1.00	-	0.71	nr	1.71
Crowders Nurseries; stakes; 1500 mm high for "Crowders Tree Tube"; driving into ground	0.50	0.05	0.75	-	0.50	nr	1.25
Crowders Nurseries; galvanized welded mesh tree guards; nailing to tree stakes (tree stakes not included)							
840 mm high x 200 mm diameter	10.25	0.25	3.75	-	10.25	nr	14.00
1200 mm high x 250 mm diameter	8.25	0.33	5.00	-	8.25	nr	13.25
1200 mm high x 300 mm diameter	9.00	0.33	5.00	-	9.00	nr	14.00
1800 mm high x 300 mm diameter	12.00	0.33	5.00	-	12.00	nr	17.00
Expandable plastic tree guards; including 25 mm softwood stakes							
500 mm high	0.59	0.17	2.50	-	0.59	nr	3.09
1.00 m high	0.65	0.17	2.50	-	0.65	nr	3.15
J Toms; "Spiral Guard" perforated PVC guards; for trees 10 to 40 mm diameter; white, black or grey							
450 mm	0.23	0.03	0.50	-	0.23	nr	0.73
600 mm	0.28	0.03	0.50	-	0.28	nr	0.79
750 mm	0.34	0.03	0.50	-	0.34	nr	0.84
J. Toms; "Sheltatree" tree guards; square; corrugated bio-degradable polypropylene							
93 mm x 93 mm x 1200 mm	1.18	0.07	1.00	-	1.18	nr	2.18
English Woodlands; Plastic Mesh Treeguards supplied in 50m rolls							
13 mm x 13 mm small mesh; roll width 60 cm; black	0.60	0.04	0.54	-	1.01	nr	1.55
13 mm x 13 mm small mesh; roll width 120 cm; black	1.13	0.06	0.88	-	1.65	nr	2.53
13 mm x 13 mm small mesh; roll width 60 cm; brown	-	0.06	0.88	-	0.41	nr	1.29
13 mm x 13 mm small mesh; roll width 120 cm; brown	1.41	0.06	0.88	-	1.96	nr	2.84
ready made tree guard in dark green fine plastic mesh, supplied flat packed	0.34	0.06	0.88	-	0.75	nr	1.63

Q PAVING/PLANTING/FENCING/SITE FURNITURE Excluding overheads and profit	PC £	Labour hours	Labour £	Plant £	Material £	Unit	Total rate £
Q31 PLANTING - cont'd							
Tree planting; tree protection - cont'd							
Tree guards of 3 nr 3.00 m x 100 mm stakes; driving 600 mm into firm ground; bracing with timber braces at top and bottom; including 3 strands barbed wire	21.44	1.00	15.00	-	21.44	nr	36.44
English Woodlands; strimmer guard in heavy duty black plastic, 225 mm high	-	0.07	1.00	-	1.98	nr	2.98
Tubex Ltd; "Treeshelter" inclusive of 25 mm stake; prices shown for quantities of 500 nr							
"Ecostart" shelter for forestry transplants and seedlings,	0.57	0.07	1.00	-	0.57	nr	1.57
0.6 m high	0.46	0.05	0.75	-	0.68	each	1.43
0.75 m high	0.56	0.05	0.75	-	0.82	each	1.57
0.9 m high	0.66	0.05	0.75	-	0.95	each	1.70
1.2 m high	1.73	0.05	0.75	-	2.09	each	2.84
1.5 m high	0.97	0.05	0.75	-	1.40	each	2.15
1 8 m high	1 17	0.05	0.75	-	1.67	each	2.42
Tubex Ltd; "Shrubshelter" inclusive of 25 mm stake; prices shown for quantities of 500 nr							
0.6 m high	1 01	0.07	1.00	-	1.23	nr	2.23
0.75 m high	1 35	0.07	1.00	-	1.61	nr	2.61
Tubex Ltd; "Quill" self supporting small diameter tree shelter							
Standard; 50 mm diameter	0.36	0.04	0.60	-	0.36	nr	0.96
Tree guards; Netlon							
"Treeguard 2"; flat sheet 450 mm wide x 1.00 m high;cutting to length on site; 40 x 20 mm mesh; average 1.00 m high	0.84	0.03	0.50	-	0.84	nr	1.34
Extra over trees for spraying with antidessicant spray; "Wiltpruf"							
selected standards; standards, light standards	2.57	0.20	3.00	-	2.57	nr	5.57
standards; heavy standards	4.29	0.25	3.75	-	4.29	nr	8.04
Hedges							
Excavating trench for hedges; depositing soil alongside trench; by machine							
300 mm deep x 300 mm wide	-	0.03	0.45	0.31	-	m	0.76
300 mm deep x 450 mm wide	-	0.05	0.68	0.47	-	m	1.14
Excavating trench for hedges; depositing soil alongside trench; by hand							
300 mm deep x 300 mm wide	-	0.12	1.80	-	-	m	1.80
300 mm deep x 450 mm wide	-	0.23	3.38	-	-	m	3.38
Setting out; notching out; excavating trench; breaking up subsoil to minimum depth 300 mm minimum 400 mm deep	-	0.25	3.75	-	-	m	3.75
Hedge planting; including backfill with excavated topsoil; PC £0.30/nr							
single row; 200 mm centres	1.50	0.06	0.94	-	1.50	m	2.44
single row; 300 mm centres	1.00	0.06	0.83	-	1.00	m	1.83
single row; 400 mm centres	0.75	0.04	0.63	-	0.75	m	1.38
single row; 500 mm centres	0.60	0.03	0.50	-	0.60	m	1.10

Q PAVING/PLANTING/FENCING/SITE FURNITURE Excluding overheads and profit	PC £	Labour hours	Labour £	Plant £	Material £	Unit	Total rate £
double row; 200 mm centres	3.00	0.17	2.50	-	3.00	m	5.50
double row; 300 mm centres	2.00	0.13	2.00	-	2.00	m	4.00
double row; 400 mm centres	1.50	0.08	1.25	-	1.50	m	2.75
double row; 500 mm centres	1.20	0.07	1.00	-	1.20	m	2.20
Extra over hedges for incorporating manure;							
at 1 m³ per 30 m	0.37	0.03	0.38	-	0.37	m	0.74
Topiary; Clifton Nurseries Ltd							
Mop head trees; trimming to ball							
1 50 m high x 450 mm diameter	75.00	0.50	7.50	-	75.00	nr	82.50
Pyramid Bay trees; trimming to shape							
2.50 m high	275.00	0.50	7.50	-	275.00	nr	282.50
Shrub planting - General							
Preamble: For preparation of planting areas see "Cultivation" at the beginning of the section on planting							
Shrub planting							
Setting out; selecting planting from holding area; loading to wheel barrows; planting as plan or as directed; distance from holding area maximum 50 m; plants 2 - 3 litre containers							
plants in groups of 100 nr minimum	-	0.01	0.17	-	-	nr	0.17
plants in groups of 10 - 100 nr	-	0.02	0.25	-	-	nr	0.25
plants in groups of 3 - 5 nr	-	0.03	0.38	-	-	nr	0.38
single plants not grouped	-	0.04	0.60	-	-	nr	0.60
Forming planting holes; in cultivated ground (cultivating not included); by mechanical auger; trimming holes by hand; depositing excavated material alongside holes							
250 diameter	-	0.03	0.50	0 04	-	nr	0.54
250 x 250 mm	-	0.04	0.60	0.08	-	nr	0.68
300 x 300 mm	-	0.08	1.13	0.10	-	nr	1.22
Hand excavation; forming planting holes; in cultivated ground (cultivating not included); depositing excavated material alongside holes							
100 mm x 100 mm x 100 mm deep; with mattock or hoe	-	0.01	0.10	-	-	nr	0.10
250 mm x 250 mm x 300 mm deep	-	0.04	0.60	-	-	nr	0.60
300 mm x 300 mm x 300 mm deep	-	0.06	0.83	-	-	nr	0.83
400 mm x 400 mm x 400 mm deep	-	0.13	1.88	-	-	nr	1.88
500 mm x 500 mm x 500 mm deep	-	0.25	3.75	-	-	nr	3.75
600 mm x 600 mm x 600 mm deep	-	0.43	6.49	-	-	nr	6.49
900 mm x 900 mm x 600 mm deep	-	1.00	15.00	-	-	nr	15.00
1.00 m x 1.00 m x 600 mm deep	-	1.23	18.45	-	-	nr	18.45
1.25 m x 1.25 m x 600 mm deep	-	1.93	28.95	-	-	nr	28.95
Hand excavation; forming planting holes; in uncultivated ground; depositing excavated material alongside holes							
100 mm x 100 mm x 100 mm deep; with mattock or hoe	-	0.03	0.38	-	-	nr	0.38
250 mm x 250 mm x 300 mm deep	-	0.06	0.83	-	-	nr	0.83
300 mm x 300 mm x 300 mm deep	-	0.06	0.94	-	-	nr	0.94
400 mm x 400 mm x 400 mm deep	-	0.25	3.75	-	-	nr	3.75
500 mm x 500 mm x 500 mm deep	-	0.33	4.88	-	-	nr	4.88
600 mm x 600 mm x 600 mm deep	-	0.55	8.25	-	-	nr	8.25

Q PAVING/PLANTING/FENCING/SITE FURNITURE Excluding overheads and profit	PC £	Labour hours	Labour £	Plant £	Material £	Unit	Total rate £
Q31 PLANTING - cont'd							
Shrub planting - cont'd							
900 mm x 900 mm x 600 mm deep	-	1.25	18.75	-	-	nr	**18.75**
1.00 m x 1.00 m x 600 mm deep	-	1.54	23.06	-	-	nr	**23.06**
1 25 m x 1.25 m x 600 mm deep	-	2.41	36.19	-	-	nr	**36.19**
Bare root planting; to planting holes (forming holes not included); including backfilling with excavated material (bare root plants not included)							
bare root 1+1; 30 - 90 mm high	-	0.02	0.25	-	-	nr	**0.25**
bare root 1+2; 90 - 120 mm high	-	0.02	0.25	-	-	nr	**0.25**
Containerised planting; to planting holes (forming holes not inIcluded); including backfilling with excavated material (shrub or ground cover not included)							
9 cm pot	-	0.01	0.15	-	-	nr	**0.15**
2 litre container	-	0.02	0.30	-	-	nr	**0.30**
3 litre container	-	0.02	0.33	-	-	nr	**0.33**
5 litre container	-	0.03	0.50	-	-	nr	**0.50**
10 litre container	-	0.05	0.75	-	-	nr	**0.75**
15 litre container	-	0.07	1.00	-	-	nr	**1.00**
20 litre container	-	0.08	1.25	-	-	nr	**1.25**
Shrub planting; 2 litre containerised plants; in cultivated ground (cultivating not included); PC £1 80/nr							
average 2 plants per m²	-	0.06	0.84	-	3.60	m²	**4.44**
average 3 plants per m²	-	0.08	1.26	-	5.40	m²	**6.66**
average 4 plants per m²	-	0.11	1.68	-	7.20	m²	**8.88**
average 6 plants per m²	-	0.17	2.52	-	10.80	m²	**13.32**
Extra over shrubs for stakes	0.50	0.02	0.25	-	0.50	nr	**0.75**
Composted bark soil conditioners; 20 m³ loads; on beds by mechanical loader; spreading and rotavating into topsoil; by machine							
50 mm thick	130.00	-	-	7.38	130.00	100m²	**137.38**
100 mm thick	260.00	-	-	11.29	273.00	100m²	**284.29**
150 mm thick	390.00	-	-	15.35	409.50	100m²	**424.85**
200 mm thick	520.00	-	-	19.34	546.00	100m²	**565.34**
Mushroom compost; 20 m³ loads; delivered not further than 25m from location; cultivating into topsoil by pedestrian operated machine							
50 mm thick	40.00	2.86	42.86	1.78	40.00	100m²	**84.64**
100 mm thick	80.00	6.05	90.73	1.78	80.00	100m²	**172.51**
150 mm thick	120.00	8.90	133.57	1.78	120.00	100m²	**255.35**
200 mm thick	160.00	12.90	193.57	1.78	160.00	100m²	**355.35**
Mushroom compost; 20 m³ loads; on beds by mechanical loader; spreading and rotavating into topsoil by tractor drawn rotavator							
50 mm thick	40.00	-	-	7.38	40.00	100 m²	**47.38**
100 mm thick	80.00	-	-	11.29	80.00	100 m²	**91.29**
150 mm thick	120.00	-	-	15.35	120.00	100 m²	**135.35**
200 mm thick	160.00	-	-	19.34	160.00	100 m²	**179.34**

Q PAVING/PLANTING/FENCING/SITE FURNITURE Excluding overheads and profit	PC £	Labour hours	Labour £	Plant £	Material £	Unit	Total rate £
Manure; 20 m³ loads; delivered not further than 25m from location; cultivating into topsoil by pedestrian operated machine							
50 mm thick	-	2.86	42.86	1.78	65.00	100 m²	109.64
100 mm thick	110 00	6.05	90.73	1.78	115.50	100 m²	208.01
150 mm thick	165 00	8.90	133.57	1.78	173.25	100 m²	308.60
200 mm thick	220.00	12.90	193.57	1.78	231.00	100 m²	426.35
Fertilizers (7:7:7); PC £0.48/kg; to beds; by hand							
35 g/m²	1.68	0.17	2.50	-	1.68	100 m²	4.18
50 g/m²	2.40	0.17	2.50	-	2.40	100 m²	4.90
70 g/m²	3.36	0.17	2.50	-	3.36	100 m²	5.86
Fertilizers; "Enmag" PC £1.11/kg slow release fertilizer; to beds; by hand							
35 g/m²	3 90	0.17	2.50	-	3.90	100 m²	6.40
50 g/m²	5 57	0.17	2.50	-	5.57	100 m²	8.07
70 g/m²	8 36	0.17	2.50	-	8.36	100 m²	10.86
Note: For machine incorporation of fertilizers and soil conditioners see "Cultivation".							
Herbaceous and groundcover planting							
Herbaceous plants; PC £1.00/nr; including forming planting holes in cultivated ground (cultivating not included); backfilling with excavated material, 1 litre containers							
average 4 plants per m² - 500 mm centres	-	0.09	1.40	-	4.00	m²	5.40
average 6 plants per m² - 408 mm centres	-	0.14	2.10	-	6.00	m²	8.10
average 8 plants per m² - 354 mm centres	-	0.19	2.80	-	8.00	m²	10.80
Note: For machine incorporation of fertilizers and soil conditioners see "Cultivation".							
Plant support netting; Bridport Gundry; on 50 mm diameter stakes; 750 mm long; driving into ground at 1.50 m centres							
green extruded plastic mesh; 125 mm square	0.36	0.04	0.60	-	0.36	m²	0.96
Bulb planting							
Bulbs; including forming planting holes in cultivated area (cultivating not included); backfilling with excavated material							
small	12.90	0.83	12.50	-	12.90	100 nr	25.40
medium	20.60	0.83	12.50	-	20.60	100 nr	33.10
large	24.50	0.91	13.64	-	24.50	100 nr	38.14
Bulbs; in grassed area; using bulb planter; including backfilling with screened topsoil or peat and cut turf plug							
small	12.90	1.67	25.00	-	12.90	100 nr	37.90
medium	20.60	1.67	25.00	-	20.60	100 nr	45.60
large	24.50	2.00	30.00	-	24.50	100 nr	54.50
Aquatic planting							
Approved hessian bags filled with soil and fertilizer; tied, slit and weighted; in pool to receive three submerged floating leafed marginal or swamp plants	-	0.10	1.50	-	0.55	nr	2.05
Aquatic plants; in prepared growing medium in pool; plant size 2 - 3 litre containerised (plants not included)	-	0.04	0.60	-	-	nr	0.60

Q PAVING/PLANTING/FENCING/SITE FURNITURE Excluding overheads and profit	PC £	Labour hours	Labour £	Plant £	Material £	Unit	Total rate £
Q31 PLANTING - cont'd							
Operations after planting							
Initial cutting back to shrubs and hedge plants; including disposal of all cuttings	-	1.00	15.00	-	-	100m²	**15.00**
Mulch; Melcourt Industries Ltd; "Bark Nuggets"; to plant beds; delivered in 70 m³ loads; maximum distance 25 m							
50 mm thick	1.45	0.03	0.40	-	1.52	m²	**1.93**
75 mm thick	2.17	0.04	0.61	-	2.29	m²	**2.89**
100 mm thick	2.90	0.05	0.81	-	3.04	m²	**3.85**
Mulch; Melcourt Industries Ltd; "Bark Nuggets"; to plant beds; delivered in 20 m³ loads; maximum distance 25 m							
50 mm thick	2.07	0.03	0.44	-	2.17	m²	**2.61**
75 mm thick	3.10	0.04	0.66	-	3.26	m²	**3.92**
100 mm thick	4.14	0.06	0.87	-	4.35	m²	**5.22**
Mulch; Melcourt Industries Ltd; "Amenity Bark"; to plant beds; delivered in 70 m³ loads; maximum distance 25 m							
50 mm thick	0.89	0.03	0.44	-	0.93	m²	**1.37**
75 mm thick	1.33	0.04	0.66	-	1.40	m²	**2.06**
100 mm thick	1.77	0.04	0.55	-	1.86	m²	**2.42**
Mulch; Melcourt Industries Ltd; "Amenity Bark"; to plant beds; delivered in 20 m³ loads; maximum distance 25 m							
50 mm thick	1.51	0.03	0.44	-	1.58	m²	**2.02**
75 mm thick	2.26	0.04	0.66	-	2.38	m²	**3.03**
100 mm thick	3.02	0.06	0.87	-	3.17	m²	**4.04**
Selective herbicides, in accordance with manufacturer's instructions; PC £26.45 / litre; by machine; application rate							
30 ml/m²	-	-	-	0.29	0.79	100m²	**1.08**
35 ml/m²	-	-	-	0.29	0.93	100m²	**1.22**
40 ml/m²	-	-	-	0.29	1.06	100m²	**1.35**
50 ml/m²	-	-	-	0.29	1.32	100m²	**1.61**
3.00 l /ha	-	-	-	28.89	79.42	ha	**108.31**
3.50 l/ha	-	-	-	28.89	92.66	ha	**121.55**
4.00 l/ha	-	-	-	28.89	105.90	ha	**134.79**
Selective herbicides; in accordance with manufacturer's instructions; PC £26.45 / litre; by hand; application rate							
30 ml/m²	-	0.17	2.50	-	0.79	100m²	**3.29**
35 ml/m²	-	0.17	2.50	-	0.93	100m²	**3.43**
40 ml/m²	-	0.17	2.50	-	1.06	100m²	**3.56**
50 ml/m²	-	0.17	2.50	-	1.32	100m²	**3.82**
3.00 l/ha	-	16.67	250.00	-	79.42	ha	**329.43**
3.50 l/ha	-	16.67	250.00	-	92.66	ha	**342.66**
4.00 l/ha	-	16.67	250.05	-	105.90	ha	**355.95**
General herbicides; in accordance with manufacturer's instructions; "Knapsack" spray application							
"Casoron G (residual)"; at 1kg/125m²	3.04	0.05	0.75	-	3.34	100m²	**4.09**
"Spasor"; at 50 ml/100 m²	0.70	0.05	0.75	-	0.70	100m²	**1.45**
"Roundup Pro"; at 40 ml /100 m²	0.16	0.05	0.75	-	0.16	100m²	**0.91**
"Super Verdone"; at 100ml/100 m²	0.60	0.05	0.75	-	0.60	100m²	**1.35**

Q PAVING/PLANTING/FENCING/SITE FURNITURE Excluding overheads and profit	PC £	Labour hours	Labour £	Plant £	Material £	Unit	Total rate £
Fertilizers; in top 150 mm of topsoil at 35 g/m²							
fertilizer (18:0:0+Mg +Fe)	5.94	0.12	1.84	-	5.94	100m²	7.77
"Enmag"	3.90	0.12	1.84	-	4.09	100m²	5.93
fertilizer (7:7:7)	1.68	0.12	1.84	-	1.68	100m²	3.52
fertilizer (20:10:10)	2.37	0.12	1.84	-	2.37	100m²	4.21
"Superphosphate"	1 57	0.12	1.84	-	1.65	100m²	3.49
"Hoof and Horn"	2.17	0.12	1.84	-	2.28	100m²	4.12
"Bone meal"	3 01	0.12	1.84	-	3.16	100m²	5.00
Fertilizers; in top 150 mm of topsoil at 70 g/m²							
fertilizer (18:0:0+Mg +Fe)	11.87	0.12	1.84	-	11.87	100m²	13.71
"Enmag"	7.80	0.12	1.84	-	8.19	100m²	10.03
fertilizer (7:7:7)	3 36	0.12	1.84	-	3.36	100m²	5.20
fertilizer (20:10:10)	4.75	0.12	1.84	-	4.75	100m²	6.58
"Superphosphate"	3 15	0.12	1.84	-	3.31	100m²	5.14
"Hoof and Horn"	4.34	0.12	1.84	-	4.56	100m²	6.39
"Bone meal"	6.02	0.12	1.84	-	6.32	100m²	8.16
Maintenance operations (Note: the following rates apply to aftercare maintenance executed as part of a landscaping contract only)							
Weeding and hand forking planted areas; including disposing weeds and debris on site; areas maintained weekly	-	-	0.06	-	-	m²	0.06
Weeding and hand forking planted areas; including disposing weeds and debris on site; areas maintained monthly	-	0.01	0.15	-	-	m²	0.15
Extra over weeding and hand forking planted areas for disposing excavated material off site; to tip not exceeding 13 km; mechanically loaded							
slightly contaminated	-	-	-	1.52	16.00	m³	17.52
rubbish	-	-	-	1.52	16.00	m³	17.52
inert material	-	-	-	1.01	10.00	m³	11.01
Mulch; Melcourt Industries Ltd; "Bark Nuggets"; to plant beds, delivered in 70 m³ loads; maximum distance 25 m							
50 mm thick	1.45	0.03	0.40	-	1.52	m²	1.93
75 mm thick	2.17	0.04	0.61	-	2.29	m²	2.89
100 mm thick	2 90	0.05	0.81	-	3.04	m²	3.85
Mulch; Melcourt Industries Ltd; "Bark Nuggets"; to plant beds; delivered in 20 m³ loads; maximum distance 25 m							
50 mm thick	2.07	0.03	0.44	-	2.17	m²	2.61
75 mm thick	3.10	0.04	0.66	-	3.26	m²	3.92
100 mm thick	4.14	0.06	0.87	-	4.35	m²	5.22
Mulch; Melcourt Industries Ltd; "Amenity Bark"; to plant beds; delivered in 70 m³ loads; maximum distance 25 m							
50 mm thick	0.89	0.03	0.44	-	0.93	m²	1.37
75 mm thick	1 33	0.04	0.66	-	1.40	m²	2.06
100 mm thick	1.77	0.04	0.55	-	1.86	m²	2.42
Mulch; Melcourt Industries Ltd; "Amenity Bark"; to plant beds; delivered in 20 m³ loads; maximum distance 25 m							
50 mm thick	1.51	0.03	0.44	-	1.58	m²	2.02
75 mm thick	2.26	0.04	0.66	-	2.38	m²	3.03
100 mm thick	3.02	0.06	0.87	-	3.17	m²	4.04

Q PAVING/PLANTING/FENCING/SITE FURNITURE Excluding overheads and profit	PC £	Labour hours	Labour £	Plant £	Material £	Unit	Total rate £
Q31 PLANTING - cont'd							
Maintenance operations (Note: the following rates apply to aftercare maintenance executed as part of a landscaping contract only) - cont'd							
Selective herbicides in accordance with manufacturer's instructions; PC £26.45 / litre, by machine; application rate							
30 ml/m²	-	-	-	0.29	0.79	100m²	1.08
35 ml/m²	-	-	-	0.29	0.93	100m²	1.22
40 ml/m²	-	-	-	0.29	1.06	100m²	1.35
50 ml/m²	-	-	-	0.29	1.32	100m²	1.61
3 00 l/ha	-	-	-	28.89	79.42	ha	108.31
3.50 l/ha	-	-	-	28.89	92.66	ha	121.55
4.00 l/ha	-	-	-	28.89	105.90	ha	134.79
Selective herbicides; in accordance with manufacturer's instructions; PC £25.70 / litre, by hand; application rate							
30 ml/m²	-	0.17	2.50	-	0.79	100m²	3.29
35 ml/m²	-	0.17	2.50	-	0.93	100m²	3.43
40 ml/m²	-	0.17	2.50	-	1.06	100m²	3.56
50 ml/m²	-	0.17	2.50	-	1.32	100m²	3.82
3.00 l/ha	-	16.67	250.00	-	79.42	ha	329.43
3 50 l/ha	-	16.67	250.00	-	92.66	ha	342.66
4.00 l/ha	-	16.67	250.00	-	105.90	ha	355.90
General herbicides; in accordance with manufacturer's instructions; "Knapsack" spray application							
"Casoron G (residual)"; at 1kg/125m²	3.04	0.05	0.75	-	3.34	100m²	4.09
"Spasor"; at 50 ml/100 m²	0.70	0.05	0.75	-	0.70	100m²	1.45
"Roundup Pro"; at 40 ml/100 m²	0.16	0.05	0.75	-	0.16	100m²	0.91
"Super Verdone"; at 100 ml/100 m²	0.60	0.05	0.75	-	0.60	100m²	1.35
Fertilizers; at 35 g/m²							
fertilizer (18:0:0+Mg +Fe)	5.94	0.12	1.84	-	5.94	100m²	7.77
"Enmag"	3.90	0.12	1.84	-	4.09	100m²	5.93
fertilizer (7:7 7)	1.68	0.12	1.84	-	1.68	100m²	3.52
fertilizer (20:10:10)	2.37	0.12	1.84	-	2.37	100m²	4.21
"Superphosphate"	1.57	0.12	1.84	-	1.65	100m²	3.49
"Hoof and Horn"	2.17	0.12	1.84	-	2.28	100m²	4.12
"Bone meal"	3.01	0.12	1.84	-	3.16	100m²	5.00
Fertilizers; at 70 g/m²							
fertilizer (18:0:0+Mg +Fe)	11.87	0.12	1.84	-	11.87	100m²	13.71
"Enmag"	7.80	0.12	1.84	-	8.19	100m²	10.03
fertilizer (7:7:7)	3.36	0.12	1.84	-	3.36	100m²	5.20
fertilizer (20:10:10)	4.75	0.12	1.84	-	4.75	100m²	6.58
"Superphosphate"	3.15	0.12	1.84	-	3.31	100m²	5.14
"Hoof and Horn"	4.34	0.12	1.84	-	4.56	100m²	6.39
"Bone meal"	6.02	0.12	1.84	-	6.32	100m²	8.16
Treating plants with growth retardant (maleic hydrazide); at rate of 16 l/ha; in accordance with manufacturer's instructions							
"Burtolin" spray; at 1.50 litres per 5 l water	1.06	0.03	0.50	-	1.06	nr	1.56

Q PAVING/PLANTING/FENCING/SITE FURNITURE Excluding overheads and profit	PC £	Labour hours	Labour £	Plant £	Material £	Unit	Total rate £
Watering planting; evenly; at a rate of 5 litre/m²							
using hand-held watering equipment	-	0.25	3.75	-	-	100m²	3.75
using sprinkler equipment and with sufficient water pressure to run 1 nr 15 m radius sprinkler	-	0.14	2.09	-	-	100m²	2.09
using movable spray lines powering 3 nr sprinkler heads with a radius of 15 m and allowing for 60% overlap (irrigation machinery costs not included)	-	0.02	0.24	-	-	100m²	0.24
Forestry planting							
Deep ploughing rough ground to form planting ridges at							
2.00 m centres	-	0.63	9.38	7.23	-	100m²	16.60
3.00 m centres	-	0.59	8.82	6.80	-	100m²	15.62
4.00 m centres	-	0.40	6.00	4.62	-	100m²	10.62
Notching plant forestry seedlings; "T" or "L" notch	26.00	0.75	11.25	-	26.00	100 nr	37.25
Turf planting forestry seedlings	26.00	2.00	30.00	-	26.00	100 nr	56.00
Selective herbicides; in accordance with manufacturer's instructions; PC £26.45 / litre; by hand; application rate							
30 ml/100m²	-	0.17	2.50	-	0.79	100m²	3.29
35 ml/100m²	-	0.17	2.50	-	0.93	100m²	3.43
40 ml/100m²	-	0.17	2.50	-	1.06	100m²	3.56
50 ml/100m²	-	0.17	2.50	-	1.32	100m²	3.82
3.00 l/ha	-	16.67	250.00	-	79.42	ha	329.43
3.50 l/ha	-	16.67	250.00	-	92.66	ha	342.66
4.00 l/ha	-	16.67	250.00	-	105.90	ha	355.90
Selective herbicides; in accordance with manufacturer's instructions; PC £26.45 / litre; by machine; application rate							
30 ml/100m²	-	-	-	0.29	0.79	100m²	1.08
35 ml/100m²	-	-	-	0.29	0.93	100m²	1.22
40 ml/100m²	-	-	-	0.29	1.06	100m²	1.35
50 ml/100m²	-	-	-	0.29	1.32	100m²	1.61
3.00 l/ha	-	-	-	28.89	79.42	ha	108.31
3.50 l/ha	-	-	-	28.89	92.66	ha	121.55
4.00 l/ha	-	-	-	28.89	105.90	ha	134.79
General herbicides; in accordance with manufacturer's instructions; "Knapsack" spray application							
"Casoron G (residual)"; at 1kg/125m²	3.04	0.05	0.75	-	3.34	100m²	4.09
"Spasor"; at 50 ml/100 m²	0.70	0.05	0.75	-	0.70	100m²	1.45
"Roundup Pro"; at 40 ml/100 m²	0.16	0.05	0.75	-	0.16	100m²	0.91
"Super Verdone"; at 100 ml/100 m²	0.60	0.05	0.75	-	0.60	100m²	1.35
Fertilizers; at 35 g/m²							
fertilizer (18:0:0+Mg + Fe)	5.94	0.12	1.84	-	5.94	100m²	7.77
"Enmag"	3.90	0.12	1.84	-	4.09	100m²	5.93
fertilizer (7:7:7)	1.68	0.12	1.84	-	1.68	100m²	3.52
fertilizer (20:10:10)	2.37	0.12	1.84	-	2.37	100m²	4.21
"Superphosphate"	1.57	0.12	1.84	-	1.65	100m²	3.49
"Hoof and Horn"	2.17	0.12	1.84	-	2.28	100m²	4.12
"Bone meal"	3.01	0.12	1.84	-	3.16	100m²	5.00

Q PAVING/PLANTING/FENCING/SITE FURNITURE Excluding overheads and profit	PC £	Labour hours	Labour £	Plant £	Material £	Unit	Total rate £
Q31 PLANTING - cont'd							
Forestry planting - cont'd							
Fertilizers; at 70 g/m²							
fertilizer (18:0:0+Mg + Fe)	11.87	0.12	1.84	-	11.87	100m²	13.71
"Enmag"	7.80	0.12	1.84	-	8.19	100m²	10.03
fertilizer (7:7:7)	3.36	0.12	1.84	-	3.36	100m²	5.20
fertilizer (20:10:10)	4.75	0.12	1.84	-	4.75	100m²	6.58
"Superphosphate"	3.15	0.12	1.84	-	3.31	100m²	5.14
"Hoof and Horn"	4.34	0.12	1.84	-	4.56	100m²	6.39
"Bone meal"	6.02	0.12	1.84	-	6.32	100m²	8.16
Tree tubes; to young trees	121.00	0.30	4.50	-	121.00	100 nr	125.50
Cleaning and weeding around seedlings; once	-	0.50	7.50	-	-	100 nr	7.50
Treading in and firming ground around seedlings planted; at 2500 per ha after frost or other ground disturbance; once	-	0.33	5.00	-	-	100 nr	5.00
Beating up initial planting; once (including supply of replacement seedlings at 10% of original planting)	2.60	0.25	3.75	-	2.60	100 nr	6.35
Work to existing planting							
Control shoots lichen and suckers on street trees; Rhone Poulenc							
"Burtolin" spray at 1.50 litres per 5 l	1.06	0.03	0.50	-	1.06	nr	1.56
Cutting and trimming ornamental hedges; to specified profiles; including cleaning out hedge bottoms; hedge cut 2 occasions per annum; by hand							
up to 2.00 m high	-	0.03	0.50	-	0.25	m	0.75
2.00 - 4.00 m high	-	0.05	0.75	1.60	0.50	m	2.85
Cutting and laying field hedges; including stakes and ethering; removing or burning all debris. (Note: Rate at which work executed varies greatly with width and height of hedge. A typical hedge could be cut and laid at rate of 7 m run per man day)	-	1.00	15.00	-	1.65	m	16.65
labour charge per man day (specialist daywork)	-	1.00	15.00	-	-	m	15.00
Trimming field hedges; to specified heights and shapes							
using cutting bar	-	5.00	75.00	-	-	100 m	75.00
using flail	-	0.20	3.00	2.52	-	100 m	5.52

Q PAVING/PLANTING/FENCING/SITE FURNITURE Excluding overheads and profit	PC £	Labour hours	Labour £	Plant £	Material £	Unit	Total rate £
Q35 LANDSCAPE MAINTENANCE							
Preamble; Long term Landscape maintenance							
Maintenance on long term contracts differs in cost from that of maintenance as part of a landscape contract. In this section the contract period is generally 3 - 5 years. Staff are generally allocated to a single project only and therefore productivity is higher whilst overhead costs are lower. Labour costs in this section are lower than the costs used in other parts of the book. Machinery is assumed to be leased over a 5 year period and written off over the same period. The costs of maintenance and consumables for the various machinery types have been included in the information that follows. Finance costs for the machinery have not been allowed for. The rates shown below are for machines working in unconfined contiguous areas. Users should adjust the times and rates if working in smaller spaces or spaces with obstructions.							
Grass Cutting - tractor mounted equipment; Keith Banyard Grounds Maintenance Ltd Using multiple-gang mower with cylindrical cutters; contiguous areas such as playing fields and the like larger than 3000 m²							
3 gang; 2.13 m cutting width	-	0.01	0.12	0.10	-	100m²	**0.22**
5 gang; 3.40 m cutting width	-	0.01	0.11	0.13	-	100 m²	**0.24**
7 gang; 4.65 m cutting width	-	0.01	0.07	0.09	-	100 m²	**0.16**
Using multiple-gang mower with cylindrical cutters; non contiguous areas such as verges and general turf areas							
3 gang	-	0.02	0.25	0.10	-	100m²	**0.35**
5 gang	-	0.01	0.16	0.13	-	100 m²	**0.29**
Using multiple-rotary mower with vertical drive shaft and horizontally rotating bar or disc cutters; contiguous areas larger than 3000 m²							
cutting grass, overgrowth or the like using flail mower or reaper	-	0.02	0.24	0.22	-	100m²	**0.46**
Collection of arisings							
22 cuts per year	-	0.02	0.20	0.21	-	100m²	**0.41**
18 cuts per year	-	0.02	0.27	0.21	-	100m²	**0.48**
12 cuts per year	-	0.03	0.40	0.21	-	100m²	**0.61**
4 cuts per year	-	0.07	0.80	0.21	-	100m²	**1.01**
Disposal of arisings							
22 cuts per year	-	0.01	0.10	0.04	0.13	100m²	**0.26**
18 cuts per year	-	0.01	0.12	0.05	0.17	100m²	**0.34**
12 cuts per year	-	0.01	0.08	0.12	0.30	100m²	**0.50**
4 cuts per year	-	0.02	0.24	0.76	0.50	100m²	**1.50**
Cutting grass, overgrowth or the like: using tractor-mounted side-arm flail mower: in areas inaccessible to alternative machine: on surface							
not exceeding 30 deg from horizontal	-	0.02	0.28	0.28	-	100m²	**0.56**
30 deg to 50 deg from horizontal	-	0.05	0.56	0.28	-	100m²	**0.84**

Q PAVING/PLANTING/FENCING/SITE FURNITURE Excluding overheads and profit	PC £	Labour hours	Labour £	Plant £	Material £	Unit	Total rate £
Q35 LANDSCAPE MAINTENANCE - cont'd							
Grass Cutting - 'Ride-On' self-propelled equipment; Norris and Gardener Grounds Maintenance							
Using ride-on multiple-cylinder mower							
3 gang; 2.13 m cutting width	-	0.01	0.12	0.10	-	100m²	0.22
5 gang; 3.40 m cutting width	-	0.01	0.08	0.13	-	100m²	0.21
Using ride-on multiple-rotary mower with horizontally rotating bar, disc or chain cutters							
cutting width 1.52 m	-	0.01	0.15	0.15	-	100m²	0.30
cutting width 1.82 m	-	0.01	0.13	0.11	-	100m²	0.24
cutting width 2.97 m	-	0.01	0.08	0.12	-	100m²	0.20
Add for using grass box/collector for removal and depositing of arisings	-	0.04	0.46	0.05	-	100m²	0.51
Grass Cutting - pedestrian operated equipment							
Using cylinder lawn mower fitted with not less than five cutting blades, front and rear rollers: on surface not exceeding 30 deg from horizontal; arisings let fly; width of cut							
51 cm	-	0.06	0.73	0.16	-	100 m	0.89
61 cm	-	0.05	0.61	0.14	-	100 m	0.75
71 cm	-	0.04	0.52	0.14	-	100 m	0.66
91 cm	-	0.03	0.41	0.14	-	100 m	0.55
Using rotary self propelled mower; width of cut							
45 cm	-	0.04	0.47	0.05	-	100m²	0.52
81 cm	-	0.02	0.26	0.06	-	100m²	0.32
91 cm	-	0.02	0.23	0.05	-	100m²	0.28
120 cm	-	0.01	0.18	0.20	-	100m²	0.38
Add for using grass box for collecting and depositing arisings							
removing and depositing arisings	-	0.05	0.60	-	-	100m²	0.60
Add for 30 to 50 deg from horizontal	-	-	-	-	-	33%	-
Add for slopes exceeding 50 deg	-	-	-	-	-	100%	-
cutting grass or light woody undergrowth: using trimmer with nylon cord or metal disc cutter: on surface							
not exceeding 30 deg from horizontal	-	0.20	2.40	0.27	-	100m²	2.67
30 -50 deg from horizontal	-	0.40	4.80	0.54	-	100m²	5.34
exceeding 50 deg from horizontal	-	0.50	6.00	0.68	-	100m²	6.67
Grass Cutting - collecting arisings							
Extra over for tractor drawn and self propelled machinery using attached grass boxes; depositing arisings							
22 cuts per year	-	0.05	0.60	-	-	100m²	0.60
18 cuts per year	-	0.08	0.90	-	-	100m²	0.90
12 cuts per year	-	0.10	1.20	-	-	100m²	1.20
4 cuts per year	-	0.25	3.00	-	-	100m²	3.00
Disposing arisings							
22 cuts per year	-	0.01	0.09	0.02	0.02	100m²	0.14
18 cuts per year	-	0.01	0.12	0.03	0.03	100m²	0.17
12 cuts per year	-	0.01	0.17	0.04	0.05	100m²	0.26
4 cuts per year	-	0.04	0.52	0.12	0.14	100m²	0.78

Q PAVING/PLANTING/FENCING/SITE FURNITURE Excluding overheads and profit	PC £	Labour hours	Labour £	Plant £	Material £	Unit	Total rate £
Harrowing							
Harrowing grassed area							
drag harrow	-	0.01	0.15	0.12	-	100m²	0.27
chain or light flexible spiked harrow	-	0.02	0.20	0.12	-	100m²	0.32
Scarifying							
A Plant Groundcare							
Sisis ARP4, including grass collection box, towed by tractor; area scarified annually	-	0.02	0.28	0.48	-	100m²	0.76
Sisis ARP4, including grass collection box, towed by tractor; area scarified two years previously	-	0.03	0.34	0.58	-	100m²	0.92
pedestrian operated self-powered equipment	-	0.07	0.84	0.65	-	100m²	1.49
add for disposal of arisings	-	0.03	0.30	0.95	2.50	100m²	3.75
By hand							
hand implement	-	0.50	6.00	-	-	100m²	6.00
add for disposal of arisings	-	0.03	0.30	0.95	2.50	100m²	3.75
Rolling							
Rolling grassed area; equipment towed by tractor; once over, using							
smooth roller	-	0.01	0.17	0.15	-	100m²	0.32
Turf Aeration							
By machine; A Plant Groundcare							
Vertidrain turf aeration equipment towed by tractor to effect a minimum penetration of 100 to 250 mm at 100 mm centres	-	0.04	0.47	1.21	-	100m²	1.67
Ryan GA 30; self propelled turf aerating equipment; to effect a minimum penetration of 100 mm at varying centres	-	0.04	0.47	1.40	-	100m²	1.87
Groundsman; pedestrian operated; self powered solid or slitting tine turf aerating equipment to effect a minimum penetration of 100 mm	-	0.14	1.68	1.33	-	100m²	3.01
Cushman core harvester; self propelled; for collection of arisings	-	0.02	0.23	0.62	-	100m²	0.85
By hand							
hand fork: to effect a minimum penetration of 100 mm and spaced 150 mm apart	-	1.33	16.00	-	-	100m²	16.00
hollow tine hand implement: to effect a minimum penetration of 100 mm and spaced 150 mm apart:	-	2.00	24.00	-	-	100m²	24.00
collection of arisings by hand	-	3.00	36.00	-	-	100m²	36.00
Turf areas; Surface treatments and Top dressing; Boughton Loam Ltd							
Apply screened topdressing to grass surfaces. Spread using Tru-Lute							
Sand soil mixes 90/10 to 50/50	0.07	-	0.02	0.03	0.07	m²	0.13
Apply screened soil 3 mm, Kettering loam to goal mouths and worn areas							
20 mm thick	0.69	0.01	0.12	-	0.83	m²	0.95
10 mm thick	0.34	0.01	0.12	-	0.41	m²	0.53

Q PAVING/PLANTING/FENCING/SITE FURNITURE Excluding overheads and profit	PC £	Labour hours	Labour £	Plant £	Material £	Unit	Total rate £
Q35 LANDSCAPE MAINTENANCE - cont'd							
Leaf Clearance; clearing grassed area of leaves and other extraneous debris							
Using equipment towed by tractor							
large grassed areas with perimeters of mature trees such as sports fields and amenity areas	-	0.01	0.15	0.27	-	100m²	0.42
large grassed areas containing ornamental trees and shrub beds	-	0.03	0.30	0.38	-	100m²	0.68
Using pedestrian operated mechanical equipment and blowers							
grassed areas with perimeters of mature trees such as sports fields and amenity areas	-	0.04	0.48	0.09	-	100m²	0.57
grassed areas containing ornamental trees and shrub beds	-	0.10	1.20	0.24	-	100m²	1.44
verges	-	0.07	0.80	0.16	-	100m²	0.96
By hand							
grassed areas with perimeters of mature trees such as sports fields and amenity areas	-	0.05	0.60	0.09	-	100m²	0.69
grassed areas containing ornamental trees and shrub beds	-	0.08	1.00	0.15	-	100m²	1.15
verges	-	1.00	12.00	1.84	-	100m²	13.84
Removal of arisings							
areas with perimeters of mature trees	-	0.01	0.08	0.08	0.20	100m²	0.36
areas containing ornamental trees and shrub beds	-	0.02	0.24	0.31	0.50	100m²	1.05
Litter Clearance							
Collection and disposal of litter from grassed area	-	0.01	0.12	-	0.06	100m²	0.18
Collection and disposal of litter from isolated grassed area not exceeding 1000 m²	-	0.04	0.48	-	0.06	100m²	0.54
Edge Maintenance							
Maintain edges where lawn abuts pathway or hard surface using;							
strimmer	-	0.01	0.06	0.01	-	m	0.07
shears	-	0.02	0.20	-	-	m	0.20
Maintain edges where lawn abuts plant bed using							
mechanical edging tool	-	0.01	0.08	0.01	-	m	0.09
shears	-	0.01	0.13	-	-	m	0.13
half moon edging tool	-	0.02	0.24	-	-	m	0.24
Tree Guards, Stakes and Ties etc.							
Adjusting existing tree tie	-	0.03	0.40	-	-	nr	0.40
Taking up single or double tree stake and ties: removing and disposing	-	0.05	0.75	-	-	nr	0.75
Pruning Shrubs							
Trimming ground cover planting							
soft groundcover; vinca ivy and the like	-	1.00	12.00	-	-	100m²	12.00
woody groundcover; cotoneaster and the like	-	1.50	18.00	-	-	100m²	18.00
Pruning massed shrub border (measure ground area)							
shrub beds pruned annually	-	0.01	0.12	-	-	m²	0.12
shrub beds pruned hard every 3 years	-	0.03	0.33	-	-	m²	0.33

Q PAVING/PLANTING/FENCING/SITE FURNITURE Excluding overheads and profit	PC £	Labour hours	Labour £	Plant £	Material £	Unit	Total rate £
Cutting off dead heads							
bush or standard rose	-	0.05	0.60	-	-	nr	0.60
climbing rose	-	0.08	1.00	-	-	nr	1.00
Pruning Roses							
bush or standard rose	-	0.05	0.60	-	-	nr	0.60
climbing rose or rambling rose: tying in as required	-	0.07	0.80	-	-	nr	0.80
Pruning Ornamental Shrub: height before pruning (increase these rates by 50% if pruning work has not been executed during the previous two years)							
not exceeding 1m	-	0.04	0.48	-	-	nr	0.48
1 to 2 m	-	0.06	0.67	-	-	nr	0.67
exceeding 2 m	-	0.13	1.50	-	-	nr	1.50
Removing excess growth etc from face of building etc: height before pruning							
not exceeding 2 m	-	0.03	0.34	-	-	nr	0.34
2 to 4 m	-	0.05	0.60	-	-	nr	0.60
4 to 6 m	-	0.08	1.00	-	-	nr	1.00
6 to 8 m	-	0.13	1.50	-	-	nr	1.50
8 to 10 m	-	0.14	1.71	-	-	nr	1.71
Removing epicormic growth from base of shrub or trunk and base of tree: any height: any diameter: number of growths							
not exceeding 10	-	0.05	0.60	-	-	nr	0.60
10 to 20	-	0.07	0.80	-	-	nr	0.80
Beds Borders and Planters							
Lifting							
bulbs	-	0.50	6.00	-	-	100 nr	6.00
tubers or corms	-	0.40	4.80	-	-	100 nr	4.80
established herbaceous plants; hoeing and depositing for replanting	-	2.00	24.00	-	-	100 nr	24.00
Temporary staking and tying in herbaceous plant	-	0.03	0.40	-	0.12	nr	0.52
cutting down spent growth of herbaceous plant; clearing arisings							
unstaked	-	0.02	0.24	-	-	nr	0.24
staked: not exceeding 4 stakes per plant: removing stakes and putting into store	-	0.03	0.30	-	-	nr	0.30
Hand weeding							
newly planted areas	-	2.00	24.00	-	-	100m²	24.00
established areas	-	0.50	6.00	-	-	100m²	6.00
Removing grasses from groundcover areas	-	3.00	36.04	-	-	100m²	36.04
Hand digging with fork: not exceeding 150 mm deep: breaking down lumps: leaving surface with a medium tilth	-	0.01	0.16	-	-	100m²	0.16
Hand digging with fork or spade to an average depth of 230 mm: breaking down lumps: leaving surface with a medium tilth	-	0.02	0.24	-	-	100m²	0.24
Hand hoeing: not exceeding 50 mm deep: Leaving surface with a medium tilth	-	0.40	4.80	-	-	100m²	4.80
Hand raking to remove stones etc: breaking down lumps: leaving surface with a fine tilth prior to planting	-	0.01	0.08	-	-	100m²	0.08
Hand weeding; planter, window box; not exceeding 1.00 m²							
ground level box	-	0.05	0.60	-	-	nr	0.60
box accessed by stepladder	-	0.08	1.00	-	-	nr	1.00

Q PAVING/PLANTING/FENCING/SITE FURNITURE Excluding overheads and profit	PC £	Labour hours	Labour £	Plant £	Material £	Unit	Total rate £
Q35 LANDSCAPE MAINTENANCE - cont'd							
Beds Borders and Planters - cont'd							
Spreading compost, mulch or processed bark to a depth of 75 mm							
on shrub bed with existing mature planting	-	0.04	0.53	-	-	m²	0.53
recently planted areas	-	0.06	0.70	-	-	m²	0.70
groundcover and herbaceous areas	-	0.08	0.90	-	-	m²	0.90
Clearing cultivated area of leaves, litter and other extraneous debris: using hand implement							
weekly maintenance	-	0.13	1.50	-	-	100m²	1.50
daily maintenance	-	0.02	0.20	-	-	100m²	0.20
Bedding							
Lifting							
bedding plants; hoeing and depositing for disposal	-	0.33	4.00	-	-	100m²	4.00
Hand digging with fork: not exceeding 150 mm deep: breaking down lumps: leaving surface with a medium tilth	-	0.01	0.16	-	-	100m²	0.16
Hand weeding							
newly planted areas	-	2.00	24.00	-	-	100m²	24.00
established areas	-	0.50	6.00	-	-	100m²	6.00
Hand digging with fork or spade to an average depth of 230 mm: breaking down lumps: leaving surface with a medium tilth		0.02	0.24	-	-	100m²	0.24
Hand hoeing: not exceeding 50 mm deep: leaving surface with a medium tilth	-	0.40	4.80	-	-	100m²	4.80
Hand raking to remove stones etc: breaking down lumps: leaving surface with a fine tilth prior to planting	-	0.01	0.08	-	-	100m²	0.08
Hand weeding; planter, window box; not exceeding 1.00 m²							
ground level box	-	0.05	0.60	-	-	nr	0.60
box accessed by stepladder	-	0.08	1.00	-	-	nr	1.00
Spreading only; compost, mulch or processed bark to a depth of 75 mm							
on shrub bed with existing mature planting	-	0.04	0.53	-	-	m²	0.53
recently planted areas	-	0.06	0.70	-	-	m²	0.70
groundcover and herbaceous areas	-	0.08	0.90	-	-	m²	0.90
Collecting bedding from nursery	-	3.00	36.00	12.61	-	100m²	48.61
setting out							
mass planting single variety	-	0.13	1.50	-	-	m²	1.50
pattern	-	0.33	4.00	-	-	m²	4.00
Planting only							
massed bedding plants	-	0.20	2.40	-	-	m²	2.40
Clearing cultivated area of leaves, litter and other extraneous debris: using hand implement							
weekly maintenance	-	0.13	1.50	-	-	100m²	1.50
daily maintenance	-	0.02	0.20	-	-	100m²	0.20

Q PAVING/PLANTING/FENCING/SITE FURNITURE Excluding overheads and profit	PC £	Labour hours	Labour £	Plant £	Material £	Unit	Total rate £
Irrigation and watering							
Hand held hosepipe; flow rate 25 litres per minute; irrigation requirement							
10 litres /m²	-	0.74	8.84	-	-	100m²	**8.84**
15 litres /m²	-	1.10	13.20	-	-	100m²	**13.20**
20 litres /m²	-	1.46	17.56	-	-	100m²	**17.56**
25 litres /m²	-	1.84	22.04	-	-	100m²	**22.04**
Hand held hosepipe; flow rate 40 litres per minute; irrigation requirement							
10 litres /m²	-	0.46	5.54	-	-	100m²	**5.54**
15 litres /m²	-	0.69	8.32	-	-	100m²	**8.32**
20 litres /m²	-	0.91	10.96	-	-	100m²	**10.96**
25 litres /m²	-	1.15	13.75	-	-	100m²	**13.75**
Hedge Cutting; field hedges cut once or twice annually							
Trimming sides and top using hand tool or hand held mechanical tools							
not exceeding 2 m high	-	0.10	1.20	0.14	-	10m²	**1.33**
2 to 4 m high	-	0.33	4.00	0.45	-	10m²	**4.45**
Hedge Cutting; ornamental							
Trimming sides and top using hand tool or hand held mechanical tools							
not exceeding 2 m high	-	0.13	1.50	0.17	-	10m²	**1.67**
2 to 4 m high	-	0.50	6.00	0.68	-	10m²	**6.67**
Hedge Cutting; reducing width; hand tool or hand held mechanical tools							
Not exceeding 2 m high							
average depth of cut not exceeding 300 mm	-	0.10	1.20	0.14	-	10m²	**1.33**
average depth of cut 300 to 600 mm	-	0.83	10.00	1.13	-	10m²	**11.12**
average depth of cut 600 to 900 mm	-	1.25	15.00	1.69	-	10m²	**16.69**
2 to 4 m high							
average depth of cut not exceeding 300 mm	-	0.03	0.30	0.03	-	10m²	**0.33**
average depth of cut 300 to 600 mm	-	0.13	1.50	0.17	-	10m²	**1.67**
average depth of cut 600 to 900 mm	-	2.50	30.00	3.38	-	10m²	**33.38**
4 to 6 m high							
average depth of cut not exceeding 300 mm	-	0.10	1.20	0.14	-	m²	**1.33**
average depth of cut 300 to 600 mm	-	0.17	2.00	0.23	-	m²	**2.23**
average depth of cut 600 to 900 mm	-	0.50	6.00	0.68	-	m²	**6.67**
Hedge Cutting; reducing width; tractor mounted hedge cutting equipment							
Not exceeding 2 m high							
average depth of cut not exceeding 300 mm	-	0.04	0.48	0.62	-	10m²	**1.10**
average depth of cut 300 to 600 mm	-	0.05	0.60	0.77	-	10m²	**1.37**
average depth of cut 600 to 900 mm	-	0.20	2.40	3.08	-	10m²	**5.48**
2 to 4 m high							
average depth of cut not exceeding 300 mm	-	0.01	0.15	0.19	-	10m²	**0.34**
average depth of cut 300 to 600 mm	-	0.03	0.30	0.38	-	10m²	**0.68**
average depth of cut 600 to 900 mm	-	0.02	0.24	0.31	-	10m²	**0.55**

Q PAVING/PLANTING/FENCING/SITE FURNITURE Excluding overheads and profit	PC £	Labour hours	Labour £	Plant £	Material £	Unit	Total rate £
Q35 LANDSCAPE MAINTENANCE - cont'd							
Hedge Cutting; reducing height; hand tool or hand held mechanical tools							
Not exceeding 2 m high							
average depth of cut not exceeding 300 mm	-	0.07	0.80	0.09	-	10m²	**0.89**
average depth of cut 300 to 600 mm	-	0.13	1.60	0.18	-	10m²	**1.78**
average depth of cut 600 to 900 mm	-	0.40	4.80	0.54	-	10m²	**5.34**
2 to 4 m high							
average depth of cut not exceeding 300 mm	-	0.03	0.40	0.05	-	m²	**0.44**
average depth of cut 300 to 600 mm	-	0.07	0.80	0.09	-	m²	**0.89**
average depth of cut 600 to 900 mm	-	0.20	2.40	0.27	-	m²	**2.67**
4 to 6 m high							
average depth of cut not exceeding 300 mm	-	0.07	0.80	0.09	-	m²	**0.89**
average depth of cut 300 to 600 mm	-	0.13	1.50	0.17	-	m²	**1.67**
average depth of cut 600 to 900 mm	-	0.25	3.00	0.34	-	m²	**3.34**
Hedge Cutting; removal and disposal of arisings							
Sweeping up and depositing arisings							
300 mm cut	-	0.05	0.60	-	-	10m²	**0.60**
600 mm cut	-	0.20	2.40	-	-	10m²	**2.40**
900 mm cut	-	0.40	4.80	-	-	10m²	**4.80**
Chipping arisings							
300 mm cut	-	0.02	0.24	0.09	-	10m²	**0.33**
600 mm cut	-	0.08	1.00	0.38	-	10m²	**1.38**
900 mm cut	-	0.20	2.40	0.92	-	10m²	**3.32**
Disposal of unchipped arisings							
300 mm cut	-	0.02	0.20	0.47	0.33	10m²	**1.01**
600 mm cut	-	0.03	0.40	0.95	0.50	10m²	**1.85**
900 mm cut	-	0.08	1.00	2.36	1.25	10m²	**4.61**
Disposal of chipped arisings							
300 mm cut	-	-	0.04	0.16	0.50	10m²	**0.70**
600 mm cut	-	0.02	0.20	0.16	1.00	10m²	**1.36**
900 mm cut	-	0.03	0.40	0.31	2.50	10m²	**3.21**
Q40 FENCING AND GATES							
Protective fencing							
Rolled chestnut pale fencing; fixing to 100 mm diameter chestnut posts; driving into firm ground at 3 m centres							
900 mm high	1.76	0.11	1.60	-	3.79	m	**5.39**
1200 mm high	2.36	0.16	2.40	-	4.42	m	**6.82**
1500 mm high; 3 strand	3.58	0.21	3.20	-	5.53	m	**8.74**
Enclosures; Earth Anchors							
"Rootfast" anchored galvanized steel enclosures post ref ADP 20-1000; 1000 mm high x 20 mm diameter with ref AA25-750 socket and padlocking ring	30.00	0.10	1.50	-	32.40	nr	**33.91**
steel cable, orange plastic coated	-	-	0.03	-	1.25	m	**1.28**

Q PAVING/PLANTING/FENCING/SITE FURNITURE Excluding overheads and profit	PC £	Labour hours	Labour £	Plant £	Material £	Unit	Total rate £
Boundary fencing; strained wire and wire mesh; H.S.Jackson & Son (Fencing)							
Strained wire fencing; concrete posts only at 2750 mm centres, 610 mm below ground; excavating holes; filling with concrete; replacing top soil; disposing surplus soil off site							
900 mm high	2.18	0.56	8.33	0.61	3.46	m	12.40
1200 mm high	2.89	0.56	8.33	0.61	4.17	m	13.11
1400 mm high	3.50	0.56	8.33	0.61	4.77	m	13.72
1800 mm high	3.95	0.78	11.67	1.52	5.22	m	18.41
2400 mm high	5.33	1.67	25.00	0.61	6.60	m	32.22
Extra over strained wire fencing for concrete straining posts with one strut; posts and struts 610 mm below ground; struts cleats, stretchers, winders, bolts, and eye bolts; excavating holes; filling to within 150 mm of ground level with concrete (1:12) - 40 mm aggregate; replacing topsoil; disposing surplus soil off site							
900 mm high	18.67	0.67	10.05	1.37	41.82	nr	53.24
1200 mm high	20.75	0.67	10.00	1.38	44.27	nr	55.65
1400 mm high	23.00	0.67	10.00	1.38	48.53	nr	59.91
1800 mm high	29.53	0.67	10.00	1.38	53.86	nr	65.24
2400 mm high	43.92	0.83	12.49	1.63	72.70	nr	86.82
Extra over strained wire fencing for concrete straining posts with two struts; posts and struts 610 mm below ground; excavating holes; filling to within 150 mm of ground level with concrete (1 12) - 40 mm aggregate; replacing topsoil; disposing surplus soil off site							
900 mm high	26.93	0.91	13.65	1.76	61.22	nr	76.63
1200 mm high	30.01	0.85	12.75	1.38	65.67	nr	79.79
1400 mm high	33.50	0.91	13.62	1.38	74.65	nr	89.64
1800 mm high	42.86	0.85	12.75	1.38	81.61	nr	95.73
2400 mm high	43.92	0.67	10.00	1.63	87.13	nr	98.75
Strained wire fencing; galvanised steel angle posts only at 2750 mm centres; 610 mm below ground; driving in							
900 mm high; 40 mm x 40 mm x 5 mm	1.33	0.03	0.45	-	1.33	m	1.78
1200 mm high; 40 mm x 40 mm x 5 mm	1.56	0.03	0.50	-	1.56	m	2.06
1400 mm high; 40 mm x 40 mm x 5 mm	1.68	0.06	0.91	-	1.68	m	2.59
1800 mm high; 40 mm x 40 mm x 5 mm	2.25	0.07	1.09	-	2.25	m	3.34
2400 mm high; 45 mm x 45 mm x 5 mm	3.09	0.12	1.82	-	3.09	m	4.91
Galvanised steel straining posts with one strut for strained wire fencing; setting in concrete							
900 mm high; 50 mm x 50 mm x 6 mm	-	-	-	-	-	nr	39.90
1200 mm high; 50 mm x 50 mm x 6 mm	-	-	-	-	-	nr	42.80
1400 mm high; 50 mm x 50 mm x 6 mm	-	-	-	-	-	nr	44.07
1800 mm high; 50 mm x 50 mm x 6 mm	-	-	-	-	-	nr	49.66
2400 mm high; 60 mm x 60 mm x 6 mm	-	-	-	-	-	nr	60.77
Galvanised steel straining posts with two struts for strained wire fencing; setting in concrete							
900 mm high; 50 mm x 50 mm x 6 mm	-	-	-	-	-	nr	56.27
1200 mm high; 50 mm x 50 mm x 6 mm	-	-	-	-	-	nr	61.18
1400 mm high; 50 mm x 50 mm x 6 mm	-	-	-	-	-	nr	62.86
1800 mm high; 50 mm x 50 mm x 6 mm	-	-	-	-	-	nr	74.17
2400 mm high; 60 mm x 60 mm x 6 mm	-	-	-	-	-	nr	91.38

Q PAVING/PLANTING/FENCING/SITE FURNITURE

Item (Excluding overheads and profit)	PC £	Labour hours	Labour £	Plant £	Material £	Unit	Total rate £
Q40 FENCING AND GATES - cont'd							
Boundary fencing; strained wire and wire mesh; H.S.Jackson & Son (Fencing) - cont'd							
Strained wire; to posts (posts not included); 3 mm galvanized wire; fixing with galvanized stirrups							
900 mm high; 2 wire	0.20	0.03	0.50	-	0.26	m	**0.76**
1200 mm high; 3 wire	0.30	0.05	0.70	-	0.36	m	**1.06**
1400 mm high; 3 wire	0.30	0.05	0.70	-	0.36	m	**1.06**
1800 mm high; 3 wire	0.30	0.05	0.70	-	0.36	m	**1.06**
Barbed wire; to posts (posts not included); 3 mm galvanised wire; fixing with galvanised stirrups							
900 mm high; 2 wire	0.20	0.07	1.00	-	0.26	m	**1.26**
1200 mm high; 3 wire	0.30	0.09	1.40	-	0.36	m	**1.76**
1400 mm high; 3 wire	0.30	0.09	1.40	-	0.36	m	**1.76**
1800 mm high; 3 wire	0.30	0.09	1.40	-	0.36	m	**1.76**
Chain link fencing; H.S.Jackson & Son (Fencing); to strained wire and posts priced separately, 3 mm galvanised wire; 51 mm mesh; galvanized steel components; fixing to line wires threaded through posts and strained with eye-bolts; posts (not included)							
900 mm high	1.92	0.07	1.00	-	1.98	m	**2.98**
1200 mm high	2.42	0.07	1.00	-	2.48	m	**3.48**
1800 mm high	3.76	0.10	1.50	-	3.82	m	**5.32**
2400 mm high	3.76	0.10	1.50	-	3.82	m	**5.32**
Chain link fencing; to strained wire and posts priced separately; 3.15 mm plastic coated galvanized wire (wire only 2.50 mm); 51 mm mesh; galvanised steel components; fencing with line wires threaded through posts and strained with eye-bolts; posts (not included) (Note: plastic coated fencing can be cheaper than galvanised finish as wire of a smaller cross-sectional area can be used)							
900 mm high	1.58	0.07	1.00	-	1.70	m	**2.70**
1200 mm high	2.09	0.07	1.00	-	2.21	m	**3.21**
1400 mm high	2.40	0.07	1.00	-	2.52	m	**3.52**
1800 mm high	-	0.13	1.88	-	3.56	m	**5.43**
Extra over strained wire fencing for cranked arms and three rows of plain barbed wire							
1800 mm high	1.13	0.06	0.83	-	1.33	m	**2.16**
As above but galvanized barbed wire and clips in lieu of plain wire							
1 row	1.24	0.02	0.25	-	1.24	m	**1.49**
2 row	1.34	0.05	0.75	-	1.34	m	**2.09**
3 row	1.44	0.05	0.75	-	1.44	m	**2.19**
Field fencing; welded wire mesh; fixed to posts and straining wires measured separately							
Cattle fence; 1143 high 114 x 300 mm at bottom to 230 x 300 mm at top	0.69	0.10	1.50	-	0.89	m	**2.39**
Sheep fence; 910 high; 140 x 300 mm at bottom to 230 x 300 mm at top	0.55	0.10	1.50	-	0.75	m	**2.25**
Deer Fence; 1905 high; 89 x 150 mm at bottom to 267 x 300 mm at top	2.48	0.13	1.88	-	2.68	m	**4.56**
Extra for concreting in posts	-	-	-	-	1.70	nr	**1.70**
Extra for straining post	-	0.75	11.25	-	12.55	nr	**23.80**

Q PAVING/PLANTING/FENCING/SITE FURNITURE Excluding overheads and profit	PC £	Labour hours	Labour £	Plant £	Material £	Unit	Total rate £
Rabbit netting; H.S Jackson and Son; Timber stakes; peeled kiln dried pressure treated; pointed; 1.8 m posts driven 900 mm into ground at 3 m centres; (line wires and netting priced separately)							
75 -100 mm stakes	-	0.25	3.75	-	0.60	m	4.35
Corner posts or straining posts 150 mm diameter. 2.3 m high set in concrete; centres to suit local conditions or changes of direction							
1 strut	10.90	1.00	15.00	3.92	16.93	each	35.85
2 strut	12.88	1.00	15.00	3.92	18.91	each	37.83
Strained wire; to posts (posts not included); 3 mm galvanized wire; fixing with galvanized stirrups							
900 mm high; 2 wire	0.20	0.03	0.50	-	0.26	m	0.76
1200 mm high; 3 wire	0.30	0.05	0.70	-	0.36	m	1.06
Rabbit netting; 31 mm 19 gauge 1050 high netting fixed to posts line wires and straining posts or corner posts all priced separately.							
900 high turned in	0.56	0.04	0.60	-	0.57	m	1.17
900 high buried 150 mm in trench	0.56	0.08	1.25	-	0.57	m	1.82
Boundary fencing; strained wire and wire mesh; H.S.Jackson & Son (Fencing) Tubular chain link fencing; galvanized; plastic-coated; 60.3 mm diameter posts at 3.0 m centres; setting 700 mm into ground; choice of ten mesh colours; including excavating holes; backfilling and removing surplus soil; with top rail only							
900 mm high	-	-	-	-	-	m	19.42
1200 mm high	-	-	-	-	-	m	20.48
1800 mm high	-	-	-	-	-	m	22.70
2000 mm high	-	-	-	-	-	m	24.08
Tubular chain link fencing; galvanised; plastic coated; 60.3 mm diameter posts at 3.0 m centres; cranked arms and 3 lines barbed wire; setting 700 mm into ground; including excavating holes; backfilling and removing surplus soil; with top rail only							
2000 mm high	-	-	-	-	-	m	25.20
1800 mm high	-	-	-	-	-	m	24.73
Boundary fencing; Steelway Fensecure "Steelway-Fensecure" tubular fencing; comprising 60.3 mm tubular posts at 3.00 m centres; setting in concrete; 35 mm top rail tied with aluminium and steel fittings; 50 mm x 50 mm x 355/2.5 mm PVC coated chain link; all components galvanized and coated in green nylon							
900 mm high	-	-	-	-	-	m	19.54
1200 mm high	-	-	-	-	-	m	20.45

Q40 FENCING AND GATES - cont'd

Q PAVING/PLANTING/FENCING/SITE FURNITURE Excluding overheads and profit	PC £	Labour hours	Labour £	Plant £	Material £	Unit	Total rate £
Boundary fencing; Steelway Fensecure - cont'd							
1400 mm high	-	-	-	-	-	m	21.45
1800 mm high	-	-	-	-	-	m	24.43
2100 mm high	-	-	-	-	-	m	32.27
2400 mm high	-	-	-	-	-	m	35.64
3050 mm high	-	-	-	-	-	m	33.87
3600 mm high	-	-	-	-	-	m	48.85
End Posts; 60.3 mm diameter; setting in concrete							
900 mm high	-	-	-	-	-	nr	19.38
1200 mm high	-	-	-	-	-	nr	20.45
1400 mm high	-	-	-	-	-	nr	21.45
1800 mm high	-	-	-	-	-	nr	24.73
2100 mm high	-	-	-	-	-	nr	26.92
2400 mm high	-	-	-	-	-	nr	28.57
3050 mm high	-	-	-	-	-	nr	33.87
3600 mm high	-	-	-	-	-	nr	36.04
Corner Posts; 60.3 mm diameter; setting in concrete							
900 mm high	-	-	-	-	-	nr	19.54
1200 mm high	-	-	-	-	-	nr	20.45
1400 mm high	-	-	-	-	-	nr	21.45
1800 mm high	-	-	-	-	-	nr	24.43
2100 mm high	-	-	-	-	-	nr	32.27
2400 mm high	-	-	-	-	-	nr	35.64
3050 mm high	-	-	-	-	-	nr	33.87
3600 mm high	-	-	-	-	-	nr	48.85
Boundary fencing; Twil Fencing Products							
"Sentinel Paladin" Welded mesh colour coated green fencing; fixing to metal posts at 3 m centres with manufacturer's fixings; setting 600 mm deep in firm ground; including excavating holes; backfilling and removing surplus excavated material							
900 mm high	-	-	-	-	13.81	m	13.81
1200 mm high	-	-	-	-	19.77	m	19.77
1800 mm high	-	-	-	-	25.00	m	25.00
Extra over welded galvanised plastic coated mesh fencing for concreting in posts	-	-	-	-	1.70	nr	1.70
Boundary fencing; Alpha Rail Limited							
"Orsogril" rectangular steel bar mesh fence panels; pleione pattern; bolting to 60 mm x 80 mm uprights at 2 m centres; mesh 62 mm x 66 mm, setting in concrete							
930 mm high panels	-	-	-	-	-	m	88.15
1326 mm high panels	-	-	-	-	-	m	122.01
1722 mm high panels	-	-	-	-	-	m	146.74
Alpha Rail Ltd; "Orsogril" rectangular steel bar mesh fence panels; pleione pattern; bolting to 80 mm x 80 mm uprights at 2 m centres; mesh 62 mm x 66 mm; setting in concrete							

Q PAVING/PLANTING/FENCING/SITE FURNITURE Excluding overheads and profit	PC £	Labour hours	Labour £	Plant £	Material £	Unit	Total rate £
930 mm high panel	-	-	-	-	-	m	90.30
1326 mm high panel	-	-	-	-	-	m	119.06
1722 mm high panel	-	-	-	-	-	m	151.31
"Orsogril" rectangular steel bar mesh fence panels, sterope pattern; bolting to 60 mm x 80 mm uprights at 2.00 m centres; mesh 62 mm x 132 mm; setting in concrete paving (paving not included)							
930 mm high panels	-	-	-	-	-	m	83.53
1326 mm high panels	-	-	-	-	-	m	109.49
1722 mm high panels	-	-	-	-	-	m	137.71
Security fencing; H.S.Jackson & Son (Fencing)							
"Barbican" galvanized steel paling fencing; on 60 mm x 60 mm posts at 3 m centres; setting in concrete							
1250 mm high	-	-	-	-	-	m	49.78
1500 mm high	-	-	-	-	-	m	53.28
2000 mm high	-	-	-	-	-	m	61.43
2500 mm high	-	-	-	-	-	m	72.17
Gates; to match "Barbican" galvanised steel paling fencing							
width 1m	-	-	-	-	-	nr	654.00
width 2 m	-	-	-	-	-	nr	662.00
width 3 m	-	-	-	-	-	nr	680.00
width 4 m	-	-	-	-	-	nr	703.00
width 8 m	-	-	-	-	-	pair	1433.00
width 9m	-	-	-	-	-	pair	1728.00
width 10 m	-	-	-	-	-	pair	1964.00
Security fencing; intruder guards							
"Viper Spike Intruder Guards"; to existing structures, including fixing bolts							
ref Viper 1; 40 mm x 5 mm x 1.2 m long with base plate	32 45	1.00	15.00	-	32.45	nr	47.45
ref Viper 3; 160 mm x 190 mm wide; U shape; to prevent intruders climbing pipes	32.08	0.50	7.50	-	32.08	nr	39.58
Security fencing; Siddall and Hilton							
"Razor Barb Concertina" ; spiral wire security barriers; fixing to 600 mm steel ground stakes							
ref RBC4M/G 455 mm diameter roll - medium barb, galvanized	0.91	0.02	0.30	-	1.03	m	1.33
ref RBC7M/G 710 mm diameter roll - medium barb, galvanized	1.09	0.02	0.30	-	1.21	m	1.51
ref RBC9M/G 965 mm diameter roll - medium barb, galvanized	1.39	0.02	0.30	-	1.51	m	1.81
3 lines "Razor Tape" Medium Barb barbed tape; on 50 mm x 50 mm mild steel angle posts; setting in concrete							
ref RTS5M/G Razor Tape medium barb, galvanized	0.32	0.21	3.16	-	7.90	m	11.06
5 lines "Razor Tape" Medium Barb barbed tape; on 50 mm x 50 mm mild steel angle posts; setting in concrete							
ref RTS5M/G Razor Tape medium barb, galvanized	0.53	0.22	3.33	-	8.11	m	11.44

Q PAVING/PLANTING/FENCING/SITE FURNITURE Excluding overheads and profit	PC £	Labour hours	Labour £	Plant £	Material £	Unit	Total rate £
Q40 FENCING AND GATES - cont'd							
Trip rails; timber; AVS Fencing Contractors							
200 mm x 38 mm softwood rails; screwing to 100 mm x 100 mm softwood posts 700 mm long; setting in concrete; at 1.00 m centres; all treated with timber preservative	10.20	0.71	10.68	-	13.59	m	**24.27**
Extra over timber trip rails for priming in lieu of timber preservative	0.10	0.08	1.25	-	0 10	m	**1.35**
Extra over timber trip rails for bolting in lieu of screwing rails to softwood posts	1.10	0.13	1.88	-	1.10	m	**2.98**
Extra over timber trip rails for galvanised 50 mm x 50 mm mild steel posts at 1.50 m centres in lieu of 100 mm x 100 mm softwood posts	3.84	0.22	3.33	-	3.84	m	**7.17**
50 mm x 150 mm "BFC" oiled hardwood rails on 100 mm x 75 mm western red cedar posts 700 mm long; setting in concrete at 1.00 m centres	15.70	0.35	5.25	-	15.70	m	**20.95**
50 mm x 150 mm oak rails on 150 mm x 200 mm oak posts 1100 mm long; driving into firm ground at 1.00 m centres	12.40	0.35	5.25	-	12.40	m	**17.65**
Trip rails; metal; Broxap Streetscene							
double row 48.3 mm internal diameter mild steel tubular rails with sleeved joints; to cast iron posts 525 mm above ground mm long; setting in concrete at 1.20 m centres; all standard painted BX 243 with BX Rail C	77.31	1.00	15.00	-	82.12	m	**97.12**
22 mm diameter mild steel rails with ferrule joints; fixing through holes in 44 mm x 13 mm mild steel standards 700 mm long; setting in concrete at 1 20 m centres; priming	22.66	1.33	20.00	-	23.02	m	**43.02**
Double 48 mm diameter mild steel rails; to 75 mm diameter mild steel "Y" posts 700 mm long; setting in concrete at 1.20 m centres; all galvanized	45.86	1.00	15.00	-	46.22	m	**61.22**
48 mm diameter galvanized tubular rails; fixing through holes in 125 mm x 125 mm oak posts 1 20 m long; setting in firm ground at 2.00 m centres	18.25	1.18	17.65	-	18.61	m	**36.26**
48 mm diameter galvanised tubular rails; fixing through holes in 100 mm x 150 mm concrete barrel posts 1.00 m long; setting in firm ground at 2.00 m centres	12.25	1.14	17.14	-	12.68	m	**29.82**
Trip rails; Townscape Products Ltd							
Hollow steel section knee rails, coated; 500 mm high; setting in concrete							
1000 mm bays	52.69	2.00	30.00	-	59.31	m	**89.30**
1200 mm bays	16.34	2.00	30.00	-	54.64	m	**84.63**

Q PAVING/PLANTING/FENCING/SITE FURNITURE Excluding overheads and profit	PC £	Labour hours	Labour £	Plant £	Material £	Unit	Total rate £
Timber fencing; AVS Fencing Supply							
Palisade fencing; 22 mm x 75 mm softwood vertical palings with flat tops; nailing to 3 nr 50 mm x 125 mm horizontal softwood cant rails; housing into 100 mm x 100 mm softwood posts with weathered tops at 3.00 m centres; setting in concrete; all treated timber							
1800 mm high	15.18	0.47	7.01	-	19.04	m	26.05
Palisade fencing; 22 mm x 75 mm oak vertical palings with flat tops; nailing to 3 nr 50 mm x 125 mm horizontal oak cant rails; housing into 100 mm x 100 mm oak posts with weathered tops at 3.00 m centres; setting in concrete							
1800 mm high	35.43	0.47	7.00	-	39.30	m	46.30
Post-and-rail fencing; 3 nr 90 mm x 38 mm softwood horizontal rails; fixing with galvanized nails to 150 mm x 75 mm softwood posts; including excavating and backfilling into firm ground at 1.80 m centres; all treated timber							
1200 mm high	9.14	0.35	5.25	-	9.19	m	14.44
Post-and-rail fencing; 3 nr 90 mm x 38 mm oak horizontal rails; fixing with galvanised nails to 150 mm x 75 mm oak posts; including excavating and backfilling into firm ground at 1.80 m centres							
1200 mm high	15.47	0.35	5.25	-	15.52	m	20.77
Post-and-rail fencing; 3 nr 90 mm x 38 mm western red cedar rails; fixing with galvanised nails to 150 mm x 75 mm hardwood posts; including excavating and backfilling into firm ground at 1.80 m centres							
1200 mm high	58.10	0.35	5.25	-	58.15	m	63.40
Morticed post-and-rail fencing; 3 nr horizontal 90 mm x 38 mm softwood rails; fixing with galvanized nails; 90 mm x 38 mm softwood centre prick posts; to 150 mm x 75 mm softwood posts; including excavating and backfilling into firm ground at 2.85 m centres; all treated timber							
1200 mm high	8.00	0.35	5.25	-	8.05	m	13.30
1350 mm high five rails	8.79	0.40	6.00	-	8.84	m	14.84
Cleft rail fencing; oak or chestnut adze tapered rails 2.80 m long; morticed into joints; to 150 mm x 100 mm softwood posts 1.98 m long; including excavating and backfilling into firm ground at 2.80 m centres							
three rails	11.01	0.28	4.20	-	11.01	m	15.21
four rails	20.05	0.35	5.25	-	20.05	m	25.30
Close boarded fencing; 2 nr softwood rails; 150 x 25 mm gravel boards; 89 mm x 19 mm softwood pales lapped 13 mm; to concrete posts; including excavating and backfilling into firm ground at 3.00 m centres							
900 mm high	8.67	0.47	7.01	-	9.94	m	16.95
1050 mm high	6.86	0.47	7.01	-	10.50	m	17.51
1200 mm high	7.46	0.47	7.01	-	11.25	m	18.26

Q PAVING/PLANTING/FENCING/SITE FURNITURE
Excluding overheads and profit

Q40 FENCING AND GATES - cont'd

Timber fencing; AVS Fencing Supply - cont'd

Item	PC £	Labour hours	Labour £	Plant £	Material £	Unit	Total rate £
Close boarded fencing; 3 nr softwood rails; 150 x 25 mm gravel boards; 89 mm x 19 mm softwood pales lapped 13 mm; to concrete posts; including excavating and backfilling into firm ground at 3 00 m centres							
1350 mm high	11.83	0.56	8.40	-	13.10	m	21.50
1650 mm high	12.93	0.56	8.40	-	14.20	m	22.60
1800 mm high	13.30	0.56	8.40	-	14.57	m	22.97
Close boarded fencing; 3 nr softwood rails; 150 x 25 mm gravel boards; 89 mm x 19 mm softwood pales lapped 13 mm; to softwood posts 100 x 100 mm; including excavating and backfilling into firm ground at 3.00 m centres.							
1350 mm high	10.52	0.56	8.40	-	11.79	m	20.19
1650 mm high	11.60	0.56	8.40	-	12.87	m	21.27
1800 mm high	12.25	0.56	8.40	-	13.53	m	21.93
Close boarded fencing; 3 nr softwood rails; 150 x 25 mm gravel boards; 89 mm x 19 mm softwood pales lapped 13 mm; to oak posts; including excavating and backfilling into firm ground at 3 00 m centres							
1350 mm high	8.91	0.56	8.40	-	14.32	m	22.73
1650 mm high	14 69	0.56	8.40	-	15.96	m	24.37
1800 mm high	15 63	0.56	8.40	-	16.90	m	25.31
Close boarded fencing; 3 nr softwood rails; 150 x 25 mm gravel boards; 89 mm x 19 mm softwood pales lapped 13 mm; to oak posts; including excavating and backfilling into firm ground at 2 47 m centres							
1800 mm high	16.82	0.68	10.16	-	18.09	m	28.25
Close boarded fencing - all oak; 2 nr 76 mm x 38 mm rectangular rails; 152 mm x 25 mm gravel boards; 89 mm x 19 mm oak featheredge boards lapped 13 mm; to oak posts; including excavating and backfilling into firm ground at 2 47 m centres							
1070 mm high	7.08	0.56	8.40	-	29.02	m	37.43
1800 mm high	34.10	0.56	8.40	-	35.38	m	43.78
Extra over close boarded fencing for 65 mm x 38 mm oak cappings	4.32	0.05	0.75	-	4.32	nr	5.07
"Hit and miss" horizontal rail fencing; 87 mm x 38 mm top and bottom rails; 100 mm x 22 mm vertical boards arranged alternately on opposite side of rails; to 100 mm x 100 mm posts; including excavating and backfilling into firm ground; setting in concrete at 1.8 m centres							
treated softwood; 1800 mm high	19.08	1.33	20.00	-	21.83	m	41.83
treated softwood; 2000 mm high	19.85	1.33	20.00	-	25.18	m	45.17
primed softwood; 1800 mm high	19.08	1.67	25.00	-	22.22	m	47.22
primed softwood; 2000 mm high	19.85	1.67	25.00	-	25.57	m	50.57
Screen fencing in horizontal wavy edge overlap panels; 1.80 m wide; to 100 mm x 100 mm preservative treated softwood posts; including excavating and backfilling into firm ground							

Q PAVING/PLANTING/FENCING/SITE FURNITURE Excluding overheads and profit	PC £	Labour hours	Labour £	Plant £	Material £	Unit	Total rate £
900 mm high	6.08	0.47	7.01	-	11.12	m	18.13
1200 mm high	6.25	0.47	7.01	-	11.76	m	18.77
1500 mm high	6.39	0.56	8.43	-	13.21	m	21.63
1800 mm high	6.53	0.56	8.43	-	13.79	m	22.21
Screen fencing for oak in lieu of softwood posts							
900 mm high	4.72	-	-	-	4.72	m	4.72
1200 mm high	5.90	-	-	-	5.90	m	5.90
1500 mm high	7.08	-	-	-	7.08	m	7.08
1800 mm high	8.21	-	-	-	8.21	m	8.21
Extra over screen fencing for 300 mm high trellis tops; slats at 100 mm centres; including additional length of posts	3.78	0.10	1.50	-	3.78	m	5.28
Timber trellis; Landscapes by Design; pressure treated softwood panels							
Timber section 38 x 19 mm at 120 mm centres at 45, 55 or 90 deg angle; framed and weathered; panel size 1.8 x 1.8 m							
fixed to timber freestanding posts 100 x 100 mm	-	-	-	-	-	m	60.00
fixed to wall on 38 mm batten	-	-	-	-	-	m	56.00
Extra for post finials	-	-	-	-	-	nr	9.60
extra for angled top to each trellis panel at 55 deg projecting 450 mm above the top frame of each panel	-	-	-	-	-	nr	12.00
extra for scalloped radius top depression 300 mm into each panel	-	-	-	-	-	nr	15.00
extra for painting with Sadolin	-	-	-	-	-	m	12.00
Concrete fencing							
Panel fencing; to precast concrete posts; in 2 m bays; setting posts 600 mm into ground; sandfaced finish							
900 mm high	11.18	0.25	3.75	-	12.86	m	16.61
1200 mm high	14.90	0.25	3.75	-	16.59	m	20.34
Panel fencing; to precast concrete posts; in 2 m bays, setting posts 750 mm into ground; sandfaced finish							
1500 mm high	18.63	0.33	5.00	-	20.31	m	25.31
1800 mm high	22.36	0.36	5.45	-	24.04	m	29.49
2100 mm high	26.08	0.40	6.00	-	27.76	m	33.76
2400 mm high	31.05	0.40	6.00	-	32.73	m	38.73
Extra over concrete panel fencing for aggregate faced one side	7.09	-	-	-	7.09	m²	7.09
Windbreak fencing							
Fencing; Netlon Ltd; "Tensar" windbreak fencing; to 100 mm diameter treated softwood posts; setting 450 mm into ground; fixing with 50 mm x 25 mm treated softwood battens nailed to posts, including excavating and backfilling into firm ground; setting in concrete at 3 m centres							
1500 mm high	3.21	0.16	2.33	-	6.49	m	8.82
2000 mm high	7.35	0.16	2.33	-	10.63	m	12.96

Q PAVING/PLANTING/FENCING/SITE FURNITURE Excluding overheads and profit	PC £	Labour hours	Labour £	Plant £	Material £	Unit	Total rate £
Q40 FENCING AND GATES - cont'd							
Ball stop fencing; Steelway Fensecure							
Ball Stop Net; 30 x 30 mm netting fixed to 60.3 diameter 12 mm solid bar lattice galvanised dual posts, top, middle and bottom rails							
4.5 m high	-	-	-	-	-	m	80.18
5.0 m high	-	-	-	-	-	m	91.53
6.0 m high	-	-	-	-	-	m	113.06
7.0 m high	-	-	-	-	-	m	133.67
8.0 m high	-	-	-	-	-	m	150.73
9.0 m high	-	-	-	-	-	m	160.29
10.0 m high	-	-	-	-	-	m	181.90
Ball Stop Net; Corner posts							
4.5 m high	-	-	-	-	-	m	86.60
5.0 m high	-	-	-	-	-	m	95.00
6.0 m high	-	-	-	-	-	m	119.73
7.0 m high	-	-	-	-	-	m	130.60
8.0 m high	-	-	-	-	-	m	148.40
9.0 m high	-	-	-	-	-	m	166.20
10 m high	-	-	-	-	-	m	184.00
Ball Stop Net; End posts							
4.5 m high	-	-	-	-	-	m	75.90
5.0 m high	-	-	-	-	-	m	90.90
6.0 m high	-	-	-	-	-	m	104.80
7.0 m high	-	-	-	-	-	m	122.60
8.0 m high	-	-	-	-	-	m	144.30
9.0 m high	-	-	-	-	-	m	158.20
10 m high	-	-	-	-	-	m	173.10
Premier Range; Ball stop fencing; 50 x 50 mm 3.55/2.55 galvanized green PVC chain link; 3.0 mm gauge straining wires; 35 mm top middle and bottom horizontal tubular rail; to 60 mm diam steel posts; setting in concrete; including all fittings							
2.75 m high	-	-	-	-	-	m	40.27
3.6 m high	-	-	-	-	-	m	57.69
5.0 m high	-	-	-	-	-	m	69.54
Premier Range; Ball stop fencing; 50 x 50 mm 3.55/3.00 galvanized and green PVC weldmesh; 3.0 mm gauge straining wires; 35 mm top middle and bottom horizontal tubular rail; to 60 mm diam steel posts; setting in concrete; including all fittings							
2.75 m high	-	-	-	-	-	m	47.25
3.6 m high	-	-	-	-	-	m	64.09
5.0 m high	-	-	-	-	-	m	79.03
Premier Range; Corner posts							
2.75 m	-	-	-	-	-	nr	76.97
3.6 m	-	-	-	-	-	nr	100.94
5.0 m	-	-	-	-	-	nr	137.73
Premier Range; End posts							
2.75 m	-	-	-	-	-	nr	60.45
3.6 m	-	-	-	-	-	nr	82.57
5.0 m	-	-	-	-	-	nr	113.71
Classic Range; Subtract for system with top and bottom rail only	-	-	-	-	-	m	4.00

Q PAVING/PLANTING/FENCING/SITE FURNITURE Excluding overheads and profit	PC £	Labour hours	Labour £	Plant £	Material £	Unit	Total rate £
Railings; Alpha Rail Ltd							
Mild steel bar railings of balusters at 115 mm centres welded to flat rail top and bottom; bays 2.00 m long; bolting to 51 mm x 51 mm hollow square section posts; setting in concrete							
galvanized; 900 mm high	-	-	-	-	-	m	41.92
galvanized; 1200 mm high	-	-	-	-	-	m	50.79
galvanized; 1500 mm high	-	-	-	-	-	m	56.44
galvanized; 1800 mm high	-	-	-	-	-	m	67.72
primed; 900 mm high	-	-	-	-	-	m	39.51
primed; 1200 mm high	-	-	-	-	-	m	47.98
primed; 1500 mm high	-	-	-	-	-	m	53.05
primed; 1800 mm high	-	-	-	-	-	m	63.21
Mild steel blunt top railings of 19 mm balusters at 130 mm centres welded to bottom rail; passing through holes in top rail and welded; top and bottom rails 40 mm x 10mm; bolting to 51 mm x 51 mm hollow square section posts; setting in concrete							
galvanized; 800 mm high	-	-	-	-	-	m	45.15
galvanized; 1000 mm high	-	-	-	-	-	m	47.41
galvanized; 1300 mm high	-	-	-	-	-	m	56.44
galvanized; 1500 mm high	-	-	-	-	-	m	60.95
primed; 800 mm high	-	-	-	-	-	m	43.45
primed; 1000 mm high	-	-	-	-	-	m	45.15
primed; 1300 mm high	-	-	-	-	-	m	53.62
primed; 1500 mm high	-	-	-	-	-	m	57.57
Metal Estate Fencing; Broxap Ltd; Mild steel flat bar angular fencing; galvanized; main posts at 5.00 m centres; fixed with 1:3:6 concrete 430 deep; intermediate posts fixed with fixing claw at 1.00 m centres.							
"Cheshire"; curved top posts and 5 nr flat rails 1200 high	29.60	0.28	4.17	-	32.00	m	36.17
"Prestbury" 5 nr plain flat or round bar 1200 high	27.20	0.28	4.17	-	29.60	m	33.77
Kissing gate; Mild steel construction, flat or angular steel;							
Galvanized	271.00	2.00	30.00	-	273.40	m	303.40
Five bar horizontal steel fencing; 1.20 m high; 38 mm x 10 mm joiner standards at 4.50 m centres; 38 mm x 8 mm intermediate standards at 900 mm centres; including excavating and backfilling into firm ground; setting in concrete							
galvanized	-	-	-	-	-	m	61.75
primed	-	-	-	-	-	m	50.75
Extra over five bar horizontal steel fencing for 76 mm diameter end and corner posts							
galvanized	-	-	-	-	-	each	67.50
primed	-	-	-	-	-	each	39.50

Q PAVING/PLANTING/FENCING/SITE FURNITURE Excluding overheads and profit	PC £	Labour hours	Labour £	Plant £	Material £	Unit	Total rate £
Q40 FENCING AND GATES - cont'd							
Metal Estate Fencing; Broxap Ltd; - cont'd							
Five bar horizontal steel fencing with mild steel copings, 38 mm square hollow section mild steel standards at 1.80 m centres; including excavating and backfilling into firm ground; setting in concrete							
galvanized; 900 mm high	-	-	-	-	-	m	80.80
galvanized; 1200 mm high	-	-	-	-	-	m	85.00
galvanized; 1500 mm high	-	-	-	-	-	m	91.00
galvanized; 1800 mm high	-	-	-	-	-	m	97.00
primed; 900 mm high	-	-	-	-	-	m	71.70
primed; 1200 mm high	-	-	-	-	-	m	73.60
primed; 1500 mm high	-	-	-	-	-	m	77.60
primed; 1800 mm high	-	-	-	-	-	m	83.50
Railings; traditional pattern; 16 mm diameter verticals at 127 mm intervals with horizontal bars near top and bottom; balusters with spiked tops; 51 x 20 mm standards; including setting 520 mm into concrete at 2.75 m centres							
primed; 1200 mm high	-	-	-	-	-	m	50.79
primed; 1500 mm high	-	-	-	-	-	m	57.57
primed; 1800 mm high	-	-	-	-	-	m	62.08
Interlaced bow-top mild steel railings; traditional park type; 16 mm diameter verticals at 80 mm intervals, welded at bottom to 50 x 10 mm flat and slotted through 38 mm x 8 mm top rail to form hooped top profile; 50 mm x 10 mm standards; setting 560 mm into concrete at 2.75 m centres							
galvanized; 900 mm high	-	-	-	-	-	m	76.76
galvanized; 1200 mm high	-	-	-	-	-	m	93.69
galvanized; 1500 mm high	-	-	-	-	-	m	111.75
galvanized; 1800 mm high	-	-	-	-	-	m	129.81
primed; 900 mm high	-	-	-	-	-	m	74.50
primed; 1200 mm high	-	-	-	-	-	m	90.30
primed; 1500 mm high	-	-	-	-	-	m	107.23
primed; 1800 mm high	-	-	-	-	-	m	124.16
Gates - General							
Preamble: Gates in fences; see specification for fencing, as gates in traditional or proprietary fencing systems are usually constructed of the same materials and finished as the fencing itself.							
Gates, hardwood; Norbury (Fencing & Building Materials) Ltd							
"Yeoman" entrance gate, five bar diamond braced, curved hanging stile, planed keruing; fixed to 150 mm x 150 mm softwood posts; inclusive of hinges and furniture							
ref 1100 040; 0.9 m wide	107.24	5.00	75.00	-	161.07	nr	236.07
ref 1100 041; 1.2 m wide	116.46	5.00	75.00	-	170.29	nr	245.29
ref 1100 042; 1.5 m wide	125.72	5.00	75.00	-	179.55	nr	254.55
ref 1100 043; 1.8 m wide	181.54	5.00	75.00	-	235.37	nr	310.37
ref 1100 044; 2.1 m wide	194.41	5.00	75.00	-	248.24	nr	323.24

Prices for Measured Works

Q PAVING/PLANTING/FENCING/SITE FURNITURE Excluding overheads and profit	PC £	Labour hours	Labour £	Plant £	Material £	Unit	Total rate £
ref 1100 045; 2.4 m wide	201.76	5.00	75.00	-	255.59	nr	330.59
ref 1100 047; 3.0 m wide	219.82	5.00	75.00	-	273.65	nr	348.65
ref 1100 048; 3.3 m wide	227.15	5.00	75.00	-	280.98	nr	355.98
ref 1100 049; 3.6 m wide	234.46	5.00	75.00	-	288.29	nr	363.29
"Yeoman" entrance gate, five bar diamond braced, curved hanging stile, planed iroko; fixed to 150 mm x 150 mm softwood posts; inclusive of hinges and furniture							
ref 1500 040; 0.9 m wide	150.90	5.00	75.00	-	204.73	nr	279.73
ref 1500 041; 1.2 m wide	167 20	5.00	75.00	-	221.03	nr	296.03
ref 1500 042; 1.5 m wide	180 09	5.00	75.00	-	233.92	nr	308.92
ref 1500 043; 1.8 m wide	283.76	5.00	75.00	-	337.59	nr	412.59
ref 1500 044; 2.1 m wide	295.56	5.00	75.00	-	349.39	nr	424.39
ref 1500 045; 2.4 m wide	305.86	5.00	75.00	-	359.69	nr	434.69
ref 1500 046; 2.7 m wide	316.15	5.00	75.00	-	369.98	nr	444.98
ref 1500 048; 3.3 m wide	340.80	5.00	75.00	-	394.63	nr	469.63
ref 1500 049; 3.6 m wide	350.88	5.00	75.00	-	404.71	nr	479.71
Hardwood field gate, five bar diamond braced, planed keruing; fixed to 150 mm x 150 mm softwood posts; inclusive of hinges and furniture							
ref 1100 100; 0.9 m wide	67 72	4.00	60.00	-	121.55	nr	181.55
ref 1100 101; 1.2 m wide	73.45	4.00	60.00	-	127.28	nr	187.28
ref 1100 102; 1.5 m wide	86 43	4.00	60.00	-	140.26	nr	200.26
ref 1100 103; 1.8 m wide	93.99	4.00	60.00	-	147.82	nr	207.82
ref 1100 104; 2.1 m wide	103.06	4.00	60.00	-	156.89	nr	216.89
ref 1100 105; 2.4 m wide	113.48	4.00	60.00	-	167.31	nr	227.31
ref 1100 106; 2.7 m wide	122.36	4.00	60.00	-	176.19	nr	236.19
ref 1100 108; 3.3 m wide	137 29	4.00	60.00	-	191.12	nr	251.12
ref 1100 109; 3.6 m wide	143.66	4.00	60.00	-	197.49	nr	257.49
Gates, softwood; H.S.Jackson & Son (Fencing)							
Timber field gates, including wrought iron ironmongery; five bar type; diamond braced; 1.80 m high, to 200 mm x 200 mm posts; setting 750 mm into firm ground							
treated softwood; width 2400 mm	68.13	10.00	150.00	-	150.74	nr	300.74
treated softwood; width 2700 mm	74.61	10.00	150.00	-	157.22	nr	307.22
treated softwood; width 3000 mm	76.41	10.00	150.00	-	159.02	nr	309.02
treated softwood; width 3300 mm	83.25	10.00	150.00	-	165.86	nr	315.86
Featherboard garden gates, including ironmongery; to 100 mm x 120 mm posts; 1 nr diagonal brace							
treated softwood; 0.9m x 1.15m high	56.44	3.00	45.00	-	81.80	nr	126.80
treated softwood; 1.0m x 1.2m high	17.73	3.00	45.00	-	81.12	nr	126.12
treated softwood; 1.0m x 1.5m high	23.67	3.00	45.00	-	88.41	nr	133.41
treated softwood; 1.0m x 1.8m high	60.26	3.00	45.00	-	94.44	nr	139.44
Picket garden gates, including ironmongery; to match picket fence; width 1000 mm; to 100 mm x 120 mm posts; 1 nr diagonal brace							
treated softwood; 950 mm high	54.85	3.00	45.00	-	80.22	nr	125.22
treated softwood; 1200 mm high	57.38	3.00	45.00	-	82.74	nr	127.74
treated softwood; 1800 mm high	65.92	3.00	45.00	-	100.11	nr	145.11

Q PAVING/PLANTING/FENCING/SITE FURNITURE Excluding overheads and profit	PC £	Labour hours	Labour £	Plant £	Material £	Unit	Total rate £
Q40 FENCING AND GATES - cont'd							
Gates, tubular steel; H.S.Jackson & Son (Fencing)							
Tubular mild steel field gates, including ironmongery; diamond braced; 1.80 m high; to tubular steel posts; setting in concrete							
galvanized; width 3000 mm	69.70	5.00	75.00	-	138.63	nr	213.63
galvanized; width 3300 mm	73.60	5.00	75.00	-	142.53	nr	217.53
galvanized; width 3600 mm	77.50	5.00	75.00	-	146.43	nr	221.43
galvanized; width 4200 mm	85.50	5.00	75.00	-	154.43	nr	229.43
Gates, sliding; H.S.Jackson & Son (Fencing)							
"Sliding Gate"; including all galvanized rails and vertical rail infill panels; special guide and shutting frame posts (Note: foundations installed by suppliers)							
access width 4.00 m; 1.5m high gates	-	-	-	-	-	nr	3135.00
access width 4.00 m; 2.0m high gates	-	-	-	-	-	nr	3179.00
access width 4.00 m; 2.5m high gates	-	-	-	-	-	nr	3222.00
access width 6.00 m; 1.5m high gates	-	-	-	-	-	nr	3582.00
access width 6.00 m; 2.0m high gates	-	-	-	-	-	nr	3635.00
access width 6.00 m; 2.5m high gates	-	-	-	-	-	nr	3686.00
access width 8.00 m; 1.5m high gates	-	-	-	-	-	nr	4082.00
access width 8.00 m; 2.0m high gates	-	-	-	-	-	nr	4143.00
access width 8.00 m; 2.5m high gates	-	-	-	-	-	nr	4201.00
access width 10.00 m; 2.0m high gates	-	-	-	-	-	nr	5034.00
access width 10.00 m; 2.5m high gates	-	-	-	-	-	nr	5109.00
Stiles and kissing gates							
Stiles; H S Jackson & Son (Fencing); "Jacksons Nr 2007" stiles; 2 nr posts; setting into firm ground; 3 nr rails; 2 nr treads	62.05	3.00	45.00	-	73.45	nr	118.45
Kissing gates; Townscape Products; in mild steel bar, fixing to fencing posts (posts not included); 1 65 m x 1.30 m x 1.00 m high	134.82	5.00	75.00	-	134.82	nr	209.82
Pedestrian guard rails; Broxap Streetscene							
Mild steel pedestrian guard rails; to BS 3049:1976; 1.00 m high with 150 mm toe space; to posts at 2.00 m centres, galvanized finish							
vertical bar infill panel	24.20	2.00	30.00	-	30.23	m	60.23
vertical bar infill with 200 mm visibility gap at top	28.60	2.00	30.00	-	34.63	m	64.63

Q PAVING/PLANTING/FENCING/SITE FURNITURE Excluding overheads and profit	PC £	Labour hours	Labour £	Plant £	Material £	Unit	Total rate £
Q50 SITE/STREET FURNITURE/EQUIPMENT							
Cattle Grids - General							
Preamble: Cattle grids are not usually prefabricated, owing to their size and weight. Specification must take into account the maximum size and weight of vehicles likely to cross the grid. Drainage and regular clearance of the grid is essential. Warning signs should be erected on both approaches to the grid. For the cost of a complete grid and pit installation see "Approximate Estimates" section (book only).							
Cattle grids; H S Jackson & Son (Fencing)							
Cattle grids; supply only							
to carry 12 tonnes evenly distributed; 2.90 m x 2 36 m grid; galvanized	565.00	-	-	-	565.00	nr	565.00
to carry 12 tonnes evenly distributed; 3.66 m x 2.50 m grid; galvanized	985.00	-	-	-	985.00	nr	985.00
Barriers - General							
Preamble: The provision of car and lorry control barriers may form part of the landscape contract. Barriers range from simple manual counterweighted poles to fully automated remote-control security gates, and the exact degree of control required must be specified. Complex barriers may need special maintenance and repair.							
Barriers: Autopa							
Manually operated pole barriers; counterbalance; to tubular steel supports; bolting to concrete foundation (foundation not included); aluminium boom; various finishes;							
clear opening up to 3.00 m	603.00	6.00	90.00	-	623.36	nr	713.36
clear opening up to 4.00 m	662.00	6.00	90.00	-	682.36	nr	772.36
clear opening 5.00	717.00	6.00	90.00	-	737.36	nr	827.36
clear opening 6.00	785.00	6.00	90.00	-	805 36	nr	895.36
clear opening 7.00	860.00	6.00	90.00	-	880.36	nr	970.36
catch pole; arm rest for all manual barriers	101.00	-	-	-	103.54	nr	103.54
Electrically operated pole barriers; enclosed fan-cooled motor; double worm reduction gear; overload clutch coupling with remote controls; aluminium boom; various finishes; (exclusive of electrical connections by electrician)							
ref BL6 "Bulwark" clear opening up to 3.00 m	1760.00	6.00	90.00	-	1780.36	nr	1870.36
ref BL6 "Bulwark" clear opening up to 4.50 m	1830.00	6.00	90.00	-	1850.36	nr	1940.36
ref BL20 "Bulwark" clear opening up to 6.00 m inclusive of catchpost	1900.00	6.00	90.00	-	2021.36	nr	2111.36

Q PAVING/PLANTING/FENCING/SITE FURNITURE Excluding overheads and profit	PC £	Labour hours	Labour £	Plant £	Material £	Unit	Total rate £
Q50 SITE/STREET FURNITURE/EQUIPMENT - cont'd							
Vehicle crash barriers - General Preamble: See Department of Environment Technical Memorandum BE5.							
Vehicle crash barriers Steel corrugated beams; untensioned Broxap Limited; effective length 3.20 m 310 mm deep x 85 mm corrugations;							
steel posts; Z-section; roadside posts	51.11	0.33	5.00	0.81	54.64	m	60.45
steel posts; Z-section; off highway	46.73	0.33	5.00	0.81	50.26	m	56.07
steel posts RSJ 760 high; for anchor fixing	57.67	0.33	5.00	-	61.17	m	66.17
steel posts RSJ 560 high; for anchor fixing to car parks	-	0.33	5.00	-	59.61	m	64.61
extra over for curved rail 6.00 m radius	18.64	-	-	-	18.64	m	18.64
Bases for street furniture Excavating; filling with no-fines cement:aggregate (1:12); bases for street furniture							
300 mm x 450 mm x 500 mm deep	-	0.75	11.25	-	5.33	nr	16.58
300 mm x 600 mm x 500 mm deep	-	1.11	16.65	-	7.11	nr	23.76
300 mm x 900 mm x 500 mm deep	-	1.67	25.05	-	10.66	nr	35.72
1750 mm x 900 mm x 300 mm deep	-	2.33	34.95	-	37.33	nr	72.28
2000 mm x 900 mm x 300 mm deep	-	2.67	40.05	-	42.66	nr	82.71
2400 mm x 900 mm x 300 mm deep	-	3.20	48.00	-	51.20	nr	99.19
2400 mm x 1000 mm x 300 mm deep	-	3.56	53.40	-	56.88	nr	110.28
Precast concrete flags; to BS 7263; to concrete bases (not included); bedding and jointing in cement:mortar (1:4)							
450 mm x 600 mm x 50 mm	7.59	1.17	17.50	-	13.05	m²	30.55
Precast concrete paving blocks; to concrete bases (not included); bedding in sharp sand; butt joints							
200 mm x 100 mm x 65 mm	8.18	0.50	7.50	-	9.92	m²	17.42
200 mm x 100 mm x 80 mm	9.06	0.50	7.50	-	10.68	m²	18.18
Engineering paving bricks; to concrete bases (not included); bedding and jointing in sulphate-resisting cement:lime:sand mortar (1:1:6)							
over 300 mm wide	10.50	0.56	8.43	-	20.84	m²	29.26
Edge restraints to pavings; haunching in concrete (1:3:6)							
200 mm x 300 mm	-	0.10	1.50	-	5.34	m	6.84
Bases; Earth Anchors Ltd "Rootfast" ancillary anchors; ref A1; 500 mm long 25 mm diameter and top strap ref F2; including bolting to site furniture (site furniture not included)	7.84	0.50	7.50	-	7.84	set	15.34
installation tool for above	19.00	-	-	-	19.00	nr	19.00
"Rootfast" ancillary anchors; ref A4; heavy duty 40 mm square fixed head anchors; including bolting to site furniture (site furniture not included)	-	0.33	5.00	-	-	set	5.00
"Rootfast" ancillary anchors; F4; vertical socket; including bolting to site furniture (site furniture not included)	10.85	0.05	0.75	-	10.85	set	11.60

Q PAVING/PLANTING/FENCING/SITE FURNITURE Excluding overheads and profit	PC £	Labour hours	Labour £	Plant £	Material £	Unit	Total rate £
"Rootfast" ancillary anchors; ref F3; horizontal socket; including bolting to site furniture (site furniture not included)	11.90	0.05	0.75	-	11.90	set	12.65
installation tools for the above	56.00	-	-	-	56.00	nr	56.00
"Rootfast" ancillary anchors; ref A3 Anchored bases; including bolting to site furniture (site furniture not included)	46.00	0.33	5.00	-	46.00	set	51.00
installation tools for the above	56.00	-	-	-	56.00	nr	56.00
Furniture/equipment - General Preamble: The following items include fixing to manufacturer's instructions; holding down bolts or other fittings and making good (excavating, backfilling and tarmac, concrete or paving bases not included)							
Dog waste bins; all-steel Earth Anchors Ltd							
HG45A, 45l, earth anchored, post mounted	145.00	0.33	5.00	-	145.00	nr	150.00
HG45A, 45l, as above with pedal operation	172.00	0.33	5.00	-	172.00	nr	177.00
Dog waste bins; cast iron Bins; Furnitubes International Ltd							
ref PED 701; "Pedigree"; post mounted cast iron dog waste bins; 1250 mm total height above ground; 400 mm square bin	635.00	0.75	11.25	-	641.03	nr	652.28
Litter bins; precast concrete in textured white or exposed aggregate finish; with wire baskets and drainage holes Bins; Marshalls Mono							
"Strada 710"; 710 mm diameter x 500 mm high	159.94	1.00	15.00	-	159.94	nr	174.94
"Newstead"; exposed aggregate finish; 760 mm high x 485 mm diameter; including wire baskets	107.83	0.67	10.00	-	107.83	nr	117.83
"Shirley" circular bins; 760 mm high x 1066 mm diameter; including coated wire baskets	409.12	1.00	15.00	6.09	409.12	nr	430.21
Bins; Neptune Outdoor Furniture							
ref SF 16 - 42l	169.00	0.50	7.50	-	169.00	nr	176.50
ref SF 14 - 100l	235.00	0.50	7.50	-	235.00	nr	242.50
Bins; Townscape Products							
"Sutton"; 750 mm high x 500 mm diameter, 70 litre capacity including GRP canopy	135.61	0.33	5.00	-	232.50	nr	237.50
"Braunton"; 750 mm high x 500 mm diameter, 70 litre capacity including GRP canopy	143.77	1.50	22.50	-	240.66	nr	263.16
Litter bins; metal; stove-enamelled perforated metal for holder and container Bins; Abacus Municipal Ltd							
"Model 621"; stove enamelled steel; including brackets; fixing to pole	47.00	0.25	3.75	-	47.00	nr	50.75
"Model 601"; stove enamelled aluminium inner containers; galvanized mild steel cradles; ragged floor mounting pieces; concreting in	85.00	0.50	7.50	-	91.03	nr	98.53
Bins; SMP Playgrounds							
ref LBT18; 710 mm high	229.00	1.00	15.00	-	237.21	nr	252.21
ref LBT28; 710 mm high	369.00	1.00	15.00	-	377.21	nr	392.21

Q PAVING/PLANTING/FENCING/SITE FURNITURE Excluding overheads and profit	PC £	Labour hours	Labour £	Plant £	Material £	Unit	Total rate £
Q50 SITE/STREET FURNITURE/EQUIPMENT - cont'd							
Litter bins; metal; stove-enamelled perforated metal for holder and container - cont'd							
Bins; Townscape Products							
"Metro"; 440 x 420 x 800 high; 62 litre capacity	444.33	0.33	5.00	-	444.33	nr	**449.33**
"Voltan" Large Round; 460 diameter x 780 high; 56 litre capacity	433.31	0.33	5.00	-	433.31	nr	**438.31**
"Voltan" Small Round with pedestal; 410 diameter x 760 high; 31 litre capacity	377.94	0.33	5.00	-	377.94	nr	**382.94**
Litter bins; all-steel							
Bins; Earth Anchors							
"Ranger", 100l Litter bin, 107l, pedestal mounted	364.00	0.50	7.50	-	364.00	nr	**371.50**
"Big Ben", 87l, steel frame and liner, Vyflex coated finish, earth anchored	235.00	1.00	15.00	-	235.00	nr	**250.00**
"Beau", 42l, steel frame and liner, Vyflex coated finish, earth anchored	188.00	1.00	15.00	-	188.00	nr	**203.00**
Bins; Townscape Products							
"Baltimore Major" with GRP canopy; 560 diameter x 960 high, 140 litre capacity	654.86	1.00	15.00	-	654.86	nr	**669.86**
Litter bins; cast Iron							
Bins; Furnitubes International Ltd							
"Wave Bin"; ref WVB 440; free standing 55 litre liners, 440 mm diameter x 850 mm high	643.00	0.50	7.50	-	646.78	nr	**654.28**
"Wave Bin" ref WVB 520; free standing 85 litre liners; 520 mm diameter x 850 mm high; cast iron plinth	716.00	0.50	7.50	-	719.78	nr	**727.28**
"Covent Garden"; ref COV 702; side opening, 500 mm square x 1050 mm high; 105 litre capacity	399.35	0.50	7.50	-	405.67	nr	**413.17**
"Covent Garden"; ref COV 803; side opening, 500 mm diameter x 1025 mm high; 85 litre capacity	361.63	0.50	7.50	-	367.95	nr	**375.45**
"Covent Garden"; ref COV 912; open top; 500 mm A/F octagonal x 820 mm high; 85 litre capacity	515.00	0.50	7.50	-	521.32	nr	**528.82**
"Albert"; ref ALB 800; open top; 400 mm diameter x 845 mm high; 55 litre capacity	359.65	0.50	7.50	-	365.97	nr	**373.47**
Bins; Broxap Streetscene							
"Chester"; pedestal mounted	253.05	1.00	15.00	-	259.37	nr	**274.37**
Bins; Townscape Products							
"York Major"; 650 diameter x 1060 high, 140 litre capacity	633.44	1.00	15.00	-	635.22	nr	**650.22**
Litter bins; timber faced; hardwood slatted casings with removable metal litter containers; ground or wall fixing							
Bins; SMP Playgrounds							
ref LBC22; 560 mm high	400.00	1.00	15.00	-	408.21	nr	**423.21**
Bins; Lister Lutyens							
"Monmouth"; 675 mm high x 450 mm wide; freestanding	105.00	0.33	5.00	-	105.00	nr	**110.00**
"Monmouth"; 675 mm high x 450 mm wide; bolting to ground (without legs)	98.00	1.00	15.00	-	106.21	nr	**121.21**

Prices for Measured Works 277

Q PAVING/PLANTING/FENCING/SITE FURNITURE Excluding overheads and profit	PC £	Labour hours	Labour £	Plant £	Material £	Unit	Total rate £
Bins; Woodscape Ltd							
Square, 580 mm x 580 mm x 950 mm high, with lockable lid	525.00	0.50	7.50	-	525.00	nr	532.50
Round, 580 mm diameter by 950 mm high, with lockable lid	495.00	0.50	7.50	-	495.00	nr	502.50
Plastic litter and grit bins; glassfibre reinforced polyester grit bins; yellow body; hinged lids							
Bins; Furnitubes International Ltd							
ref Q6; grit and salt bins; 170 litre capacity	245.46	0.25	3.75	-	245.46	nr	249.21
ref Q11; grit and salt bins; 310 litre capacity	269.82	0.25	3.75	-	269.82	nr	273.57
Bins; Wybone; Victoriana glass fibre, cast iron effect litter bins, including lockable liner							
ref LBV/2; 521 mm x 521 mm x 673 mm high; open top; square shape; 0.078 m³ capacity; with lockable liner	198.58	1.00	15.00	-	204.90	nr	219.90
ref LVC/3; 457 mm diameter x 648 mm high; open top; drum shape; 0.084 m³ capacity; with lockable liner	215.05	1.00	15.00	-	221.37	nr	236.37
Bins; Amberol Ltd; floor standing double walled to accept stabilizing ballast							
"Enviro Bin"; 150 litre	145.00	0.33	5.00	-	145.00	nr	150.00
"Westminster" hooded; 90 litre	95.00	0.33	5.00	-	95.00	nr	100.00
"Westminster" 90 litre	70.00	0.33	5.00	-	70.00	nr	75.00
Bins; Amberol Ltd; pole mounted							
Envirobin 50 litre top emptying with liner	60.00	0.50	7.50	-	60.00	nr	67.50
Lifebuoy stations							
Lifebuoy stations; Earth Anchors Limited							
"Rootfast" lifebuoy station complete with post ref AP44; SOLAS approved lifebouys 590 mm and lifeline	148.00	2.00	30.00	-	157.42	nr	187.42
installation tool for above	-	-	-	-	72.00	nr	72.00
Outdoor seats; concrete framed - General							
Preamble: Prices for the following concrete framed seats and benches with hardwood slats include for fixing (where necessary) by bolting into existing paving or concrete bases (bases not included) or building into walls or concrete foundations (walls and foundations not included)							
Outdoor seats; concrete framed							
Outdoor seats; Marshalls Mono							
"Kelvin" benches; free-standing; 2.00 m long	196.74	0.50	7.50	-	196.74	nr	204.24
"Kelvin" benches; below-ground fixing; 2.00 m long	196.74	2.00	30.00	-	208.80	nr	238.80
"Kelvin" seats; concrete with hardwood slats; 2.00 m long	258.44	2.00	30.00	-	270.50	nr	300.50
Outdoor seats; Townscape Products							
"Oxford" benches; 1800 mm x 430 mm x 440 mm	201.76	2.00	30.00	-	213.82	nr	243.82
"Maidstone" seats; 1800 mm x 610 mm x 785 mm	301.73	2.00	30.00	-	313.79	nr	343.79

Q PAVING/PLANTING/FENCING/SITE FURNITURE Excluding overheads and profit	PC £	Labour hours	Labour £	Plant £	Material £	Unit	Total rate £
Q50 SITE/STREET FURNITURE/EQUIPMENT - cont'd							
Outdoor seats; metal framed - General Preamble: Metal framed seats with hardwood backs to various designs can be bolted to ground anchors.							
Outdoor seats; metal framed "Orsogril" shaped steel grill seats; to steel frame; galvanized and plastic coated; in a range of colours							
"Dream"; 1.8 m long, backless	266.00	2.00	30.00	-	278.06	nr	**308.06**
"Condor"; 1.8 m long	340.00	3.00	45.00	-	352.06	nr	**397.06**
"Dream"; 1.8 m long	435.00	2.00	30.00	-	447.06	nr	**477.06**
Outdoor seats; Furnitubes International Ltd							
ref NS 6; "Newstead"; steel standards with iroko slats; 1.80 m long	251.50	2.00	30.00	-	271.12	nr	**301.12**
ref NEB 6; "New Forest Single Bench"; cast iron standards with iroko slats; 1.83 m long	370.00	2.00	30.00	-	389.62	nr	**419.62**
ref EA 6; "Eastgate"; cast iron standards with iroko slats; 1.86 m long	345.00	2.00	30.00	-	364.62	nr	**394.62**
ref NE 6; "New Forest Seat"; cast iron standards with iroko slats; 1.83 m long	495.00	2.00	30.00	-	514.62	nr	**544.62**
Outdoor seats; Geometric Furniture; Teak bench with back and armrests							
ref 8204; two seater; 1.38 m	286.00	2.00	30.00	-	305.62	nr	**335.62**
ref 8207; two seater; 1.50 m	295.10	2.00	30.00	-	314.72	nr	**344.72**
ref 8206; three seater; 1.85 m	534.30	2.00	30.00	-	553.92	nr	**583.92**
ref 8202; three seater; 1.95 m	400.40	2.00	30.00	-	420.02	nr	**450.02**
ref 8205; four seater; 2.40 m	642.20	2.00	30.00	-	661.82	nr	**691.82**
Outdoor seats; Columbia Cascade Ltd; perforated metal bench seats, range of 8 nr colours							
ref 2604-6; back-to-back metal benches; 1800 mm x 1500 mm x 860 mm high	1295.00	3.00	45.00	-	1307.65	nr	**1352.65**
Outdoor seats; Orchard Street Furniture; "Bramley"; broad iroko slats to steel frame							
1.20 m long	161.44	2.00	30.00	-	174.09	nr	**204.09**
1.80 m long	198.94	2.00	30.00	-	211.59	nr	**241.59**
2.40 m long	229.79	2.00	30.00	-	242.44	nr	**272.44**
Outdoor seats; Orchard Street Furniture; "Laxton"; narrow iroko slats to steel frame							
1.20 m long	182.45	2.00	30.00	-	195.10	nr	**225.10**
1.80 m long	224.20	2.00	30.00	-	236.85	nr	**266.85**
2.40 m long	254.79	2.00	30.00	-	267.44	nr	**297.44**
Outdoor seats; Orchard Street Furniture; "Lambourne"; iroko slats to cast iron frame							
1.80 m long	473.94	2.00	30.00	-	486.59	nr	**516.59**
2.40 m long	557.71	2.00	30.00	-	570.36	nr	**600.36**
Outdoor seats; SMP Playgrounds; "Datchet", steel and timber							
backless benches; ref OFDB6; 1.80 m long	287.00	2.00	30.00	-	299.65	nr	**329.65**
bench seats with backrests; ref OFDS6; 1.80 m long	387.00	2.00	30.00	-	399.65	nr	**429.65**
picnic tables with attached seats; ref OFDT6; 1.80 m long	674.00	2.00	30.00	-	686.65	nr	**716.65**

Q PAVING/PLANTING/FENCING/SITE FURNITURE Excluding overheads and profit	PC £	Labour hours	Labour £	Plant £	Material £	Unit	Total rate £
Outdoor seats; Broxap Streetscene; metal and timber							
"Manchester"	352.80	2.00	30.00	-	365.45	nr	**395.45**
"Eastgate"	352.80	2.00	30.00	-	365.45	nr	**395.45**
"Serpent"	398.00	2.00	30.00	-	410.65	nr	**440.65**
"Rotherham"	529.00	2.00	30.00	-	541.65	nr	**571.65**
Outdoor seats; Earth Anchors; "Forest-Saver", steel frame and recycled slats							
bench, 1.8m	171.00	1.00	15.00	-	171.00	nr	**186.00**
seat, 1.8m	265.00	1.00	15.00	-	265.00	nr	**280.00**
Outdoor seats; Earth Anchors; "Forest-Saver", ci frame and recycled slats							
"Evergreen" bench, 1.8m	345.00	1.00	15.00	-	345.00	nr	**360.00**
"Evergreen" seat, 1.8m	419.00	1.00	15.00	-	419.00	nr	**434.00**
Outdoor seats; all-steel							
Outdoor seats; Earth Anchors, "Ranger"							
bench, 1.8m	238.00	1.00	15.00	-	238.00	nr	**253.00**
seat, 1.8m	389.00	1.00	15.00	-	389.00	nr	**404.00**
Outdoor seats; all timber							
Outdoor seats; Lister Lutyens; "Mendip"; teak							
1.524 m long	411.00	1.00	15.00	-	434.39	nr	**449.39**
1.829 m long	434.00	1.00	15.00	-	457.39	nr	**472.39**
2.438 m long	562.00	1.00	15.00	-	585.39	nr	**600.39**
Outdoor seats; Lister Lutyens; "Sussex"; hardwood							
1.5 m	170.00	2.00	30.00	-	193.39	nr	**223.39**
Outdoor seats; Woodscape; solid hardwood							
seat type "3" with back; 2.00 m long; freestanding	580.00	2.00	30.00	-	603.39	nr	**633.39**
seat type "3" with back; 2.00 m long; building in	620.00	4.00	60.00	-	643.39	nr	**703.39**
seat type "4" with back; 2.00 m long; fixing to wall	345.00	3.00	45.00	-	356.34	nr	**401.34**
seat type "5" with back; 2.50 m long; freestanding	685.00	2.00	30.00	-	708.39	nr	**738.39**
seat type "4"; 2.00 m long; fixing to wall	270.00	2.00	30.00	-	293.39	nr	**323.39**
seat type "5" with back; 2.00 m long; freestanding	595.00	2.00	30.00	-	618.39	nr	**648.39**
seat type "4" with back; 2.50 m long; fixing to wall	435.00	3.00	45.00	-	446.34	nr	**491.34**
seat type "5" with back; building in	635.00	2.00	30.00	-	658.39	nr	**688.39**
seat type "5" with back; 2.50 m long; building in	725.00	3.00	45.00	-	748.39	nr	**793.39**
bench type "1"; 2.00 m long; freestanding	455.00	2.00	30.00	-	478.39	nr	**508.39**
bench type "1"; 2.00 m long; building in	495.00	4.00	60.00	-	518.39	nr	**578.39**
bench type "2"; 2.00 m long; freestanding	440.00	2.00	30.00	-	463.39	nr	**493.39**
bench type "2"; 2.00 m long; building in	480.00	4.00	60.00	-	503.39	nr	**563.39**
bench type "2"; 2.50 m long; freestanding	505.00	2.00	30.00	-	528.39	nr	**558.39**
bench type "2"; 2.50 m long; building in	545.00	4.00	60.00	-	568.39	nr	**628.39**
bench type "2"; 2.00 m long overall; curved to 5 m radius, building in	640.00	4.00	60.00	-	663.39	nr	**723.39**

Q PAVING/PLANTING/FENCING/SITE FURNITURE Excluding overheads and profit	PC £	Labour hours	Labour £	Plant £	Material £	Unit	Total rate £
Q50 SITE/STREET FURNITURE/EQUIPMENT - cont'd							
Outdoor seats; all timber - cont'd							
Outdoor seats; Orchard Street Furniture; "Allington"; all iroko							
1 20 m long	254.26	2.00	30.00	-	277.65	nr	307.65
1.80 m long	292.29	2.00	30.00	-	315.68	nr	345.68
2.40 m long	355.59	2.00	30.00	-	378.98	nr	408.98
Outdoor seats; SMP Playgrounds; "Lowland range"; all timber							
bench seats with backrest; ref OFDS6; 1.50 m long	387.00	2.00	30.00	-	399.65	nr	429.65
Outdoor seats; plastic							
Outdoor seats; Sarena Plastics; "Benchmark GRP" benches; range of colours							
1830 mm x 254 mm x 457 mm high	135.00	2.00	30.00	-	147 65	nr	177.65
Outdoor seats; tree benches/seats							
Tree bench; Neptune Outdoor Furniture, "Beaufort" Hexagonal, Timber							
SF34-05A, 500 mm diameter	620.00	0.50	7.50	-	620.00	nr	627.50
SF34-15A, 1500 mm diameter	754.00	0.50	7.50	-	754.00	nr	761.50
Tree seat; Neptune Outdoor Furniture, "Beaufort" Hexagonal, Timber, with back							
SF32-10A, 720 mm diameter	905.00	0.50	7.50	-	905.00	nr	912.50
SF32-20A, 1720 mm diameter	1057.00	0.50	7.50	-	1057.00	nr	1064.50
Street furniture ranges							
Townscape Products; "Belgrave"; natural grey concrete							
bollards; 250 mm diameter x 500 mm high	49.98	1.00	15.00	-	53.15	nr	68.15
seats; 1800 mm x 600 mm x 736 mm high	240.94	2.00	30.00	-	240.94	nr	270.94
Abacus Municipal; "300" range of street furniture; hardwood seats without arms							
ref 304H; 1220 mm long	211.00	2.00	30.00	-	223.65	nr	253.65
ref 306H; 1830 mm long	234.00	2.00	30.00	-	246.65	nr	276.65
ref 308H; 2440 mm long	326.00	2.00	30.00	-	338.65	nr	368.65
Abacus Municipal; "300" range of street furniture; hardwood double seats without arms							
ref 304DH; 1220 mm long	346.00	2.00	30.00	-	358.65	nr	388.65
ref 306DH; 1830 mm long	418.00	2.00	30.00	-	430.65	nr	460.65
ref 308DH; 2440 mm long	583.00	2.00	30.00	-	595.65	nr	625.65
Picnic benches - General							
Preamble: The following items include for fixing to ground in to manufacturer's instructions or concreting in.							
Picnic tables and benches							
Picnic tables and benches; Abacus Municipal							
ref 306SP; softwood	379.00	2.00	30.00	-	391.65	nr	421.65
ref 306HP; hardwood	441.00	2.00	30.00	-	453.65	nr	483.65

Q PAVING/PLANTING/FENCING/SITE FURNITURE Excluding overheads and profit	PC £	Labour hours	Labour £	Plant £	Material £	Unit	Total rate £
Picnic tables and benches; SMP Playgrounds ref OFLTH; "Datchet" Special needs Picnic Table	894.00	2.00	30.00	-	906.65	nr	936.65
ref OFDT6; "Datchet"; 1.80 m; steel and timber; with attached seats	674.00	2.00	30.00	-	686.65	nr	716.65
Picnic tables and benches; Broxap Streetscene "Eastgate" picnic unit	548.00	2.00	30.00	-	560.65	nr	590.65
Picnic tables and benches; Woodscape Ltd Table and benches built in, 2 m long	1455.00	2.00	30.00	-	1467.65	nr	1497.65
Market Prices of Containers							
Plant containers; terracotta							
Capital Garden Products							
Large Pot LP63; weathered terracotta - 1170 x 1600 dia.	-	-	-	-	584.66	nr	584.66
Large Pot LP38; weathered terracotta - 610 x 970 dia.	-	-	-	-	254.85	nr	254.85
Large Pot LP23; weathered terracotta - 480 x 580 dia.	-	-	-	-	134.59	nr	134.59
Manhole Cover Planter 3022; 760 x 560 x 210 high	-	-	-	-	93.79	nr	93.79
Apple Basket 2215; 380 x 560 dia.	-	-	-	-	96.80	nr	96.80
Indian style Shimmer Pot 2322; 585 x 560 dia.	-	-	-	-	211.45	nr	211.45
Indian style Shimmer Pot 1717; 430 x 430 dia.	-	-	-	-	109.24	nr	109.24
Indian style Shimmer Pot 1314; 330 x 355 dia.	-	-	-	-	70.94	nr	70.94
Plant containers; faux lead							
Capital Garden Products							
Trough 2508 'Tudor Rose'; 635 x 215 x 230 high	-	-	-	-	64.87	nr	64.87
Tub 2004 'Elizabethan'; 508 mm square	-	-	-	-	89.10	nr	89.10
Tub 1513 'Elizabethan'; 380 mm square	-	-	-	-	69.51	nr	69.51
Tub 1601 'Tudor Rose'; 420 x 395 dia.	-	-	-	-	61.40	nr	61.40
Plant containers; window boxes							
Capital Garden Products							
Adam Design AD5401; 1370 x 270 x 210 h	-	-	-	-	100.42	nr	100.42
Wheatsheaf Design WH54; 1370 x 270 x 210 h	-	-	-	-	100.42	nr	100.42
Oakleaf Design OK24; 610 x 370 x 350 h	-	-	-	-	71.71	nr	71.71
Swag Design SW2402; 610 x 200 x 210 h	-	-	-	-	71.71	nr	71.71
Plant containers; timber							
Plant containers; Hardwood; Neptune Outdoor Furniture							
'Beaufort' T38-4D, 1500 x 1500 x 900 mm high	-	-	-	-	845.00	nr	845.00
'Beaufort' T38-3C, 1000 x 1500 x 700 mm high	-	-	-	-	629.00	nr	629.00
'Beaufort' T38-2A, 1000 x 500 x 500 mm high	-	-	-	-	328.00	nr	328.00
'Kara' T42-4D, 1500 x 1500 x 900 mm high	-	-	-	-	912.00	nr	912.00
'Kara' T42-3C, 1000 x 1500 x 700 mm high	-	-	-	-	679.00	nr	679.00
'Kara' T42-2A, 1000 x 500 x 500 mm high	-	-	-	-	354.00	nr	354.00
Plant containers; Hardwood; Woodscape Ltd Square, 900 x 900 x 420 mm high	-	-	-	-	244.80	nr	244.80

Q PAVING/PLANTING/FENCING/SITE FURNITURE Excluding overheads and profit	PC £	Labour hours	Labour £	Plant £	Material £	Unit	Total rate £
Q50 SITE/STREET FURNITURE/EQUIPMENT - cont'd							
Plant containers; precast concrete							
Plant containers; Marshalls Mono							
"Strada"; 710 mm circular planter	120.85	2.00	30.00	30.36	120.85	nr	**181.21**
"Elba"; 970 mm diameter x 470 mm high	131.11	2.00	30.00	30.36	131.11	nr	**191.47**
"Shirley"; 1066 mm diameter x 760 mm high	409.12	2.00	30.00	30.36	409.12	nr	**469.48**
"Boulevard" bases; 2000 mm diameter x 260 mm high	287.71	2.00	30.00	15.18	287.71	nr	**332.89**
"Boulevard" bases; 1200 mm diameter x 260 mm high	299.90	1.00	15.00	15.18	299.90	nr	**330.08**
"Boulevard" rings; 1200 mm diameter x 235 mm high	201.23	1.00	15.00	15.18	201.23	nr	**231.41**
"Boulevard" rings; 2000 mm diameter x 235 mm high	248.90	1.50	22.50	22.77	248.90	nr	**294.17**
Cycle holders							
Cycle holders; George Fischer Sales; "Velop A" galvanized steel							
ref R; fixing to wall or post; making good	26.50	1.00	15.00	-	39.15	nr	**54.15**
ref SR(V); fixing in ground; making good	28.50	1.00	15.00	-	41.15	nr	**56.15**
Cycle holders; Townscape Products							
"Guardian" cycle holders; tubular steel frame; setting in concrete, 1250 mm x 550 mm x 775 mm high; making good	215.01	1.00	15.00	-	227.66	nr	**242.66**
"Penny" cycle stands; 600 mm diameter; exposed aggregate bollards with 8 nr cycle holders; in galvanized steel; setting in concrete; making good	450.07	1.00	15.00	-	462.72	nr	**477.72**
Cycle holders; Mawrob & Co; combined cycle racks and shelters; corrugated aluminium roof; on angle iron frame; galvanized							
ref MW/AV10; for 10 nr cycles, semi-vertical; 1.22 m wide x 3.05 m long x 2.15 m high	460.00	10.00	150.00	-	485.30	nr	**635.30**
ref MW/AW10; for 10 nr cycles; horizontal; 2.13 m wide x 3.05 m long x 2.15 m high	449.00	10.00	150.00	-	474.30	nr	**624.30**
Cycle holders; Broxap Mawrob							
cycle parking stand; BX11 6045-6, pillar mounted double sided rail of front wheel supports; holds twelve cycles	359.00	4.00	60.00	-	369.18	nr	**429.18**
Directional signage; cast aluminium							
Signage; Furnitubes International Ltd							
ref FFL 1; "Lancer"; cast aluminium finials	69.95	0.07	1.00	-	69.95	nr	**70.95**
ref FAA IS; arrow end type cast aluminium directional arms, single line; 90 mm wide	118.90	2.00	30.00	-	118.90	nr	**148.90**
ref FAA ID; arrow end type cast aluminium directional arms; double line; 145 mm wide	152.45	0.13	2.00	-	152.45	nr	**154.45**
ref FAA IS; arrow end type cast aluminium directional arms; treble line; 200 mm wide	188.50	0.20	3.00	-	188.50	nr	**191.50**
ref FCK1 211 G; "Kingston"; composite standard root columns	552.00	2.00	30.00	-	567.48	nr	**597.48**

Prices for Measured Works

Q PAVING/PLANTING/FENCING/SITE FURNITURE Excluding overheads and profit	PC £	Labour hours	Labour £	Plant £	Material £	Unit	Total rate £
Q50 SITE/STREET FURNITURE/EQUIPMENT - cont'd							
Excavating; for bollards and barriers; by hand							
Holes for bollards							
400 mm x 400 mm x 400 mm; disposing off site	-	0.42	6.25	-	0.77	nr	7.02
600 mm x 600 mm x 600 mm; disposing off site	-	0.97	14.55	-	2.16	nr	16.71
Concrete bollards - General							
Preamble. Precast concrete bollards are available in a very wide range of shapes and sizes. The bollards listed here are the most commonly used sizes and shapes; manufacturer's catalogues should be consulted for the full range. Most manufacturers produce bollards to match their suites of street furniture, which may include planters, benches, litter bins and cycle stands. Most parallel sided bollards can be supplied in removable form, with a reduced shank, precast concrete socket and lifting hole to permit removal with a bar.							
Concrete bollards							
Marshalls Mono; cylinder; straight or tapered; 200 mm - 400 mm diameter; plain grey concrete; setting into firm ground (excavating and backfilling not included)							
"Richmond" 535 mm high above ground	57.93	2.00	30.00	-	69.06	nr	99.06
"Bridgford" 915 mm high above ground	52.11	2.00	30.00	-	63.24	nr	93.24
"Truro" 760 mm high above ground	132.49	2.00	30.00	-	138.47	nr	168.47
Marshalls Mono; cylinder; straight or tapered, 200 mm - 400 mm diameter; white concrete; setting into firm ground (excavating and backfilling not included)							
"Bridgford" 915 mm high above ground	86.48	2.00	30.00	-	92.46	nr	122.46
"Richmond" 535 mm high above ground	94.80	2.00	30.00	-	100.78	nr	130.78
"Truro" 760 mm high above ground	155.22	2.00	30.00	-	161.20	nr	191.20
Marshalls Mono; cylinder; straight or tapered; 200 mm - 400 mm diameter; exposed aggregate; setting into firm ground (excavating and back filling not included)							
"Worcester" 495 mm high above ground	87.86	2.00	30.00	-	93.84	nr	123.84
"Bridgford" 915 mm high above ground	65.42	2.00	30.00	-	71.40	nr	101.40
"Truro" 760 mm high above ground	149.40	2.00	30.00	-	155.38	nr	185.38
Marshalls Mono; cylinder; straight or tapered; 200 mm - 400 mm diameter; reflective finish; setting into firm ground (excavating and backfilling not included)							
"Richmond" 535 mm high above ground	130.55	2.00	30.00	-	136.63	nr	166.63
"Bridgford" 915 mm high above ground	137.21	2.00	30.00	-	143.19	nr	173.19
"Wexham" 760 mm high above ground	172.96	2.00	30.00	-	184.09	nr	214.09
Marshalls Mono; "Millbank" bollards; exposed aggregate with ribbed top							
230 mm diameter x 1000 mm high	111.98	2.00	30.00	-	117.96	nr	147.96
Extra over concrete bollards for chain fixing rings	5.20	-	-	-	5.20	nr	5.20

Q PAVING/PLANTING/FENCING/SITE FURNITURE Excluding overheads and profit	PC £	Labour hours	Labour £	Plant £	Material £	Unit	Total rate £
Q50 SITE/STREET FURNITURE/EQUIPMENT - cont'd							
Concrete bollards - cont'd							
Precast concrete verge markers; various shapes; 450 mm high							
plain grey concrete	24.91	1.00	15.00	-	29.18	nr	44.18
white concrete	27.46	1.00	15.00	-	31.73	nr	46.73
exposed aggregate	30.49	1.00	15.00	-	34.76	nr	49.76
Extra over concrete bollards for removable bollards (any size)	16.13	-	-	-	16.13	nr	16.13
Precast concrete spur stones; 450 mm diameter x 750 mm high; plain white concrete							
half-section	108.52	1.00	15.00	-	112.34	nr	127.34
three-quarter section	132.25	1.00	15.00	-	136.07	nr	151.07
Other bollards							
Marshalls Mono; hollow section steel bollards with cast capping units and 300 mm root; stove enamel finish							
ref MSF 601 168 mm diameter; 900 mm high	161.70	2.00	30.00	-	167.40	nr	197.40
ref MSF 602 168 mm diameter; 900 mm high	154.01	2.00	30.00	-	159.71	nr	189.71
ref MSF 605 114 mm diameter; 925 mm high	169.65	2.00	30.00	-	175.35	nr	205.35
Hinged or removable bollards; Marshalls Mono; powder coated; steel tube with locking device; base plate; bolting to ground, or socket 267 mm deep in ground							
"Brunel" removable bollards	314.05	2.00	30.00	-	325.65	nr	355.65
"Telford" removable bollards	244.87	2.00	30.00	-	256.47	nr	286.47
Removable parking posts; Dee-Organ "Spacekeeper"; ref 3014203; folding plastic coated galvanized steel parking posts with key; reflective bands; including 300 mm x 300 mm x 300 mm concrete foundations and fixing bolts; 850 mm high	114.22	2.00	30.00	-	125.82	nr	155.82
"Spacesaver"; ref 1508101; hardwood removable bollards including key and base-plate; 150 mm x 150 mm; 600 mm high	116.86	2.00	30.00	-	120.90	nr	150.90
"Spacesaver" ref 1506101; hardwood removable bollards including key and base-plate; 150 mm x 150 mm; 800 mm high	130 04	2.00	30.00	-	134.08	nr	164.08
Cast iron bollards - General							
Preamble: The following bollards are particularly suitable for conservation areas. Logos for civic crests can be incorporated to order.							
Cast iron bollards							
Bollards; Furnitubes International Ltd (excavating and backfilling not included)							
"Doric Round"; 920 mm high x 170 mm diameter	101.00	2.00	30.00	-	106.70	nr	136.70
"Gunner round"; 750 mm high x 165 mm diameter	92.00	2.00	30.00	-	97.70	nr	127.70
"Manchester Round"; 975 mm high; 225 mm square base	112.00	2.00	30.00	-	117.70	nr	147.70
"Cannon"; 1140 mm x 210 mm diameter	146.00	2.00	30.00	-	151.70	nr	181.70
"Kenton"; heavy duty galvanized steel; 900 mm high; 350 mm diameter	330.00	2.00	30.00	-	335.70	nr	365.70

Q PAVING/PLANTING/FENCING/SITE FURNITURE Excluding overheads and profit	PC £	Labour hours	Labour £	Plant £	Material £	Unit	Total rate £
Cast iron bollards with rails - General Preamble: The following cast iron bollards are suitable for conservation areas.							
Cast iron bollards with rails Cast iron posts with steel tubular rails; Broxap Streetscene; setting into firm ground (excavating not included)							
"Sheffield"; 450 mm high; one rail, type A	50.00	1.50	22.50	-	52.54	nr	75.04
"Southbank"; 1150 mm high above ground; four rails, type C	85.00	1.00	15.00	-	87.54	nr	102.54
"Mersey"; 1085 mm high above ground; two rails, type D	91.80	1.50	22.50	-	94.34	nr	116.84
"Promenade"; 1150 mm high above ground; square; three rails, type C	99.80	1.50	22.50	-	102.34	nr	124.84
"Type A" mild steel tubular rail including connector	3.97	0.03	0.50	-	4.87	m	5.37
"Type C" mild steel tubular rail including connector	5.15	0.03	0.50	-	6.35	m	6.85
"Type D" mild steel tubular rail including connector	6.61	-	-	-	8.49	m	8.49
Timber bollards - General **Timber bollards** Woodscape Ltd; Durable Hardwood							
RP 250/1500; 250 mm diameter x 1500 mm long	172.85	1.00	15.00	-	173.10	nr	188.10
SP 250/1500; 250 mm square x 1500 mm long	172.85	1.00	15.00	-	173.10	nr	188.10
SP 150/1200; 150mm square x 1200 mm long	62.25	1.00	15.00	-	62.50	nr	77.50
SP 125/750; 125 mm square x 750 mm long	34.35	1.00	15.00	-	34.60	nr	49.60
RP 125/750; 125 mm diameter x 750 mm long	34.35	1.00	15.00	-	34.60	nr	49.60
Deterrent bollards Semi-mountable vehicle deterrent and kerb protection bollards; Furnitubes International Ltd (excavating and backfilling not included)							
"Bell decorative"	312.00	2.00	30.00	-	320.11	nr	350.11
"Half bell"	400.00	2.00	30.00	-	408.11	nr	438.11
"Full bell"	486.00	2.00	30.00	-	494.11	nr	524.11
"Three quarter bell"	496.16	2.00	30.00	-	504.27	nr	534.27
Security Bollards Security bollards; Furnitubes International Ltd (excavating and backfilling not included)							
"Gunner"; reinforced with steel insert and tie bars; 750 mm high above ground; 600 mm below ground	1053.00	2.00	30.00	-	1061.55	nr	1091.55
"Burr Bloc Type 6" removable steel security bollard; 750 mm high above ground; 410 mm x 285 mm	409.50	2.00	30.00	-	428.73	nr	458.73
"Burr 7 - Burr Post Type 8"; circular hollow steel post; heavy duty anti-Ram Raid; galvanised; 900 mm high above ground; 170 mm diameter	222.45	2.00	30.00	-	241.68	nr	271.68

Prices for Measured Works

Q PAVING/PLANTING/FENCING/SITE FURNITURE Excluding overheads and profit	PC £	Labour hours	Labour £	Plant £	Material £	Unit	Total rate £
Q50 SITE/STREET FURNITURE/EQUIPMENT - cont'd							
Tree grilles; Cast Iron							
Cast iron tree grilles; Furnitubes International							
ref GS 1070 Greenwich; two part; 1000 mm square; 700 mm diameter tree hole	107.50	2.00	30.00	-	107.50	nr	137.50
ref GC 1270 Greenwich; two part, 1200 mm diameter; 700 mm diameter tree hole	155.50	2.00	30.00	-	155.50	nr	185.50
ref GCF 1026 Greenwich; steel tree grille frame for GC 1045, one part	75.40	2.00	30.00	-	75.40	nr	105.40
ref GSF 1226 Greenwich; steel tree grille frame for GC 1270 and GC 1245, one part	84.10	2.00	30.00	-	84.10	nr	114.10
Cast iron tree grilles; Marshalls Mono							
ref MSF310 grille; two part; 1200 x 1200 mm, 440 mm diameter hole; primed	333.45	3.00	45.00	-	343.97	nr	388.97
ref MSF310B grille; two part; 1200 x 1200 mm, 440 mm diameter hole ; painted black	367.66	3.00	45.00	-	378.18	nr	423.18
ref GFB310 frame; for bolt fixing; primed	147.06	1.00	15.00	-	147.06	nr	162.06
ref GFS310 frame; for spike fixing; primed	124.22	1.00	15.00	-	124.22	nr	139.22
Cast iron tree grilles; Townscape Products							
"Baltimore" 1200 mm square x 460 mm diameter tree hole	343.53	2.00	30.00	-	343.53	nr	373.53
"Baltimore" Hexagonal, maximum width 1440 mm nominal, 600 mm diameter tree hole	625.31	2.00	30.00	-	625.31	nr	655.31
Note: Care must be taken to ensure that tree grids and guards are removed when trees grow beyond the specified diameter of guard							
Playground equipment - General Preamble: The range of equipment manufactured or available in the UK is so great that comprehensive coverage would be impossible, especially as designs, specifications and prices change fairly frequently. The following information should be sufficient to give guidance to anyone designing or equipping a playground. In comparing prices note that only outline specification details are given here and that other refinements which are not mentioned may be the reason for some difference in price between two apparently identical elements. The fact that a particular manufacturer does not appear under one item heading does not necessarily imply that he does not make it. Landscape designers are advised to check that equipment complies with ever more stringent safety standards before specifying.							
Playground equipment - Installation The rates below include for installation of the specified equipment by the manufacturers. Most manufactures will offer an option to install the equipment supplied by them.							

Q PAVING/PLANTING/FENCING/SITE FURNITURE
Excluding overheads and profit

Item	PC £	Labour hours	Labour £	Plant £	Material £	Unit	Total rate £
Swings - General							
Preamble: Prices for the following vary considerably. Those given represent the middle of the range and include multiple swings with tubular steel frames and timber or tyre seats; ground fixing and priming only.							
Swings							
Swings; Wicksteed Ltd							
traditional swings; 1850 mm high; 1 bay; 2 seat	-	-	-	-	-	nr	1418.00
traditional swings; 1850 mm high; 2 bay; 4 seat	-	-	-	-	-	nr	2264.00
traditional swings; 2450 mm high; 1 bay; 2 seat	-	-	-	-	-	nr	1368.00
traditional swings; 2450 mm high; 2 bay; 4 seat	-	-	-	-	-	nr	2133.00
traditional swings; 3050 mm high; 1 bay; 2 seat	-	-	-	-	-	nr	1448.00
traditional swings; 3050 mm high; 2 bay; 4 seat	-	-	-	-	-	nr	2236.00
single arch swings; cradle safety seats; 1850 mm high	-	-	-	-	-	nr	1113.00
single arch swings; flat rubber safety seats; 2450 mm high	-	-	-	-	-	nr	1099.00
double arch swings; flat rubber safety seats; 2450 mm high	-	-	-	-	-	nr	1342.00
Swings; Lappset UK Ltd							
ref 020414; swing frame	-	-	-	-	-	nr	885.00
Swings; Kompan Ltd							
ref M947-52; double swings	-	-	-	-	-	nr	2477.00
ref M948-52; double swings	-	-	-	-	-	nr	2098.00
ref M951; "Sunflower" swings	-	-	-	-	-	nr	1134.00
Slides							
Slides; Wicksteed Ltd							
"Pedestal" slides; 3.40 m	-	-	-	-	-	nr	1928.00
"Pedestal" slides; 4.40 m	-	-	-	-	-	nr	2499.00
"Pedestal" slides; 5.80 m	-	-	-	-	-	nr	2904.00
"Embankment" slides; 3.40 m	-	-	-	-	-	nr	1478.00
"Embankment" slides; 4.40 m	-	-	-	-	-	nr	1915.00
"Embankment" slides; 5.80 m	-	-	-	-	-	nr	2434.00
"Embankment" slides; 7.30 m	-	-	-	-	-	nr	3092.00
"Embankment" slides; 9.10 m	-	-	-	-	-	nr	3868.00
"Embankment" slides; 11.00 m	-	-	-	-	-	nr	4629.00
"Mini" slides	-	-	-	-	-	nr	938.00
Slides; Lappset UK Ltd							
ref 142015; slide	-	-	-	-	-	nr	1910.00
ref 141115; "Jumbo" slide	-	-	-	-	-	nr	2589.00
Slides; Kompan Ltd							
ref M³51; slides	-	-	-	-	-	nr	1926.00
ref M³22; slide and cave	-	-	-	-	-	nr	2262.00
Moving equipment - General							
Preamble: The following standard items of playground equipment vary considerably in quality and price; the following prices are middle of the range.							
Moving equipment							
Roundabouts; Wicksteed Ltd							
"Roll-the-Barrel"	-	-	-	-	-	nr	605.00
"Pineapple Whirl"	-	-	-	-	-	nr	2760.00
"Honeycomb Whirl"	-	-	-	-	-	nr	3040.00

Q PAVING/PLANTING/FENCING/SITE FURNITURE Excluding overheads and profit	PC £	Labour hours	Labour £	Plant £	Material £	Unit	Total rate £
Q50 SITE/STREET FURNITURE/EQUIPMENT - cont'd							
Playground equipment - cont'd							
Seesaws							
Seesaws; Lappset UK Ltd							
ref 010300; seesaws	-	-	-	-	-	nr	618.00
ref 010236; seesaws	-	-	-	-	-	nr	1337.00
Seesaws; Wicksteed Ltd							
"Non Bump Seesaw"	-	-	-	-	-	nr	1580.00
"Jolly Gerald"	-	-	-	-	-	nr	1698.00
"Rocking Rockette"	-	-	-	-	-	nr	2269.00
"Rocking Horse"	-	-	-	-	-	nr	2577.00
Play sculptures - General Preamble: Many variants on the shapes of playground equipment are available, simulating spacecraft, trains, cars, houses etc., and these designs are frequently changed. The basic principles remain constant but manufacturer's catalogues should be checked for the latest styles.							
Play sculptures							
Wooden animals; SMP Playgrounds							
pig	-	-	-	-	-	nr	440.00
donkey	-	-	-	-	-	nr	460.00
rhino	-	-	-	-	-	nr	620.00
giraffe	-	-	-	-	-	nr	630.00
camel	-	-	-	-	-	nr	650.00
hippo	-	-	-	-	-	nr	950.00
elephant	-	-	-	-	-	nr	1240.00
Climbing equipment and Play structures; General Preamble: Climbing equipment generally consists of individually designed modules. Play structures generally consist of interlinked and modular pieces of equipment and sculptures. These may consist of climbing, play and skill based modules, nets and various other activities. Both are set into either safety surfacing or defined sand pit areas. The equipment below outlines a range from various manufacturers. Individual catalogues should be consulted in each instance. Safety areas should be allowed round all equipment.							
Climbing equipment and Play structures							
Climbing equipment and Play structures; Kompan Ltd							
ref MQ1002; "Mosaic" combination system	-	-	-	-	-	nr	4321.00
ref OK1103; "Oasis" combination system	-	-	-	-	-	nr	3912.00
Mosaiq ref MQ 1001	-	-	-	-	-	nr	3739.00
ref MQ2004; "Mosaic" combination system	-	-	-	-	-	nr	6753.00
ref OK3100; "Oasis" combination system	-	-	-	-	-	nr	15901.00
ref M480 "Castle	-	-	-	-	-	nr	22468.00

Q PAVING/PLANTING/FENCING/SITE FURNITURE Excluding overheads and profit	PC £	Labour hours	Labour £	Plant £	Material £	Unit	Total rate £
Climbing equipment; Wicksteed Ltd							
"Junior Commando Bridge"	-	-	-	-	-	nr	827.00
"In Line 2"	-	-	-	-	-	nr	1992.00
"In Line 3"	-	-	-	-	-	nr	2776.00
"Pentagon"	-	-	-	-	-	nr	4155.00
"Open 5"	-	-	-	-	-	nr	4443.00
"Fantasy Funrun" set of 17 units	-	-	-	-	-	nr	14229.00
Climbing equipment; Lappset UK Ltd							
ref 120254; "Climbing Frame"	-	-	-	-	-	nr	2135.00
ref 122457; "Playhouse"	-	-	-	-	-	nr	2610.00
ref 120100; "Activity Tower"	-	-	-	-	-	nr	7401.00
ref 122124; "Tower and Climbing Frame"	-	-	-	-	-	nr	8941.00
SMP Playgrounds							
Action Pack - Cape Horn; 8 module	-	-	-	-	-	nr	7360.00
Spring equipment							
Spring based equipment for 1 - 8 year olds; Kompan Ltd							
ref M101; "Crazy Hen"	-	-	-	-	-	nr	486.00
ref M128; "Crazy Daisy"	-	-	-	-	-	nr	675.00
ref M141; "Crazy Seesaw"	-	-	-	-	-	nr	1270.00
ref M155; "Quartet Seesaw"	-	-	-	-	-	nr	960.00
ref M199; "Crazy Springboard"	-	-	-	-	-	nr	1688.00
Spring based equipment for under 12's; Lappset UK Ltd							
ref 010443; "Rocking Horse"	-	-	-	-	-	nr	680.00
Sand pits							
Market Prices							
Play Pit sand; Boughton Loam Ltd	-	-	-	-	38.70	m³	38.70
Kompan Ltd							
" Basic 500" 2900 mm x 1570 mm x 310 mm deep	-	-	-	-	-	nr	751.00
Flagpoles							
Flagpoles; Harrison Flagpoles; in glass fibre; smooth white finish; terylene halyards with mounting accessories; setting in concrete; to manufacturers recommendations (excavating not included)							
6.00 m high, plain	151.00	3.00	45.00	-	171.36	nr	216.36
6.00 m high, hinged baseplate, external halyard	192.00	4.00	60.00	-	212.36	nr	272.36
6.00 m high, hinged baseplate, internal halyard	289.00	4.00	60.00	-	309.36	nr	369.36
10.00 m high, plain	286.00	5.00	75.00	-	306.36	nr	381.36
10.00 m high, hinged baseplate, external halyard	699.00	5.00	75.00	-	719.36	nr	794.36
10.00 m high, hinged baseplate, internal halyard	407.00	5.00	75.00	-	427.36	nr	502.36
15.00 m high, plain	654.00	6.00	90.00	-	674.36	nr	764.36
15.00 m high, hinged baseplate, external halyard	699.00	6.00	90.00	-	719.36	nr	809.36
15.00 m high, hinged baseplate, internal halyard	880.00	6.00	90.00	-	900.36	nr	990.36
Flagpoles; Harrison Flagpoles; tapered hollow steel section; base plate flange; galvanized; white painted finish; lowering gear; including bolting to concrete (excavating not included)							
7.00 m high	125.00	3.00	45.00	-	144.34	nr	189.34
10.00 m high	286.00	3.50	52.50	-	308.20	nr	360.70
13.00 m high	310.00	6.00	90.00	-	336.41	nr	426.41

Q PAVING/PLANTING/FENCING/SITE FURNITURE Excluding overheads and profit	PC £	Labour hours	Labour £	Plant £	Material £	Unit	Total rate £
Q50 SITE/STREET FURNITURE/EQUIPMENT - cont'd							
Flagpoles - cont'd							
Flagpoles; Harrison Flagpoles; in glass fibre; smooth white finish, terylene halyards for wall mounting							
3.0 m pole	171.00	2.00	30.00	-	171.00	nr	**201.00**
Flagpoles; Harrison Flagpoles; parallel galvanised steel banner flagpole - guide price							
6 m high	800.00	4.00	60.00	-	820.36	nr	**880.36**
Sports equipment; Edwards Sports Products							
Tennis courts; steel posts and fixings for hardcourt							
Round	130.00	1.00	15.00	-	130.00	set	**145.00**
Square	176.90	1.00	15.00	-	176.90	set	**191.90**
Tennis nets; not including posts or fixings							
ref 5021; "Matchplay"	59.40	3.00	45.00	-	66.19	set	**111.19**
ref 5075; "Club"	45.43	3.00	45.00	-	52.22	set	**97.22**
ref 5001; "Championship"	67.60	3.00	45.00	-	74.39	set	**119.39**
Wooden football goal posts with sockets; ref 2721; including nets							
ref 2026; "Club" nets	374.40	4.00	60.00	-	381.19	set	**441.19**
ref 2037; "Medium Club" nets	364.00	0.14	2.16	-	552.48	set	**554.64**
Steel football goal posts with sockets; ref 2760; including nets							
ref 2026; "Club" net	364.00	4.00	60.00	-	370.79	set	**430.79**
ref 2037; "Medium Club" net	358.28	4.00	60.00	-	365.07	set	**425.07**
Junior wooden football goal posts with sockets; ref 2741; including 2 mm nets; ref 2150	303.68	4.00	60.00	-	310.47	set	**370.47**
Telescopic tubular steel rugby goal posts with sockets							
ref 2800; 7.30 m high	394.16	6.00	90.00	-	412.25	set	**502.25**
ref 2802; 10.70 m high	565.04	6.00	90.00	-	583.13	set	**673.13**
Hockey goal posts with sockets; including "Club" nets							
ref 2851; steel posts	190.61	0.14	2.16	-	331.97	set	**334.13**
backboards 450 mm	244.61	-	-	-	244.61	set	**244.61**
ref 2830; wooden posts	106.87	4.00	60.00	-	113.66	set	**173.66**
backboards 150 mm practice	90.85	-	-	-	90.85	set	**90.85**
Football mini goal complete with posts, net and fittings							
12' x 6'; aluminium and steel	343.20	4.00	60.00	-	349.99	set	**409.99**
12' x 6'; all steel	290.68	3.00	45.00	-	297.47	set	**342.47**

Q PAVING/PLANTING/FENCING/SITE FURNITURE Excluding overheads and profit	PC £	Labour hours	Labour £	Plant £	Material £	Unit	Total rate £
Clothes line fittings							
Rotary outdoor clothes line; Hills Industries Ltd; mild steel tube; to concrete base (base not included)							
"Airdry"; ref 4/40	31.41	2.00	30.00	-	34.80	nr	**64.80**
"Airdry"; ref 3/30	24.75	2.00	30.00	-	28.14	nr	**58.14**
"Portadry"; ref 3/35	37.32	2.00	30.00	-	40.71	nr	**70.71**
"Supadry"; ref 4/50	82.83	2.00	30.00	-	86.22	nr	**116.22**
"Supadry"; ref 3/50	65.08	2.00	30.00	-	68.47	nr	**98.47**
"Builders Special"	138.52	2.00	30.00	-	141.91	nr	**171.91**
"Handiline"; ref 4/45	76.66	2.00	30.00	-	80.05	nr	**110.05**
"Handiline"; ref 3/35	53.77	2.00	30.00	-	57.16	nr	**87.16**
Pair of precast concrete clothes posts 2.65 m long; 125 mm x 125 mm at base tapering to 75 mm; including setting in concrete (1:2:4); base size 300 mm x 450 mm x 600 mm deep; including excavating and disposing off site	23.46	3.00	45.00	-	35.52	nr	**80.52**

R DISPOSAL SYSTEMS Excluding overheads and profit	PC £	Labour hours	Labour £	Plant £	Material £	Unit	Total rate £
R12 DRAINAGE BELOW GROUND							
Silt pits and inspection chambers							
Excavating pits; starting from ground level; by machine							
maximum depth not exceeding 1.00 m	-	0.50	7.50	15.18	-	m^3	22.68
maximum depth not exceeding 2.00 m	-	0.50	7.50	22.77	-	m^3	30.27
maximum depth not exceeding 4.00 m	-	0.50	7.50	45.54	-	m^3	53.04
Disposal of excavated material; depositing on site in permanent spoil heaps; average 50 m	-	0.04	0.63	1.57	-	m^3	2.20
Filling to excavations; obtained from on site spoil heaps; average thickness not exceeding 0.25 m	-	0.13	2.00	4.05	-	m^3	6.05
Surface treatments; compacting; bottoms of excavations	-	0.05	0.75	-	-	m^2	0.75
Earthwork support; distance between opposing faces not exceeding 2.00 m							
maximum depth not exceeding 1.00 m	-	0.20	3.00	-	4.11	m^2	7.11
maximum depth not exceeding 2.00 m	-	0.30	4.50	-	4.11	m^2	8.61
maximum depth not exceeding 4.00 m	-	0.67	10.00	-	1.84	m^2	11.83
Silt pits and inspection chambers; in situ concrete							
Beds; plain in situ concrete; 11.50 N/mm² - 40 mm aggregate							
thickness not exceeding 150 mm	-	1.00	15.00	-	94.24	m^3	109.24
thickness 150 mm - 450 mm	-	0.67	10.00	-	94.24	m^3	104.24
Benchings in bottoms; plain in situ concrete; 25.50 N/mm² - 20mm aggregate							
thickness 150 mm - 450 mm	-	2.00	30.00	-	94.24	m^3	124.24
Isolated cover slabs; reinforced in situ concrete; 21 00 N/mm² - 20mm aggregate							
thickness not exceeding 150 mm	94 24	4.00	60.00	-	94.24	m^3	154.24
Fabric reinforcement; BS 4483; A193 (3.02 kg/m²) in cover slabs	3.99	0.06	0.94	-	4.39	m^2	5.33
Formwork to reinforced in situ concrete; isolated cover slabs							
soffits; horizontal	-	3.28	49.20	-	11.60	m^2	60.80
height not exceeding 250 mm	-	0.97	14.55	-	7.25	m	21.80
Silt pits and inspection chambers; precast concrete units							
Precast concrete inspection chamber units; BS 5911; bedding, jointing and pointing in cement mortar (1:3); 600 mm x 450 mm internally							
600 mm deep	30.43	6.00	90.00	-	34.75	nr	124.75
900 nm deep	37.66	7.00	105.00	-	42.16	nr	147.16
Drainage chambers; 1200 mm x 750 mm reducing to 600 mm x 600 mm; no base unit; depth of invert							
1050 mm deep	228.13	9.00	135.00	-	239.76	nr	374.76
1650 mm deep	332.97	11.00	165.00	-	350.77	nr	515.77
2250 mm deep	437.81	12.50	187.50	-	460.61	nr	648.11
Cover slabs for chambers or shaft sections; heavy duty							
900 mm diameter internally	41.70	0.67	9.99	-	41.70	nr	51.69
1050 mm diameter internally	50.70	2.00	30.00	10.12	50.70	nr	90.82
1200 mm diameter internally	62.48	1.00	15.00	10.12	62.48	nr	87.60

R DISPOSAL SYSTEMS Excluding overheads and profit	PC £	Labour hours	Labour £	Plant £	Material £	Unit	Total rate £
1500 mm diameter internally	117.09	1.00	15.00	10.12	117.09	nr	142.21
1800 mm diameter internally	136.52	2.00	30.00	22.77	136.52	nr	189.29
Brickwork							
Walls to manholes; common bricks; PC £160.00 /1000; in cement mortar (1:3)							
one brick thick	21.60	3.00	45.00	-	29.79	m²	74.79
one and a half brick thick	32.40	4.00	60.00	-	44.69	m²	104.69
two brick thick projection of footing or the like	43.20	4.80	72.00	-	59.58	m²	131.58
Walls to manholes; engineering bricks; PC £260.00 /1000; in cement mortar (1:3)							
one brick thick	31.20	3.00	45.00	-	39.87	m²	84.87
one and a half brick thick	46.80	4.00	60.00	-	59.80	m²	119.81
two brick thick projection of footing or the like	62.40	4.80	72.00	-	79.74	m²	151.74
Extra over common or engineering bricks in any mortar for fair face; flush pointing as work proceeds; English bond walls or the like	-	0.13	2.00	-	-	m²	2.00
In situ finishings; cement:sand mortar (1:3); steel trowelled; 13 mm one coat work to manhole walls; to brickwork or blockwork base;							
over 300 mm wide	-	0.80	12.00	-	2.37	m²	14.37
Building into brickwork; ends of pipes; making good facings or renderings							
small	-	0.20	3.00	-	-	nr	3.00
large	-	0.30	4.50	-	-	nr	4.50
extra large	-	0.40	6.00	-	-	nr	6.00
extra large; including forming ring arch cover	-	0.50	7.50	-	-	nr	7.50
Cast iron inspection chambers; to BS 437; Drainage Systems; bolted flat covers; bedding in cement mortar (1:3); mechanical coupling joints							
100 mm x 100 mm							
one branch each side	175.17	2.20	33.00	-	178.04	nr	211.04
two branches each side	319.08	2.40	36.00	-	320.30	nr	356.30
100 mm x 150 mm							
one branch each side	222.25	1.80	27.00	-	222.49	nr	249.49
150 mm x 150 mm							
one branch	229.09	1.50	22.50	-	229.33	nr	251.83
one branch each side	264.14	2.75	41.25	-	264.38	nr	305.63
Step irons; to BS 1247; Drainage Systems; malleable cast iron; galvanized; building into joints							
General purpose pattern; for one brick walls	4.28	0.17	2.55	-	4.28	nr	6.83
Best quality vitrified clay half section channels; Hepworth Plc; bedding and jointing in cement:mortar (1:2)							
Channels; straight							
100 mm	2.95	0.80	12.00	-	6.35	m	18.35
150 mm	5.19	1.00	15.00	-	8.59	m	23.59
225 mm	13.01	1.35	20.25	-	16.41	m	36.66
300 mm	27.38	1.80	27.00	-	30.78	m	57.78
Bends; 15, 30, 45 or 90 degrees							
100 mm bends	2.99	0.75	11.25	-	4.69	nr	15.94
150 mm bends	5.16	0.90	13.50	-	7.72	nr	21.22
225 mm bends	20.04	1.20	18.00	-	23.44	nr	41.44
300 mm bends	61.28	1.10	16.50	-	65.54	nr	82.04

R DISPOSAL SYSTEMS Excluding overheads and profit	PC £	Labour hours	Labour £	Plant £	Material £	Unit	Total rate £
R12 DRAINAGE BELOW GROUND - cont'd							
Best quality vitrified clay three quarter section channels; Hepworth Plc; bedding and jointing in cement:mortar (1:2)							
Branch bends; 115, 140 or 165 degrees; left or right hand							
100 mm	2.99	0.75	11.25	-	4.69	nr	15.94
150 mm	5.16	0.90	13.50	-	7.72	nr	21.22
Intercepting traps							
Vitrified clay; inspection arms; brass stoppers; iron levers; chains and staples; galvanized; staples cut and pinned to brickwork; cement:mortar (1:2) joints to vitrified clay pipes and channels; bedding and surrounding in concrete; 11.50 N/mm² - 40 mm aggregate; cutting and fitting brickwork; making good facings							
100 mm inlet; 100 mm outlet	35.54	3.00	45.00	-	64.82	nr	109.82
150 mm inlet; 150 mm outlet	51.26	2.00	30.00	-	86.32	nr	116.32
Excavating trenches; using 3 tonne tracked excavator; to receive pipes; grading bottoms; earthwork support; filling with excavated material to within 150 mm of finished surfaces and compacting; completing fill with topsoil; disposal of surplus soil							
Services not exceeding 200 mm nominal size							
average depth of run not exceeding 0.50 m	0.52	0.12	1.80	1.15	0.63	m	3.58
average depth of run not exceeding 0.75 m	0.52	0.16	2.44	1.59	0.63	m	4.67
average depth of run not exceeding 1.00 m	0.52	0.28	4.25	2.78	0.63	m	7.66
average depth of run not exceeding 1.25 m	0.52	0.38	5.75	3.74	0.52	m	10.01
Granular beds to trenches; lay granular material, to trenches excavated separately, to receive pipes (not included)							
300 wide x 100 thick							
reject sand	1.90	0.05	0.75	0.26	2.09	m	3.10
reject gravel	2.27	0.05	0.75	0.26	2.38	m	3.39
shingle 40 mm aggregate	2.61	0.05	0.75	0.26	2.74	m	3.75
sharp sand	2.62	0.05	0.75	0.26	2.88	m	3.89
300 wide x 150 thick							
reject sand	2.85	0.08	1.13	0.39	3.13	m	4.65
reject gravel	3.40	0.08	1.13	0.39	3.57	m	5.08
shingle 40 mm aggregate	3.92	0.08	1.13	0.39	4.11	m	5.62
sharp sand	3.93	0.08	1.13	0.39	4.32	m	5.84
Excavating trenches; using 3 tonne tracked excavator; to receive pipes; grading bottoms; earthwork support; filling with imported granular material type 2 and compacting; disposal of surplus soil							

Prices for Measured Works

R DISPOSAL SYSTEMS Excluding overheads and profit	PC £	Labour hours	Labour £	Plant £	Material £	Unit	Total rate £
Services not exceeding 200 mm nominal size							
average depth of run not exceeding 0.50 m	4.37	0.09	1.30	0.80	6.17	m	8.28
average depth of run not exceeding 0.75 m	6.56	0.11	1.62	1.01	9.26	m	11.89
average depth of run not exceeding 1.00 m	8.75	0.14	2.10	1.31	12.35	m	15.76
average depth of run not exceeding 1.25 m	10.94	0.23	3.43	2.23	15.44	m	21.10
Excavating trenches, using 3 tonne tracked excavator, to receive pipes; grading bottoms; earthwork support; filling with concrete, ready mixed ST2 ; disposal of surplus soil							
Services not exceeding 200 mm nominal size							
average depth of run not exceeding 0.50 m	10.13	0.11	1.60	0.41	12.19	m	14.20
average depth of run not exceeding 0.75 m	15.20	0.13	1.95	0.52	18.28	m	20.74
average depth of run not exceeding 1.00 m	20.27	0.17	2.50	0.69	24.37	m	27.56
average depth of run not exceeding 1.25 m	25.33	0.23	3.38	1.03	30.47	m	34.88
Earthwork Support; providing support to opposing faces of excavation; moving along as work proceeds; A Plant Acrow							
Maximum depth not exceeding 2.00 m							
distance between opposing faces not exceeding 2.00 m	-	0.80	12.00	15.39	-	m	27.39
Clay pipes and fittings: to BS EN295:1:1991:Hepworth Plc; Supersleeve							
100 mm clay pipes; polypropylene slip coupling; in trenches (trenches not included)							
laid straight	1.83	0.25	3.75	-	3.18	m	6.93
short runs under 3.00 m	1.83	0.31	4.69	-	3.18	m	7.86
Extra over 100 mm clay pipes for							
bends; 15-90 degree; single socket	5.82	0.25	3.75	-	5.82	nr	9.57
junction; 45 or 90 degree; double socket	12.71	0.25	3.75	-	12.71	nr	16.46
slip couplings polypropylene	2.16	0.08	1.25	-	2.16	nr	3.41
gully with "P" trap; 100 mm; 154 mm x 154 mm plastic grating	47.29	1.00	15.00	-	68.23	nr	83.23
150 mm vitrified clay pipes; polypropylene slip coupling; in trenches (trenches not included)							
laid straight	7.30	0.30	4.50	-	7.30	m	11.80
short runs under 3.00 m	5.62	0.33	5.00	-	5.62	m	10.62
Extra over 150 mm vitrified clay pipes for							
bends; 15 - 90 degree	8.13	0.28	4.20	-	12.05	nr	16.25
junction; 45 or 90 degree; 100 x 150	15.56	0.40	6.00	-	23.40	nr	29.40
junction; 45 or 90 degree; 150 x 150	17.08	0.40	6.00	-	24.92	nr	30.92
slip couplings polypropylene	3.92	0.05	0.75	-	3.92	nr	4.67
taper pipe 100 -150 mm	12.24	0.50	7.50	-	12.24	nr	19.74
taper pipe 150 -225 mm	30.05	0.50	7.50	-	30.05	nr	37.55
socket adaptor; connection to traditional pipes and fittings	7.89	0.33	4.95	-	11.81	nr	16.76
Accessories in clay							
150 mm access pipe	41.91	0.50	7.50	-	49.75	nr	57.25
150 mm rodding eye	39.98	0.50	7.50	-	44.43	nr	51.93
gully with "P" traps; 150 mm; 154 mm x 154 mm plastic grating	47.29	1.00	15.00	-	52.99	nr	67.99

R DISPOSAL SYSTEMS Excluding overheads and profit	PC £	Labour hours	Labour £	Plant £	Material £	Unit	Total rate £
R12 DRAINAGE BELOW GROUND - cont'd							
PVC-u pipes and fittings; to BS EN1401; **Wavin Plastics Ltd; OsmaDrain**							
110 mm PVC-u pipes; in trenches (trenches not included)							
laid straight	3.98	0.08	1.20	-	3.98	m	5.18
short runs under 3.00 m	3.98	0.12	1.80	-	4.57	m	6.37
Extra over 110 mm PVC-u pipes for							
bends, short radius	9.14	0.25	3.75	-	9.14	nr	12.89
bends; long radius	15.86	0.25	3.75	-	15.86	nr	19.61
junctions; equal; double socket	10.91	0.25	3.75	-	10.91	nr	14.66
slip couplings	3.18	0.25	3.75	-	3.18	nr	6.93
adaptors to clay	7.91	0.50	7.50	-	8.08	nr	15.58
160 mm PVC-u pipes; in trenches (trenches not included)							
laid straight	7.80	0.08	1.20	-	7.80	m	9.00
short runs under 3.00 m	9.92	0.12	1.80	-	11.27	m	13.07
Extra over 160 mm PVC-u pipes for							
socket bend double 90 or 45 degrees	23.29	0.20	3.00	-	23.29	nr	26.29
socket bend double 15 or 30 degrees	22.95	0.20	3.00	-	22.95	nr	25.95
socket bend single 87.5 or 45 degrees	18.24	0.20	3.00	-	18.24	nr	21.24
socket bend single 15 or 30 degrees	18.24	0.20	3.00	-	18.24	nr	21.24
bends, short radius	6.36	0.25	3.75	-	15.50	nr	19.25
bends; long radius	39.39	0.25	3.75	-	39.39	nr	43.14
junctions; single	39.30	0.33	5.00	-	39.30	nr	44.30
pipe coupler	3.18	0.05	0.75	-	3.18	nr	3.93
slip couplings PVC-u	6.42	0.05	0.75	-	6.42	nr	7.17
adaptors to clay	22.59	0.50	7.50	-	22.76	nr	30.26
level invert reducer	10.76	0.50	7.50	-	10.76	nr	18.26
spigot	13.48	0.20	3.00	-	13.48	nr	16.48
Accessories in PVC-u							
110 mm screwed access cover	-	-	-	-	6.42	nr	6.42
110 mm rodding eye	21.20	0.50	7.50	-	25.65	nr	33.15
gully with "P" traps; 110 mm; 154 mm x 154 mm grating	36.63	1.00	15.00	-	38.41	nr	53.41
inspection chambers; 450 mm diameter; 960 mm deep; heavy duty round covers and frames; double seal recessed with 6 nr 110 mm outlets/inlets	125.23	1.00	15.00	-	127.90	nr	142.90
Kerbs; to gullies; in one course Class B engineering bricks; to 4 nr sides; rendering in cement:mortar (1:3); dished to gully gratings	2.08	1.00	15.00	-	2.92	nr	17.92
Gullies Concrete; Hepworth Plc							
Concrete road gullies; to BS 5911; trapped with rodding eye and stoppers; 450 mm diameter x 1.07 m deep	30.50	6.00	90.00	-	55.14	nr	145.14
Gullies Vitrified Clay; Hepworth Plc; **bedding in concrete; 11.50 N/mm² - 40 mm aggregate**							
Vitrified clay yard gullies (mud); trapped; domestic duty (up to 1 tonne)							
RGP5; 100 mm outlet; 100 mm dia; 225 mm internal width 585 mm internal depth	58.98	3.50	52.50	-	59.63	nr	112.13
RGP7; 150 mm outlet; 100 mm dia; 225 mm internal width 585 mm internal depth	55.64	3.50	52.50	-	56.29	nr	108.79

R DISPOSAL SYSTEMS Excluding overheads and profit	PC £	Labour hours	Labour £	Plant £	Material £	Unit	Total rate £
Vitrified clay yard gullies (mud); trapped; medium duty (up to 5 tonnes)							
100 mm outlet; 100 mm dia; 225 mm internal width 585 mm internal depth	83.37	3.50	52.50	-	84.02	nr	136.52
150 mm outlet; 100 mm dia; 225 mm internal width 585 mm internal depth	83.37	3.50	52.50	-	84.02	nr	136.52
Combined filter and silt bucket for Yard gullies 225 mm wide	20.74	-	-	-	20.74	nr	20.74
Vitrified clay road gullies; trapped with rodding eye,							
100 mm outlet; 300 mm internal dia, 600 mm internal depth	51.77	3.50	52.50	-	52.42	nr	104.92
150 mm outlet; 300 mm internal dia, 600 mm internal depth	53.01	3.50	52.50	-	53.66	nr	106.16
150 mm outlet; 400 mm internal dia, 750 mm internal depth	61.48	3.50	52.50	-	62.13	nr	114.63
150 mm outlet; 450 mm internal dia, 900 mm internal depth	-	3.50	52.50	-	83.84	nr	136.34
Hinged Gratings and frames for gullies; Alloy;							
135 mm for 100 mm dia gully;	-	-	-	-	7.60	nr	7.60
193 mm for 150 mm dia gully;	-	-	-	-	13.15	nr	13.15
120 mm x 120 mm	-	-	-	-	4.67	nr	4.67
150 mm x 150 mm	-	-	-	-	8.45	nr	8.45
230 mm x 230 mm	-	-	-	-	15.46	nr	15.46
316 mm x 316 mm	-	-	-	-	40.99	nr	40.99
Hinged Gratings and frames for gullies; Cast Iron;							
265 mm for 225 mm dia gully;	-	-	-	-	26.25	nr	26.25
150 mm x 150 mm	-	-	-	-	8.45	nr	8.45
230 mm x 230 mm	-	-	-	-	15.46	nr	15.46
316 mm x 316 mm	-	-	-	-	40.99	nr	40.99
Universal Gully Trap PVC-u; Wavin Plastics Ltd; OsmaDrain system; bedding in concrete; 11.50 N/mm² - 40 mm aggregate							
Universal gully fitting; comprising gully trap only							
110 mm outlet; 110 mm dia; 205 mm internal depth	7.75	3.50	52.50	-	8.40	nr	60.91
Vertical inlet hopper c\w plastic grate							
272 x 183 mm	10.42	0.25	3.75	-	10.42	nr	14.17
Sealed access hopper							
110 x 110 mm	17.37	0.25	3.75	-	17.37	nr	21.12
Universal Gully PVC-u; Wavin Plastics Ltd; OsmaDrain system; Accessories to Universal gully trap							
Hoppers; backfilling with clean granular material; tamping; surrounding in lean mix concrete							
plain hopper with 110 spigot 150 mm long	8.53	0.40	6.00	-	8.89	nr	14.89
vertical inlet hopper with 110 spigot 150 mm long	10.42	0.40	6.00	-	10.42	nr	16.42
sealed access hopper with 110 spigot 150 mm long	17.37	0.40	6.00	-	17.37	nr	23.37
plain hopper; solvent weld to trap	5.82	0.40	6.00	-	5.82	nr	11.82
vertical inlet hopper; solvent wed to trap	7.67	0.40	6.00	-	7.67	nr	13.67
sealed access cover; PVC-U	8.85	0.10	1.50	-	8.85	nr	10.35

Prices for Measured Works

R DISPOSAL SYSTEMS Excluding overheads and profit	PC £	Labour hours	Labour £	Plant £	Material £	Unit	Total rate £
R12 DRAINAGE BELOW GROUND - cont'd							
Gullies PVC-u; Wavin Plastics Ltd; **OsmaDrain system; bedding in concrete;** **11.50 N/mm² - 40 mm aggregate**							
Bottle Gully; providing access to the drainage system for cleaning							
bottle gully; 228 x 228 x 317 mm deep	17.67	0.50	7.50	-	18.20	nr	25.70
sealed access cover; PVC-u 217 x 217 mm	11.39	0.10	1.50	-	11.39	nr	12.89
grating; ductile iron 215 x 215 mm	8.79	0.10	1.50	-	8.79	nr	10.29
bottle gully riser; 325 mm	2.04	0.50	7.50	-	3.20	nr	10.70
Yard Gully; trapped 300 mm diameter 600 mm deep; including catchment bucket and ductile iron cover and frame, medium duty loading							
305 mm diameter 600 mm deep	114.09	2.50	37.50	-	117.27	nr	154.77
Kerbs to Gullies							
One course Class B engineering bricks to 4 nr sides; rendering in cement:mortar (1:3); dished to gully gratings							
150 mm x 150 mm	1 04	0.33	5.00	-	1.61	nr	6.61
Access Covers and Frames; Jones of **Oswestry; bedding frame in cement** **mortar (1:3); cover in grease and sand;** **clear opening sizes**							
Access Covers and frames; "Suprabloc"; to paved areas; filling with blocks cut and fitted to match surrounding paving; BS standard size manholes available							
pedestrian weight; 300 mm x 300 mm	110.43	2.50	37.50	-	116.30	nr	153.80
pedestrian weight; 450 mm x 450 mm	127.62	2.50	37.50	-	127.62	nr	165.12
pedestrian weight; 450 mm x 600 mm	137.19	2.50	37.50	-	143.30	nr	180.80
light vehicular weight; 300 mm x 300 mm	110.43	2.40	36.00	-	116.30	nr	152.30
light vehicular weight; 450 mm x 450 mm	110.43	3.00	45.00	-	117.15	nr	162.15
light vehicular weight; 450 mm x 600 mm	127.62	4.00	60.00	-	133.49	nr	193.49
heavy vehicular weight; 300 mm x 300 mm	128.08	3.00	45.00	-	134.19	nr	179.19
heavy vehicular weight; 450 mm x 600 mm	159 13	3.00	45.00	-	165.85	nr	210.85
heavy vehicular weight; 600 mm x 600 mm	178.70	4.00	60.00	-	188.48	nr	248.48
Extra over "Suprabloc" manhole frames and covers for							
filling recessed manhole covers with brick paviors; PC £305.00 /1000	12.20	1.00	15.00	-	13.92	m²	28.92
filling recessed manhole covers with vehicular paving blocks; PC £8.18/m²	8.18	0.75	11.25	-	8.68	m²	19.93
filling recessed manhole covers with concrete paving flags; PC £9.62/m²	9.61	0.35	5.25	-	9.91	m²	15.16
Access covers and frames to BS 497; **Saint -Gobain Pipelines; bedding frame** **in cement mortar (1:3); cover in grease** **and sand; light duty; clear opening sizes**							
Grade A; double triangular; coated							
600 mm diameter	119.14	3.00	45.00	-	121.49	nr	166.49
675 mm diameter	203.83	3.00	45.00	-	206.18	nr	251.18

R DISPOSAL SYSTEMS Excluding overheads and profit	PC £	Labour hours	Labour £	Plant £	Material £	Unit	Total rate £
Grade A; coated; single seal solid top							
450 mm x 450 mm	65.18	1.50	22.50	-	67.00	nr	89.50
450 mm x 600 mm	60.18	1.80	27.00	-	62.71	nr	89.71
600 mm x 600 mm	127.90	2.00	30.00	-	131.14	nr	161.14
Grade A; coated; double seal solid top							
450 mm x 450 mm	111.99	1.50	22.50	-	113.81	nr	136.31
600 mm x 450 mm	120.60	1.80	27.00	-	123.13	nr	150.13
600 mm x 600 mm	170.00	2.00	30.00	-	173.24	nr	203.24
Grade A; coated; single seal recessed							
450 mm x 450 mm	71.78	1.50	22.50	-	73.60	nr	96.10
600 mm x 600 mm	127.11	2.00	30.00	-	130.35	nr	160.35
Grade A; coated; double seal recessed							
450 mm x 600 mm	142.07	1.80	27.00	-	144.60	nr	171.60
600 mm x 600 mm	204.94	2.00	30.00	-	208.18	nr	238.18
Access covers and frames; to BS 497; Drainage Systems; bedding frame in cement mortar (1:3); cover in grease and sand; medium duty; clear opening sizes							
Grade B; coated							
"Standard" 600 mm diameter single seal	91.11	3.00	45.00	-	93.46	nr	138.46
"Standard" 600 x 450 mm single seal	96.26	3.00	45.00	-	98.61	nr	143.61
"Bricast" 600 x 600 mm diameter; single seal	85.15	3.00	45.00	-	87.50	nr	132.50
Grade B; recessed							
"Trucast" 600 x 450 mm single seal recessed	181.10	3.00	45.00	-	183.45	nr	228.45
R13 LAND DRAINAGE							
Ditching; clear and bottom ditch not exceeding 1.50 m deep; trimming back vegetation; disposing to spoil heaps; by machine							
Up to 1.50 m wide at top	-	18.00	270.00	62.06	-	100 m	332.06
1.50 - 2.50 m wide at top	-	26.00	390.00	93.10	-	100 m	483.10
2.50 - 4.00 m wide at top	-	42.00	630.00	144.82	-	100 m	774.82
Ditching; clear and bottom ditch not exceeding 1.50 m deep; trimming back vegetation; disposing to spoil heaps; by hand							
Up to 1.50 m wide at top	-	15.00	225.00	6.75	-	100 m	231.75
1.50 - 2.50 m wide at top	-	27.00	405.00	12.15	-	100 m	417.15
2.50 - 4.00 m wide at top	-	42.00	630.00	18.90	-	100 m	648.90
Ditching; excavating and forming ditch and bank to given profile (normally 45 degrees); in loam or sandy loam; by machine							
Width 300 mm							
depth 600 mm	-	3.70	55.50	38.27	-	100 m	93.77
depth 900 mm	-	5.20	78.00	53.79	-	100 m	131.79
depth 1200 mm	-	7.20	108.00	74.48	-	100 m	182.48
depth 1500 mm	-	9.40	141.00	85.08	-	100 m	226.08

R DISPOSAL SYSTEMS Excluding overheads and profit	PC £	Labour hours	Labour £	Plant £	Material £	Unit	Total rate £
R13 LAND DRAINAGE - cont'd							
Ditching; excavating and forming ditch and bank to given profile (normally 45 degrees); in loam or sandy loam; by machine - cont'd							
Width 600 mm							
depth 600 mm	-	-	-	152.77	-	100 m	152.77
depth 900 mm	-	-	-	226.69	-	100 m	226.69
depth 1200 mm	-	-	-	308.00	-	100 m	308.00
depth 1500 mm	-	-	-	443.52	-	100 m	443.52
Width 900 mm							
depth 600 mm	-	-	-	289.80	-	100 m	289.80
depth 900 mm	-	-	-	510.60	-	100 m	510.60
depth 1200 mm	-	-	-	686.14	-	100 m	686.14
depth 1500 mm	-	-	-	850.08	-	100 m	850.08
Width 1200 mm							
depth 600 mm	-	-	-	455.40	-	100 m	455.40
depth 900 mm	-	-	-	676.20	-	100 m	676.20
depth 1200 mm	-	-	-	895.62	-	100 m	895.62
depth 1500 mm	-	-	-	1147.61	-	100 m	1147.61
Width 1500 mm							
depth 600 mm	-	-	-	561.66	-	100 m	561.66
depth 900 mm	-	-	-	856.15	-	100 m	856.15
depth 1200 mm	-	-	-	1138.50	-	100 m	1138.50
depth 1500 mm	-	-	-	1426.92	-	100 m	1426.92
Extra for ditching in clay	-	-	-	-	-	20%	-
Ditching; excavating and forming ditch and bank to given profile (normal 45 degrees); in loam or sandy loam; by hand							
Width 300 mm							
depth 600 mm	-	36.00	540.00	-	-	100 m	540.00
depth 900 mm	-	42.00	630.00	-	-	100 m	630.00
depth 1200 mm	-	56.00	840.00	-	-	100 m	840.00
depth 1500 mm	-	70.00	1050.00	-	-	100 m	1050.00
Width 600 mm							
depth 600 mm	-	56.00	840.00	-	-	100 m	840.00
depth 900 mm	-	84.00	1260.00	-	-	100 m	1260.00
depth 1200 mm	-	112.00	1680.00	-	-	100 m	1680.00
depth 1500 mm	-	140.00	2100.00	-	-	100 m	2100.00
Width 900 mm							
depth 600 mm	-	84.00	1260.00	-	-	100 m	1260.00
depth 900 mm	-	126.00	1890.00	-	-	100 m	1890.00
depth 1200 mm	-	168.00	2520.00	-	-	100 m	2520.00
depth 1500 mm	-	210.00	3150.00	-	-	100 m	3150.00
Width 1200 mm							
depth 600 mm	-	112.00	1680.00	-	-	100 m	1680.00
depth 900 mm	-	168.00	2520.00	-	-	100 m	2520.00
depth 1200 mm	-	224.00	3360.00	-	-	100 m	3360.00
depth 1500 mm	-	280.00	4200.00	-	-	100 m	4200.00
Extra for ditching in clay -	-	-	-	-		50%	
Extra for ditching in compacted soil -	-	-	-	-		90%	

R DISPOSAL SYSTEMS Excluding overheads and profit	PC £	Labour hours	Labour £	Plant £	Material £	Unit	Total rate £
Piped ditching							
Jointed concrete pipes; to BS 5911 pt.3; including bedding, haunching and topping with 150 mm concrete; 11.50 N/mm² - 40 mm aggregate; to existing ditch							
300 mm diameter	13.15	0.67	10.00	6.07	31.80	m	47.87
450 mm diameter	21.38	0.67	10.00	6.07	49.36	m	65.43
600 mm diameter	31.63	0.67	10.00	6.07	71.35	m	87.42
900 mm diameter	74.00	1.00	15.00	9.11	93.05	m	117.16
Jointed concrete pipes; to BS 5911 pt.1 class S; including bedding, haunching and topping with 150 mm concrete; 11.50 N/mm² - 40 mm aggregate; to existing ditch							
1200 mm diameter	120.93	1.00	15.00	9.11	145.86	m	169.96
Extra over jointed concrete pipes for bends 300 mm radius	564.96	0.67	10.00	7.59	572.56	nr	590.15
Extra over jointed concrete pipes for single junctions	61.08	0.67	10.00	7.59	68.68	nr	86.27
Concrete road gullies; to BS 5911; trapped; cement:mortar (1:2) joints to concrete pipes; bedding and surrounding in concrete; 11.50 N/mm² - 40 mm aggregate; 450 mm diameter x 1.07 m deep; rodding eye; stoppers	30.50	6.00	90.00	-	47.55	nr	137.55
Mole drainage; White Horse Contractors Ltd							
Drain by mole plough; 50 mm diameter mole set at depth of 450 mm in parallel runs							
1.20 m centres	-	-	-	-	-	100 m²	11.19
1.20 m centres	-	-	-	-	-	ha	557.36
1.50 m centres	-	-	-	-	-	100 m²	10.28
1.50 m centres	-	-	-	-	-	ha	504.00
2.00 m centres	-	-	-	-	-	100 m²	9.35
2.00 m centres	-	-	-	-	-	ha	462.93
2.50 m centres	-	-	-	-	-	100 m²	8.31
2.50 m centres	-	-	-	-	-	ha	412.26
3.00 m centres	-	-	-	-	-	100 m²	7.27
3.00 m centres	-	-	-	-	-	ha	370.81
Drain by mole plough; 75 mm diameter mole set at depth of 450 mm in parallel runs							
1.20 m centres	-	-	-	-	-	100 m²	13.36
1.20 m centres	-	-	-	-	-	ha	667.92
1.50 m centres	-	-	-	-	-	100 m²	12.17
1.50 m centres	-	-	-	-	-	ha	601.07
2.00 m centres	-	-	-	-	-	100 m²	11.76
2.00 m centres	-	-	-	-	-	ha	569.10
2.50 m centres	-	-	-	-	-	100 m²	9.83
2.50 m centres	-	-	-	-	-	ha	492.87
3.00 m centres	-	-	-	-	-	100 m²	10.28
3.00 m centres	-	-	-	-	-	ha	428.39

R DISPOSAL SYSTEMS Excluding overheads and profit	PC £	Labour hours	Labour £	Plant £	Material £	Unit	Total rate £
R13 LAND DRAINAGE - cont'd							
Sand slitting; White Horse Contractors Ltd							
Drainage slits; at 2.00 m centres; using slitting machine; backfilling to 75 mm of surface with pea gravel; blind with sharp sand							
250 mm depth	-	-	-	-	-	100 m²	25.29
300 mm depth	-	-	-	-	-	100 m²	28.42
400 mm depth	-	-	-	-	-	100 m²	34.16
450 mm depth	-	-	-	-	-	100 m²	36.70
250 mm depth	-	-	-	-	-	ha	6500.00
300 mm depth	-	-	-	-	-	ha	6750.00
400 mm depth	-	-	-	-	-	ha	7000.00
Trenchless drainage system; White Horse Contractors Ltd; Insert perforated pipes by means of laser graded deep plough machine; backfill with gravel (not included)							
laterals 60 mm depth 700 mm	-	-	-	-	-	m	1.45
main 100 mm depth 900 mm	-	-	-	-	-	m	2.05
main 160 mm depth 1000 mm	-	-	-	-	-	m	2.55
Gravel backfill							
laterals 60 mm	-	-	-	-	-	m²	3.10
main 110 mm	-	-	-	-	-	m²	5.40
main 160 mm	-	-	-	-	-	m²	4.60
Agricultural drainage; calculation table							
For calculation of drainage per hectare, the following table can be used: rates show the lengths of drains per unit and not the value.							
Lateral drains							
4.00 m centres	-	-	-	-	-	m/ha	2500.00
6.00 m centres	-	-	-	-	-	m/ha	1670.00
8.00 m centres	-	-	-	-	-	m/ha	1250.00
10.00 m centres	-	-	-	-	-	m/ha	1000.00
15.00 m centres	-	-	-	-	-	m/ha	650.00
25.00 m centres	-	-	-	-	-	m/ha	400.00
30.00 m centres	-	-	-	-	-	m/ha	330.00
Main drains (per hectare)							
1 nr (at 100 m centres)	-	-	-	-	-	m/ha	100.00
2 nr (at 50 m centres)	-	-	-	-	-	m/ha	200.00
3 nr (at 33.333 m centres)	-	-	-	-	-	m/ha	300.00
4 nr (at 25 m centres)	-	-	-	-	-	m/ha	400.00
Agricultural drainage; excavating							
Removing 150 mm depth of topsoil; 300 mm wide; depositing beside trench; by machine	-	1.50	22.50	15.52	-	100 m	38.02
Removing 150 mm depth of topsoil; 300 mm wide; depositing beside trench; by hand	-	8.00	120.00	-	-	100 m	120.00
Disposing on site; to spoil heaps; by machine							
not exceeding 100 m distance	-	0.07	0.99	2.30	-	m³	3.29
average 100 - 150 m distance	-	0.08	1.19	2.77	-	m³	3.95
average 150 - 200 m distance	-	0.09	1.39	3.23	-	m³	4.62

R DISPOSAL SYSTEMS Excluding overheads and profit	PC £	Labour hours	Labour £	Plant £	Material £	Unit	Total rate £
Removing excavated material from site to tip not exceeding 13 km; mechanically loaded excavated material and clean hardcore rubble	-	-	-	1.01	10.00	m³	11.01
Agricultural drainage; White Horse Contractors Ltd excavating trenches (with minimum run of 500 m) by trenching machine; including disposing subsoil to spoil heaps not exceeding 100 m							
Width 150 mm							
depth 450 mm	-	5.04	75.60	84.01	-	100 m	159.61
depth 600 mm	-	5.79	86.80	98.51	-	100 m	185.31
depth 750 mm	-	6.53	97.94	112.88	-	100 m	210.83
Width 225 mm							
depth 450 mm	-	4.67	70.13	79.43	-	100 m	149.56
depth 600 mm	-	6.60	99.00	117.97	-	100 m	216.97
depth 750 mm	-	7.42	111.38	135.52	-	100 m	246.90
Width 300 mm							
depth 600 mm	-	7.33	109.95	137.87	-	100 m	247.82
depth 750 mm	-	8.17	122.55	158.17	-	100 m	280.72
depth 900 mm	-	9.00	135.00	178.07	-	100 m	313.07
depth 1000 mm	-	9.72	145.80	193.76	-	100 m	339.56
Width 375 mm							
depth 600 mm	-	8.67	130.05	165.40	-	100 m	295.45
depth 750 mm	-	9.58	143.70	188.38	-	100 m	332.08
depth 900 mm	-	10.50	157.50	211.74	-	100 m	369.24
depth 1000 mm	-	11.78	176.70	236.96	-	100 m	413.66
Agricultural drainage; excavating for drains; by backacter excavator JCB C3X Sitemaster; including disposing spoil to spoil heaps not exceeding 100 m							
Width 150 mm							
depth 450 mm	-	0.50	7.50	79.93	-	100 m	87.43
depth 600 mm	-	0.67	10.00	101.51	-	100 m	111.51
depth 700 mm	-	0.78	11.70	121.04	-	100 m	132.74
depth 900 mm	-	1.00	15.00	129.49	-	100 m	144.49
Width 225 mm							
depth 450 mm	-	0.75	11.25	89.53	-	100 m	100.78
depth 600 mm	-	1.00	15.00	114.31	-	100 m	129.31
depth 700 mm	-	1.17	17.55	136.02	-	100 m	153.57
depth 900 mm	-	1.50	22.50	163.88	-	100 m	186.38
depth 1000 mm	-	1.67	25.05	215.95	-	100 m	241.00
Width 300 mm							
depth 450 mm	-	1.00	15.00	100.04	-	100 m	115.04
depth 700 mm	-	1.56	23.40	155.56	-	100 m	178.96
depth 900 mm	-	2.00	30.00	199.78	-	100 m	229.78
depth 1000 mm	-	2.22	33.30	221.89	-	100 m	255.19
depth 1200 mm	-	2.67	40.05	266.50	-	100 m	306.55
depth 1500 mm	-	3.33	49.95	332.84	-	100 m	382.79
Width 375 mm							
depth 450 mm	-	1.25	18.75	124.83	-	100 m	143.58
depth 700 mm	-	1.94	29.10	194.14	-	100 m	223.24
depth 900 mm	-	2.50	37.50	249.65	-	100 m	287.15
depth 1000 mm	-	2.78	41.70	277.41	-	100 m	319.11
depth 1200 mm	-	3.33	49.95	332.84	-	100 m	382.79
depth 1500 mm	-	4.17	62.55	416.11	-	100 m	478.66

R DISPOSAL SYSTEMS Excluding overheads and profit	PC £	Labour hours	Labour £	Plant £	Material £	Unit	Total rate £
R13 LAND DRAINAGE - cont'd							
Agricultural drainage; excavating for drains; by backacter excavator JCB C3X Sitemaster; including disposing spoil to spoil heaps not exceeding 100 m - cont'd							
Width 450 mm							
depth 450 mm	-	1.50	22.50	162.97	-	100 m	185.47
depth 700 mm	-	2.33	34.95	253.44	-	100 m	288.39
depth 900 mm	-	3.00	45.00	325.93	-	100 m	370.93
depth 1000 mm	-	3.33	49.95	362.29	-	100 m	412.24
depth 1200 mm	-	4.00	60.00	434.78	-	100 m	494.78
depth 1500 mm	-	5.00	75.00	543.33	-	100 m	618.33
depth 2000 mm	-	6.67	100.05	724.66	-	100 m	824.71
depth 2500 mm	-	8.33	124.95	905.62	-	100 m	1030.57
Width 600 mm							
depth 450 mm	-	6.63	99.45	254.67	-	100 m	354.12
depth 600 mm	-	8.84	132.60	339.56	-	100 m	472.16
depth 700 mm	-	10.31	154.65	396.03	-	100 m	550.68
depth 900 mm	-	11.05	165.75	424.45	-	100 m	590.20
depth 1000 mm	-	14.73	220.95	565.81	-	100 m	786.76
depth 1200 mm	-	17.68	265.20	679.12	-	100 m	944.32
depth 1500 mm	-	22.10	331.50	848.91	-	100 m	1180.41
depth 2000 mm	-	29.46	441.90	1131.62	-	100 m	1573.52
depth 2500 mm	-	36.83	552.45	1414.71	-	100 m	1967.16
depth 3000 mm	-	36.83	552.45	1414.71	-	100 m	1967.16
Width 900 mm							
depth 450 mm	-	6.63	99.45	254.67	-	100 m	354.12
depth 600 mm	-	8.84	132.60	339 56	-	100 m	472.16
depth 700 mm	-	10.31	154.65	396.03	-	100 m	550.68
depth 900 mm	-	11.05	165.75	424.45	-	100 m	590.20
depth 1000 mm	-	14.73	220.95	565.81	-	100 m	786.76
depth 1200 mm	-	17.68	265.20	679.12	-	100 m	944.32
depth 1500 mm	-	22.10	331.50	848.91	-	100 m	1180.41
depth 2000 mm	-	29.46	441.90	1131.62	-	100 m	1573.52
depth 2500 mm	-	36.83	552.45	1414.71	-	100 m	1967.16
depth 3000 mm	-	36.83	552.45	1414.71	-	100 m	1967.16
Agricultural drainage; excavating for drains; by 7 tonne tracked excavator; including disposing spoil to spoil heaps not exceeding 100 m							
Width 600 mm							
depth 450 mm	-	4.49	67.35	146.79	-	100 m	214.14
depth 600 mm	-	5.99	89.85	195.83	-	100 m	285.68
depth 700 mm	-	6.99	104.85	228.52	-	100 m	333.37
depth 900 mm	-	8.98	134.70	293.57	-	100 m	428.27
depth 1000 mm	-	9.98	149.70	326.27	-	100 m	475.97
depth 1200 mm	-	11.98	179.70	391.65	-	100 m	571.35
depth 1500 mm	-	14.97	224.55	489.40	-	100 m	713.95
depth 2000 mm	-	19.97	299.55	652.86	-	100 m	952.41
depth 2500 mm	-	24.96	374.40	958.76	-	100 m	1333.16
depth 3000 mm	-	29.95	449.25	979.13	-	100 m	1428.38

R DISPOSAL SYSTEMS Excluding overheads and profit	PC £	Labour hours	Labour £	Plant £	Material £	Unit	Total rate £
Width 700 mm							
depth 450 mm	-	5.24	78.60	171.31	-	100 m	249.91
depth 600 mm	-	6.99	104.85	228.52	-	100 m	333.37
depth 700 mm	-	8.15	122.25	266.44	-	100 m	388.69
depth 900 mm	-	10.48	157.20	342.61	-	100 m	499.81
depth 1000 mm	-	11.65	174.75	380.86	-	100 m	555.61
depth 1200 mm	-	13.98	209.70	457.03	-	100 m	666.73
depth 1500 mm	-	17.47	262.05	571.13	-	100 m	833.18
depth 2000 mm	-	23.29	349.35	761.40	-	100 m	1110.75
depth 2500 mm	-	29.11	436.68	1118.56	-	100 m	1555.24
depth 3000 mm	-	34.94	524.10	1142.26	-	100 m	1666.36
Width 900 mm							
depth 600 mm	-	9.89	148.35	323.32	-	100 m	471.67
depth 700 mm	-	11.54	173.10	377.27	-	100 m	550.37
depth 900 mm	-	14.84	222.60	485.40	-	100 m	708.00
depth 1000 mm	-	16.48	247.20	538.76	-	100 m	785.96
depth 1200 mm	-	19.78	296.70	646.65	-	100 m	943.35
depth 1500 mm	-	24.73	370.95	808.47	-	100 m	1179.42
depth 2000 mm	-	32.97	494.55	1077.86	-	100 m	1572.41
depth 2500 mm	-	41.21	618.15	1582.96	-	100 m	2201.11
depth 3000 mm	-	49.45	741.75	1616.62	-	100 m	2358.37
Width 1000 mm							
depth 600 mm	-	11.64	174.60	380.53	-	100 m	555.13
depth 700 mm	-	13.59	203.85	444.28	-	100 m	648.13
depth 900 mm	-	17.47	262.05	571.13	-	100 m	833.18
depth 1000 mm	-	19.41	291.15	634.55	-	100 m	925.70
depth 1200 mm	-	23.29	349.35	761.40	-	100 m	1110.75
depth 1500 mm	-	29.11	436.65	951.66	-	100 m	1388.31
depth 2000 mm	-	38.81	582.15	1268.78	-	100 m	1850.93
depth 2500 mm	-	48.52	727.80	1863.75	-	100 m	2591.55
depth 3000 mm	-	58.22	873.30	1903.33	-	100 m	2776.63
Width 1500 mm							
depth 600 mm	-	17.47	262.05	571.13	-	100 m	833.18
depth 700 mm	-	20.38	305.70	666.26	-	100 m	971.96
depth 900 mm	-	26.20	393.00	856.53	-	100 m	1249.53
depth 1000 mm	-	29.11	436.65	951.66	-	100 m	1388.31
depth 1200 mm	-	34.93	523.95	1141.93	-	100 m	1665.88
depth 1500 mm	-	43.67	655.05	1427.66	-	100 m	2082.71
depth 2000 mm	-	58.22	873.30	1903.33	-	100 m	2776.63
depth 2500 mm	-	72.78	1091.70	2795.63	-	100 m	3887.33
depth 3000 mm	-	87.33	1309.95	2854.99	-	100 m	4164.94
Agricultural drainage; excavating for drains; by hand, including disposing spoil to spoil heaps not exceeding 100 m							
Width 150 mm							
depth 450 mm	-	22.03	330.45	-	-	100 m	330.45
depth 600 mm	-	29.38	440.70	-	-	100 m	440.70
depth 700 mm	-	34.27	514.05	-	-	100 m	514.05
depth 900 mm	-	44.06	660.90	-	-	100 m	660.90
Width 225 mm							
depth 450 mm	-	33.05	495.75	-	-	100 m	495.75
depth 600 mm	-	44.06	660.90	-	-	100 m	660.90
depth 700 mm	-	51.41	771.15	-	-	100 m	771.15
depth 900 mm	-	66.10	991.50	-	-	100 m	991.50
depth 1000 mm	-	73.44	1101.60	-	-	100 m	1101.60

R DISPOSAL SYSTEMS Excluding overheads and profit	PC £	Labour hours	Labour £	Plant £	Material £	Unit	Total rate £
R13 LAND DRAINAGE - cont'd							
Agricultural drainage; excavating for drains; by hand, including disposing spoil to spoil heaps not exceeding 100 m - cont'd							
Width 300 mm							
depth 450 mm	-	44.06	660.90	-	-	100 m	**660.90**
depth 600 mm	-	58.75	881.25	-	-	100 m	**881.25**
depth 700 mm	-	68.54	1028.10	-	-	100 m	**1028.10**
depth 900 mm	-	88.13	1321.95	-	-	100 m	**1321.95**
depth 1000 mm	-	97.92	1468.80	-	-	100 m	**1468.80**
Width 375 mm							
depth 450 mm	-	55.08	826.20	-	-	100 m	**826.20**
depth 600 mm	-	73.44	1101.60	-	-	100 m	**1101.60**
depth 700 mm	-	85.68	1285.20	-	-	100 m	**1285.20**
depth 900 mm	-	110.16	1652.40	-	-	100 m	**1652.40**
depth 1000 mm	-	122.40	1836.00	-	-	100 m	**1836.00**
Width 450 mm							
depth 450 mm	-	66.10	991.50	-	-	100 m	**991.50**
depth 600 mm	-	88.13	1321.95	-	-	100 m	**1321.95**
depth 700 mm	-	102.82	1542.30	-	-	100 m	**1542.30**
depth 900 mm	-	132.19	1982.85	-	-	100 m	**1982.85**
depth 1000 mm	-	146.88	2203.20	-	-	100 m	**2203.20**
Width 600 mm							
depth 450 mm	-	88.13	1321.95	-	-	100 m	**1321.95**
depth 600 mm	-	117.50	1762.50	-	-	100 m	**1762.50**
depth 700 mm	-	137.09	2056.35	-	-	100 m	**2056.35**
depth 900 mm	-	176.26	2643.90	-	-	100 m	**2643.90**
depth 1000 mm	-	195.84	2937.60	-	-	100 m	**2937.60**
Width 900 mm							
depth 450 mm	-	132.19	1982.85	-	-	100 m	**1982.85**
depth 600 mm	-	176.26	2643.90	-	-	100 m	**2643.90**
depth 700 mm	-	205.63	3084.45	-	-	100 m	**3084.45**
depth 900 mm	-	264.38	3965.70	-	-	100 m	**3965.70**
depth 1000 mm	-	293.76	4406.40	-	-	100 m	**4406.40**
Earthwork Support; moving along as work proceeds							
Maximum depth not exceeding 2.00 m							
distance between opposing faces not exceeding 2.00 m	-	0.80	12.00	15.39	-	m	**27.39**
Agricultural drainage; pipe laying; Hepworth; Agricultural clay drain pipes; to BS 1196; 300 mm length; butt joints; in straight runs							
75 mm diameter	237.10	8.00	120.00	-	243.02	100 m	**363.02**
100 mm diameter	407.59	9.00	135.00	-	417.78	100 m	**552.78**
150 mm diameter	836.50	10.00	150.00	-	857.41	100 m	**1007.41**
Extra over clay drain pipes for filter-wrapping pipes with "Terram" or similar filter fabric							
"Terram 700"	0.20	0.04	0.60	-	0.20	m²	**0.80**
"Terram 1000"	0.22	0.04	0.60	-	0.22	m²	**0.82**

R DISPOSAL SYSTEMS Excluding overheads and profit	PC £	Labour hours	Labour £	Plant £	Material £	Unit	Total rate £
Junctions between drains in clay pipes							
75 mm x 75 mm	8.63	0.25	3.75	-	9.05	nr	12.80
100 mm x 100 mm	10.78	0.25	3.75	-	11.46	nr	15.21
100 mm x 150 mm	13.27	0.25	3.75	-	14.29	nr	18.04
Outfalls; to ditches or culverts; setting in weak concrete							
75 mm diameter clay pipes	1.01	1.00	15.00	-	3.53	nr	18.53
100 mm diameter clay pipes	1.81	1.00	15.00	-	5.01	nr	20.01
150 mm diameter clay pipes	3.86	1.00	15.00	-	7.13	nr	22.13
Wavin Plastics Ltd; flexible plastic perforated pipes in trenches (not included); to a minimum depth of 450 mm (couplings not included)							
"OsmaDrain"; flexible plastic perforated pipes in trenches (not included); to a minimum depth of 450 mm (couplings not included)							
60 mm diameter; available in 150 m coil	23.50	2.00	30.00	-	24.09	100 m	54.09
80 mm diameter; available in 100 m coil	35.50	2.00	30.00	-	36.39	100 m	66.39
100 mm diameter; available in 100 m coil	62.50	2.00	30.00	-	64.06	100 m	94.06
160 mm diameter; available in 35 m coil	151.00	2.00	30.00	-	154.78	100 m	184.78
"WavinCoil"; Plastic pipe junctions							
60 mm x 60 mm	1.35	0.05	0.75	-	1.35	nr	2.10
80 mm x 80 mm	1.48	0.05	0.75	-	1.48	nr	2.23
100 mm x 100 mm	1.66	0.05	0.75	-	1.66	nr	2.41
80 mm x 60 mm	1.43	0.05	0.75	-	1.43	nr	2.17
100 mm x 60 mm	1.52	0.05	0.75	-	1.52	nr	2.27
100 mm x 80 mm	1.58	0.05	0.75	-	1.58	nr	2.33
125 mm x 60 mm	1.68	0.05	0.75	-	1.68	nr	2.42
125 mm x 100 mm	1.74	0.05	0.75	-	1.74	nr	2.49
160 mm x 100 mm	5.15	0.05	0.75	-	5.15	nr	5.90
160 mm x 160 mm	4.28	0.05	0.75	-	4.28	nr	5.03
"WavinCoil"; couplings for flexible pipes							
60 mm diameter	0.52	0.03	0.50	-	0.52	nr	1.02
80 mm diameter	0.61	0.03	0.50	-	0.61	nr	1.11
100 mm diameter	0.68	0.03	0.50	-	0.68	nr	1.17
160 mm diameter	0.91	0.03	0.50	-	0.91	nr	1.41

R DISPOSAL SYSTEMS Excluding overheads and profit	PC £	Labour hours	Labour £	Plant £	Material £	Unit	Total rate £
R13 LAND DRAINAGE - cont'd							
Market prices of backfilling materials							
Sand	18.99	-	-	-	22.79	m³	22.79
Gravel rejects	22.68	-	-	-	24.95	m³	24.95
Topsoil; allowing for 20% settlement	8.33	-	-	-	10.00	m³	10.00
Agricultural drainage; backfilling trench after laying pipes with gravel rejects or similar; blind filling with ash or sand; topping with 150 mm topsoil from dumps not exceeding 100 m; by machine							
Width 150 mm							
depth 450 mm	-	3.30	49.50	49.38	119.47	100 m	218.35
depth 600 mm	-	4.30	64.50	64.56	173.92	100 m	302.98
depth 750 mm	-	4.96	74.40	74.58	204.52	100 m	353.50
depth 900 mm	-	6.30	94.50	94.92	272.56	100 m	461.98
Width 225 mm							
depth 450 mm	-	4.95	74.25	74.07	184.54	100 m	332.87
depth 600 mm	-	6.45	96.75	96.84	260.98	100 m	454.57
depth 750 mm	-	7.95	119.25	119.61	337.63	100 m	576.50
depth 900 mm	-	9.45	141.75	142.38	414.07	100 m	698.20
Width 375 mm							
depth 450 mm	-	8.25	123.75	123.45	307.43	100 m	554.64
depth 600 mm	-	10.75	161.25	161.40	434.90	100 m	757.55
depth 750 mm	-	13.25	198.75	199.35	562.58	100 m	960.69
depth 900 mm	-	15.75	236.25	237.30	690.05	100 m	1163.60
Agricultural drainage; backfilling trench after laying pipes with gravel rejects or similar, blind filling with ash or sand, topping with 150 mm topsoil from dumps not exceeding 100 m; by hand							
Width 150 mm							
depth 450 mm	-	18.63	279.45	-	115.72	100 m	395.17
depth 600 mm	-	24.84	372.60	-	173.92	100 m	546.52
depth 750 mm	-	31.05	465.75	-	185.78	100 m	651.53
depth 900 mm	-	37.26	558.90	-	272.56	100 m	831.46
Width 225 mm							
depth 450 mm	-	27.94	419.10	-	184.54	100 m	603.64
depth 600 mm	-	37.26	558.90	-	260.98	100 m	819.88
depth 750 mm	-	46.57	698.55	-	337.63	100 m	1036.18
depth 900 mm	-	55.89	838.35	-	414.07	100 m	1252.42
Width 375 mm							
depth 450 mm	-	46.57	698.55	-	307.43	100 m	1005.98
depth 600 mm	-	61.10	916.50	-	434.90	100 m	1351.40
depth 750 mm	-	77.63	1164.45	-	562.58	100 m	1727.03
depth 900 mm	-	93.15	1397.25	-	690.05	100 m	2087.30

R DISPOSAL SYSTEMS Excluding overheads and profit	PC £	Labour hours	Labour £	Plant £	Material £	Unit	Total rate £
Catchwater or french drains; 100 mm diameter non-coilable perforated plastic pipes; to BS4962; including straight jointing; pipes laid with perforations uppermost; lining trench; wrapping pipes with filter fabric							
Width 300 mm							
depth 450 mm	680.91	9.10	136.50	31.88	1028.44	100 m	1196.82
depth 600 mm	684.72	9.84	147.60	43.11	1150.27	100 m	1340.98
depth 750 mm	688.71	10.60	159.00	54.65	1272.31	100 m	1485.95
depth 900 mm	692.61	11.34	170.10	65.88	1394.24	100 m	1630.23
depth 1000 mm	695.21	11.84	177.60	73.47	1475.53	100 m	1726.60
depth 1200 mm	700.41	12.84	192.60	88.65	1638.11	100 m	1919.36
Width 450 mm							
depth 450 mm	686.76	10.44	156.60	79.54	1211.34	100 m	1447.48
depth 600 mm	692.61	11.34	170.10	65.88	1394.24	100 m	1630.23
depth 750 mm	698.46	12.46	186.90	82.88	1577.14	100 m	1846.93
depth 900 mm	704.31	13.60	204.00	100.19	1760.05	100 m	2064.23
depth 1000 mm	708.21	14.34	215.10	111.42	1881.98	100 m	2208.50
depth 1200 mm	716.01	15.84	237.60	134.19	2125.85	100 m	2497.64
depth 1500 mm	727.71	18.10	271.50	168.50	2491.65	100 m	2931.65
depth 2000 mm	747.21	21.84	327.60	225.27	3101.33	100 m	3654.20
Width 600 mm							
depth 450 mm	692.61	11.24	168.60	65.88	1394.24	100 m	1628.73
depth 600 mm	700.41	12.84	192.60	88.65	1638.11	100 m	1919.36
depth 750 mm	708.21	14.34	215.10	111.42	1881.98	100 m	2208.50
depth 900 mm	716.01	15.84	237.60	134.19	2125.85	100 m	2497.64
depth 1000 mm	723.44	16.84	252.60	149.37	2279.49	100 m	2681.46
depth 1200 mm	731.61	18.84	282.60	179.73	2613.59	100 m	3075.92
depth 1500 mm	747.21	21.84	327.60	420.27	3101.33	100 m	3849.20
depth 2000 mm	773.21	19.84	297.60	407.43	3914.22	100 m	4619.26
Width 900 mm							
depth 450 mm	704.31	13.60	204.00	100.19	1760.05	100 m	2064.23
depth 600 mm	716.01	7.92	118.80	134.19	2125.85	100 m	2378.84
depth 750 mm	727.71	9.05	135.75	168.50	2491.65	100 m	2795.90
depth 900 mm	739.41	10.17	152.55	202.50	2857.46	100 m	3212.51
depth 1000 mm	747.21	10.92	163.80	225.27	3101.33	100 m	3490.40
depth 1200 mm	762.81	12.42	186.30	270.81	3589.07	100 m	4046.18
depth 1500 mm	786.21	14.67	220.05	339.12	4320.67	100 m	4879.84
depth 2000 mm	825.21	18.92	283.80	517.97	5540.02	100 m	6341.79
depth 2500 mm	864.21	22.17	332.55	631.82	6759.37	100 m	7723.74
depth 3000 mm	903.21	25.92	388.80	680.67	7978.71	100 m	9048.18
Catchwater or french drains; 160 mm diameter non-coilable perforated plastic pipes; to BS4962; including straight jointing; pipes laid with perforations uppermost; lining trench; wrapping pipes with filter fabric							
Width 600 mm							
depth 450 mm	1272.62	11.20	168.00	63.76	1967.10	100 m	2198.85
depth 600 mm	1280.42	12.70	190.50	86.53	2210.96	100 m	2487.99
depth 750 mm	1288.22	14.20	213.00	109.30	2454.83	100 m	2777.13
depth 900 mm	1296.02	15.70	235.50	132.07	2698.70	100 m	3066.27
depth 1000 mm	1301.22	16.70	250.50	147.25	2861.28	100 m	3259.03
depth 1200 mm	1311.62	18.70	280.50	177.69	3186.44	100 m	3644.64

R DISPOSAL SYSTEMS Excluding overheads and profit	PC £	Labour hours	Labour £	Plant £	Material £	Unit	Total rate £
R13 LAND DRAINAGE - cont'd							
Catchwater or french drains; 160 mm diameter non-coilable perforated plastic pipes; to BS4962; including straight jointing; pipes laid with perforations uppermost; lining trench; wrapping pipes with filter fabric - cont'd							
depth 1500 mm	1327.22	21.70	325.50	223.15	3674.18	100 m	**4222.83**
depth 2000 mm	1353.22	26.70	400.50	299.05	4487.08	100 m	**5186.62**
depth 2500 mm	1379.22	31.70	475.50	374.95	5299.98	100 m	**6150.42**
depth 3000 mm	1405.22	36.70	550.50	450.85	6112.87	100 m	**7114.22**
Width 900 mm							
depth 450 mm	704.31	13.60	204.00	100.19	1760.05	100 m	**2064.23**
depth 600 mm	716.01	15.84	237.60	134.19	2125.85	100 m	**2497.64**
depth 750 mm	727.71	18.10	271.50	168.50	2491.65	100 m	**2931.65**
depth 900 mm	739.41	20.34	305.10	202.50	2857.46	100 m	**3365.06**
depth 1000 mm	747.21	21.84	327.60	225.27	3101.33	100 m	**3654.20**
depth 1200 mm	762.81	24.84	372.60	270.81	3589.07	100 m	**4232.48**
depth 1500 mm	786.21	29.34	440.10	339.12	4320.67	100 m	**5099.89**
depth 2000 mm	825.21	36.84	552.60	452.97	5540.02	100 m	**6545.59**
depth 2500 mm	864.21	44.34	665.10	566.82	6759.37	100 m	**7991.29**
depth 3000 mm	903.21	51.84	777.60	680.67	7978.71	100 m	**9436.98**
Catchwater or french drains; Exxon Geochemical Polymers; "Filtram" filter drain; in trenches (trenches not included); comprising filter fabric, liquid conducting core and 110 mm uPVC slitpipes; all in accordance with manufacturer's instructions; backfilling with clean broken stone 40 mm minimum; top 300 mm of fill to be gravel rejects or quarry waste							
Width 600 mm							
depth 1000 mm	721.21	16.84	252.60	149.37	2288.43	100 m	**2690.40**
depth 1200 mm	731.61	18.84	282.60	179.73	2613.59	100 m	**3075.92**
depth 1500 mm	747.21	21.84	327.60	420.27	3101.33	100 m	**3849.20**
depth 2000 mm	773.21	19.84	297.60	407.43	3914.22	100 m	**4619.26**
Width 900 mm							
depth 450 mm	704.31	13.60	204.00	100.19	1760.05	100 m	**2064.23**
depth 600 mm	716.01	15.84	237.60	134.19	2125.85	100 m	**2497.64**
depth 750 mm	727.71	18.10	271.50	168.50	2491.65	100 m	**2931.65**
depth 900 mm	739.41	20.34	305.10	202.50	2857.46	100 m	**3365.06**
depth 1000 mm	747.21	21.84	327.60	225.27	3101.33	100 m	**3654.20**
depth 1200 mm	762.81	24.84	372.60	270.81	3589.07	100 m	**4232.48**
depth 1500 mm	786.21	29.34	440.10	339.12	4320.67	100 m	**5099.89**
depth 2000 mm	825.21	37.84	567.60	517.97	5540.02	100 m	**6625.59**
depth 2500 mm	864.21	44.34	665.10	631.82	6759.37	100 m	**8056.29**
depth 3000 mm	903.21	51.84	777.60	745.67	7978.71	100 m	**9501.98**

R DISPOSAL SYSTEMS Excluding overheads and profit	PC £	Labour hours	Labour £	Plant £	Material £	Unit	Total rate £
Outfalls							
Reinforced concrete outfalls to water course; flank walls; for 150 mm drain outlets; overall dimensions							
900 mm x 1050 mm x 900 mm high	-	-	-	-	-	m	424.00
Soakaway design based on BS EN 752-4. Flat rate hourly rainfall = 50 mm /hr and assumes 100 impermeability of the run-off area. A storage capacity of the soakaway should be 1/3 of the hourly rainfall; Formulae for calculating soakaway depths are provided in the publications mentioned below and in the memoranda section of this publication. The design of soakaways is dependent on amongst other factors, soil conditions, permeability, groundwater level and runoff. The definitive documents for design of soakaways are CIRIA 156 and BRE Digest 365 dated September 1991; The suppliers of the systems below will assist through their technical divisions. Excavation earthwork support of pits and disposal not included.							
Excavating; mechanical							
To reduce levels							
maximum depth not exceeding 1.00 m; JCB sitemaster	-	0.05	0.75	1.52	-	m³	2.27
maximum depth not exceeding 1.00 m; 360 Tracked excavator	-	0.04	0.60	1.19	-	m³	1.79
maximum depth not exceeding 2.00 m; 360 Tracked excavator	-	0.06	0.90	1.78	-	m³	2.68
Disposal; mechanical							
Excavated material; off site; to tip not exceeding 13 km; mechanically loaded.							
inert material	-	0.03	0.50	1.01	10.00	m³	11.51
Insitu concrete ring beam foundations to base of soakaway; 300 mm wide x 250 deep; poured on or against earth or unblinded hardcore							
internal diameters of rings							
900 mm	18.88	4.00	60.00	-	20.77	nr	80.77
1200 mm	25.17	4.50	67.50	-	27.69	nr	95.19
1500 mm	31.46	5.00	75.00	-	34.60	nr	109.60
2400 mm	50.34	5.50	82.50	-	55.37	nr	137.87

R DISPOSAL SYSTEMS Excluding overheads and profit	PC £	Labour hours	Labour £	Plant £	Material £	Unit	Total rate £
R13 LAND DRAINAGE - cont'd							
Concrete soakaway rings; Milton Pipes Ltd; perforations and step irons to concrete rings at manufacturers recommended centres; placing of concrete ring soakaways to in situ concrete ring beams (1:3:6) (not included); filling and surrounding base with gravel 225 deep (not included);							
Ring diameter 900 mm							
1.00 deep; volume 636 litres	31.00	1.25	18.75	13.01	99.95	nr	131.71
1.50 deep; volume 954 litres	46.50	1.65	24.75	34.34	147.85	nr	206.94
2.00 deep; volume 1272 litres	62.00	1.65	24.75	34.34	195.75	nr	254.84
Ring diameter 1200 mm							
1.00 deep; volume 1131 litres	46.00	3.00	45.00	31.22	125.77	nr	201.99
1.50 deep; volume 1696 litres	69.00	4.50	67.50	46.82	184.97	nr	299.30
2.00 deep; volume 2261 litres	92.00	4.50	67.50	46.82	244.17	nr	358.50
2.50 deep; volume 2827 litres	115.00	6.00	90.00	62.43	276.77	nr	429.21
Ring diameter 1500 mm							
1.00 deep; volume 1767 litres	84.00	4.50	67.50	31.22	183.13	nr	281.85
1.50 deep; volume 2651 litres	126.00	6.00	90.00	43.70	268.93	nr	402.63
2.00 deep; volume 3534 litres	168.00	6.00	90.00	43.70	354.73	nr	488.43
2.50 deep; volume 4418 litres	115.00	7.50	112.50	78.04	345.53	nr	536.07
Ring diameter 2400 mm							
1.00 deep; volume 4524 litres	214.00	6.00	90.00	31.22	347.33	nr	468.55
1.50 deep; volume 6786 litres	321.00	7.50	112.50	43.70	513.33	nr	669.53
2.00 deep; volume 9048 litres	428.00	7.50	112.50	43.70	683.13	nr	839.33
2.50 deep; volume 11310 litres	535.00	9.00	135.00	78.04	849.13	nr	1062.17
Extra over for							
250 mm depth chamber ring	-	-	-	-	-	100 %	
500 mm depth chamber ring	-	-	-	-	-	50 %	
Cover slabs to soakaways							
heavy duty precast concrete							
900 diameter	53.00	1.00	15.00	15.61	53.00	nr	83.61
1200 diameter	68.00	1.00	15.00	15.61	68.00	nr	98.61
1500 diameter	108.00	1.00	15.00	15.61	108.00	nr	138.61
2400 diameter	383.00	1.00	15.00	15.61	383.00	nr	413.61
Step irons to concrete chamber rings.	19.20	-	-	-	19.20	m	19.20
Extra over soakaways for filter wrapping with a proprietary filter membrane							
900 mm diameter x 1.00 m deep	1.23	1.00	15.00	-	1.53	nr	16.53
900 mm diameter x 2.00 m deep	2.45	1.50	22.50	-	3.07	nr	25.57
1050 mm diameter x 1.00 m deep	1.43	1.50	22.50	-	1.79	nr	24.29
1050 mm diameter x 2.00 m deep	2.86	2.00	30.00	-	3.57	nr	33.57
1200 mm diameter x 1.00 m deep	1.63	2.00	30.00	-	2.04	nr	32.04
1200 mm diameter x 2.00 m deep	3.27	2.50	37.50	-	4.08	nr	41.58
1500 mm diameter x 1.00 m deep	2.04	2.50	37.50	-	2.55	nr	40.05
1500 mm diameter x 2.00 m deep	4.07	2.50	37.50	-	5.09	nr	42.59
1800 mm diameter x 1.00 m deep	2.45	3.00	45.00	-	3.06	nr	48.06
1800 mm diameter x 2.00 m deep	4.90	3.25	48.75	-	6.12	nr	54.87

R DISPOSAL SYSTEMS Excluding overheads and profit	PC £	Labour hours	Labour £	Plant £	Material £	Unit	Total rate £
Gravel surrounding to concrete ring soakaway 40 mm aggregate backfilled to vertical face of soakaway wrapped with geofabric (not included) 250 thick	26.10	0.20	3.00	6.24	26.10	m³	35.34
Backfilling to face of soakaway; **Carefully compacting as work proceeds** arising from the excavations average thickness exceeding 0.25 m; depositing in layers 150 mm maximum thickness	-	0.03	0.50	3.12	-	m³	3.62
"Aquacell" soakaway; Wavin Plastics Ltd; **Preformed polypropylene soakaway infiltration crate units; to trenches; surrounded by geotextile and 40 mm aggregate laid 100 thick; in trenches (Excavation, disposal and backfilling not included)** 1.00 x 500 x 400; internal volume 190 litres							
4 crates; 2.00 x 1.00 x 400; 760 litres	108.00	1.60	24.00	12.75	142.92	nr	179.67
8 crates; 2.00 x 1.00 x 800; 1520 litres	216.00	3.20	48.00	16.39	259.68	nr	324.08
12 crates; 6.00 x 500 x 800; 2280 litres	324.00	4.80	72.00	34.00	410.04	nr	516.05
16 crates; 4.00 x 1.00 x 800; 3040 litres	432.00	6.40	96.00	30.36	509.28	nr	635.64
20 crates; 5.00 x 1.00 x 800; 3800 litres	540.00	8.00	120.00	30.06	620.79	nr	770.85
30 crates; 15.00 x 1.00 x 400; 5700 litres	810.00	12.00	180.00	81.97	1015.69	nr	1277.66
60 crates; 15.00 x 1.00 x 800; 11400 litres	1620.00	20.00	300.00	107.17	1894.79	nr	2301.96
Geofabric surround to Aquacell units; **Terram Ltd** "Terram" synthetic fibre filter fabric; to face of concrete rings (not included); anchoring whilst backfilling (not included) "Terram 1000", 0.70 mm thick; mean water flow 50 litre/m²/s	0.43	0.05	0.75	-	0.52	m²	1.27

S PIPED SUPPLY SYSTEMS Excluding overheads and profit	PC £	Labour hours	Labour £	Plant £	Material £	Unit	Total rate £
S10 COLD WATER							
Blue MDPE polythene pipes; type 50; for cold water services; with compression fittings; bedding on 100 mm DOT type 1 granular fill material							
Pipes							
20 mm diameter	0.70	0.08	1.20	-	2.27	m	3.47
25 mm diameter	0.90	0.08	1.20	-	2.48	m	3.68
32 mm diameter	0.81	0.08	1.20	-	2.38	m	3.58
50 mm diameter	1.40	0.10	1.50	-	2.99	m	4.49
60 mm diameter	3.03	0.10	1.50	-	4.66	m	6.16
Hose union bib taps; to BS 5412; including fixing to wall; making good surfaces							
15 mm	10 56	0.75	11.25	-	12.34	nr	23.59
22 mm	14.90	0.75	11.25	-	17.57	nr	28.82
Stopcocks; to BS 1010; including fixing to wall; making good surfaces							
15 mm	6.26	0.75	11.25	-	8.04	nr	19.29
22 mm	9.76	0.75	11.25	-	11.54	nr	22.79
Standpipes; to existing 25mm water mains							
1.00 m high	-	-	-	-	-	nr	85.00
Hose junction bib taps; to standpipes							
19 mm	-	-	-	-	-	nr	33.00
S14 IRRIGATION							
Leaky Pipe Systems Ltd; " Leaky Pipe"; moisture leaking pipe irrigation system.							
Main supply pipe inclusive of machine excavation; exclusive of connectors							
20 mm LDPE Polytubing	0.84	0.05	0.75	0.57	0.86	m	2.18
16 mm LDPE Polytubing	0.56	0.05	0.75	0.57	0.57	m	1.89
Water filters and cartridges							
No 10; 20 mm	-	-	-	-	41.90	nr	41.90
Big Blue and RR30 cartridge; 25 mm	-	-	-	-	107.20	nr	107.20
Water filters and pressure regulator sets. Complete assemblies							
No 10; Flow rate 3.1 - 82 litres per minute	-	-	-	-	71.50	nr	71.50
Leaky pipe hose; placed 150 mm sub surface for turf irrigation							
Distance between laterals 350 mm; excavation and backfilling priced separately							
LP12L low leak	3.65	0.04	0.60	-	3.65	m²	4.25
LP12H high leak	3 16	0.04	0.60	-	3.16	m²	3.76
LP12UH ultra high leak	3.85	0.04	0.60	-	3.85	m²	4.45
Leaky pipe hose; laid to surface for landscape irrigation. Distance between laterals 600 mm							
LP12L low leak	2.12	0.03	0.38	-	2.18	m²	2.55
LP12H high leak	1.84	0.03	0.38	-	1.89	m²	2.26
LP12UH ultra high leak	2.24	0.03	0.38	-	2.30	m²	2.67
Leaky pipe hose; laid to surface for landscape irrigation. Distance between laterals 900 mm							
LP12L low leak	1.42	0.02	0.25	-	1.46	m²	1.71
LP12H high leak	1.23	0.02	0.25	-	1.26	m²	1.51
LP12UH ultra high leak	1.50	0.02	0.25	-	1.54	m²	1.79

S PIPED SUPPLY SYSTEMS Excluding overheads and profit	PC £	Labour hours	Labour £	Plant £	Material £	Unit	Total rate £
Leaky pipe hose; laid to surface for tree irrigation laid around circumference of tree pit.							
LP12L low leak	1.96	0.13	1.88	-	2 15	nr	4.03
LP12H high leak	1.70	0.13	1.88	-	1.87	nr	3.74
LP12UH ultra high leak	2.07	0.13	1.88	-	2.27	nr	4.15
Accessories							
Automatic multi-station controller stations inclusive of connections	289.00	2.00	30.00	-	289.00	nr	319.00
Solenoid valves inclusive of wiring and connections to a multi-station controller; nominal distance from controller 25 m	48.00	0.50	7.50	-	398.00	nr	405.50

S15 FOUNTAINS/WATER FEATURES

Lakes and ponds - General

Preamble: The pressure of water against a retaining wall or dam is considerable, and where water retaining structures form part of the design of water features, the landscape architect is advised to consult a civil engineer. Artificially contained areas of water in raised reservoirs over 25,000 m³ have to be registered with the local authority, and their dams will have to be covered by a civil engineer's certificate of safety.

Typical linings - General

Preamble: In addition to the traditional methods of forming the linings of lakes and ponds in puddled clay or concrete, there are a number of lining materials available. They are mainly used for reservoirs but can also help to form comparatively economic water features especially in soil which is not naturally water retentive. Information on the construction of traditional clay puddle ponds can be obtained from the British Trust for Conservation Volunteers, 36 St. Mary's Street, Wallingford, Oxfordshire. OX10 0EU. Tel: (01491) 39766. The cost of puddled clay ponds depends on the availability of suitable clay, the type of hand or machine labour that can be used, and the use to which the pond is to be put.

S PIPED SUPPLY SYSTEMS Excluding overheads and profit	PC £	Labour hours	Labour £	Plant £	Material £	Unit	Total rate £
S15 FOUNTAINS/WATER FEATURES - cont'd							
Lake liners; Fairwater Water Gardens Ltd; to evenly graded surface of excavations (excavating not included); all stones over 75 mm; removing debris; including all welding and jointing of liner sheets							
Geotextile underlay; inclusive of spot welding to prevent dragging							
to water features	-	-	-	-	-	m²	1.80
to lakes or large features	-	-	-	-	-	1000m²	1520.00
Butyl rubber liners; "Varnamo" inclusive of site vulcanising							
0.75 mm thick	-	-	-	-	-	m²	5.70
0.75 mm thick	-	-	-	-	-	1000m²	5530.00
1.00 mm thick	-	-	-	-	-	m²	6.45
1.00 mm thick	-	-	-	-	-	1000m²	6280.00
Lake liners; Landline Ltd; "Landflex" or "Alkorplan" geomembranes; to prepared surfaces (surfaces not included); all joints fully welded; installation by Landline employees							
"Landflex HC" polyethylene geomembranes							
0.50 mm thick	-	-	-	-	-	1000m²	2800.00
0 75 mm thick	-	-	-	-	-	1000m²	3200.00
1 00 mm thick	-	-	-	-	-	1000m²	3600.00
1.50 mm thick	-	-	-	-	-	1000m²	4100.00
2.00 mm thick	-	-	-	-	-	1000m²	4550.00
"Alkorplan PVC" geomembranes							
0.80 mm thick	-	-	-	-	-	1000 m	4500.00
1.20 mm thick	-	-	-	-	-	1000 m	6750.00
Lake liners; Rawell Marketing; to prepared graded surfaces (surfaces not included); maximum slope 2 horizontal:1 vertical; in accordance with manufacturer's instructions "Rawmat" bentonite sheets; 200 mm lap joints; spreading 150 mm thick screened approved topsoil; "Rawmat" to be bedded in anchor trenches round perimeter (excavation and material to trenches not included)	774.90	4.00	60.00	24.64	929.88	100 m²	1014.52
Lake liners; Monarflex Polyethylene lake and reservoir lining system; welding on site by Monarflex technicians (surface preparation and backfilling not included)							
"Blackline"; 500 micron	-	-	-	-	-	100 m²	393.00
"Blackline"; 750 micron	-	-	-	-	-	100 m²	526.00
"Blackline"; 1000 micron	-	-	-	-	-	100 m²	560.00

S PIPED SUPPLY SYSTEMS Excluding overheads and profit	PC £	Labour hours	Labour £	Plant £	Material £	Unit	Total rate £
Operations over surfaces of lake liners							
Dug ballast; evenly spread over excavation already brought to grade							
150 mm thick	407.10	2.00	30.00	49.28	447.81	100 m²	527.09
200 mm thick	542.80	3.00	45.00	73.92	597.08	100 m²	716.00
300 mm thick	814.20	3.50	52.50	86.24	895.62	100 m²	1034.36
Imported topsoil; evenly spread over excavation							
100 mm thick	116.60	1.50	22.50	36.96	139.92	100 m²	199.38
150 mm thick	174.90	2.00	30.00	49.28	209.88	100 m²	289.16
200 mm thick	233.20	3.00	45.00	73.92	279.84	100 m²	398.76
Blinding existing subsoil with 50 mm sand	131.05	1.00	15.00	49.28	144.16	100 m²	208.44
Topsoil from excavation; evenly spread over excavation							
100 mm thick	-	-	-	36.96	-	100 m²	36.96
200 mm thick	-	-	-	49.28	-	100 m²	49.28
300 mm thick	-	-	-	73.92	-	100 m²	73.92
Extra over for screening topsoil using a "Powergrid screener; removing debris	-	-	-	3.22	0.50	m³	3.72
Waterfall construction; Fairwater Water Gardens Ltd;							
Stone placed on top of butyl liner; securing with concrete and dressing to form natural rock pools and edgings							
Portland stone m³ rate	-	-	-	-	-	m³	518.00
Portland stone tonne rate	-	-	-	-	-	tonne	288.00
Balancing tank; blockwork construction; inclusive of recirculation pump and pond level control; 110 mm balancing pipe to pond; waterproofed with butyl rubber membrane; pipework mains water top-up and overflow							
450 x 600 x 1000 mm	-	-	-	-	-	nr	923.00
extra for pump							
2000 gallons per hour; submersible	-	-	-	-	-	nr	195.00
Ornamental pools - General							
Preamble: Small pools may be lined with one of the materials mentioned under lakes and ponds, or may be in rendered brickwork, puddled clay or, for the smaller sizes, fibreglass. Most of these tend to be cheaper than waterproof concrete. Basic prices for various sizes of concrete pools are given in the approximate estimates section (book only). Prices for excavation, grading, mass concrete, and precast concrete retaining walls are given in the relevant sections. The manufacturers should be consulted before specifying the type and thickness of pool liner, as this depends on the size, shape and proposed use of the pool. The manufacturer's recommendation on foundations and construction should be followed.							

S PIPED SUPPLY SYSTEMS Excluding overheads and profit	PC £	Labour hours	Labour £	Plant £	Material £	Unit	Total rate £
S15 FOUNTAINS/WATER FEATURES - cont'd							
Ornamental pools							
Pool liners; to 50 mm sand blinding to excavation (excavating not included); all stones over 50 mm; removing debris from surfaces of excavation; including all welding and jointing of liner sheets							
black polythene; 1000 gauge	-	-	-	-	-	m²	2.18
blue polythene; 1000 gauge	-	-	-	-	-	m²	2.30
coloured PVC; 1500 gauge	-	-	-	-	-	m²	2.88
black PVC; 1500 gauge	-	-	-	-	-	m²	2.55
black butyl; 0.75 mm thick	-	-	-	-	-	m²	11.31
black butyl; 1.00 mm thick	-	-	-	-	-	m²	15.57
black butyl; 1.50 mm thick	-	-	-	-	-	m²	24.76
Fine gravel; 100 mm; evenly spread over area of pool; by hand	2.75	0.13	1.88	-	2.75	m²	4.63
Selected topsoil from excavation; 100 mm; evenly spread over area of pool; by hand	-	0.13	1.88	-	-	m²	1.88
Extra over selected topsoil for spreading imported topsoil over area of pool; by hand	11.66	-	-	-	13.99	m³	13.99
Pool surrounds and ornament; Haddonstone Ltd; Portland Bath or Terracotta cast stone							
Pool surrounds; Installed to pools or water feature construction priced separately. Surrounds and copings to 112.5 mm internal brickwork							
C4HS; half small pool surround; internal diameter 1780 mm; kerb features continuous moulding enriched with ovolvo and palmette designs; inclusive of plinth and integral conch shell vases flanked by dolphins	905.52	16.00	240.00	-	917.42	nr	1157.42
C4S; small pool surround as above but with full circular construction; internal diameter 1780 mm	1650.19	48.00	720.00	-	1665.32	nr	2385.32
C4M; medium pool surround; internal diameter 2705 mm; inclusive of plinth and integral vases	2042.52	48.00	720.00	-	2074.24	nr	2794.24
C4XL; extra large pool surround; internal diameter 5450 mm	4085.04	140.00	2100.00	-	4122.86	nr	6222.86
Pool centre pieces and fountains; inclusive of plumbing and pumps							
HC350; Lotus bowl; 1830 wide with C1700 triple dolphin fountain; C180 pedestal	3111.49	8.00	120.00	-	3114.22	nr	3234.22
C251; Gothic Fountain and Gothic Upper Base A350; freestanding fountain	566.81	4.00	60.00	-	569.54	nr	629.54
HC521; Romanesque Fountain; freestanding bowl with self circulating fountain; filled with cobbles; 815 mm diameter; 348 mm high	276.60	2.00	30.00	-	332.73	nr	362.73
C300; Lion fountain 610 mm high on fountain base C305; 280 mm high	234.03	2.00	30.00	-	236.76	nr	266.76

S PIPED SUPPLY SYSTEMS Excluding overheads and profit	PC £	Labour hours	Labour £	Plant £	Material £	Unit	Total rate £
Wall Fountains, Watertanks and Fountains							
Capital Garden Products							
Lion Wall Fountain F010; 965 x 940 x 520	-	-	-	-	253.22	nr	253.22
Dolphin Wall Fountain F001; 735 x 510	-	-	-	-	78.64	nr	78.64
Dutch Master Wall Fountain F012; 740 x 410	-	-	-	-	76.25	nr	76.25
Tank 2801, James II design; 725 x 725 x 760 h - 308 litres	-	-	-	-	207.42	nr	207.42
James II Tank and Fountain 4901bp; 725 x 1245 x 760 h - 557 litres	-	-	-	-	474.30	nr	474.30
Pool Fountain F005; Crane Catching Fish; 635 high	-	-	-	-	187.53	nr	187.53
Pool Fountain F004; Mannekin Pis; 340 high	-	-	-	-	55.69	nr	55.69
Fountain kits; typical prices of submersible units comprising fountain pumps, fountain nozzles, underwater spotlights, nozzle extension armatures, underwater terminal boxes and electrical control panels							
Single aerated white foamy water columns; ascending jet 70 mm diameter; descending water up to four times larger; jet height adjustable between 1.00 m and 1.70 m	3600.00	-	-	-	4170.00	nr	4170.00
Single aerated white foamy water columns, ascending jet 110 mm diameter, descending water up to four times larger; jet height adjustable between 1.50 m and 3.00 m	6300.00	-	-	-	7180.02	nr	7180.02

V ELEC. SUPPLY/POWER/LIGHTING SYSTEMS Excluding overheads and profit	PC £	Labour hours	Labour £	Plant £	Material £	Unit	Total rate £
V41 STREET AREA FLOODLIGHTING							
Street area floodlighting - Generally Preamble: There are an enormous number of luminaires available which are designed for small scale urban and garden projects. The designs are continually changing and the landscape designer is advised to consult the manufacturer's latest catalogue. Most manufacturers supply light fittings suitable for column, bracket, bulkhead, wall or soffit mounting. Highway lamps and columns for trafficked roads are not included in this section as the design of highway lighting is a very specialised subject outside the scope of most landscape contracts. The IP reference number refers to the waterproof properties of the fitting; the higher the number the more waterproof the fitting. Most items can be fitted with time clocks or PIR controls.							
Market Prices of Lamps Lamps							
42w TCT	-	-	-	-	16.48	nr	16.48
125w MBFU	-	-	-	-	20.11	nr	20.11
75w MBIF	-	-	-	-	57.02	nr	57.02
70w HQI-TS	-	-	-	-	57 02	nr	57.02
70w SON	-	-	-	-	54.00	nr	54.00
70w SONT	-	-	-	-	54.00	nr	54.00
100w SONT	-	-	-	-	65.79	nr	65.79
150w SONT	-	-	-	-	65.79	nr	65.79
10/13w TCD	-	-	-	-	9.83	nr	9.83
28w 2D	-	-	-	-	10.19	nr	10.19
100w GLS\E27	-	-	-	-	0.87	nr	0.87
Wall brackets and lanterns; including fixing to wall; (lamps, electric wiring, connections or related fixtures such as switch gear and time clock mechanisms not included unless otherwise indicated) Wall brackets and lanterns; Sugg Lighting							
762 mm "Westminster"; on large ornate bracket	783.75	0.50	7.50	-	786.42	nr	793.92
533 mm "Westminster"; on 1372 mm bow bracket	747 17	0.50	7.50	-	749.84	nr	757.34
406 mm "Classic Globe" on "Universal" plinth	339.63	0.50	7.50	-	342.30	nr	349.80
356 mm "Windsor" in "Abbey" cradle bracket	339.63	0.50	7.50	-	342.30	nr	349.80
305 mm "Windsor" in "Contract" cradle bracket	245.57	0.50	7.50	-	248.24	nr	255.74
406 mm "Windsor" on "Abbey" bracket	397.10	0.50	7.50	-	399.77	nr	407.27
Wall brackets and lanterns; The Woodhouse Company							
ref SX711; 90 mm diameter x 570 mm high x 205mm projection; TC-D26W; to IP 44	267.00	0.50	7.50	-	269.67	nr	277.17
ref SX442; "Alpha" lantern cone-shaped light; aluminium dome; 630 mm diameter x 710 mm high x 815 mm projection; HSE 70W; to IP 44	811.00	0.50	7.50	-	813.67	nr	821.17

V ELEC. SUPPLY/POWER/LIGHTING SYSTEMS Excluding overheads and profit	PC £	Labour hours	Labour £	Plant £	Material £	Unit	Total rate £
ref SX462; "Delta" lantern half-globe light; shallow metal dome; 540 mm diameter; 815 mm projection o/a; HME 125W; to IP 44	644.00	0.50	7.50	-	646.67	nr	654.17
ref SX 472; "Saturn" lantern cylindrical light; flat metal dome; 660 mm high; 915 mm projection o/a; HSE 80W; to IP 44	744.00	0.50	7.50	-	746.67	nr	754.17
ref SX 476; "Saturn" lantern cylindrical light; flat metal dome; 490 mm high; 915 mm projection o/a; HSE 80W; to IP 44	628.00	0.50	7.50	-	630.67	nr	638.17
ref SX120; "Kugel" plain spherical light; clear; opal; 300 mm diameter; 236 projection; TC-D 13W; to IP 44	140.00	0.50	7.50	-	142.67	nr	150.17
Marlin Lighting; Skyline 76 mm diameter column mounting; 196 mm diameter 677 mm high to IP64; Black							
ref SKYC 4206P; 42w TCT	226.66	1.00	15.00	-	226.66	nr	241.66
ref SKYC 7506H; 75w MBIF	226.66	1.00	15.00	-	226.66	nr	241.66
ref SKYC 1006N; 70w SONT	226.66	1.00	15.00	-	226.66	nr	241.66
ref SKYC 1006H; 100w MBIF	244.97	1.00	15.00	-	244.97	nr	259.97
ref SKYC 1006N; 100w SONT	244.97	1.00	15.00	-	244.97	nr	259.97
ref SKYC 1506N; 150w SONT	276.37	1.00	15.00	-	276.37	nr	291.37
Marlin Lighting; Skyline 76 mm diameter column mounting; 196 mm diameter 677 mm high to IP64; Aluminium							
ref SKYC 4212P; 42w TCT	226.66	1.00	15.00	-	226.66	nr	241.66
ref SKYC 7512H; 75w MBIF	226.66	1.00	15.00	-	226.66	nr	241.66
ref SKYC 7012N; 70w SONT	226.66	1.00	15.00	-	226.66	nr	241.66
ref SKYC 12512M; 125w MBFU	244.97	1.00	15.00	-	244.97	nr	259.97
ref SKYC 1012H; 100w MBIF	244.97	1.00	15.00	-	244.97	nr	259.97
ref SKYC 1012N; 100w SONT	244.97	1.00	15.00	-	244.97	nr	259.97
ref SKYC 1512N; 150w SONT	276.37	1.00	15.00	-	276.37	nr	291.37
Marlin Lighting; Skyline; Optional canopies							
Canopy 1 ref SKYAH106; black or aluminium	-	-	-	-	56.98	nr	56.98
Canopy 2 ref SKYAH206; black or aluminium	-	-	-	-	113.94	nr	113.94
Wall brackets and lanterns; Noral							
"Nova III"; aluminium hooded globe on swan neck; 330 mm diameter x 460 mm projection; 18W SL; to IP 44	129.00	0.50	7.50	-	131.67	nr	139.17
"Nove IV"; aluminium hooded globe on swan neck; 420 mm diameter x 359 mm high x 588 mm projection; 18W SL; to IP 44	207.00	0.50	7.50	-	209.67	nr	217.17
"Expo IV" polycarbonate globe in aluminium cradle; 433 mm diameter x 523 mm high x 547 mm projection o/a; 26W SL; to IP 44	268.00	0.50	7.50	-	270.67	nr	278.17

V ELEC. SUPPLY/POWER/LIGHTING SYSTEMS Excluding overheads and profit	PC £	Labour hours	Labour £	Plant £	Material £	Unit	Total rate £
V41 STREET AREA FLOODLIGHTING - cont'd							
Bulkhead and canopy fittings; including fixing to wall and light fitting (lamp, final painting, electric wiring, connections or related fixtures such as switch gear and time clock mechanisms not included unless otherwise indicated)							
Bulkhead and canopy fittings; Louis Poulsen "Nyhavn Wall" small domed top conical shade with rings; copper wall lantern; finished untreated copper to achieve verdigris finish; also available in white aluminium; 310 mm diameter shade; with wall mounting arm; to IP 44	347.11	0.50	7.50	-	349.78	nr	357.28
"Skot Wall" bulkhead of diecast aluminium; diffuser of clear or opal polycarbonate; masked diffuser available for directional lighting; 239 mm diameter; to IP 66	115.47	0.50	7.50	-	118.14	nr	125.64
Bulkhead and canopy fittings; Sugg Lighting							
"Princess" backlamp in brass	120.00	0.50	7.50	-	122.67	nr	130.17
"Victoria" backlamp in copper	200.00	0.50	7.50	-	202.67	nr	210.17
"Palace" backlamp in brass	-	0.50	7.50	-	202.67	nr	210.17
"Windsor" backlamp	150.00	0.50	7.50	-	152.67	nr	160.17
Bulkhead fittings; Marlin Lighting; "MONITOR"; surface mounted; vandal resistant; range of finishes; to IP55							
ref MON386; 38w 2D	122.59	1.00	15.00	-	122.59	nr	137.59
ref MON 706; 70w SON; lamp with internal ignitor	139.50	1.00	15.00	-	139.50	nr	154.50
ref MON386HFM³; 38w 2D with integral emergency gear 3 hour	268.01	1.00	15.00	-	268.01	nr	283.01
Bulkhead fittings; Marlin Lighting; "STERLING"; surface mounted; vandal resistant; black to IP55							
ref SGD 286; 28w 2D	91.97	1.00	15.00	-	91.97	nr	106.97
ref SGD 286M³; 28w 2D; with integral HPF/ and 3hr emergency gear	218.76	1.00	15.00	-	218.76	nr	233.76
Bulkhead fittings; Marlin Lighting; "VEDO"; to IP55; white							
ref SIM 6709-01 18w TCD; oval	79.79	1.00	15.00	-	79.79	nr	94.79
ref SIM 6749-01 18w TCD; bis	90.00	1.00	15.00	-	90.00	nr	105.00
ref SIM 6759-01 18w TCD;hood	95.77	1.00	15.00	-	95.77	nr	110.77
ref SIM 6219-01 18wTCD; oval angled	191.65	1.00	15.00	-	191.65	nr	206.65
Bulkhead fittings; Marlin Lighting; "WIP"; to IP65; surface/recessed bulkhead; integral gear; aluminium black							
ref SIM 4349-09 18w TCD	104.63	1.00	15.00	-	104.63	nr	119.63
ref SIM 4343-12; recommended wall recessing box	13.05	1.00	15.00	-	13.05	nr	28.05
ref SIM 4350-09; bollard housing	173.84	1.00	15.00	-	173.84	nr	188.84
ref SIM 4372; bollard base plate	24.33	0.08	1.25	-	24.33	nr	25.58
ref SIM 4345-09; 45 deg directional grille	16.58	0.08	1.25	-	16.58	nr	17.83
ref SIM 4344-09; standard grille	12.56	0.08	1.25	-	12.56	nr	13.81
ref SIM V0049; polycarbonate diffuser	6.76	0.08	1.25	-	6.76	nr	8.01

V ELEC. SUPPLY/POWER/LIGHTING SYSTEMS Excluding overheads and profit	PC £	Labour hours	Labour £	Plant £	Material £	Unit	Total rate £
Canopy fittings; Marlin Lighting; "FLAT"; to IP65; flat, vandal resistant, under canopy downlight; adjustable tilt; integral/remote gear; toughened 15mm glass; aluminium recessing stirrup and removable anti-glare louvre							
ref SIM 4442-09 70w CDMT/HQIT c/w remote control gear	369.58	1.00	15.00	-	369.58	nr	384.58
ref SIM 4449-01 18w TCT	197.36	1.00	15.00	-	197.36	nr	212.36
ref SIM 4441-09 75w PAR 30	158.92	1.00	15.00	-	158.92	nr	173.92
Floodlighting; ground, wall or pole mounted; including fixing, (lamp, final painting, electric wiring, connections or related fixtures such as switch gear and time clock mechanisms not included unless otherwise indicated)							
Floodlighting; Outdoor Lighting (OLS)							
"Compact Flood"; ref FL-01; multi purpose ground / spike mounted wide angle floodlight; 70w HIT or SON; to IP 65	225.00	1.00	15.00	-	230.48	nr	245.48
"Morph 38"; ref SP-07; mains voltage spotlight; 60/80/120 w PAR 38; to IP 54	42.23	1.00	15.00	-	47.71	nr	62.71
Floodlight Accessories; Outdoor Lighting (OLS)							
"ES"; earth spike for Compact Flood	37.00	-	-	-	37.00	nr	37.00
"ES"; earth spike for Morph 38	5.15	-	-	-	5.15	nr	5.15
"L" Louvre for Compact Flood	44.00	-	-	-	44.00	nr	44.00
"L38"; Louvre for Morph 38	19.05	-	-	-	19.05	nr	19.05
"C ", Cowl for Compact Flood	28.00	-	-	-	28.00	nr	28.00
"BD "; Barn doors for Compact Flood	60.73	-	-	-	60.73	nr	60.73
Large area/pitch floodlighting; CU Phosco;							
ref FL444 1000w SON-T; floodlights with lamp and loose gear; narrow asymmetric beam	338.61	1.00	15.00	-	338.61	nr	353.61
ref FL444 2.0kw MBIOS; floodlight with lamp and loose gear; projector beam	589.37	1.00	15.00	-	589.37	nr	604.37
Large area floodlighting; CU Phosco;							
ref FL345/G/100S; floodlight with lamp and integral gear	240.16	1.00	15.00	-	240.16	nr	255.16
ref FL345/G/250S; floodlight with lamp and integral gear	242.47	1.00	15.00	-	242.47	nr	257.47
ref FL345/G/400 MBI; floodlight with lamp and integral gear	263.66	1.00	15.00	-	263.66	nr	278.66
Small area floodlighting; CU Phosco;							
ref FL325/G/50S; floodlight with lamp and integral gear	239.08	1.00	15.00	-	239.08	nr	254.08
ref FL325/G/110S; floodlight with lamp and integral gear	237.89	1.00	15.00	-	237.89	nr	252.89
ref FL325/G/80 MBF; floodlight with lamp and integral gear	224.41	1.00	15.00	-	224.41	nr	239.41
Pole clamp	19.98	-	-	-	19.98	nr	19.98
ref B76; spigot cap	6.80	-	-	-	6.80	nr	6.80
"D" wall bracket	14.58	-	-	-	14.58	nr	14.58

V ELEC. SUPPLY/POWER/LIGHTING SYSTEMS Excluding overheads and profit	PC £	Labour hours	Labour £	Plant £	Material £	Unit	Total rate £
V41 STREET AREA FLOODLIGHTING - cont'd							
Spotlights for uplighting and for illuminating signs and notice boards, statuary and other features); ground, wall or pole mounted; including fixing, light fitting and priming (lamp, final painting, electric wiring, connections or related fixtures such as switch gear and time clock mechanisms not included); All mains voltage (240v) unless otherwise stated							
Spotlights; Outdoor Lighting (OLS)							
Maxispotter" miniature solid copper spotlight designed to patternate and age naturally; with mounting bracket; and internal anti glare louvre; low voltage halogen reflector; 20/35/50w; to IP56;	67.98	1.00	15.00	-	73.46	nr	**88.46**
"WeeBee Spot"; ref SP-05 for Low voltage halogen reflector; 20/35/50w; c/w integral transformer and wall mounting box; to IP 65	78.95	1.00	15.00	-	84.43	nr	**99.43**
Spotlight Accessories; Outdoor Lighting (OLS)							
"ES"; earth spike for Maxispotter	6.18	-	-	-	6.18	nr	**6.18**
"ES"; earth spike for WeeBee Spot	14.70	-	-	-	14.70	nr	**14.70**
"ES"; earth spike for Morph 20	5.15	-	-	-	5.15	nr	**5.15**
Spotlighters, uplighters and cowl lighting; Marlin Lighting; TECHNO - SHORT ARM; head adjustable 130 deg rotation 350 deg; projection 215 mm on 210 mm base plate; 355 high; integral gear; PG16 cable gland; black							
ref SIM 3518-09; 150w CDMT/HQIT	337.90	1.00	15.00	-	337.90	nr	**352.90**
ref SIM 3514 -09; 70w SONT	332.21	1.00	15.00	-	332.21	nr	**347.21**
ref SIM 3517-09; 70w CDMT/HQIT	321.42	1.00	15.00	-	321.42	nr	**336.42**
Recessed uplighting; including walk/drive over fully recessed uplighting; excavating, ground fixing, concreting in and making good surfaces (electric wiring, connections or related fixtures such as switch gear and time-clock mechanisms not included unless otherwise stated) (Note: Transformers will power multiple lights dependent on the distance between the light units) All mains voltage (240v) unless otherwise stated							
Recessed uplighting; Outdoor lighting (OLS); "WeeBee Up"; ultra-compact recessed halogen uplight in diecast aluminium and S/Steel top plate and toughened safety glass; 20/35w; complete with installation sleeve; low voltage requires transformer; to IP 67							
ref BU-01;	91.67	2.00	30.00	-	97.15	nr	**127.15**

V ELEC. SUPPLY/POWER/LIGHTING SYSTEMS Excluding overheads and profit	PC £	Labour hours	Labour £	Plant £	Material £	Unit	Total rate £
Recessed uplighting; Outdoor lighting (OLS); BU-Pharo; Pharo uplighters; 266 mm diameter; diecast aluminium with stainless steel top plate 10 mm toughened safety glass; 2000 kg drive over; integral control gear; to IP67							
BU-Pharo; HIT Metal Halide; white light; spot or flood or wall wash distribution	309.00	1.50	22.50	-	314.48	nr	336.98
HSE-E High pressure sodium; golden light; spot or flood or wall wash distribution	319.30	1.50	22.50	-	324.78	nr	347.28
HME-Mercury Vapour; cool white; flood distribution	278.10	1.50	22.50	-	283.58	nr	306.08
TC-T Compact fluorescent; white low power consumption; flood distribution	277.08	1.50	22.50	-	282.56	nr	305.06
Tungsten halogen 150w; flood distribution	267.80	1.50	22.50	-	273.28	nr	295.78
Accessories for Pharo uplighters; Outdoor Lighting(OLS);							
"RG" Rockguard	56.65	-	-	-	56.65	nr	56.65
"IS" Stainless steel installation sleeve	55.62	-	-	-	55.62	nr	55.62
"L" Anti glare louvre	22.66	-	-	-	22.66	nr	22.66
Recessed uplighting; Outdoor lighting (OLS); Nimbus 125-150 mm diameter uplighters; manufactured from diecast aluminium with stainless steel top plate 8 mm toughened safety glass; 2000 kg drive over; integral control gear; to IP67							
BU-02; 95 mm deep; 20/35/50w low voltage halogen; requires remote transformer	97.85	2.00	30.00	-	103.33	nr	133.33
BU-03; 172 mm deep; 20/35/50w low voltage halogen; with integral transformer	134.93	2.00	30.00	-	140.41	nr	170.41
Accessories for Nimbus uplighters; Outdoor Lighting(OLS);							
"IS" Stainless steel installation sleeve	32.45	-	-	-	32.45	nr	32.45
"BR" Brass top plate surcharge	17.51	-	-	-	17.51	nr	17.51
Wall recessed; Marlin Lighting; "EOS" range ; integral gear; asymmetric reflector; toughened reeded glass; to IP55							
ref SIM 4629-09 10w TCD "MINI EOS" 145 x 90 mm	76.83	1.00	15.00	-	76.83	nr	91.83
ref SIM 4640-09 "MEGA EOS" 80w MBFU 310 x 310 mm	355.00	1.00	15.00	-	355.00	nr	370.00
ref SIM 4639-09 "MEGA EOS" Wall recessing box	12.98	0.25	3.75	-	12.98	nr	16.73
ref SIM 4619-09 26w TCD "RECTANGULAR EOS" 270 x 145 mm	125.84	1.00	15.00	-	125.84	nr	140.84
ref SIM 4532-12 "RECTANGULAR EOS" Wall recessing box	12.98	0.25	3.75	-	12.98	nr	16.73
ref SIM 4628-09 12v 20w QT9 "MINI EOS" 145 x 90 mm	120.65	1.00	15.00	-	120.65	nr	135.65
ref SIM 4623-12 "MINI EOS" Wall recessing box	7.42	0.25	3.75	-	7.42	nr	11.17

V ELEC. SUPPLY/POWER/LIGHTING SYSTEMS Excluding overheads and profit	PC £	Labour hours	Labour £	Plant £	Material £	Unit	Total rate £
V41 STREET AREA FLOODLIGHTING - cont'd							
Recessed uplighting; including walk/drive over fully recessed uplighting; excavating, ground fixing, concreting in and making good surfaces (electric wiring, connections or related fixtures such as switch gear and time-clock mechanisms not included unless otherwise stated) (Note: Transformers will power multiple lights dependent on the distance between the light units) All mains voltage (240v) unless otherwise stated - cont'd							
Buried Uplighter; Marlin Lighting "ZIP"; round glass with round trim; withstands loads to 1000 kg; to IP67; aluminium							
ref SIM4848-09 35w QR CB51; round with round glass	91.79	1.00	15.00	-	91.79	nr	106.79
ref MTX 300; 35 -300VA; standard; intelligent transformer; stabilizes output to provide constant colour and prolong lamp life	238.90	1.00	15.00	-	238.90	nr	253.90
ref MTX 300DIM; 35 -300VA; as above but dimmable	293.45	1.00	15.00	-	293.45	nr	308.45
Buried Uplighter; Marlin Lighting; "SPARK" drive over 3000 kg load; various window options to IP67; including recessing tube and P16 cable gland;							
ref SIM 4836 35wCDMT; one window	433.80	1.00	15.00	-	433.80	nr	448.80
ref SIM 4826-09 35w CDMT; two windows	433.80	1.00	15.00	-	433.80	nr	448.80
ref SIM 4816-09 35w CDMT; twelve window	433.80	1.00	15.00	-	433.80	nr	448.80
Underwater Lighting							
Underwater Lighting; Marlin Lighting							
ref SIM 3651; "Sub" submersible light; 12v 300w PAR56	580.37	2.00	30.00	-	580.37	nr	610.37
ref SIM 3641; "Mini-Sub" submersible light; 12v 50w PAR36	418.14	2.00	30.00	-	418.14	nr	448.14
ref MTX 300; 35 -300VA; standard; intelligent transformer; stabilizes output to provide constant colour and prolong lamp life	238.90	1.00	15.00	-	238.90	nr	253.90
ref MTX 300DIM; 35 -300VA; as above but dimmable	293.45	1.00	15.00	-	293.45	nr	308.45
ref SIM 3654; Filter guard / holder for "Sub"	108.39	0.35	5.25	-	108.39	nr	113.64
ref SIM 3644; Filter guard / holder for "Mini-Sub"	68.81	0.25	3.75	-	68.81	nr	72.56
ref SIM 3657; colour filter for "Sub"	42.13	0.08	1.25	-	42.13	nr	43.38
ref SIM 3648; colour filter for "Mini-Sub"	21.54	0.08	1.25	-	21.54	nr	22.79
Underwater Lighting; Outdoor lighting (OLS) UW-05 "Minipower" cast bronze underwater floodlight c\w mounting bracket; 50w low voltage reflector lamp; requires remote transformer.	180.00	1.00	15.00	-	180.00	nr	195.00

V ELEC. SUPPLY/POWER/LIGHTING SYSTEMS Excluding overheads and profit	PC £	Labour hours	Labour £	Plant £	Material £	Unit	Total rate £
Low-level lighting; positioned to provide glare free light wash to pathways steps and terraces; including forming post holes, concreting in, making good to surfaces (final painting, electric wiring, connections or related fixtures such as switch gear and time clock mechanisms not included)							
Low-level lighting; Outdoor Lighting (OLS)							
"Sentry"; ref AM-08; single-sided bollard type sculptural pathway light; aluminum; 670 mm high; for 35w HIT; to IP 65	324.45	2.00	30.00	-	329.93	nr	359.93
"Footliter"; ref GL-05; low voltage pathlighter; 311 mm high; 6.00 m light distribution; in aluminium or solid copper; c\w earth spike; requires remote transformer; 20w halogen; to IP 44 - copper finish	69.01	2.00	30.00	-	74.49	nr	104.49
"Bricklight"; ref ST-03; for 9w PL lamp; diecast aluminium for recessing to walls; conforms to standard brick size; to IP 65	51.50	2.00	30.00	-	56.98	nr	86.98
Low-level lighting; The Woodhouse Company; pathway fittings; two aluminium posts; horizontal light							
ref SX731; 1627 mm long x 1000 mm high; including information plate; 58W SL; to IP 44	760.00	2.50	37.50	-	766.03	nr	803.53
ref SX731; 1327 mm long x 1000 mm high; including information plate; 36W SL; to IP 44	681.00	2.50	37.50	-	687.03	nr	724.53
ref SX730; 1627 mm long x 1000 mm high; 58W SL; to IP 44	652.00	2.50	37.50	-	658.03	nr	695.53
ref SX732; 702 mm long x 1000 mm high; including information plate; T18W; to IP 44	595.00	2.50	37.50	-	601.03	nr	638.53
ref SX730; 1327 mm long x 1000 mm high; 36W SL; to IP 44	584.00	2.50	37.50	-	590.03	nr	627.53
Lighted bollards; including excavating, ground fixing, concreting in and making good surfaces (lamp, final painting, electric wiring, connections or related fixtures such as switch gear and time-clock mechanisms not included) (Note: All illuminated bollards must be earthed); heights given are from ground level to top of bollards							
Lighted bollards; Louis Poulsen							
"Orbiter" Vandal resistant; head of cast aluminium; domed top; anti glare rings; pole extruded aluminium; diffuser clear UV stabilised polycarbonate; powder coated; 1040 mm high; 255 mm diameter; with root or base plate; IP44	390.37	2.50	37.50	-	396.40	nr	433.90
"Waterfront" solidly proportioned; head of cast silumin; domed top; symmetrical distribution; pole extruded aluminium sandblasted or painted white; internal diffuser clear UV stabilised polycarbonate; 865 mm high; diameter 260 mm; IP55	410.00	2.50	37.50	-	416.03	nr	453.53

V ELEC. SUPPLY/POWER/LIGHTING SYSTEMS Excluding overheads and profit	PC £	Labour hours	Labour £	Plant £	Material £	Unit	Total rate £
V41 STREET AREA FLOODLIGHTING - cont'd							
Lighted bollards; including excavating, ground fixing, concreting in and making good surfaces (lamp, final painting, electric wiring, connections or related fixtures such as switch gear and time-clock mechanisms not included) (Note: All illuminated bollards must be earthed); heights given are from ground level to top of bollards - cont'd							
"Bysted" concentric louvred bollard; head cast iron; post COR-TEN steel; externally untreated to provide natural aging effect of uniform oxidised red surface finish; internal painted white; lamp diffuser rings of clear polycarbonate; 1130 mm high 280 mm diameter: to IP 44	594 89	2.50	37.50	-	600.92	nr	638.42
"Centurion" heavy duty bollard; 2 sided light distribution from stepped reflector; cast aluminium; 900 mm high; width 240 x 160 mm; to IP 44	397 58	2.50	37.50	-	403.61	nr	441.11
"Nyhavn" split level overhanging canopy type luminaire head of cast silumin; diameter 565 mm; sandblasted or powder coated; diffuser clear UV stabilized polycarbonate; post of extruded aluminium; diameter 220 mm; 1235 mm high; diameter to IP44	515 00	2.50	37.50	-	521.03	nr	558.53
Lighted bollards; Marlin Lighting; Stepped Cone Sterling Bollard; to IP65; black or aluminium							
ref SBC 2606; 26w TCT	439.67	2.00	30.00	-	439.67	nr	469.67
ref SBC 8006; 50/80w MBFU	472 09	2.00	30.00	-	472.09	nr	502.09
SBC 5006 ; 50w SON	483.63	2.00	30.00	-	483.63	nr	513.63
SBC 7006; 70w SON	483.63	2.00	30.00	-	483.63	nr	513.63
Lighted bollards; Sugg Lighting							
"Cannon" bollards; flange mounted in cast aluminium with slotted apertures	339 63	2.50	37.50	-	345.66	nr	383.16
Reproduction street lanterns and columns; including excavating, concreting in, backfilling and disposing of spoil, or fixing to ground or wall, making good surfaces, light fitting and priming (final painting, electric wiring, connections or related fixtures such as switch gear and time-clock mechanisms not included) (Note: lanterns up to 14 in and columns up to 7 ft are suitable for residential lighting)							
Reproduction street lanterns and columns; Sugg Lighting; reproduction lanterns hand made to original designs; all with ES lamp holder							
762 mm "Westminster"	540.00	0.50	7.50	-	542.67	nr	550.17
533 mm "Westminster"	450.00	0.50	7.50	-	452.67	nr	460.17
432 mm "Guildhall"	400.00	0.50	7.50	-	402.67	nr	410.17
432 mm "Grosvenor"	300.00	0.50	7.50	-	302.67	nr	310.17
406 mm "Classic Globe"	292.60	0.50	7.50	-	295.27	nr	302.77

V ELEC. SUPPLY/POWER/LIGHTING SYSTEMS Excluding overheads and profit	PC £	Labour hours	Labour £	Plant £	Material £	Unit	Total rate £
"Windsor"; large, with door, IP54	280.00	0.50	7.50	-	282.67	nr	290.17
"Windsor"; medium, with door, IP54	240.00	0.50	7.50	-	242.67	nr	250.17
"Windsor"; small, with door, IP54	220.00	0.50	7.50	-	222.67	nr	230.17
Reproduction street lanterns and columns; D.W.Windsor; reproduction polished copper, natural copper or black stove enamelled lanterns with lead or gunmetal ornament; no IP rating given (Note: Diamond optic variable reflectors can be fitted to lanterns to prevent light pollution)							
"Westminster Boulevard"; 462 mm diameter x 910 mm high; SL25	471.00	0.50	7.50	-	473.67	nr	481.17
"Windsor Boulevard"; 520 mm square x 960 mm high; SL25	451.00	0.50	7.50	-	453.67	nr	461.17
"Westminster Avenue"; 348 mm diameter x 668 mm high; SL25	308.00	0.50	7.50	-	310.67	nr	318.17
"Windsor Court"; 310 mm square x 603 mm high; SL25	258.00	0.50	7.50	-	260.67	nr	268.17
"Westminster Garden"; 304 mm diameter x 558 mm high; SL25	238.00	0.50	7.50	-	240.67	nr	248.17
"Windsor Terrace"; 240 mm square x 446 mm high; SL18	198.00	0.50	7.50	-	200.67	nr	208.17
Reproduction street lanterns and columns; D W Windsor; aluminium metalwork powder coated black lanterns with lead or gunmetal ornament; no IP rating given (Note: Diamond optic variable reflectors can be fitted to lanterns to prevent light pollution)							
"Kingston 500" globe with ornate belt; 555 mm diameter; SL25	404.00	0.50	7.50	-	406.67	nr	414.17
"Kingston 400" globe with ornate belt; 440 mm diameter; SL25	281.00	0.50	7.50	-	283.67	nr	291.17
"Kingston 300" globe with ornate belt; 340 mm diameter; SL18	222.00	0.50	7.50	-	224.67	nr	232.17
Reproduction brackets and suspensions							
Reproduction brackets and suspensions; Sugg Lighting							
"Bow" bracket	334.40	2.00	30.00	-	337.07	nr	367.07
"Large ornate"	188.10	2.00	30.00	-	190.77	nr	220.77
"Large swan neck"	161.97	2.00	30.00	-	164.64	nr	194.64
"Medium ornate"	156.75	2.00	30.00	-	159.42	nr	189.42
"Universal" bracket/pedestal	130.63	2.00	30.00	-	133.30	nr	163.30
"Medium swan neck"	130.63	2.00	30.00	-	133.30	nr	163.30
"Decorated P" bracket	78.38	2.00	30.00	-	81.05	nr	111.05
"3-arm" suspension; general	78.38	2.00	30.00	-	81.05	nr	111.05
"Plaza" bracket	62.70	2.00	30.00	-	65.37	nr	95.37
"Universal" plinth pedestal	47.02	2.00	30.00	-	49.69	nr	79.69
"Frog"; general	31.35	2.00	30.00	-	34.02	nr	64.02
Reproduction brackets and suspensions; D. W. Windsor; reproduction brackets, pedestals, and suspensions; of cast iron or composite construction (Note: short brackets can only be used with small lanterns)							
"Edinburgh Hoop" bracket; 760 mm high x 1000 mm projection	221.00	2.00	30.00	-	223.67	nr	253.67
"Edinburgh Scroll" bracket; 760 mm high x 1000 mm projection	195.00	2.00	30.00	-	197.67	nr	227.67

V ELEC. SUPPLY/POWER/LIGHTING SYSTEMS Excluding overheads and profit	PC £	Labour hours	Labour £	Plant £	Material £	Unit	Total rate £
V41 STREET AREA FLOODLIGHTING - cont'd							
Reproduction brackets and suspensions - cont'd							
"Worcester" ornate pedestal; 76 mm diameter shaft x 642 mm high	96.00	2.00	30.00	-	98.67	nr	128.67
"Large Plain Cast"; 345 mm high x 450 mm projection	62.00	2.00	30.00	-	64.67	nr	94.67
"Wells 75" ornate pedestal; 50 mm diameter shaft x 785 mm high	73.00	2.00	30.00	-	75.67	nr	105.67
"Wells" ornate pedestal; 50 mm diameter shaft x 465 mm high	55.00	2.00	30.00	-	57.67	nr	87.67
"Small Ornate"; 233 mm high x 318 mm projection	55.00	2.00	30.00	-	57.67	nr	87.67
"Small Plain Cast"; 296 mm high x 300 mm projection	48.00	2.00	30.00	-	50.67	nr	80.67
"London fluted"; 273 mm high	47.00	2.00	30.00	-	49.67	nr	79.67
Reproduction lighting columns							
Reproduction lighting columns; Sugg Lighting							
2133 mm "Royal Exchange" cast aluminium columns with flanges; triple	1672.00	8.00	120.00	-	1672.00	nr	1792.00
3505 mm "Constitution Hill" cast aluminium columns with flanges	1567.50	8.00	120.00	-	1567.50	nr	1687.50
2133 mm "Royal Exchange" cast aluminium columns with flanges; single	1254.00	6.00	90.00	-	1254.00	nr	1344.00
3505 mm "Aylesbury" cast aluminium columns	836.00	8.00	120.00	-	836.00	nr	956.00
3.00 m - 5.00 m "Harborne" combined steel/iron columns with roots	522.50	8.00	120.00	-	530.04	nr	650.04
2438 mm "Barley Sugar" cast aluminium columns	532.95	6.00	90.00	-	532.95	nr	622.95
3.00 m - 5.00 m "Cannonbury"; combined steel/iron columns with flanges	470.25	8.00	120.00	-	470.25	nr	590.25
3048 mm "Trafalgar" cast aluminium columns with flanges	438.90	6.00	90.00	-	438.90	nr	528.90
2743 mm "Napier" cast aluminium columns with flanges	386.65	6.00	90.00	-	407.01	nr	497.01
Reproduction lighting columns; D. W. Windsor; reproduction columns of mild steel, cast iron, cast aluminium, or composite construction (Note: columns under 7ft are only suitable for small lanterns; refurbished original lamp posts are also available)							
"Cambridge 5M" tubular base and ladder bar; 4400 mm high	436.00	8.00	120.00	-	451.08	nr	571.08
"Cambridge 4M" tubular base and ladder bar; 3400 mm high	412.00	8.00	120.00	-	427.08	nr	547.08
"Chester 5M" tubular plain base; 4700 mm high	225.00	8.00	120.00	-	232.54	nr	352.54
"Chester 4M" tubular plain base; 3700 mm high	221.00	8.00	120.00	-	228.54	nr	348.54
"London fluted" tapered with ladder bar; 1805 mm high	177.00	4.00	60.00	-	192.08	nr	252.08
"Worcester" tubular base and ladder bar; 2350 mm high	158.00	4.00	60.00	-	165.54	nr	225.54
"Wells" tubular base and ladder bar; 1395 mm high	80.00	4.00	60.00	-	87.54	nr	147.54

V ELEC. SUPPLY/POWER/LIGHTING SYSTEMS Excluding overheads and profit	PC £	Labour hours	Labour £	Plant £	Material £	Unit	Total rate £
Precinct lighting lanterns; ground, wall or pole mounted; including fixing and light fitting (lamps, poles, brackets, final painting, electric wiring, connections or related fixtures such as switch gear and time clock mechanisms not included)							
Precinct lighting lanterns; Louis Poulsen							
"Nyhavn Park" side entry 90 degree curved arm mounted; steel rings over domed top; housing cast silumin sandblasted with integral gear; protected by UV stabilized clear polycarbonate diffuser; IP55	504.70	1.00	15.00	-	504.70	nr	519.70
"Kipp"; bottom entry pole-top; shot blasted or lacquered Hanover design award; hinged diffuser for simple maintenance; indirect lighting technique ensures low glare; IP55	272.95	1.00	15.00	-	272.95	nr	287.95
"Bjarne Bech"; luminaire head of galvanized sheet steel; domed head over cylindrical light shield; internal reflector surfaces painted white; IP23	279.13	1.00	15.00	-	279.13	nr	294.13
Precinct lighting lanterns; Marlin Lighting; "Sphereline"; clear or opal; vandal resistant; integral HPF gear; finish - black							
ref 8031 XX; 300 mm diameter opal polycarbonate globe; 10/13w TCD; to IP 44	57.63	1.00	15.00	-	57.63	nr	72.63
Precinct lighting lanterns; Marlin Lighting; "Sphereline"; accessories and fittings;							
ref 8030 PB26 "Boulevard" pole top 2 globe bracket	313.73	0.50	7.50	-	313.73	nr	321.23
ref 8030 B26; "Boulevard" 2 globe wall bracket; 460 projection	272.67	0.50	7.50	-	272.67	nr	280.17
ref 8030 B26; "Boulevard" wall bracket; 570 projection	182.81	0.50	7.50	-	182.81	nr	190.31
ref 8030 A3; pole top 3 globe bracket	188.41	0.50	7.50	-	188.41	nr	195.91
ref 8030 A2; pole top 2 globe bracket	137.01	0.50	7.50	-	137.01	nr	144.51
ref 8030 W16; wall bracket 90 degree; 350 mm projection	39.04	0.50	7.50	-	39.04	nr	46.54
ref 8035 L6; plinth; 400 mm overall height	35.19	0.50	7.50	-	35.19	nr	42.69
ref PC766; pole top spigot for 76 mm diameter pole	30.89	0.50	7.50	-	30.89	nr	38.39
ref 8000 WB6; conduit mounting box	24.20	0.50	7.50	-	24.20	nr	31.70
ref 8000 PC13; optional louvrestack	20.57	0.50	7.50	-	20.57	nr	28.07
Precinct lighting lanterns; Sugg Lighting							
large suspended "Superseal Rochester"; with 150W SON-T lamp	420.00	1.00	15.00	-	420.00	nr	435.00
medium suspended "Superseal Rochester"; with 70W SON-T lamp	340.00	1.00	15.00	-	340.00	nr	355.00
small suspended "Street Rochester"; with 70W SON-E lamp	280.00	1.00	15.00	-	280.00	nr	295.00
Precinct lighting lanterns; Noral							
"Nova III"; ref N1; aluminium hooded globe on swan neck; 330 mm diameter x 460 mm projection on fluted base; tubular post 2640 mm high o/a; 18W SL; to IP 44	288.00	8.00	120.00	-	295.54	nr	415.54
"Nova III"; ref N2; double; aluminium hooded globe on swan neck; 330 mm diameter x 460 mm projection on fluted base; tubular post 2640 mm high o/a; 18W SL; to IP 44	404.00	8.00	120.00	-	411.54	nr	531.54

V ELEC. SUPPLY/POWER/LIGHTING SYSTEMS Excluding overheads and profit	PC £	Labour hours	Labour £	Plant £	Material £	Unit	Total rate £
V41 STREET AREA FLOODLIGHTING - cont'd							
Precinct lighting lanterns; ground, wall or pole mounted; including fixing and light fitting (lamps, poles, brackets, final painting, electric wiring, connections or related fixtures such as switch gear and time clock mechanisms not included) - cont'd							
Precinct lighting lanterns; The Woodhouse Company							
ref SX441; double; "Alpha" lantern; aluminium dome; 630 mm diameter x 710 mm high; HSE 70W; to IP 44	2019.00	5.00	75.00	-	2019.00	nr	2094.00
ref SX471; double; "Saturn" lantern; flat metal dome; 660 mm high; HSE 80W; to IP 44	1865.00	4.00	60.00	-	1865.00	nr	1925.00
ref SX460; "Delta" lantern; shallow metal dome; 540 mm diameter; HME 125W; to IP 44	644.00	4.00	60.00	-	644.00	nr	704.00
ref SX440; "Alpha" lantern; aluminium dome; 630 mm diameter x 710 mm high; HSE 70W; to IP 44	642.00	4.00	60.00	-	642.00	nr	702.00
ref SX470; "Saturn" lantern; flat metal dome; 660 mm high; HSE 80W; to IP 44	551.00	4.00	60.00	-	551.00	nr	611.00
ref SX475; "Saturn" lantern; flat metal dome; 490 mm high; HSE 80W; to IP 44	451.00	4.00	60.00	-	451.00	nr	511.00
ref SX170 "Kugel" opal; 650 mm diameter on 2000 mm high steel pole; TC-D 13W; to IP 44	394.00	8.00	120.00	-	395.51	nr	515.51
ref SX112; "Kugel" opal; 300 mm diameter on 2000 mm high steel pole; TC-D 13W; to IP 44	202.00	8.00	120.00	-	209.54	nr	329.54
ref SX111; "Kugel" opal; 300 mm diameter on 1300 mm high steel pole; TC-D 13W; to IP 44	174.00	8.00	120.00	-	181.54	nr	301.54
Precinct lighting columns; including excavating, concreting, backfilling and disposing of spoil (final painting, electric wiring, connections or related fixtures such as switch gear and time-clock mechanisms not included)							
Precinct lighting columns; Marlin Lighting; "Skyline" tubular steel poles							
ref SPR 750 A6; waisted; 5.0 m high	179.46	6.00	90.00	-	208.77	nr	298.77
ref SPR 740 A6; waisted; 4.0 m high	155.53	6.00	90.00	-	184.84	nr	274.84
ref SPR 730 A6; waisted; 3.0 m high	131.60	6.00	90.00	-	160.91	nr	250.91
ref SPS 735 A6; straight; 3.5 m high	101.69	6.00	90.00	-	131.00	nr	221.00
Precinct lighting columns; Marlin Lighting; "Sphereline" tubular steel poles							
ref PS 735 A6; 76 mm diameter x 3.5 m high	161.88	6.00	90.00	-	169.42	nr	259.42
ref PR 730 A6; straight; 3.0 m high	166.97	6.00	90.00	-	196.28	nr	286.28
ref PR 740 A6; straight; 4.0 m high	174.82	6.00	90.00	-	204.13	nr	294.13
ref PR 750 A6; straight; 5.0 m high	182.67	6.00	90.00	-	211.98	nr	301.98
Precinct lighting columns; Sugg Lighting							
3.00 - 5.00 m steel columns with decorative embellishments	261.25	8.00	120.00	-	268.79	nr	388.79
3.00 - 5.00 m steel columns with contract embellishments	209.00	8.00	120.00	-	216.54	nr	336.54
Precinct lighting columns; CU Phosco							
"Sterling" tapered aluminium; 3000 mm high	115.00	8.00	120.00	-	122.54	nr	242.54
"Cub" octagonal steel; 6000 mm high	105.98	8.00	120.00	-	113.52	nr	233.52
"Cub" octagonal steel; 5000 mm high	79.06	8.00	120.00	-	86.60	nr	206.60

V ELEC. SUPPLY/POWER/LIGHTING SYSTEMS Excluding overheads and profit	PC £	Labour hours	Labour £	Plant £	Material £	Unit	Total rate £
Precinct lighting columns; The Woodhouse Company; galvanized straight tubular steel pole							
ref SX012 12-0; 76 mm diameter x 4000 mm high	171.00	6.00	90.00	-	178.54	nr	268.54
ref SX012 10-0; 90 mm diameter x 4000 mm high	154.00	6.00	90.00	-	161.54	nr	251.54
ref SX010 12-0; 76 mm diameter x 3000 mm high	147.00	6.00	90.00	-	154.54	nr	244.54
ref SX010 10-0; 90 mm diameter x 3000 mm high	131.00	6.00	90.00	-	138.54	nr	228.54
Precinct lighting columns; The Woodhouse Company; galvanized stepped tubular steel pole							
ref SX055 12-0; 76 mm - 159 mm diameter x 4000 mm high	293.00	6.00	90.00	-	300.54	nr	390.54
ref SX055 10-0; 90 mm - 159 mm diameter x 4000 mm high	274.00	6.00	90.00	-	281.54	nr	371.54
ref SX053 12-0; 76 mm - 159 mm diameter x 3000 mm high	274.00	6.00	90.00	-	281.54	nr	371.54
ref SX053 10-0; 90 mm - 159 mm diameter x 3000 mm high	259.00	6.00	90.00	-	266.54	nr	356.54
Transformers for low voltage lighting; Outdoor Lighting (OLS); distance from electrical supply 25 m; trenching and backfilling measured separately							
Boxed Transformer							
50 Va for maximum 50 w of lamp	48.92	2.00	30.00	-	70.03	nr	100.04
100Va for maximum 100 w of lamp	57.28	2.00	30.00	-	78.39	nr	108.40
150 Va for maximum 150 w of lamp	60.61	2.00	30.00	-	81.72	nr	111.72
200 Va for maximum 200 w of lamp	71.43	2.00	30.00	-	92.54	nr	122.55
250 Va for maximum 50 w of lamp	75.34	2.00	30.00	-	96.45	nr	126.46
Buried transformers							
50 Va for maximum 50 w of lamp	57.24	2.00	30.00	-	78.35	nr	108.35
Electrical Accessories							
Feeder pillar; ornamental bollard housing for exterior sockets and meters							
"Cannon" style; with 1 nr door	235.13	8.00	120.00	-	249.27	nr	369.27
"Fluted" with 2 nr doors	517.27	8.00	120.00	-	531.41	nr	651.41
Installation; electric cable in trench 500 mm deep (trench not included); all in accordance with IEE regulations							
Light duty 600 volt grade armoured							
3 core 2.5 mm	0.87	0.01	0.08	-	0.87	m	0.94
Twin and earth PVC cable in plastic conduit (conduit not included)							
2.50 mm²	-	0.01	0.15	-	0.14	m	0.29
4.00 mm²	-	0.01	0.15	-	0.43	m	0.58
16 mm heavy gauge high impact PVC conduit	-	0.03	0.50	-	3.08	m	3.58
Main switch and fuse unit; 30A	-	-	-	-	-	nr	37.92
Weatherproof switched socket; 13A	-	2.00	30.00	-	15.38	nr	45.38

ESSENTIAL READING FROM SPON PRESS

MANUFACTURED SITES
Rethinking the Post-Industrial Landscape

**Edited by Niall Kirkwood,
Harvard Graduate School of Design, USA**

' This volume offers an excellant overview of the environmental laws that govern site clean up as well as case studies in landscape restoration.'
Architectural Record

Manufactured Sites focuses on the legacy of industrial production and pollutants on the contemporary landscape and their influence on new scientific research, innovative site technologies and progressive site design. The coming quarter century will have to address the legacy of late nineteenth and twentieth-century industrialization. This area of research and development will become increasingly important as diminishing 'greenfield' sites, urban renewal and economic growth will require the reuse and renewal of manufactured sites.

This book presents innovative environmental, engineering and design approaches along with ongoing research and built projects of international significance. Topics included are soil remediation, phytoremediation brownfield development, structural soil, landfill capping and revegetation and bioengineering for soil and water quality. Ground breaking approaches to the design of contaminated land including public parks and commercial sites are outlined.

April 2001: 246x189: 256pp
154 b+w photos
Hb: 0-415-24365-3: £45.00

To Order: Tel: +44 (0) 8700 768853, or +44 (0) 1264 343071 Fax: +44 (0) 1264 343005, or Post: Spon Press Customer Services, Thomson Publishing Services, Cheriton House, Andover, Hants, SP10 5BE, UK Email: book.orders@tandf.co.uk.

For a complete listing of all our titles visit:
www.sponpress.com

8
Approximate Estimates

Prices in this section are based upon the Prices for Measured Work, but allow for incidentals which would normally be measured separately in a Bill of Quantities. They do not include for Preliminaries which are priced elsewhere in this book.

Items shown as sub-contract or specialist rates would normally include the specialists overhead and profit. All other items which could fall within the scope of works of general landscape and external works contractors would not include profit

Based on current commercial rates, profits of 15% to 35% may be added to these rates to indicate the likely "with profit" values of the tasks below. The variation quoted above is dependent on the sector in which the works are taking place – domestic, public or commercial.

ESSENTIAL READING FROM SPON PRESS

Guidelines for Landscape and Visual Impact Assessment

Second Edition

Edited by The Landscape Institute and the Institute of Environmental Management and Assessment, Co-ordinated by Sue Wilson, Landscape consultant, UK

'For any self-respecting landscape consultant, *Guidelines for Landscape and Visual Impact Assessment* is an essential reference book. The book is well-constructed, concise, compact and well-made and illustrated.'
The Architects' Journal on the first edition

This new edition of these good practice guidelines provides advice on assessing the landscape and the visual impacts of development projects. Issues covered by the guidelines include:

- integration of landscape and visual issues into the development process
- the need for a transparent approach to landscape and visual impact assessment;
- describes the baseline conditions;
- determining the magnitude and significance of impacts;
- reviews the landscape and visual components of an EIS (Environmental Impact Statement).

The guidelines will enable developers and decision makers to be aware of the issues to be addressed by this type of assessment, as well as providing environmental consultants with a framework for good practice.

April 2002: 246x189: 176pp
75 colour illustrations and 6 line drawings
Hb: 0-415-23185-X: £35.00

To Order: Tel: +44 (0) 8700 768853, or +44 (0) 1264 343071 Fax: +44 (0) 1264 343005, or Post: Spon Press Customer Services, Thomson Publishing Services, Cheriton House, Andover, Hants, SP10 5BE, UK Email: book.orders@tandf.co.uk

For a complete listing of all our titles visit:
www.sponpress.com

PRELIMINARIES	Unit	Total rate £
Excluding Overheads and Profit		
8.1 PRELIMINARIES		
CONTRACT ADMINISTRATION AND MANAGEMENT		
Prepare and maintain Health and Safety file, prepare risk and Coshh assessments, method statements, works programmes before the start of works on site; Project value:		
£35000.00	nr	**200.00**
£75000.00	nr	**300.00**
£100000.00	nr	**400.00**
£200000.00 - £500000.00	nr	**1000.00**
SITE SETUP AND SITE ESTABLISHMENT		
Costs of initial setup of Single site cabin and 200 m² of site area for storage of vehicles plant etc. Inclusive of Site office, secure container, padlocks safety fencing electricity and water and toilet facilities.		
Site establishment	nr	**411.40**
Weekly hire rates on the above	nr	**53.00**
Supply and erect temporary protective fencing and remove after planting		
Cleft chestnut paling 1.20 m high 75 larch posts at 3 m centres	100 m	**709.00**
Cleft chestnut paling 1.50 m high 75 larch posts at 3 m centres	100 m	**832.00**

DEMOLITION AND SITE CLEARANCE Excluding Overheads and Profit	Unit	Total rate £
8.2 DEMOLITION AND SITE CLEARANCE		
Break up plain concrete slab and remove to licensed tip not exceeding 13 km from site:		
150 thick	m²	5.96
200 thick	m²	7.94
300 thick	m²	11.91
Clear away light fencing and gates (chain link, chestnut paling, light boarded fence or similar) and remove to licensed tip	100 m	81.54
Clear existing scrub vegetation, including shrubs and hedges and burn on site; spread ash on specified area		
by machine		
light scrub and grasses	100 m²	14.06
undergrowth brambles and heavy weed growth	100 m²	31.72
small shrubs	100 m²	31.72
As above but remove vegetation to licensed tip not exceeding 13 km from site	100 m²	24.37
by hand	100 m²	53.40
As above but remove vegetation to licensed tip not exceeding 13 km from site	100 m²	41.51
Fell trees on site, grub up roots, all by machine and remove debris to licensed tip not exceeding 13 km from site:		
trees 600 girth	each	181.77
trees 1.5 - 3.00 m girth	each	596.68
trees over 3.00 m girth	each	1306.32

GROUNDWORK	Unit	Total rate £
Excluding Overheads and Profit		
8.3 GROUNDWORK		
Excavate trenches for foundations; trenches 225 mm deeper than specified foundation thickness. Pour plain concrete foundations GEN 1 10 N/mm and to thickness as described; Disposal of excavated material off site:		
Foundations		
250 mm wide x 250 mm deep	m³	109.42
300 mm wide x 300 mm deep	m³	106.60
400 mm wide x 600 mm deep	m³	106.60
Cut and strip by machine turves 50 thick		
load to dumper or palette by hand, stack on site		
not exceeding 100 m travel to stack	100 m²	37.05
As above but all by hand, including barrowing	100 m²	328.40
Prices for excavating and reducing to levels are for work in light or medium soils; multiplying factors for other soils are as follows:		
clay	x 1 5	-
compact Gravel	x 1.2	-
soft chalk	x 2.0	-
hard rock	x 3 0	-
Remove topsoil average depth 300 mm, deposit in spoil heaps not exceeding 100 m travel to heap, treat once with paraquat-diquat weedkiller, turn heap once during storage, all by machine	m³	3.04
As above but all by hand	m³	30.42
Excavate to reduce levels and remove spoil to dump not exceeding 100 m travel, all by machine:		
0 25 m deep average	m²	1.63
0 30 m deep	m²	1.84
1 0 m deep using 21 tonne 360 tracked excavator	m²	5.10
1 0 m deep using JCB	m²	5.58
As above but all by hand:		
0.10 m deep average	m²	6.47
0 20 m deep average	m²	13.43
0.30 m deep average	m²	60.59
1 0 m deep average	m²	89.07
Extra for carting spoil to licensed tip offsite	m³	11.51

GROUNDWORK Excluding Overheads and Profit	Unit	Total rate £
8.3 GROUNDWORK – cont'd		
Spread excavated material to levels in layers not exceeding 150 mm, using scraper blade or bucket		
average thickness 100 mm	m²	0.18
As above but with imported topsoil	m²	1.70
average thickness 200 mm	m²	0.36
As above but with imported topsoil	m²	3.40
average thickness 250 mm	m²	0.45
As above but with imported topsoil	m²	4.25
Extra for work to banks exceeding 30 slope	30%	-
Rip subsoil using approved sub-soiling machine where shown on drawings to a depth of 600 mm below topsoil at 600 mm centres, all tree roots and debris over 150 x 150 mm to be removed. Plough uncultivated ground where shown on drawings to a depths as shown		
100 mm	100 m²	3.41
200 mm	100 m²	3.54
300 mm	100 m²	3.75
400 mm	100 m²	5.47
TRENCHES; NOT SMM7		
Non SMM7 Rates for Foundations		
Excavate trenches for foundations; trenches 225 mm deeper than specified foundation thickness. Pour plain concrete foundations GEN 1 10 N/mm and to thickness as described; Disposal of excavated material off site:		
Foundations		
250 mm wide x 250 mm deep	m	8.03
300 mm wide x 300 mm deep	m	11.10
400 mm wide x 600 mm deep	m	27.53
600 mm wide x 400 mm deep	m	28.60

GROUND STABILIZATION
Excluding Overheads and Profit

8.4 GROUND STABILIZATION

Description	Unit	Total rate £
Excavate fixing trench for erosion control mat 300 x 300 mm. Place Tensar mat to trench and backfill with excavated material	m	7.41
Grade existing bank to even falls, remove all roots, stones and debris over 150 mm in any dimension. Lay 20 mm thick open texture erosion control Tensar mat with 100 mm laps, fixed with 8 x 400 mm steel pegs at 1.0 m centres. Sow with low maintenance grass at 35 g/m², spread imported topsoil to BS3882 25 mm thick incorporating medium grade sedge peat at 3 kg/m² and fertilizer at 30 g/m², water lightly.	m²	5.37
Bring bank to grade, remove all roots, stones and debris over 150 mm in any dimension. Lay Greenfix biodegradable pre-seeded erosion control mat, fixed with 6 x 300 mm steel pegs at 1.0 m centres.	100 m²	314.50
Bring surface of sub-grade to given levels and falls, remove all roots, stones and debris over 150 mm in any dimension, spread topsoil from dump on site 150 mm thick and consolidate lightly. Lay Tensar Mat erosion control mat with 100 mm laps, spread 25 mm imported topsoil with fertilizer and suitable erosion control seed.	100 m²	531.44
Bring surface of sub-grade to given levels and falls, remove all roots, stones and debris over 150 mm in any dimension, spread topsoil from dump on site and consolidate lightly. Lay Tensar Mat erosion control mat with 100 mm laps, spread 25 mm imported topsoil with fertilizer and suitable erosion control seed.	m²	5.06
Excavate trench and lay foundation concrete 1:3:6 600 x 300 mm deep		
By machine	m	22.84
By hand	m	26.28
Excavate trench 750 mm deep and lay concrete foundation 600 wide x 300 deep Construct Forticrete precast hollow concrete block wall with 450 mm below ground laid all in accordance with manufacturer's instructions. Fix reinforcing bar 12 mm as work proceeds Fill blocks with concrete 1:3:6 as work proceeds.		
1800 high		
Type 256; 400 x 225 x 256 mm	m	161.07
Type 190; 400 x 225 x 190 mm	m	131.92
As above but 2400 high		
Type 256; 400 x 225 x 256 mm	m	223.51
Type 190; 400 x 225 x 190 mm	m	180.91

GROUND STABILIZATION Excluding Overheads and Profit	Unit	Total rate £
On foundation measured above, supply and RCC precast concrete "L" shaped units, constructed all in accordance with manufacturer's instructions. Backfill with approved excavated material compacted as the work proceeds.		
1250 mm high x 1000 mm wide	m	122.85
2400 mm high x 1000 mm wide	m	193.49
3750 mm high x 1000 mm wide	m	389.97
Excavate trench to receive foundation 300 mm deep, place plain concrete foundation 150 thick in 11.50 N/mm² concrete (sulphate-resisting cement) construct precast concrete retaining wall and backfill with excavated spoil behind units.		
Anda-crib Maxi dowel-less system, average 4 m high	m	587.38
Anda-crib Mini dowel-less system, average 1.5 m high	m	189.59
Construct bank revetment, bring bank to grade by machine, lay regulating layer of quarry scalpings, lay filter membrane, lay 100 mm drainage layer of broken stone or approved hardcore 28 - 10 mm size, blind with 25 mm sharp sand, and lay grass concrete surface (price does not include edgings or toe beams),fill with approved topsoil and fertilizer at 35 g/m², seed with dwarf rye based grass seed mix:		
Grasscrete in situ reinforced concrete surfacing GC 1, 100 thick	m²	65.29
Grasscrete in situ reinforced concrete surfacing GC 2, 150 thick	m²	69.74
Grassblock 103 open matrix blocks 406 x 406 x 103	m²	66.06
Construct revetment of Maccaferri Limited Reno mattress gabions laid on firm level ground, tightly packed with broken stone or concrete and securely wired, all in accordance with manufacturers' instructions:		
gabions 6 m x 2 m x 0.17 m, one course	m²	60.87
gabions 6 m x 2 m x 0.17 m, two courses	m²	83.99

Approximate Estimates

GROUND STABILIZATION Excluding Overheads and Profit	Unit	Total rate £
Grade bank by mechanical scraper, lay soil-reinforcing fabric properly lapped and fixed with 200 mm steel pegs at 500 mm centres, spread approved topsoil 25 mm thick incorporating fertilizer at 35 g/m², sow with low maintenance grass at 35 g/m²:		
Wyretex No 8 polypropylene wire reinforced fabric	m²	5.14
unseeded "Eromat Light"	m²	2.44
seeded "Covamat Standard"	m²	3.05
Netlon Erosion mat for erosion control	m²	3.82
Construct ha-ha; excavate ditch 1200 mm deep x 900 mm wide at bottom battered one side 45° slope, excavate for foundation to wall and place GEN 1 concrete foundation 150 x 500 mm, construct one and a half brick wall (Brick PC £300.00/1000) battered 10° from vertical, laid in cement:lime:sand (1:1:6) mortar in English garden wall bond, precast concrete coping weathered and throated set 150 mm above ground on high side, rake bottom and sides of ditch and seed with low maintenance grass at 35 g/m²:		
wall 600 high	m	202.80
wall 900 high	m	250.03
wall 1200 high	m	308.69
Timber Log Retaining Walls Excavate trench 300 wide to one third of the finished height of the retaining walls below. Lay 100 mm hardcore. Fix machine rounded logs set in concrete 1:3:6. Remove excavated material from site. Fix geofabric to rear of timber logs. Backfill with previously excavated material set aside in position. All works by machine		
100 mm diameter logs		
500 mm high (constructed from 1.80 m lengths)	m	71.24
1.20 mm high (constructed from 1.80 m lengths)	m	118.10
1.60 mm high (constructed from 2.40 m lengths)	m	167.11
2.00 mm high (constructed from 3.00 m lengths)	m	213.62
As above but 150 mm diameter logs		
1.20 mm high (constructed from 1.80 m lengths)	m	160.12
1.60 mm high (constructed from 2.40 m lengths)	m	221.68
2.00 mm high (constructed from 3.00 m lengths)	m	280.29
2.40 mm high (constructed from 3.60 m lengths)	m	324.21

IN SITU CONCRETE Excluding Overheads and Profit	Unit	Total rate £
8.5 IN SITU CONCRETE		
Mix concrete on site; aggregates delivered in 20 tonne loads; deliver mixed concrete to location by barrow distance 25 m		
1:3:6	m³	86.18
1:2:4	m³	90.97
As above but aggregates delivered in 1 tonne bags		
1:3:6	m³	111.38
1:2:4	m³	116.18
As above but ready mixed concrete		
10 N/mm²	m³	108.40
15 N/mm²	m³	115.64
Excavate foundation trench mechanically, remove spoil offsite, lay 1:3:6 site mixed concrete foundations. Distance from mixer 25 m. Depth of trench to be 225 mm deeper than foundation to allow for 3 underground brick courses priced separately. Foundation size:		
200 mm deep x 400 mm wide	m	13.89
300 mm deep x 500 mm wide	m	24.03
400 mm deep x 400 mm wide	m	24.65
400 mm deep x 600 mm wide	m	37.08
600 mm deep x 600 mm wide	m	53.94
As above but hand excavation and disposal to spoil heap 25m by barrow and off site by grab lorry		
200 mm deep x 400 mm wide	m	22.78
300 mm deep x 500 mm wide	m	38.78
400 mm deep x 400 mm wide	m	38.08
400 mm deep x 600 mm wide	m	49.90
600 mm deep x 600 mm wide	m	81.41
Excavate foundation trench mechanically, remove spoil offsite, lay ready mixed concrete GEN1 discharged directly from delivery lorry to location. Depth of trench to be 225 mm deeper than foundation to allow for 3 underground brick courses priced separately. Foundation size:		
200 mm deep x 400 mm wide	m	11.41
300 mm deep x 500 mm wide	m	19.38
400 mm deep x 400 mm wide	m	19.69
400 mm deep x 600 mm wide	m	29.64
600 mm deep x 600 mm wide	m	42.78

BRICK/BLOCK WALLING	Unit	Total rate £
Excluding Overheads and Profit		
8.6 BRICK/BLOCK WALLING		
Excavate foundation trench 500 deep, remove spoil to dump off site, lay site mixed concrete foundations 1:3:6 350 x 150 thick, construct half brick wall with one brick piers at 2.0 m centres, laid in cement:lime:sand (1:1:6) mortar with flush joints, fair face one side, DPC two courses underground, engineering brick in cement:sand (1:3) mortar, coping of headers on end:		
Wall 900 high above DPC		
in engineering brick (class B) - £260.00/1000	m	128.24
in sandfaced facings - £300.00/1000	m	137.55
in rough stocks - £450.00/1000	m	146.01
Excavate foundation trench 400 deep, remove spoil to dump off site, lay GEN 1 concrete foundations 450 wide x 250 thick, construct one brick wall with one and a half brick piers at 3.0 m centres, all in English Garden Wall bond, laid in cement:lime:sand (1:1:6) mortar with flush joints, fair face one side, DPC two courses engineering brick in cement:sand (1:3) mortar, precast concrete coping 152 x 75 mm:		
Wall 900 high above DPC		
in engineering brick (class B) - £260.00/1000	m	218.24
in sandfaced facings - £300.00/1000	m	226.12
in rough stocks - £450.00/1000	m	242.94
Wall 1200 high above DPC		
in engineering brick (class B) - £260.00/1000	m	241.84
in sandfaced facings - £300.00/1000	m	251.09
in rough stocks - £450.00/1000	m	310.82
Wall 1800 high above DPC		
in engineering brick (class B) - £260.00/1000	m	303.18
in sandfaced facings - £300.00/1000	m	364.33
in rough stocks - £450.00/1000	m	392.76
Excavate foundation trench 450 deep, remove spoil to dump off site, lay GEN 1 concrete foundations 600 x 300 thick, construct one and a half brick wall with two thick brick piers at 3.0 m centres, all in English Garden Wall bond, laid in cement:lime:sand (1:1:6) mortar with flush joints, fair face one side, DPC two courses engineering brick in cement:sand (1:3) mortar, coping of headers on edge:		
Wall 900 high above DPC		
in engineering brick (class B) - £260.00/1000	m	446.57
in sandfaced facings - £300.00/1000	m	226.12
in rough stocks - £450.00/1000	m	242.94

BRICK/BLOCK WALLING Excluding Overheads and Profit	Unit	Total rate £
Wall 1200 high above DPC		
in engineering brick (class B) - £260.00/1000	m	**241.84**
in sandfaced facings - £300.00/1000	m	**251.09**
in rough stocks - £450.00/1000	m	**310.82**
Wall 1800 high above DPC		
in engineering brick (class B) - £260.00/1000	m	**303.18**
in sandfaced facings - £300.00/1000	m	**364.33**
in rough stocks - £450.00/1000	m	**392.76**

ROADS AND PAVINGS

Excluding Overheads and Profit

8.7 ROADS AND PAVINGS

KERBS AND EDGINGS

	Unit	Total rate £
Excavate and construct concrete foundation 300 mm wide x 150 mm deep. Lay precast concrete kerb units to BS 7263: Part 1, laid straight bedded and haunched both sides in semi-dry concrete, slump 35 mm maximum, haunching one side		
125 mm high x 125 mm thick bullnosed type BN	m	14.90
125 x 255 mm; ref HB2; SP	m	16.11
150 x 305 mm; ref HB1	m	19.67
On main paving base previously laid, supply and lay precast concrete channel units to BS 7263: Part 1, 255 mm wide x 125 mm high dished type CD, bedded in 1:3 lime:sand mortar, jointed in cement:sand mortar 1:3	m	12.95
Excavate and construct concrete foundation 600 mm wide x 200 mm deep. Lay channel of five courses of Class B engineering bricks to depths and falls bricks to be laid as headers along the channel, bedded in 1:3 lime:sand mortar bricks close jointed in cement:sand mortar 1:3	m	52.42
Excavate and lay precast concrete edging on concrete foundation and haunching one side 1:2:4 including all necessary formwork; to one side of straight path		
50 x 150	m	18.09
50 x 200	m	21.18
50 x 250	m	22.30
As above but to both sides of straight paths		
50 x 150	m	36.29
50 x 200	m	42.48
50 x 250	m	44.70
Timber edgings; Softwood		
150 x 38	m	4.72
150 x 50	m	4.99
Timber edgings hardwood		
150 x 38	m	9.98
150 x 50	m	11.94

ROADS AND PAVINGS Excluding Overheads and Profit	Unit	Total rate £
KERBS AND EDGINGS – cont'd		
Edge restraint to block paving; Excavate for groundbeam, lay Concrete 1:2:4 200 wide x 150mm deep. Lay butt jointed block header course on 50 mm thick sharp sand bed. Inclusive of haunching one side; butt joints		
200 x 100 x 60; laid to header course - £161.00/1000	m	11.26
200 x 100 x 60; laid to stretcher course - £161.00/1000	m	10.17
As above but clay bricks with mortar joints		
215 x 112 x 50 laid to header course - £280.00/1000	m	13.53
215 x 112 x 50 laid to stretcher course - £280.00/1000	m	12.55
ROADS		
Excavate weak points of excavation by hand and fill with well rammed Type 1 granular material;		
100 thick	m²	10.35
200 thick	m²	15.30
Excavate road bed and dispose excavated material off site, bring to grade and fill hollows with weak concrete, lay reinforcing mesh A142 lapped and joined, lay 150 mm thick in situ reinforced concrete roadbed 21 N/mm², with 15 impregnated fibreboard expansion joints and polysulphide based sealant at 50 m centres, 150 hardcore blinded with ash or sand, kerbs 155 x 255 to BS 7263 on both sides, including foundations haunched one side in 11.5 N/mm² concrete with all necessary formwork		
Falls, crossfalls and cambers not exceeding 15 degrees		
4.90 m wide	m	186.31
6.10 m wide	m	252.07
7.32 m wide	m	266.08

ROADS AND PAVINGS	Unit	Total rate £
Excluding Overheads and Profit		
Excavate 350 mm for pathways or roadbed to receive surface 100 mm thick priced separately, bring to grade, lay 100 well-rolled hardcore; lay 150 mm Type 1 granular material; lay precast edgings kerbs BS 340; 50 x 150 on both sides, including foundations haunched one side in 11.5 N/mm² concrete with all necessary formwork and disposal off site		
Work to falls, crossfalls and cambers not exceeding 15 degrees:		
1.50 m wide	m	107.84
2.00 m wide	m	121.61
3.00 m wide	m	164.25
4.00 m wide	m	206.89
5.00 m wide	m	249.53
6.00 m wide	m	292.18
7.00 m wide	m	334.82
CAR PARKS		
Excavate 350 for pathways or roadbed to receive surface 100 mm thick priced separately, bring to grade, lay 100 well-rolled hardcore; lay 150 mm Type 1 granular material; lay kerbs BS 7263; 125 x 255 on both sides, including foundations haunched one side in 11.5 N/mm² concrete with all necessary formwork		
Work to falls, crossfalls and cambers not exceeding 15 degrees :		
1.50 m wide	m	104.28
2.00 m wide	m	126.45
3.00 m wide	m	169.09
4.00 m wide	m	211.73
5.00 m wide	m	254.37
6.00 m wide	m	297.02
7.00 m wide	m	339.66
As above but excavation 450 mm deep and base of Type 1 at 250 mm thick		
1.50 m wide	m	113.81
2.00 m wide	m	137.92
3.00 m wide	m	186.30
4.00 m wide	m	234.69
5.00 m wide	m	283.07
6.00 m wide	m	331.45
7.00 m wide	m	379.83

ROADS AND PAVINGS Excluding Overheads and Profit	Unit	Total rate £
CAR PARKS – cont'd		
To excavated and prepared base above, lay roadbase of 40 size dense bitumen macadam to BS 4987, 70 thick, lay wearing course of 10 size dense bitumen macadam 30 thick, mark out car parking bays 5.0 m x 2.4 m with thermoplastic road paint: surfaces all mechanically laid		
per bay 5.0 m x 2.4 m	each	161.23
gangway	m²	11.01
As above but stainless steel road studs 100 x 100 to BS 873:pt.4, two per bay in lieu of thermoplastic paint:		
per bay 5.0 m x 2.4 m	each	166.72
Car park as above but with interlocking concrete blocks 200 x 100 x 80 mm; grey		
per bay 5.0 m x 2.4 m	each	206.40
gangway	m²	17.20
Car park as above but with interlocking concrete blocks 200 x 100 x 80 mm; colours		
per bay 5.0 m x 2.4 m	each	224.52
gangway	m²	18.71
Car park as above but with interlocking concrete blocks 200 x 100 x 60 mm; grey		
per bay 5.0 m x 2.4 m	each	195.24
gangway	m²	16.27
Car park as above but with interlocking concrete blocks 200 x 100 x 60 mm; colours		
per bay 5.0 m x 2.4 m	each	203.76
gangway	m²	16.98
Car park as above but with Marshalls Mono Hexpot 292 x 338 x 100 laid on 50 sand, fill with topsoil and peat (5:1) and fertilizer at 35 g/m², seed with dwarf rye grass at 35 g/m²:		
per bay 5.0 m x 2.4 m	each	392.04
gangway	m²	32.67
Car park as above but with lay Grass Concrete Grasscrete in situ continuously reinforced cellular surfacing, including expansion joints at 10 m centres, fill with topsoil and peat (5:1) and fertilizer at 35 g/m², seed with dwarf rye grass at 35 g/m²:		
GC2, 150 thick for HGV traffic including dust carts:		
per bay 5.0 m x 2.4 m	each	353.16
gangway	m²	29.43
GC1, 100 thick for cars and light traffic		
per bay 5.0 m x 2.4 m	each	299.76
gangway	m²	24.98

ROADS AND PAVINGS Excluding Overheads and Profit	Unit	Total rate £
CAR PARKING FOR DISABLED PEOPLE		
To excavated and prepared base above, lay roadbase of 20 size dense bitumen macadam to BS 4987, 80 thick, lay wearing course of 10 size dense bitumen macadam 30 thick, mark out car parking bays with thermoplastic road paint:		
per bay 5.80 m x 3.25 m (ambulant)	each	207.54
per bay 6.55 m x 3.80 m (wheelchair)	each	274.03
gangway	m²	11.01
Car park as above but with interlocking concrete blocks 200 x 100 x 60 mm; grey, but mark out bays		
per bay 5.80 m x 3.25 m (ambulant)	each	312.20
per bay 6.55 m x 3.80 m (wheelchair)	each	411.18
gangway	m²	16.27
PLAY AREA FOR BALL GAMES		
Excavate for playground, bring to grade, lay 225 consolidated hardcore, blind with 100 type 1, lay macadam base of 40 size dense bitumen macadam to 75 thick, lay wearing course 30 thick		
Over 1000 m²	100 m²	2824.39
400 m² -1000 m²	100 m²	2929.39
INTERLOCKING BLOCK PAVING		
Edge restraint to block paving; Excavate for groundbeam, lay Concrete 1:2:4 200 wide x 150mm deep. Lay butt jointed block header course on 50 mm thick sharp sand bed. Inclusive of haunching one side; butt joints		
200 x 100 x 60; laid to header course - £161.00/1000	m	11.26
200 x 100 x 60; laid to stretcher course - £161.00/1000	m	10.17
As above but clay bricks with mortar joints		
215 x 112 x 50 laid to header course - £280.00/1000	m	13.53
215 x 112 x 50 laid to stretcher course - £280.00/1000	m	12.55

ROADS AND PAVINGS Excluding Overheads and Profit	Unit	Total rate £
INTERLOCKING BLOCK PAVING – cont'd		
Excavate ground and bring to levels, treat substrate with total herbicide, supply and lay granular fill Type 1 150 thick laid to falls and compacted, supply and lay brick pavers to BS 6677 in herring-bone pattern, laid on 50 compacted sharp sand, bedded in 10 loose sand, vibrated, joints filled with loose sand and vibrated, all excluding edgings or kerbs measured separately		
Blocks 200 x 100 x 60	m²	26.18
200 x 100 x 80	m²	26.70
Reduce levels, lay 150 granular material Type 1, lay 200 x 100 x 60 vehicular block paving to 90 degree herringbone pattern, on 50 compacted sand bed, vibrated, jointed in sand and vibrated, excavate and lay precast concrete edging 50 x 150 to BS 7263, on concrete foundation 1:2:4		
1.0 m wide clear width between edgings	m	60.05
1.5 m wide clear width between edgings	m	72.79
2.0 m wide clear width between edgings	m	85.55
As above but blocks laid 45 degree herringbone pattern including cutting edging blocks:		
1.0 m wide clear width between edgings	m	65.00
1.5 m wide clear width between edgings	m	77.74
2.0 m wide clear width between edgings	m	90.50
As above but all excavation by hand disposal off site by grab		
1.0 m wide clear width between edgings	m	123.98
1.5 m wide clear width between edgings	m	143.91
2.0 m wide clear width between edgings	m	164.30

ROADS AND PAVINGS	Unit	Total rate £
Excluding Overheads and Profit		
BRICK PAVING		
Excavate ground and bring to levels, treat substrate with total herbicide, dig out weak spots and fill with compacted DoT Type 1 fill, supply and lay granular fill DoT Type 1 150 thick laid to falls and compacted, supply and lay brick pavers to BS 6677 in herring-bone pattern, laid on 50 mm compacted sharp sand, bedded in 10 mm loose sand, vibrated, joints filled with loose sand and vibrated, all excluding edgings or kerbs measured separately		
Pedestrian		
215 x 102.5 x 65 thick - £280.00/1000	m²	66.33
215 x 102.5 x 50 - £295.00/1000	m²	62.33
As above but to trafficked areas on 250mm thick Type 1 base		
215 x 102.5 x 65 thick - £280.00/1000	m²	71.11
215 x 102.5 x 50 - £295.00/1000	m²	67.12
Pedestrian areas as above but excavation by hand, barrowed to spoil heap maximum distance 25 m and removal off site by grab		
215 x 102.5 x 65 thick - £280.00/1000	m²	108.80
215 x 102.5 x 50 - £295.00/1000	m²	104.15
BASES FOR PAVING		
Excavate ground and reduce levels, to receive 38 mm thick slab and 25 mm mortar bed; dispose of excavated material off site; treat substrate with total herbicide		
Lay granular fill Type 1 150 thick laid to falls and compacted.		
All by machine	m²	8.54
All by hand except disposal by grab	m²	22.49
Lay 1:2:4 concrete base 150 thick laid to falls		
All by machine	m²	22.87
All by hand except disposal by grab	m²	42.42
As above but with concrete base 150 deep but inclusive of 150 mm compacted hardcore		
By Machine	m²	29.44
As above but inclusive of reinforcement fabric A142		
350 mm deep	m²	33.56

ROADS AND PAVINGS Excluding Overheads and Profit	Unit	Total rate £
FLAG PAVING TO PEDESTRIAN AREAS		
Prices are inclusive of all mechanical excavation and disposal		
Supply and lay precast concrete flags; **Excavate ground and reduce levels,** **treat substrate with total herbicide,** **supply and lay granular fill Type 1 150** **thick laid to falls and compacted,** Standard precast concrete flags to to BS 7263 bedded and jointed in lime:sand mortar (1:3).		
450 x 450 x 70 chamfered	m²	29.86
450 x 450 x 50 chamfered	m²	28.88
600 x 300 x 50	m²	26.86
600 x 450 x 50	m²	26.84
600 x 600 x 50	m²	24.50
750 x 600 x 50	m²	24.18
900 x 600 x 50	m²	23.73
Coloured flags bedded and jointed in lime:sand mortar (1:3).		
600 x 600 x 50	m²	28.44
450 x 450 x 70 chamfered	m²	36.88
600 x 600 x 50	m²	28.71
400 x 400 x 65	m²	36.42
750 x 600 x 50	m²	27.96
900 x 600 x 50	m²	26.99
Marshalls Saxon; textured concrete flags; reconstituted Yorkstone in colours, butt jointed bedded in lime:sand mortar (1:3).		
300 x 300 x 35	m²	51.61
450 x 450 x 45	m²	44.04
600 x 300 x 35	m²	41.00
600 x 600 x 50	m²	39.27
450 x 450 x 35	m²	36.17
Tactile flags; Marshalls, Blister Tactile pavings; red or buff; for blind pedestrian guidance laid to designed pattern		
450 x 450	m²	34.94
400 x 400	m²	37.21

ROADS AND PAVINGS Excluding Overheads and Profit	Unit	Total rate £
PEDESTRIAN DETERRENT PAVING.		
Excavate ground and bring to levels, treat substrate with total herbicide, supply and lay granular fill Type 1 150 mm thick laid to falls and compacted, supply and lay precast deterrent paving units bedded in lime:sand mortar (1:3) and jointed in lime:sand mortar (1:3).		
Marshalls Mono		
Lambeth pyramidal paving 600 x 600 x 75	m²	80.14
Contour undulating plain paving 498 x 498 x 125	m²	49.37
Townscape		
Abbey square cobble pattern pavings; reinforced 600 x 600 x 60	m²	45.96
Geoset chamfered studs 600 x 600 x 60	m²	42.71
IMITATION YORKSTONE PAVINGS		
As above but imitation Yorkstone paving laid to coursed patterns		
Marshalls Heritage square or rectangular		
300 x 300 x 38	m²	48.95
600 x 300 x 30	m²	41.29
600 x 450 x 38	m²	40.32
450 x 450 x 38	m²	38.64
600 x 600 x 38	m²	36.57
As above but laid to Random rectangular patterns		
Various sizes selected from the above	m²	45.32
Imitation Yorkstone laid random rectangular as above but on concrete base 150 thick		
By machine	m²	59.65
By hand	m²	73.60
NATURAL STONE SLAB PAVING		
Excavate ground by machine and reduce levels, to receive 65 mm thick slab and 25 mm mortar bed; dispose of excavated material off site; treat substrate with total herbicide, lay granular fill Type 1 150 thick laid to falls and compacted. Lay to random rectangular pattern on 25 mm mortar bed		
new riven slabs laid random rectangular	m²	79.72
reclaimed Cathedral grade riven slabs laid random rectangular	m²	111.41
new slabs sawn 6 sides laid random rectangular	m²	85.46
new slabs, 3 sizes, sawn 6 sides laid to coursed pattern	m²	89.36

ROADS AND PAVINGS Excluding Overheads and Profit	Unit	Total rate £
NATURAL STONE SLAB PAVING – cont'd		
As above but excavation by hand disposal by grab		
new riven slabs laid random rectangular	m²	93.12
reclaimed Cathedral grade riven slabs laid random rectangular	m²	124.81
new slabs sawn 6 sides laid random rectangular	m²	98.86
new slabs, 3 sizes, sawn 6 sides laid to coursed pattern	m²	102.76
Excavate ground by machine and reduce levels, to receive 65 mm thick slab and 25 mm mortar bed; dispose of excavated material off site; treat substrate with total herbicide, lay granular fill Type 1 150 thick laid to falls and compacted.		
new riven slabs laid random rectangular	m²	79.72
reclaimed Cathedral grade riven slabs laid random rectangular	m²	111.41
new slabs sawn 6 sides laid random rectangular	m²	85.46
new slabs, 3 sizes, sawn 6 sides laid to coursed pattern	m²	89.36
As above but excavation by hand disposal by grab		
new riven slabs laid random rectangular	m²	93.12
reclaimed Cathedral grade riven slabs laid random rectangular	m²	124.81
new slabs sawn 6 sides laid random rectangular	m²	98.86
new slabs, 3 sizes, sawn 6 sides laid to coursed pattern	m²	102.76
Yorkstone paving as above but laid to 1:2:4 concrete base 150 thick, excavation and disposal by machine		
new riven slabs laid random rectangular	m²	93.50
reclaimed Cathedral grade riven slabs laid random rectangular	m²	125.19
new slabs sawn 6 sides laid random rectangular	m²	99.24
new slabs, 3 sizes, sawn 6 sides laid to coursed pattern	m²	103.14
As above but excavation by hand disposal by grab		
new riven slabs laid random rectangular	m²	113.05
reclaimed Cathedral grade riven slabs laid random rectangular	m²	144.74
new slabs sawn 6 sides laid random rectangular	m²	118.79
new slabs, 3 sizes, sawn 6 sides laid to coursed pattern	m²	122.69

ROADS AND PAVINGS Excluding Overheads and Profit	Unit	Total rate £
GRANITE SETT PAVING - PEDESTRIAN		
Excavate ground and bring to levels, lay 100 hardcore to falls, compacted with 5 tonne roller, blind with compacted Type 1 50 mm thick, lay 100 concrete 1:2:4, supply and lay granite setts 100 x 100 x 100 mm bedded in cement:sand mortar (1:3) 25 mm thick minimum, close butted and jointed in fine sand, all excluding edgings or kerbs measured separately		
Setts laid to bonded pattern and jointed		
new setts 100 x 100 x 100	m²	126.70
secondhand cleaned 100 x 100 x 100	m²	137.37
Setts laid in curved pattern		
new setts 100 x 100 x 100	m²	131.65
secondhand cleaned 100 x 100 x 100	m²	142.32
GRANITE SETT PAVING TRAFFICKED AREAS		
Excavate ground and bring to levels, lay 100 hardcore to falls, compacted with 5 tonne roller, blind with compacted Type 1 50 mm thick, lay 150 site mixed concrete 1:2:4 reinforced with steel fabric to BS 4483 ref: A 142, supply and lay granite setts 100 x 100 x 100 mm bedded in cement:sand mortar (1:3) 25 mm thick minimum, close butted and jointed in fine sand, all excluding edgings or kerbs measured separately		
Site mixed concrete		
new setts 100 x 100 x 100	m²	132.77
secondhand cleaned 100 x 100 x 100	m²	143.44
Ready mixed concrete		
new setts 100 x 100 x 100	m²	128.18
secondhand cleaned 100 x 100 x 100	m²	138.85
CONCRETE PAVING		
Pedestrian Areas		
Excavate to reduce levels, lay 100 mm Type 1, lay 100 PAV 1 air entrained concrete, joints at max width of 6.0m cut out and sealed with sealant to BS 5212. Inclusive of all formwork and stripping	m²	53.62
Trafficked Areas		
Excavate to reduce levels, lay 150 mm Type 1 lay 150 mm PAV 2 40 N/mm² air entrained concrete, reinforced with steel mesh to BS 4483 200 x 200 square at 2.22kg/m², joints at max width of 6.0m cut out and sealed with sealant to BS 5212	m²	69.80

ROADS AND PAVINGS Excluding Overheads and Profit	Unit	Total rate £
BEACH COBBLE PAVING		
Excavate ground and bring to levels and fill with compacted Type 1 fill 100 thick, lay GEN 1 concrete base 100 thick, supply and lay cobbles individually laid by hand bedded in cement:sand mortar (1:3) 25 thick minimum, dry grout with 1:3 cement:sand grout, brush off surplus grout and water in, sponge of cobbles as work proceeds all excluding formwork edgings or kerbs measured separately		
Scottish beach cobbles 200 - 100 mm	m²	77.03
Kidney flint cobbles 100 - 75 mm	m²	82.28
Scottish beach cobbles 75 - 50 mm	m²	93.78
CONCRETE SETT PAVING		
Excavate ground and bring to levels, supply and lay 150 granular fill Type 1 laid to falls and compacted, supply and lay setts bedded in 50 sand and vibrated, joints filled with dry sand and vibrated, all excluding edgings or kerbs measured separately		
Marshalls Mono; Tegula precast concrete setts		
random sizes 60 thick	m²	36.89
single size 60 thick	m²	35.14
random size 80 thick	m²	39.52
single size 80 thick	m²	37.76
Concrete setts Blanc de Bierges:		
140 x 140 x 80	m²	77.70
210 x 140 x 80	m²	74.70
70 x 70 x 70	m²	82.78
GRASS CONCRETE PAVING		
Reduce levels, lay 150 mm Type 1 granular fill compacted, supply and lay precast grass concrete blocks on 20 sand and level by hand, fill blocks with 3mm sifted topsoil and pre-seeding fertilizer at 50g/m², sow with perennial ryegrass/chewings fescue see seed at 35g/m²		
Grass Concrete		
GB103 406 x 406 x 103	m²	28.48
GB83 406 x 406 x 83	m²	26.90
Extra for geotextile fabric underlayer	m²	0.68

ROADS AND PAVINGS
Excluding Overheads and Profit

	Unit	Total rate £
Firepaths		
Excavate to reduce levels, lay 300 mm well rammed hardcore, blinded with 100 mm type 1, supply and lay Marshalls Mono "Grassguard 180" precast grass concrete blocks on 50 sand and level by hand, fill blocks with sifted topsoil and pre-seeding fertilizer at 50 g/m²		
firepath 3.8 m wide	m	174.89
firepath 4.4 m wide	m	202.48
firepath 5.0 m wide	m	230.22
turning areas	m²	46.04
Reduce levels, lay granular fill Type 1, 150 thick, lay precast concrete cellular blocks on 20 sharp sand, fill with topsoil and Growmore fertilizer at 50 g/m², sow with dwarf rye grass seed at 35 g/m²		
Marshalls Mono Hexpot: 360 x 415 x 100	m²	33.65
Marshalls Mono Hexpot: 292 x 338 x 100	m²	36.61
Charcon Hard Landscaping Grassgrid: 366 x 274 x 100	m²	25.36
SLAB/BRICK PATHS		
Stepping stone path inclusive of hand excavation and mechanical disposal, 100 mm Type 1 and sand blinding. Slabs to comply with BS 7263 laid 100 mm apart to existing turf		
600 wide; 600 x 600 x 50 slabs		
natural finish	m	28.16
coloured	m	30.29
exposed aggregate	m	34.47
900 wide; 600 x 900 x 50 slabs		
natural finish	m	32.47
coloured	m	35.66
exposed aggregate	m	41.94
Pathway inclusive of mechanical excavation and disposal, 100 mm Type 1 and sand blinding. Slabs to comply with BS 7263 close butted		
Straight butted path 900 wide; 600 x 900 x 50 slabs		
natural finish	m	14.91
coloured	m	17.04
exposed aggregate	m	21.22
Straight butted path 1200 wide; double row; 600 x 900 x 50 slabs, laid stretcher bond		
natural finish	m	26.81
coloured	m	31.54
exposed aggregate	m	40.84

ROADS AND PAVINGS Excluding Overheads and Profit	Unit	Total rate £
SLAB/BRICK PATHS – cont'd		
Pathway inclusive of mechanical excavation and disposal, 100 mm Type 1 and sand blinding. Slabs to comply with BS 7263 close butted – cont'd		
Straight butted path 1200 wide, one row of 600 x 600 x 50, one row 600 x 900 x 50 slabs;		
natural finish	m	26.35
coloured	m	30.51
exposed aggregate	m	45.79
Straight butted path 1500 wide, slabs of 600 x 900 x 50, and 600 x 600 x 50 slabs, laid to bond		
natural finish	m	32.68
coloured	m	37.26
exposed aggregate	m	45.79
Straight butted path 1800 wide, two rows of 600 x 900 x 50 slabs, laid bonded		
natural finish	m	35.31
coloured	m	47.19
exposed aggregate	m	61.28
Straight butted path 1800 wide, three rows of 600 x 900 x 50 slabs, laid stretcher bond		
natural finish	m	39.06
coloured	m	50.94
exposed aggregate	m	65.03
Brick paved paths : Bricks 215 x 112.5 x 65 with mortar joints. Prices are inclusive of excavation, disposal off site, 100 hardcore, 100 1:2:4 concrete bed, edgings of brick 215 wide, haunched; all jointed in cement:lime:sand mortar (1:1:6).		
Path 1015 wide laid stretcher bond; edging course of headers		
rough stocks - £400.00/1000	m	107.78
engineering brick - £280.00/1000	m	102.91
Path 1015 wide laid stack bond:		
rough stocks - £400.00/1000	m	111.53
engineering brick - £280.00/1000	m	106.66
Path 1115 wide laid header bond:		
rough stocks - £400.00/1000	m	116.54
engineering brick - £280.00/1000	m	111.67
Path 1330 wide laid basketweave bond		
rough stocks - £400.00/1000	m	133.88
engineering brick - £280.00/1000	m	127.49
Path 1790 wide laid basketweave bond		
rough stocks - £400.00/1000	m	172.32
engineering brick - £280.00/1000	m	127.49

ROADS AND PAVINGS Excluding Overheads and Profit	Unit	Total rate £
Brick paved paths; Bricks paviors 200 x 100 x 50 chamfered edge with butt joints. Prices are inclusive of excavation, 100 mm Type 1, and 50 mm sharp sand; jointing in kiln dried sand brushed in. Exclusive of edge restraints		
Path 1000 wide laid stretcher bond; edging course of headers		
rough stocks - £400.00/1000	m	104.22
engineering brick - £280.00/1000	m	102.91
Path 1330 wide laid basketweave bond		
rough stocks - £400.00/1000	m	133.88
engineering brick - £280.00/1000	m	127.49
Path 1790 wide laid basketweave bond		
rough stocks - £400.00/1000	m	172.32
engineering brick - £280.00/1000	m	127.49
GRAVEL PATHS		
Reduce levels and remove spoil to dump on site, lay 150 hardcore well rolled, lay 25 mm sand blinding and geofabric, lay 50 gravel watered and rolled, excavate and timber edge 150 x 38 to both sides of straight paths:		
1.0 m wide	m	19.12
1.5 m wide	m	26.06
2.0 m wide	m	29.70
As above but Breedon Gravel		
1.0 m wide	m	25.14
1.5 m wide	m	34.01
2.0 m wide	m	42.90

ROADS AND PAVINGS Excluding Overheads and Profit	Unit	Total rate £
BARK PAVING		
Excavate to reduce levels, remove all topsoil to dump on site, treat area with herbicide. Lay 150 mm clean hardcore, blind with sand to BS 882 Grade C, lay 0.7 mm geotextile filter fabric water flow 50 l/m²/sec. Supply and fix treated softwood edging boards 50 x 150 mm to bark area on hardcore base extended 150 mm beyond the bark area, boards fixed with galvanized nails to treated softwood posts 750 x 50 x 75 mm driven into firm ground at 1.0 m centres. Edging boards to finish 25 mm above finished bark surface. Tops of posts to be flush with edging boards and once weathered.		
Supply and lay 100 mm Melcourt conifer walk chips 10-40 mm size.		
1.00 m wide	m	56.69
2.00 m wide	m	94.78
3.00 m wide	m	133.62
4.00 m wide	m	170.97
Extra over for wood fibre	m²	-0.05
Extra over for hardwood chips	m²	0.07
FOOTPATHS		
Excavate footpath to reduce level, remove arisings to tip on site maximum distance 25 m. Lay Type 1 granular fill 100 thick, lay base course of 28 size dense bitumen macadam 50 thick, wearing course of 10 size dense bitumen macadam 20 thick: Timber edge 150 x 38 mm.		
Areas over 1000 m²		
1.0 m wide	m	24.10
1.5 m wide	m	32.77
2.0 m wide	m	40.81
Areas 400 m² - 1000 m²		
1.0 m wide	m	26.45
1.5 m wide	m	36.30
2.0 m wide	m	55.75

SPECIAL SURFACES FOR SPORT/PLAYGROUNDS	Unit	Total rate £
Excluding Overheads and Profit		
8.8 SPECIAL SURFACES FOR SPORT/PLAYGROUNDS		
BOWLING GREEN CONSTRUCTION; **Baylis Landscape Contractors**		
Bowling Green; Complete Excavate 300 mm deep and grade to level. Excavate and install 100 mm land drain to perimeter and backfill with shingle. Install 60 mm land drain to surface at 4.5 m centres backfilled with shingle. Install 50 m non perforated pipe 50 m long. Install 100 mm compacted and levelled drainage stone, blind with grit and sand. Spread 150 mm imported 70:30 sand:soil accurately compacted and levelled. Lay bowling green turf and top dress twice luted into surface. Exclusive of perimeter ditches and bowls protection		
6 rink green 38.4 x 38.4 m	each	48300.00
Install "Toro" automatic bowling green irrigation system with pump, tank, controller, electrics, pipework and 8 nr "Toro 780" sprinklers. Excluding pumphouse (optional)	each	7500.00
Supply and install to perimeter of green, "Sportsmark" preformed bowling green ditch channels		
"Ultimate Design 99" steel reinforced concrete channel, 600 mm long section	each	7085.00
"Ultimate GRC " glass reinforced concrete channel, 1.2 m long section	each	9875.00
"Ultimate Design 2001" medium density rotational moulded channel. 1 m long section with integral bowls protection	each	9875.00
Supply and fit Bowls Protection material to rear hitting face of "Ultimate" channels 1 and 2 above		
"Curl Grass" artificial grass, 0.45 m wide	each	1950.00
"Astroturf" artificial grass, 0.45 m wide	each	3250.00
50 mm rubber bowls bumper (2 rows)	each	3650.00
Bowls protection ditch liner laid loose, 300 mm wide	each	1250.00
JOGGING TRACK		
Excavate track 250 deep, treat substrate with Casoron G4, lay filter membrane, lay 100 depth gravel waste or similar, lay 100 mm compacted gravel	100 m²	1814.28
Extra for treated softwood edging 50 x 150 on 50 x 50 posts, both sides	m	7.38

SPECIAL SURFACES FOR SPORT/PLAYGROUNDS Excluding Overheads and Profit	Unit	Total rate £
PLAYGROUNDS		
Excavate playground area, and dispose of arisings to tip; lay Type 1 granular fill, lay macadam surface two coat work 80 thick; base course of 28 size dense bitumen macadam 50 thick, wearing course of 10 size dense bitumen macadam 30 thick		
Areas over 1000 m²		
Excavation 180 mm base 100 mm thick	m²	17.85
Excavation 225 mm base 150 mm thick	m²	20.24
Areas 400 m² -1000 m²		
Excavation 180 mm base 100 mm thick	m²	20.24
Excavation 225 mm base 150 mm thick	m²	22.63
Excavate playground area to given levels and falls, remove soil off site and backfill with compacted Type 1 granular fill. Lay ready mixed concrete to fall 2% in all direction		
base 150 mm thick; surface 100 mm thick	m²	18.27
base 150 mm thick; surface 150 mm thick	m²	23.35
SAFETY SURFACING; Baylis Landscape Contractors		
Excavate ground and reduce levels, to receive 38 mm thick slab and 25 mm mortar bed; dispose of excavated material off site; treat substrate with total herbicide; Lay granular fill Type 1 150 thick laid to falls and compacted. lay macadam base 40 mm thick; Supply and lay "Ruberflex" wet pour safety system to thicknesses as specified		
All by machine except macadam by hand		
Black		
15 mm thick	100 m²	4084.90
35 mm thick	100 m²	5384.90
60 mm thick	100 m²	6584.90
Coloured		
15 mm thick	100 m²	7384.90
35 mm thick	100 m²	7884.90
60 mm thick	100 m²	8684.90
All by hand except disposal by grab		
Black		
15 mm thick	100 m²	5553.77
35 mm thick	100 m²	6853.77
60 mm thick	100 m²	8053.77
Coloured		
15 mm thick	100 m²	8853.77
35 mm thick	100 m²	9353.77
60 mm thick	100 m²	10153.77

SPECIAL SURFACES FOR SPORT/PLAYGROUNDS Excluding Overheads and Profit	Unit	Total rate £
Excavate playground area to 450 depth, lay 150 broken stone or clean hardcore, lay filter membrane, lay bark surface: Melcourt Industries		
Playbark 10/50 - 300 thick	100 m²	3006.69
Playbark 8/25 - 300 thick	100 m²	2800.69
SPORTSGROUND CONSTRUCTION; **Baylis Landscape Contractors** Plain sports pitches; site clearance, grading and drainage not included. **Cultivate ground and grade to levels, apply pre-seeding fertilizer at 900 kg/ha, apply pre-seeding selective weedkiller, seed in two operations with sports pitch type grass seed at 350 kg/ha, harrow and roll lightly, including initial cut. Size**		
association football, senior 114 m x 72 m	each	3177.50
association football, junior 106 m x 58 m	each	2562.50
rugby union pitch 156 m x 81 m	each	4663.75
rugby league pitch 134 m x 60 m	each	3075.00
hockey pitch 95 m x 60 m	each	2357.50
shinty pitch 186 m x 96 m	each	6560.00
men's lacrosse pitch 100 m x 55 m	each	2255.00
women's lacrosse pitch 110 m x 73 m	each	3280.00
target archery ground 150 m x 50 m	each	2665.00
cricket outfield 160 m x 142 m	each	8302.50
cycle track outfield 160 m x 80 m	each	5125.00
polo ground 330 m x 220 m	each	24497.50
Cricket square; excavate to depth of 150 mm, pass topsoil through 6 mm screen, return and mix evenly with imported marl or clay loam, bring to accurate levels, apply pre-seeding fertilizer at 50 g/m , apply selective weedkiller, seed with cricket square type g grass seed at 50 g/m², rake in and roll lightly, erect and remove temporary protective chestnut fencing, allow for initial cut and watering three times: 22.8 m x 22.8 m	each	13325.00

PREPARATION FOR PLANTING /TURFING Excluding Overheads and Profit	Unit	Total rate £
8.9 PREPARATION FOR PLANTING /TURFING		
CULTIVATION BY HAND		
Spread only and lightly consolidate topsoil brought from spoil heap in layers not exceeding 150, grade to specified levels, remove stones over 25mm, treat with paraquat-diquat weedkiller, all by hand		
100 mm thick	100 m²	300.06
150 mm thick	100 m²	450.09
300 mm thick	100 m²	900.18
450 mm thick	100 m²	1350.27
As above but inclusive of loading to location by barrow maximum distance 25 m; finished topsoil depth		
100 mm thick	100 m²	534.06
150 mm thick	100 m²	801.09
300 mm thick	100 m²	1602.18
450 mm thick	100 m²	2403.27
500 mm thick	100 m²	2668.80
600 mm thick	100 m²	3199.86
As above but loading to location by barrow maximum distance 100 m; finished topsoil depth		
100 mm thick	100 m²	675.06
150 mm thick	100 m²	1012.59
300 mm thick	100 m²	2025.18
450 mm thick	100 m²	3037.77
500 mm thick	100 m²	3373.80
600 mm thick	100 m²	4045.86
Extra to the above for incorporating mushroom compost at 50 mm/m² into the top 150 mm of topsoil; by hand	100 m²	107.00
Extra to above for imported topsoil PC 11.00 m³ allowing for 20% settlement		
100 mm thick	100 m²	100.00
150 mm thick	100 m²	150.00
300 mm thick	100 m²	300.00
450 mm thick	100 m²	450.00
500 mm thick	100 m²	500.00
600 mm thick	100 m²	600.00
750 mm thick	100 m²	750.00
1.00 m thick	100 m²	1000.00

PREPARATION FOR PLANTING /TURFING Excluding Overheads and Profit	Unit	Total rate £
CULTIVATION BY MACHINE		
Treat area with systemic non selective herbicide 1 month before starting cultivation operations. Rip up subsoil using subsoiling machine to a depth of 250 below topsoil at 1.20 m centres in light to medium soils, rotavate to 200 deep in two passes, cultivate with chain harrow, roll lightly, clear stones over 50mm		
by tractor	100 m²	8.14
As above but carrying out operations in clay or compacted gravel	100 m²	8.72
As above but ripping by tractor rotavation by pedestrian operated rotavator, clearance and raking by hand, herbicide application by knapsack sprayer	100 m²	31.52
As above but carrying out operations in clay or compacted gravel	100 m²	43.43
Spread and lightly consolidate topsoil brought from spoil heap not exceeding 100 m, in layers not exceeding 150, grade to specified levels, remove stones over 25, treat with paraquat-diquat weedkiller, all by machine		
100 mm thick	100 m²	58.78
150 mm thick	100 m²	88.47
300 mm thick	100 m²	176.94
450 mm thick	100 m²	265.09
Extra to above for imported topsoil PC £11.00 m³ allowing for 20% settlement		
100 mm thick	100 m²	100.00
150 mm thick	100 m²	150.00
300 mm thick	100 m²	300.00
450 mm thick	100 m²	450.00
500 mm thick	100 m²	500.00
600 mm thick	100 m²	600.00
750 mm thick	100 m²	750.00
1.00 m thick	100 m²	1000.00
Extra for incorporating mushroom compost at 50 mm/m² into the top 150 mm of topsoil (compost delivered in 20 m³) loads		
manually spread, mechanically rotavated	100 m²	84.64
mechanically spread and rotavated	100 m²	47.38
Extra for incorporating manure at 50 mm/m² into the top 150 mm of topsoil loads		
manually spread, mechanically rotavated 20m³ loads	100 m²	109.64
mechanically spread and rotavated 60 m³ loads	100 m²	99.64

SEEDING AND TURFING Excluding Overheads and Profit	Unit	Total rate £
8.10 SEEDING AND TURFING		
Bring top 200 of topsoil to a fine tilth using tractor drawn implements, remove stones over 25 by mechanical stone rake and bring to final tilth by harrow, apply pre-seeding fertilizer at 50 g/m² and work into top 50 during final cultivation, seed with certified grass seed in two operations, roll seedbed lightly after sowing.		
general amenity grass at 35 g/m²; BSH A3	100 m²	35.45
general amenity grass at 35 g/m²; Johnsons Taskmaster	100 m²	27.35
shaded areas; BSH A6 at 50 g/m²	100 m²	43.19
Motorway and road verges; Perryfields Pro 120 25-35 g/m²	100 m²	31.75
Bring top 200 of topsoil to a fine tilth using pedestrian operated rotavator, remove stones over 25, apply pre-seeding fertilizer at 50 g/m² and work into top 50 during final hand cultivation, seed with certified grass seed in two operations, rake and roll seedbed lightly after sowing,		
general amenity grass at 35 g/m²; BSH A3	100 m²	84.96
general amenity grass at 35 g/m²; Johnsons Taskmaster	100 m²	76.86
shaded areas; BSH A6 at 50 g/m²	100 m²	92.70
Motorway and road verges; Perryfields Pro 120 25-35 g/m²	100 m²	81.26
Extra to above for using imported topsoil spread by machine		
100 minimum depth	100 m²	167.00
150 minimum depth	100 m²	250.00
Extra for slopes over 30	50 %	-
Extra to above for using imported topsoil spread by hand; maximum distance for transporting soil 100 m		
100 minimum depth	m²	440.06
150 minimum depth	m²	660.09
Extra for slopes over 30	50 %	-
Extra for mechanically screening top 25 of topsoil through 6 mm screen and spreading on seedbed, debris carted to dump on site not exceeding 100 m	m³	4.21

SEEDING AND TURFING Excluding Overheads and Profit	Unit	Total rate £
Bring top 200 of topsoil to a fine tilth, remove stones over 50, apply pesticide at 100 g/m², apply pre-emergent weedkiller in accordance with manufacturer's instructions, apply pre-seeding fertilizer at 50 g/m² and work into top 50 of topsoil during final cultivation, seed with certified grass seed in two operations, harrow and roll seedbed lightly after sowing, level with mechanical lute		
outfield grass at 350 kg/ha - Perryfields Pro 40	ha	8546.00
outfield grass at 350 kg/ha - Perryfields Pro 70	ha	8196.00
sportsfield grass at 300 kg/ha - Johnsons Sportsmaster	ha	8086.00
As above but excluding pesticide and lute operations		
low maintenance grass at 350 kg/ha	ha	4383.00
verge mixture grass at 150 kg/ha	ha	3667.00
Extra for wild flora mixture at 30 kg/ha		
BSH WSF 75kg /ha	ha	3225.00
Extra for slopes over 30	50%	-
Cut existing turf to 1.0 x 1.0 x 0.5 m turves, roll up and move to stack not exceeding 100 m		
by pedestrian operated machine, roll up and stack by hand	100 m²	58.00
all works by hand	100 m²	162.50
Extra for boxing and cutting turves	100 m²	2.95
Bring top 200 of topsoil to a fine tilth in 2 passes, remove stones over 25 by and bring to final tilth, apply pre-seeding fertilizer at 50 g/m² and work into top 50 during final cultivation, roll turf bed lightly		
using tractor drawn implements and mechanical stone rake	m²	0.23
cultivation by pedestrian rotavator, all other operations by hand	m²	0.51
As above but bring turf from stack not exceeding 100 m, lay turves to stretcher bond using plank barrow runs, firm turves using wooden turf beater.		
using tractor drawn implements and mechanical stone rake	m²	1.51
cultivation by pedestrian rotavator, all other operations by hand	m²	1.67
As above but including imported turf; Rowlawn Medallion		
using tractor drawn implements and mechanical stone rake	m²	3.46
cultivation by pedestrian rotavator, all other operations by hand	m²	3.62

SEEDING AND TURFING Excluding Overheads and Profit	Unit	Total rate £
8.10 SEEDING AND TURFING – cont'd		
Bring top 200 of topsoil to a fine tilth – etc. – cont'd		
Extra over to all of the above for watering on two occasions and carrying out initial cut		
by ride on triple mower	100 m²	0.93
by pedestrian mower	100 m²	3.89
by pedestrian mower; box cutting	100 m²	4.35
Extra for using imported topsoil spread and graded by machine		
25 minimum depth	m²	0.48
75 minimum depth	m²	1.25
100 minimum depth	m²	1.67
150 minimum depth	m²	2.50
Extra for using imported topsoil spread and graded by hand; distance of barrow run 25 m		
25 minimum depth	m²	1.18
75 minimum depth	m²	2.89
100 minimum depth	m²	3.65
150 minimum depth	m²	5.48
Extra for work on slopes over 30 including pegging with 200 galvanized wire pins	m²	0.39
Inturf Big Roll **Supply, deliver in one consignment, fully prepare the area and install in Big Roll format Inturf 553; a turfgrass comprising dwarf perennial ryegrass, smooth stalked rneadowgrass and fescues; installation by tracked machine**		
Preparation by machine	m²	2.67
cultivation by pedestrian rotavator, all other operations by hand	m²	2.99
Erosion control Cultivation and preparation operations by pedestrian rotavator as above by pedestrian rotavator, all other operations by hand; Laying only to prepared ground; Greenkeeper reinforced turf on banks netting including pegging with 200 pegs at 500 centres	m²	7.96
On ground previously cultivated, bring area to level, treat with herbicide. Lay 20 mm thick open texture erosion control mat with 100 mm laps, fixed with 8 x 400 mm steel pegs at 1.0 m centres. Sow with low maintenance grass suitable for erosion control on slopes at 35 g/m². Spread imported topsoil 25 mm thick and fertilizer at 35 g/m², water lightly using sprinklers on two occasions.	m²	5.08
As above but hand-watering by hose pipe maximum distance from mains supply 50 m	m²	5.15

AFTERCARE MAINTENANCE OF GRASSED AREAS Excluding Overheads and Profit	Unit	Total rate £
8.11 AFTERCARE MAINTENANCE OF GRASSED AREAS		
Maintenance executed as part of a landscape construction contract.		
Grass cutting; fine turf, using pedestrian guided machinery; arisings boxed and disposed of off site		
per occasion	100 m²	4.01
per annum 26 cuts	m²	1.05
per annum 18 cuts	m²	0.73
Grass cutting; standard turf, using self propelled 3 gang machinery		
per occasion	100 m²	0.33
per annum 26 cuts	m²	0.09
per annum 18 cuts	m²	0.06
Maintenance for one year; recreation areas, parks, amenity grass areas, using tractor drawn machinery:		
per occasion	ha	24.71
per annum 26 cuts	ha	642.46
per annum 18 cuts	ha	444.78
Aerate ground with spiked aerator, apply weedkiller once, apply spring/summer fertilizer once, apply autumn/winter fertilizer once, cut grass, 16 cuts, sweep up leaves twice.		
as part of a landscape contract, defects liability	ha	1747.36
as part of a long term maintenance contract	ha	1337.00

PLANTING Excluding Overheads and Profit	Unit	Total rate £
8.12 PLANTING		
HEDGE PLANTING		
Excavate trench for hedge 300 wide x 450 deep by machine, deposit spoil alongside, and plant hedging plants in single row at 200 centres, and backfill with excavated material incorporating organic manure at 1 m^3 per 5 m^3, and carry out initial cut, including delivery of plants from nursery:		
bare root hedging plants		
PC - £0.30	100 m	465.02
PC - £0.60	100 m	615.02
PC - £1.00	100 m	815.02
PC - £1.20	100 m	915.02
PC - £1.50	100 m	1065.02
As above but two rows of hedging plants at 300 centres staggered rows:		
PC - £0.30	100 m	621.02
PC - £0.60	100 m	821.02
PC - £1.00	100 m	1087.68
PC - £1.50	100 m	1421.02
As above but all by hand		
two rows of hedging plants at 300 centres staggered rows		
PC - £0.30	100 m	812.00
PC - £0.60	100 m	1012.00
PC - £1.00	100 m	1278.66
PC - £1.50	100 m	1612.00
TREE PLANTING		
Excavate tree pit by hand, fork over bottom of pit, plant tree with roots well spread out, backfill with excavated material, incorporating organic manure at 1 m^3 per 3 m^3 of soil, one tree stake and two ties; tree pits square in sizes shown		
Light standard bare root tree in pit; PC £10.00		
600 x 600 deep	each	35.33
900 x 900 deep	each	47.08
Standard tree bare root tree in pit; PC £13.95		
600 x 600 deep	each	43.38
900 x 600 deep	each	50.47
900 x 900 deep	each	59.28
Standard root balled tree in pit; PC £16.00		
600 x 600 deep	each	46.63
900 x 600 deep	each	53.72
900 x 900 deep	each	62.53
Selected standard bare root tree, in pit; PC £18.80		
900 x 900 deep	each	66.83
1.00 m x 1.00 m deep	each	75.23

Approximate Estimates

PLANTING Excluding Overheads and Profit	Unit	Total rate £
Selected standard root ball tree in pit; PC £30.00		
900 x 900 deep	each	77.74
1.00 m x 1.00 m deep	each	114.12
Heavy standard bare root tree, in pit; PC £39.50		
900 x 900 deep	each	91.28
1.00 m x 1.00 m deep	each	104.60
1.20 m x 1.00 m deep	each	122.11
Heavy standard root ball tree, in pit; PC £50.00		
900 x 900 deep	each	100.80
1.00 m x 1.00 m deep	each	114.12
1.20 m x 1.00 m deep	each	131.63
Extra heavy standard bare root tree in pit; PC £49.00		
1.00 m x 1.00 deep	each	116.65
1 20 m x 1.00 m deep	each	159.48
1.50 m x 1.00 m deep	each	167.10
Extra heavy standard root ball tree in pit, PC £67.00		
1.00 m x 1.00 deep	each	137.65
1.20 m x 1.00 m deep	each	180.48
1.50 m x 1.00 m deep	each	188.10
Excavate tree pit by machine, fork over bottom of pit, plant tree with roots well spread out, backfill with excavated material, incorporating organic manure at 1 m³ per 3 m³ of soil, one tree stake and two ties; tree pits square in sizes shown		
Light standard bare root tree in pit; PC £10.00		
600 x 600 deep	each	31.70
900 x 900 deep	each	42.07
Standard tree bare root tree in pit; PC £13.95		
600 x 600 deep	each	39.75
900 x 600 deep	each	42.13
900 x 900 deep	each	46.77
Standard root balled tree in pit; PC £16.00		
600 x 600 deep	each	43.00
900 x 600 deep	each	45.38
900 x 900 deep	each	50.02
Selected standard bare root tree, in pit; PC £18.80		
900 x 900 deep	each	54.32
1 00 m x 1.00 m deep	each	63.20
Selected standard root ball tree in pit; PC £30.00		
900 x 900 deep	each	77.74
1.00 m x 1.00 m deep	each	102.09
Heavy standard bare root tree, in pit; PC £39.50		
900 x 900 deep	each	91.28
1.00 m x 1.00 m deep	each	92.57
1.20 m x 1.00 m deep	each	104.78
Heavy standard root ball tree, in pit; PC £50.00		
900 x 900 deep	each	100.80
1.00 m x 1.00 m deep	each	102.09
1.20 m x 1.00 m deep	each	114.30

PLANTING Excluding Overheads and Profit	Unit	Total rate £
TREE PLANTING – cont'd		
Excavate tree pit by machine, fork over bottom of pit, plant tree with roots well spread out, backfill with excavated material, incorporating organic manure at 1 m³ per 3 m³ of soil, one tree stake and two ties; tree pits square in sizes shown – cont'd		
Extra heavy standard bare root tree in pit; PC £49.00		
1.00 m x 1.00 deep	each	104.62
1.20 m x 1.00 m deep	each	116.83
1.50 m x 1.00 m deep	each	140.21
Extra heavy standard root ball tree in pit; PC £67.00		
1.00 m x 1.00 deep	each	125.62
1.20 m x 1.00 m deep	each	137.83
1.50 m x 1.00 m deep	each	161.21
SEMI MATURE TREE PLANTING		
Excavate tree pit deep by machine, fork over bottom of pit, plant rootballed tree Acer platanoides "Emerald Queen" using telehandler where necessary, backfill with excavated material, incorporating Melcourt Topgrow bark/manure mixture at 1 m³ per 3 m³ of soil, Platypus underground guying system; tree pits 1500 x 1500 x 1500 mm deep inclusive of Platimats; excavated material not backfilled to treepit spread to surrounding area		
16 - 18 cm girth - £69.00	each	244.62
18 - 20 cm girth - £77.00	each	261.80
20 - 25 cm girth - £130.00	each	362.32
25 - 30 cm girth - £179.00	each	466.24
30 - 35 cm girth - £220.00	each	667.14
As above but treepits 2.00 x 2.00 x 1.5 m deep		
40 - 45 cm girth - £715.00	each	977.18
45 - 50 cm girth - £805.00	each	1062.25
55 - 60 cm girth - £1265.00	each	1273.36
67 - 70 cm girth - £1800.00	each	1615.90
75 - 80 cm girth - £2400.00	each	2170.74
Extra to the above for imported topsoil moved 25 m from tipping area and disposal off site of excavated material		
Tree pits 1500 x 1500 x1500 m deep		
16 - 18 cm girth	each	91.00
18 - 20 cm girth	each	86.53
20 - 25 cm girth	each	82.45
25 - 30 cm girth	each	78.89
30 - 35 cm girth	each	75.73

PLANTING	Unit	Total rate £
Excluding Overheads and Profit		
SHRUBS GROUND COVERS AND BULBS		
Excavate planting holes on 250 x 250 mm x 300 deep to area previously ripped and rotavated; excavated material left alongside planting hole		
by mechanical auger		
250 mm centres (16 plants per m²)	m²	8.64
300 mm centres (11.11 plants per m²)	m²	6.00
400 mm centres (6.26 plants per m²)	m²	3.38
450 mm centres (4.93 plants per m²)	m²	2.66
500 mm centres (4 plants per m²)	m²	2.16
600 mm centres (2.77 plants per m²)	m²	1.50
750 mm centres (1.77 plants per m²)	m²	0.96
900 mm centres (1.23 plants per m²)	m²	0.66
1.00 m centres (1 plants per m²)	m²	0.54
1.50 m centres (0.44 plants per m²)	m²	0.24
As above but excavation by hand		
250 mm centres (16 plants per m²)	m²	8.64
300 mm centres (11.11 plants per m²)	m²	6.00
400 mm centres (6.26 plants per m²)	m²	3.38
450 mm centres (4.93 plants per m²)	m²	2.66
500 mm centres (4 plants per m²)	m²	2.16
600 mm centres (2.77 plants per m²)	m²	1.50
750 mm centres (1.77 plants per m²)	m²	0.96
900 mm centres (1.23 plants per m²)	m²	0.66
1.00 m centres (1 plants per m²)	m²	0.54
1.50 m centres (0.44 plants per m²)	m²	0.24
Clear light vegetation from planting area and remove to dump on site, dig planting holes, plant whips with roots well spread out, backfill with excavated topsoil, including one 38 x 38 treated softwood stake, two tree ties, and mesh guard 750 high. Planting matrix 1.5 m x 1.5 m. Allow for beating up once at 10% of original planting, cleaning and weeding round whips once, applying weedkiller once at 35 gm/m², applying fertilizer once at 35 gm/m², using the following mix of whips, bare rooted:		
plant bare root plants average price 0.28 p each to a required matrix		
plant mix as above	100 m²	135.07
plant bare root plants average price 0.75 p each to a required matrix		
plant mix as above	100 m²	145.67

PLANTING Excluding Overheads and Profit	Unit	Total rate £
SHRUBS GROUND COVERS AND BULBS – cont'd		
Cultivate and grade shrub bed, bring top 300 mm of topsoil to a fine tilth, incorporating Mushroom compost at 50 mm and Enmag slow release fertilizer; rake and bring to given levels, remove all stones and debris over 50 mm, dig planting holes average 300 x 300x 300 mm deep. Supply and plant specified shrubs in quantities as shown below, backfill with excavated material as above. Water to field capacity and mulch 50 mm bark chips 20-40 mm size. Water and weed regularly for 12 months and replace failed plants. Shrubs -3L PC £2.20; Ground covers - 9 cm PC £1.20		
100 % shrub area 300 centres		
300 mm centres	100 m²	4535.82
400 mm centres	100 m²	2693.23
500 mm centres	100 m²	1840.37
600 mm centres	100 m²	1377.07
100 % groundcovers		
200 mm centres	100 m²	5770.92
300 mm centres	100 m²	2743.70
400 mm centres	100 m²	1685.89
500 mm centres	100 m²	1195.67
Groundcover 30% / Shrubs 70% at the distances shown below		
200mm / 300 mm	100 m²	4853.78
300 mm /400 mm	100 m²	2683.66
300 mm /500 mm	100 m²	2101.76
400 mm/ 500 mm	100 m²	1777.66
Groundcover 50% / Shrubs 50% at the distances shown below		
200mm / 300 mm	100 m²	5157.20
300 mm /400 mm	100 m²	2714.85
300 mm /500 mm	100 m²	2298.62
400 mm/ 500 mm	100 m²	1758.87
Cultivate ground by machine and rake to level; plant bulbs as shown; bulbs PC 24.50 /100		
15 bulbs per m²	100 m²	596.15
25 bulbs per m²	100 m²	977.55
50 bulbs per m²	100 m²	1931.05
Cultivate ground by machine and rake to level; plant bulbs as shown; bulbs PC 12.90 /100		
15 bulbs per m²	100 m²	405.05
25 bulbs per m²	100 m²	659.05
50 bulbs per m²	100 m²	1294.05

PLANTING	Unit	Total rate £
Excluding Overheads and Profit		
Form holes in grass areas and plant bulbs using bulb planter, backfill with organic manure and turf plug: Bulbs PC £13.00/100		
15 bulbs per m²	100 m²	568.50
25 bulbs per m²	100 m²	947.50
50 bulbs per m²	100 m²	1895.00
PLANTING PLANTERS		
To brick planter, coat insides with 2 coats RIW liquid asphaltic composition; fill with 50 mm shingle and cover with geofabric; fill with screened topsoil incorporating 25% Topgrow compost and Enmag		
Planters 1 00 m deep		
1.00 x 1.00	nr	96.29
1.00 x 2.00	nr	163.29
1.00 x 3.00	nr	230.72
Planters 1.50 m deep		
1.00 x 1.00	nr	144.16
1 00 x 2 00	nr	209.94
1.00 x 3.00	nr	345.44
TREE GRILLES		
Supply and fix cast iron tree grille 1200 x 1200 with 440 mm diameter tree hole complete with irrigation inlet and locking points, on matching steel angle paving support frame with flanges laid on paving base. PC £309.84	each	388.97

FENCING Excluding Overheads and Profit	Unit	Total rate £
8.13 FENCING		
BALLSTOP FENCING		
Supply and erect plastic coated 30 x 30 mm netting fixed to 60.3 diameter 12 mm solid bar lattice galvanized dual posts, top, middle and bottom rails with 3 horizontal rails on 60.3 mm dia. nylon coated tubular steel posts at 3.0 m centres and 60.3 mm dia. straining posts with struts at 50 m centres set 750 mm into FND2 concrete footings 300 x 300 x 600 mm deep. Include framed chain link gate 900 x 1800 mm high to match complete with hinges and locking latch.		
4500 high	100 m	8018.00
5000 high	100 m	9153.00
6000 high	100 m	11306.00
CATTLE GRID		
Excavate pit 4.0 m x 3.0 m x 500, dispose of spoil on site, lay concrete base (C7P) 100 thick on 150 hardcore, excavate trench and lay 100 dia. clay agricultural drain outlet, form concrete sides to pit 150 thick (C15P) including all formwork, construct supporting walls one brick thick in engineering brick laid in cement mortar (1:3) at 400 mm centres, install cattle grid by H S Jackson & Son (Fencing), erect side panels on steel posts set in concrete (1:3:6), erect 2 nr warning signs standard DoT pattern	each	2349.29
Erect stock gate, ms tubular field gate, diamond braced 1.8 m high hung on tubular steel posts set in concrete (C7P), complete with ironmongery, all galvanized		
width 3.00 m	each	213.63
width 4.20 m	each	229.43

FENCING Excluding Overheads and Profit	Unit	Total rate £
CONCRETE FENCES		
Supply and erect precast concrete post and panel fence in 2 m bays, panels to be shiplap profile, aggregate faced one side, posts set 600 mm in ground in concrete.		
1500 mm high	m	25.31
1800 mm high	m	29.49
2100 mm high	m	33.76
Extra over for exposed aggregate faced panels	m²	7.09
DEER STOCK RABBIT FENCING		
Construct rabbit-stop fencing, erect galvanized wire netting, mesh 31, 900 above ground, 150 below ground turned out and buried, on 75 dia. treated timber posts 1.8 m long driven 700 into firm ground at 4.0 m centres, netting clipped to top and bottom straining wires 2.63 mm diameter, straining post 150 mm diameter, x 2.3 m long driven into firm ground at 50 m intervals		
turned in 150 mm	100 m	629.51
buried 150 mm	100 m	694.51
Construct deer-stop fencing, erect 5 no. 4 dia. plain galvanized wires and 5 no. 2 ply galvanized barbed wires at 150 spacing, on 45 x 45 x 5mm angle iron posts 2.4 m long driven into firm ground at 3.0 m centres, driven into firm ground at 3.0 m centres, with timber droppers 25 x 38 x 1.0 m long at 1.5 m centres	100 m	1105.99
Supply and erect forestry fencing of three lines of 3 mm plain galvanized wire stapled to 1700 x 65 mm dia treated softwood posts at 5.0 m centres with 1850 x 100 mm dia straining posts and 1600 x 80 mm dia struts at 50.0 m centres driven into firm ground	100 m	-
CHESTNUT PALE FENCING		
Erect chestnut pale fencing, cleft chestnut pales, two lines galvanized wire, galvanized tying wire, treated softwood posts at 3.0 m centres and straining posts and struts at 50 m centres driven into firm ground:		
900 high, posts 75 dia. x 1200 long	m	5.39
1200 high, posts 75 dia. x 1500 long	m	6.82

FENCING Excluding Overheads and Profit	Unit	Total rate £
CHAIN LINK FENCING		
Supply and erect chain link fencing. Form post holes and erect concrete posts and straining posts with struts at 50 m centres all set in 1:3:6 concrete. Fix line wires and fix 3 mm galvanized wire 50 mm chainlink fencing.		
900 mm high	m	17.20
1200 mm high	m	18.76
1800 mm high	m	21.40
2400 mm high	m	40.34
As above but with plastic coated 3.15 gauge galvanized wire mesh		
900 mm high	m	16.92
1200 mm high	m	18.49
1800 mm high	m	21.51
Extra for additional concrete straining posts with 1 strut set in concrete		
900 high	each	53.24
1200 high	each	55.65
1400 high	each	59.91
1800 high	each	65.24
2400 high	each	86.82
Extra for additional concrete straining posts with 2 struts set in concrete		
900 high	each	76.63
1200 high	each	79.79
1400 high	each	89.64
1800 high	each	95.73
2400 high	each	98.75
Chain link fencing 3 mm galvanized as above but with mild steel angle posts and straining posts instead of concrete posts		
900 mm high	m	6.32
1200 mm high	m	7.48
1800 mm high	m	10.71
2400 mm high	m	12.51
Extra for additional angle iron straining posts with 1 strut set in concrete		
900 high	each	39.90
1200 high	each	42.80
1400 high	each	44.07
1800 high	each	49.66
2400 high	each	60.77
Extra for additional angle iron straining posts with 2 struts set in concrete		
900 high	each	56.27
1200 high	each	61.18
1400 high	each	62.86
1800 high	each	74.17
2400 high	each	91.38

FENCING
Excluding Overheads and Profit

Description	Unit	Total rate £
Construct timber rail, horizontal hit and miss type, rails 150 x 25, posts 100 x 100 at 1.8 m centres, twice stained with coloured wood preservative, including excavation for posts and concreting into ground (C7P):		
In treated softwood		
1800 mm high	m	41.83
In primed softwood		
1800 mm high	m	47.22
Erect close boarded timber fence in treated softwood, pales 89 x 19 lapped, 152 x 25 gravel boards, two 76 x 38 rectangular rails:		
Concrete posts 100 x 100 at 3.0 m centres set into ground in 1:3:6 concrete		
900 mm high	m	16.95
1350 mm high	m	21.50
1800 mm high	m	22.97
As above but with oak posts		
1350 mm high	m	22.73
1800 mm high	m	25.31
As above but with softwood posts		
1350 mm high	m	20.19
1650 mm high	m	21.27
1800 mm high	m	21.93
Erect post and rail fence, three horizontal rails 100 x 38, fixed with galvanized nails to 100 x 100 posts driven into firm ground:		
In treated softwood		
1200 mm high	m	14.44
In oak or chestnut		
1200 mm high	m	20.77
Construct post and rail fence to BS 1722: Part 7 Type MPR 13/4 1300 mm high with five rails 87 x 38 mm. Rails morticed into treated softwood posts 2100 75 x 150 mm at 3.0 m centres set 700 mm into firm ground with intermediate prick posts 1800 x 87 x 38		
1300 high	m	14.84
Construct cleft oak rail fence, with rails 300 mm minimum girth tennoned both ends. 125 x 100 mm treated softwood posts double mortised for rails, corner posts 125 x 125 mm, driven into firm ground at 2.5 m centres.		
3 rails	m	15.21
4 rails	m	25.30

FENCING
Excluding Overheads and Profit

RAILINGS

Erect ms bar railing of 19 mm balusters at 115 mm centres welded to ms top and bottom rails 40 x 10, bays 2.0 m long, bolted to 51 x 51 ms hollow section posts set in C15P concrete, all metal work galvanized after fabrication:

	Unit	Total rate £
900 mm high	m	41.92
1200 mm high	m	57.61
1500 mm high	m	63.26

Supply and erect mild steel pedestrian guard rail Class A to BS 3049, panels 1000 mm high x 2000 mm wide with 150 mm toe space and 200 mm visibility gap at top, rails to be rectangular hollow sealed section 50 x 30 x 2.5 mm, vertical support 25 x 19 mm central between intermediate and top rail. Posts to be set 300 mm into paving base. All components factory welded and factory primed for painting on site — m — 64.63

SECURITY FENCING

Supply and erect chainlink fence, 51 mm x 3 mm mesh, with line wires and stretcher bars bolted to concrete posts at 3.0 m centres and straining posts at 10 m centres. Posts set in concrete 450 x 450 x 33% of height of post deep. Fit straight extension arms of 45 x 45 x 5 mm steel angle with three lines of barbed wire and droppers. All metalwork to be factory hot-dip galvanized for painting on site.

	Unit	Total rate £
900 high	m	20.70
1200 high	m	23.22
1800 high	m	31.31

As above but with PVC coated 3.15 mm mesh (diameter of wire 2.5 mm)

	Unit	Total rate £
900 high	m	20.42
1200 high	m	22.95
1800 high	m	31.42

Add to fences above for base of fence to be fixed with hairpin staples cast into concrete ground beam 1:3:6 site mixed concrete; mechanical excavation disposal to on site spoil heaps

	Unit	Total rate £
125 x 225 mm deep.	m	4.14

Add to fences above for straight extension arms of 45 x 45 x 5 mm steel angle with three lines of barbed wire and droppers. — m — 2.16

FENCING Excluding Overheads and Profit	Unit	Total rate £
Supply and erect palisade security fence Jacksons "Barbican" 2500 mm high with rectangular hollow section steel pales at 150 mm centres on three 50 x 50 x 6 mm rails. Rails bolted to 80 x 60 mm posts set in concrete 450 x 450 x 750 mm deep at 2750 mm centres, tops of pales to pointes and set at 45 degree angle. All metalwork to be hot-dip factory galvanized for painting on site.	m	72.17
Supply and erect single gate to match above complete with welded hinges and lock.		
1000 mm wide	each	654.00
4.0 mm wide	each	703.00
8.00 mm wide	pair	1433.00
Supply and erect Orsogril proprietary welded steel mesh panel fencing on steel posts set 750 mm deep in concrete foundations 600 x 600 mm. Supply and erect proprietary single gate 2.0 m wide to match fencing		
930 high	100 m	8815.00
1326 high	100 m	12201.00
1722 high	100 m	14674.00
TRIP RAILS		
Erect trip-rail, treated softwood rail 200 x 38, screwed to 100 x 100 oak posts with non-corroding screws, including excavating and setting posts in concrete (C7P) with central drainage hole, posts at 1.5 m centres, top of rail 450 above ground	m	24.27
Erect trip rail of galvanized steel tube 38 internal dia. with sleeved joint fixed to 38 dia. steel posts as above 700 long, set in C7P concrete at 1.20 m centres, metalwork primed and painted two coats metal preservative paint	m	84.63

Approximate Estimates

STREET FURNITURE Excluding Overheads and Profit	Unit	Total rate £
8.14 STREET FURNITURE		
BENCHES/SEATS/BOLLARDS		
Supply and install proprietary powder coated steel parking posts and bases to specified colour as shown on drawing, fold down top locking type, complete with keys and instructions. Posts and bases set in concrete footing 200 x 200 x 300 mm deep.	each	162.84
Supply and install 10 no. proprietary precast concrete "Millbank" bollards 355 mm high above ground x 405 mm deep, 500 mm dia. where shown on drawing, white spar exposed aggregate finish, bedded in concrete base 750 dia, x 300 mm deep.	10 nr	1549.80
In grassed area excavate for base 2500 x 1575 mm and lay 100 mm hardcore, 100 mm concrete, brick pavers in stack bond bedded in 25 mm cement:lime:sand mortar. Supply and fix where shown on drawing proprietary seat, hardwood slats on black powder coated steel frame, bolted down with 4 no. 24 x 90 mm recessed hex-head stainless steel anchor bolts set into concrete.	set	1179.08
Supply and fix to wall where shown on drawing wall-mounted proprietary litter bins powder coated timber slatted, open top with metal liner, fixed with 2 no. stainless steel recessed hex-head anchor bolts 16 x 60 mm.	each	423.21
Supply and fix cycle stand 1250 m long x 550 mm long of 60.3 mm black powder coated hollow steel sections to BS 4948:Part 2, one-piece with rounded top corners, set 250 mm into paving.	each	242.66
Supply and locate in position proprietary precast concrete planters, 970 mm diameter planter 470 high with white spar exposed aggregate finish. Drainage holes to be positioned away from footway in direction of paving fall with white spar exposed aggregate finish. Fill with topsoil placed over 50 mm shingle and terram; plant with assorted 5 litre and 3 litre shrubs at to provide instant effect	each	271.38

DRAINAGE	Unit	Total rate £
Excluding Overheads and Profit		
8.15 DRAINAGE		
AGRICULTURAL DRAINAGE		
Excavate and form ditch and bank with 45 degree sides in light to medium soils; all widths taken at bottom of ditch:		
300 wide x 600 deep	100 m	93.77
600 wide x 900 deep	100 m	226.69
1.20 m wide x 900 deep	100 m	676.20
1.50 m wide x 1.20 m deep	100 m	1138.50
Clear and bottom existing ditch average 1.50 m deep, trim back vegetation and remove debris to licensed tip not exceeding 13 km, lay jointed concrete pipes; to BS 5911 pt.1 class S; including bedding, haunching and topping with 150 mm concrete; 11.50 N/mm² - 40 mm aggregate; back fill with approved spoil from site.		
pipes 300 dia.	100 m	5513.96
pipes 450 dia.	100 m	7253.93
pipes 600 dia.	100 m	9430.53
SUB SOIL DRAINAGE; BY MACHINE		
Main drain; Remove 150 topsoil and deposit alongside trench, excavate drain trench by machine and lay flexible perforated drain, lay bed of gravel rejects 100 mm. backfill with gravel rejects or similar to within 150 of finished ground level, complete fill with topsoil, remove surplus spoil to approved dump on site:		
Main drain 160 mm in supplied in 35 m lengths		
450 mm deep	100 m	541.93
600 mm deep	100 m	655.09
900 mm deep	100 m	1107.38
Extra for couplings	each	1.41
As above but with 100 mm main drain supplied in 100 m lengths		
450 mm deep	100 m	437.86
600 mm deep	100 m	546.57
900 mm deep	100 m	738.55
Extra for couplings	each	1.17

DRAINAGE Excluding Overheads and Profit	Unit	Total rate £
SUB SOIL DRAINAGE; BY MACHINE – cont'd		
Laterals to mains above; herringbone pattern; excavation and back filling as above; inclusive of connecting lateral to main drain.		
160 mm pipe to 450 mm deep trench		
laterals at 1.0 m centres	100 m²	541.93
laterals at 2.0 m centres	100 m²	270.96
laterals at 3.0 m centres	100 m²	178.85
laterals at 5.0 m centres	100 m²	108.39
laterals at 10.0 m centres	100 m²	54.20
160 mm pipe to 600 mm deep trench		
laterals at 1.0 m centres	100 m²	655.09
laterals at 2.0 m centres	100 m²	327.55
laterals at 3.0 m centres	100 m²	216.18
laterals at 5.0 m centres	100 m²	131.02
laterals at 10.0 m centres	100 m²	65.51
160 mm pipe to 900 mm deep trench		
laterals at 1.0 m centres	100 m²	1107.38
laterals at 2.0 m centres	100 m²	553.69
laterals at 3.0 m centres	100 m²	365.45
laterals at 5.0 m centres	100 m²	221.48
laterals at 10.0 m centres	100 m²	110.74
Extra for 160/160 mm couplings connecting laterals to main drain		
laterals at 1.0 m centres	10 m	50.30
laterals at 2.0 m centres	10 m	25.15
laterals at 3.0 m centres	10 m	16.75
laterals at 5.0 m centres	10 m	10.06
laterals at 10.0 m centres	10 m	5.03
100 mm pipe to 450 mm deep trench		
laterals at 1.0 m centres	100 m²	437.86
laterals at 2.0 m centres	100 m²	218.93
laterals at 3.0 m centres	100 m²	144.50
laterals at 5.0 m centres	100 m²	87.57
laterals at 10.0 m centres	100 m²	43.79
100 mm pipe to 600 mm deep trench		
laterals at 1.0 m centres	100 m²	546.57
laterals at 2.0 m centres	100 m²	273.29
laterals at 3.0 m centres	100 m²	180.37
laterals at 5.0 m centres	100 m²	109.31
laterals at 10.0 m centres	100 m²	54.66
100 mm pipe to 900 mm deep trench		
laterals at 1.0 m centres	100 m²	738.55
laterals at 2.0 m centres	100 m²	369.28
laterals at 3.0 m centres	100 m²	243.72
laterals at 5.0 m centres	100 m²	147.71
laterals at 10.0 m centres	100 m²	73.86
Extra for 160/100 mm junctions connecting laterals to main drain		
laterals at 1.0 m centres	10 m	59.00
laterals at 2.0 m centres	10 m	29.50
laterals at 3.0 m centres	10 m	19.65
laterals at 5.0 m centres	10 m	29.50
laterals at 10.0 m centres	10 m	5.90

DRAINAGE	Unit	Total rate £
Excluding Overheads and Profit		
80 mm pipe to 450 mm deep trench		
laterals at 1.0 m centres	100 m²	410.19
laterals at 2.0 m centres	100 m²	205.10
laterals at 3.0 m centres	100 m²	135.37
laterals at 5.0 m centres	100 m²	82.04
laterals at 10.0 m centres	100 m²	41.02
80 mm pipe to 600 mm deep trench		
laterals at 1.0 m centres	100 m²	518.90
laterals at 2.0 m centres	100 m²	259.46
laterals at 3.0 m centres	100 m²	171.24
laterals at 5.0 m centres	100 m²	103.78
laterals at 10.0 m centres	100 m²	51.89
80 mm pipe to 900 mm deep trench		
laterals at 1.0 m centres	100 m²	710.88
laterals at 2.0 m centres	100 m²	355.45
laterals at 3.0 m centres	100 m²	234.59
laterals at 5.0 m centres	100 m²	142.18
laterals at 10.0 m centres	100 m²	71.09
Extra for 100/80 mm junctions connecting laterals to main drain		
laterals at 1.0 m centres	10 m	23.30
laterals at 2.0 m centres	10 m	11.65
laterals at 3.0 m centres	10 m	7.76
laterals at 5.0 m centres	10m	4.66
laterals at 10.0 m centres	10m	2.33
SUB SOIL DRAINAGE; BY HAND		
Main drain; Remove 150 topsoil and deposit alongside trench, excavate drain trench by machine and lay flexible perforated drain, lay bed of gravel rejects 100 mm. backfill with gravel rejects or similar to within 150 of finished ground level, complete fill with topsoil, remove surplus spoil to approved dump on site:		
Main drain 160 mm in supplied in 35 m lengths		
450 mm deep	100 m	1404.17
600 mm deep	100 m	1785.56
900 mm deep	100 m	2457.98
Extra for couplings	each	1.41
As above but with 100 mm main drain supplied in 100 m lengths		
450 mm deep	100 m	939.68
600 mm deep	100 m	1201.28
900 mm deep	100 m	1706.42
Extra for couplings	each	1.17

DRAINAGE Excluding Overheads and Profit	Unit	Total rate £
SUB SOIL DRAINAGE; BY HAND – cont'd		
Laterals to mains above; herringbone pattern; excavation and back filling as above; inclusive of connecting lateral to main drain.		
160 mm pipe to 450 mm deep trench		
laterals at 1.0 m centres	100 m²	1404.17
laterals at 2.0 m centres	100 m²	702.09
laterals at 3.0 m centres	100 m²	463.38
laterals at 5.0 m centres	100 m²	280.84
laterals at 10.0 m centres	100 m²	140.42
160 mm pipe to 600 mm deep trench		
laterals at 1.0 m centres	100 m²	1785.56
laterals at 2.0 m centres	100 m²	892.78
laterals at 3.0 m centres	100 m²	589.24
laterals at 5.0 m centres	100 m²	357.12
laterals at 10.0 m centres	100 m²	178.56
160 mm pipe to 900 mm deep trench		
laterals at 1.0 m centres	100 m²	2457.98
laterals at 2.0 m centres	100 m²	1228.99
laterals at 3.0 m centres	100 m²	811.13
laterals at 5.0 m centres	100 m²	491.59
laterals at 10.0 m centres	100 m²	245.80
Extra for 160/160 mm couplings connecting laterals to main drain		
laterals at 1.0 m centres	10 m	50.30
laterals at 2.0 m centres	10 m	25.15
laterals at 3.0 m centres	10 m	16.75
laterals at 5.0 m centres	10 m	10.06
laterals at 10.0 m centres	10 m	5.03
100 mm pipe to 450 mm deep trench		
laterals at 1.0 m centres	100 m²	939.68
laterals at 2.0 m centres	100 m²	469.84
laterals at 3.0 m centres	100 m²	310.10
laterals at 5.0 m centres	100 m²	187.93
laterals at 10 m centres	100 m²	93.98
100 mm pipe to 600 mm deep trench		
laterals at 1.0 m centres	100 m²	1201.28
laterals at 2.0 m centres	100 m²	600.64
laterals at 3.0 m centres	100 m²	396.42
laterals at 5.0 m centres	100 m²	240.25
laterals at 10.0 m centres	100 m²	120.13
100 mm pipe to 900 mm deep trench		
laterals at 1.0 m centres	100 m²	1706.42
laterals at 2.0 m centres	100 m²	853.21
laterals at 3.0 m centres	100 m²	563.12
laterals at 5.0 m centres	100 m²	341.28
laterals at 10.0 m centres	100 m²	170.65
Extra for 160/100 mm couplings connecting laterals to main drain		
laterals at 1.0 m centres	10 m	59.00
laterals at 2.0 m centres	10 m	29.50
laterals at 3.0 m centres	10 m	1.95
laterals at 5.0 m centres	10 m	1.18
laterals at 10.0 m centres	10 m	0.59

DRAINAGE Excluding Overheads and Profit	Unit	Total rate £
80 mm pipe to 450 mm deep trench		
laterals at 1.0 m centres	100 m²	912.01
laterals at 2.0 m centres	100 m²	456.01
laterals at 3.0 m centres	100 m²	300.97
laterals at 5.0 m centres	100 m²	182.40
laterals at 10.0 m centres	100 m²	91.21
80 mm pipe to 600 mm deep trench		
laterals at 1.0 m centres	100 m²	1173.61
laterals at 2.0 m centres	100 m²	586.81
laterals at 3.0 m centres	100 m²	387.29
laterals at 5.0 m centres	100 m²	234.72
laterals at 10.0 m centres	100 m²	117.36
80 mm pipe to 900 mm deep trench		
laterals at 1.0 m centres	100 m²	1678.75
laterals at 2.0 m centres	100 m²	839.38
laterals at 3.0 m centres	100 m²	553.99
laterals at 5.0 m centres	100 m²	335.75
laterals at 10.0 m centres	100 m²	167.88
Extra for 100/80 mm couplings connecting laterals to main drain		
laterals at 1.0 m centres	10 m	23.30
laterals at 2.0 m centres	10 m	11.65
laterals at 3.0 m centres	10 m	7.76
laterals at 5.0 m centres	10 m	4.66
laterals at 10.0 m centres	10 m	2.33
DRAIN BY MOLE PLOUGH		
50 dia mole at 450 deep	-	-
1.20 m centres	100 m²	11.19
1.50 m centres	100 m²	10.28
2.00 m centres	100 m²	9.35
2.50 m centres	100 m²	8.31
3.00 m centres	100 m²	7.27
75 dia. mole at 600 deep		
1.20 m centres	100 m²	13.36
1.50 m centres	100 m²	12.17
2.00 m centres	100 m²	11.76
2.50 m centres	100 m²	9.83
3.00 m centres	100 m²	10.28

DRAINAGE Excluding Overheads and Profit	Unit	Total rate £
SURFACE WATER DRAINAGE		
Excavate trench by excavator to 450 deep, lay 100 vitrified clay drain with butt joints, bedding Class B, backfill with excavated material screened to remove stones over 40, backfill to be laid in layers not exceeding 150, top with 150 mm topsoil remove surplus material to approved dump on site not exceeding 100m, final level of fill to allow for settlement.	100 m	1380.78
Excavate hole, supply and set in concrete (C10P) vitrified clay trapped mud (dirt) gully with rodding eye to BS 65, complete with galvanized bucket and cast iron hinged locking grate and frame, flexible joint to pipe. Connect to drainage system with flexible joints	each	136.52
Excavate pit for inspection chamber 1500 x 1500 x 1500 deep, including earthwork support and disposal of spoil to dump on site not exceeding 100m, lay concrete (1:2:4) base 1500 dia. x 200 thick, 110 vitrified clay channels, benching in concrete (1:3:6) allowing one outlet and two inlets for 110 dia pipe, construct inspection chamber with 1200 x 1200 x 1200 mm 1 brick thick walls of engineering brick Class B, backfill with excavated material, complete with 2 no. cast iron step irons, supply and fix light duty precast concrete cover slab.	each	781.97
Supply and fix vitrified clay interceptor trap 100 mm inlet; 100 mm outlet; to manhole bedded in 10 aggregate concrete (C10P), complete with brass stopper and chain.	each	109.82
Connect to drainage system with flexible joints.	each	13.86
Excavate hole and lay 100 concrete base (1:3:6) 150 x150 to suit given invert level of drain, supply and connect trapped precast concrete gully 450 dia x 1.07 m deep with 160 outlet to BS 5911, set in concrete surround, connect to vitrified clay drainage system with flexible joints, supply and fix straight bar dished top cast iron grating and frame, bedded in cement:sand mortar (1:3)	each	246.43
Excavate trench by excavator 600 deep, lay 110 PVC-U drainpipe to BS 4660, Type 2 bedding, backfill to 150 mm above pipe with gravel rejects, lay non woven geofabric and fill with topsoil to ground level.	100 m	972.57
Excavate pit for inspection chamber, supply and install cast iron inspection chamber unit 650 deep 100 x 150 mm one branch each side, bedded in Type 1 granular material.	each	272.17
Excavate hole and lay 100 concrete (C20P) base 150 x 150 to suit given invert level of drain, supply and connect trapped PVC-U gully complete with cast iron grate and frame, connect to drainage system, backfill with DoT Type 1 granular fill.	each	64.75

DRAINAGE Excluding Overheads and Profit	Unit	Total rate £
SOAKAWAYS		
Construct soakaway from perforated concrete rings; Excavation, casting insitu concrete ring beam base; and filling with gravel 250mm deep; placing perforated concrete rings; surrounding with geofabric and backfilling with 250 mm granular surround and excavated material; step irons and cover slab; inclusive of all earthwork retention and disposal offsite of surplus material		
900 mm diameter		
1.00 m deep	nr	428.28
2.00 m deep	nr	608.67
1200 mm diameter		
1.00 m deep	nr	544.52
2.00 m deep	nr	819.28

IRRIGATION Excluding Overheads and Profit	Unit	Total rate £
8.16 IRRIGATION		
LEAKY PIPE IRRIGATION		
Works by machine; Main supply and connection to laterals; Excavate trench for main or ring main 450 mm deep; Supply and lay pipe; backfill and lightly compact trench;		
20 mm LDPE	100 m	161.79
16 mm LDPE	100 m	161.50
Works by hand; Main supply and connection to laterals; Excavate trench for main or ring main 450 mm deep; Supply and lay pipe; backfill and lightly compact trench;		
20 mm LDPE	100 m	785.00
16 mm LDPE	100 m	756.00
Turf area irrigation; laterals to mains; to cultivated soil; excavate trench 150 deep using hoe or mattock; lay moisture leaking pipe laid 150 mm below ground at centres of 350 mm		
low leak	100 m²	485.00
high leak	100 m²	436.00
Landscape area irrigation; laterals to mains; moisture leaking pipe laid to the surface of irrigated areas at 600 mm centres		
low leak	100 m²	255.00
high leak	100 m²	226.00
Landscape area irrigation; laterals to mains; moisture leaking pipe laid to the surface of irrigated areas at 900 mm centres		
low leak	100 m²	171.00
high leak	100 m²	151.00
Multistation controller	each	319.00
Solenoid valves connected to automatic controller.	each	405.50

WATER FEATURES	Unit	Total rate £
Excluding Overheads and Profit		
8.17 WATER FEATURES		
LAKES AND PONDS		
Excavate for lake average depth 1.0 m, allow for bringing to specified levels, reserve topsoil, remove spoil to approved dump on site, remove all stones and debris over 75, lay 500 micron sheet including welding all joints and seams by specialist. Screen and replace topsoil 200 mm thick.		
Prices are for lakes of regular shape:		
under 1 ha	ha	95603.00
1-5 ha	ha	93053.00
As above but with 1000 micron		
under 1 ha	ha	112303.00
1-5 ha	ha	149053.00
As above but with butyl rubber sheet. Prices are for lakes of regular shape:		
0.75 mm thick	ha	111603.00
1.00 mm thick	ha	119103.00
Extra for removing spoil to tip	m^3	11.51
Extra for 25 sand blinding to lake bed	100 m^2	104.22
Extra for screening topsoil	m^2	0.74
Extra for spreading imported topsoil	m^2	398.76
Plant aquatic plants in lake topsoil:		
Aponogeton distachyum - £210.00/100	100	270.00
Acorus calamus - £169.00/100	100	270.00
Butomus umbellatus - £169.00/100	100	270.00
Typha latifolia - £169.00/100	100	270.00
Nymphaea - £425.00/100	100	270.00

TIMBER DECKING Excluding Overheads and Profit	Unit	Total rate £
8.18 TIMBER DECKING		
BOARD WALK		
Timber decking; support structure of timber joists for decking laid on blinded base (measured separately)		
joists 50 x 150 mm	10 m²	247.90
joists 50 x 200 mm	10 m²	280.90
joists 50 x 250 mm	10 m²	312.30
As above but inclusive of timber decking boards in yellow cedar 142 mm wide x 42 mm thick		
joists 50 x 150 mm	10 m²	717.70
joists 50 x 200 mm	10 m²	750.70
joists 50 x 250 mm	10 m²	782.10
As above but boards 141 x 26 mm thick		
joists 50 x 150 mm	10 m²	622.40
joists 50 x 200 mm	10 m²	280.90
joists 50 x 250 mm	10 m²	312.30
As above but decking boards in red cedar 131 mm wide x 42 mm thick		
joists 50 x 150 mm	10 m²	741.50
joists 50 x 200 mm	10 m²	774.50
joists 50 x 250 mm	10 m²	805.90
As above but boards 141 x 26 mm thick		
joists 50 x 150 mm	10 m²	623.10
joists 50 x 200 mm	10 m²	656.10
joists 50 x 250 mm	10 m²	687.50
As above but in Western Hemlock 88 x 38 mm thick		
joists 50 x 150 mm	10 m²	669.50
joists 50 x 200 mm	10 m²	702.50
joists 50 x 250 mm	10 m²	733.90
Add to all of the above for handrails fixed to posts 100 x 100 x 1370 high		
square balusters at 100 centres	m	108.27
square balusters at 300 centres	m	83.22
turned balusters at 100 centres	m	127.47
turned balusters at 300 centres	m	89.57

LIGHTING	Unit	Total rate £
Excluding Overheads and Profit		
8.19 LIGHTING		
Excavate trench 600 mm deep; supply and install 3 core 2.5 mm armoured cable and backfill with excavated material		
by machine	m	4.86
by hand	m	15.72
Supply and install Marlin Sphereline lighting units at locations shown on drawings: 3.0 m tapered aluminium columns to BS 5649, set 300 mm into paving, 500 mm dia. clear acrylic globe luminaires to IP54 complete with control gear and electrical supply, installed all in accordance with IEE regulations.	each	676.14
Supply and install 10 no. floodlighting units at locations 20 m apart; 5.0 m galvanized painted tubular steel columns to BS 5649, set 450 mm into paving, low light pollution aluminium luminaires to BS 4533:102.5 to IP65 complete with control gear and electrical supply, installed all in accordance with IEE regulations inclusive of electrical connections	each	691.82
Supply and install 30 no. illuminated bollards at locations shown on drawings: 900 mm high blue powder coated steel bollard to BS 873: Part 3 set 450 mm into ground, louvered polycarbonate luminaire to IP54, complete with control gear and electrical supply, installed all in accordance with IEE regulations.		
PC of lighting unit £395.00	each	462.90
PC of lighting unit £610.00	each	667.42
Supply and install 4 no. 12 volt ground mounted recessed brick lights at locations 5 metres apart; black powder coated diecast aluminium casing 220 mm dia. to IP55 recessed to paving, polycarbonate diffuser, complete with control gear and electrical supply, installed all in accordance with IEE regulations.	set	2127.83

ESSENTIAL READING FROM SPON PRESS

External Works Roads and Drainage
A Practical Guide
Phil Pitman

External Works Roads and Drainage: A Practical Guide bridges the gap between theory and practice in building, construction and civil engineering, providing practical guidance, and the knowledge required 'on the job'.

This comprehensive book includes sections on legislation, environmental issues, surface water, highway and foul drainage design, road and pavement design, and external works. The book is well illustrated, with flow charts to help guide you through the design process, bullet points and tables.

Written in an accessible style, this guide is an essential reference for engineers, architects, designers, technicians and students looking to increase their design and project planning knowledge.

'The text provides practical input which can be of benefit to both student and professional alike.' **Building Engineer**

2001: 234x156: 328pp:
7 line drawings, 30 tables
Pb: 0419257608: £29.99

To Order: Tel: +44 (0) 8700 768853, or +44 (0) 1264 343071 Fax: +44 (0) 1264 343005, or Post: Spon Press Customer Services, Thomson Publishing Services, Cheriton House, Andover, Hants, SP10 5BE, UK Email: book.orders@tandf.co.uk

For a complete listing of all our titles visit:
www.sponpress.com

9
Tables and Memoranda

QUICK REFERENCE CONVERSION TABLES

	Imperial			Metric	
1. LINEAR					
0.039	in	1	mm		25.4
3.281	ft	1	metre		0.305
1.094	yd	1	metre		0.914
2. WEIGHT					
0.020	cwt	1	kg		50.802
0.984	ton	1	tonne		1.016
2.205	lb	1	kg		0.454
3. CAPACITY					
1.760	pint	1	litre		0.568
0.220	gal	1	litre		4.546
4. AREA					
0.002	in^2	1	mm^2		645.16
10.764	ft^2	1	m^2		0.093
1.196	yd^2	1	m^2		0.836
2.471	acre	1	ha		0.405
0.386	mile^2	1	km^2		2.59
5. VOLUME					
0.061	in^3	1	cm^3		16.387
35.315	ft^3	1	m^3		0.028
1.308	yd^3	1	m^3		0.765
6. POWER					
1.310	HP	1	kW		0.746

CONVERSION FACTORS - METRIC TO IMPERIAL

Multiply Metric	Unit	By	To obtain Imperial	Unit
Length				
kilometre	km	0.6214	statute mile	ml
metre	m	1.0936	yard	yd
centimetre	cm	0.0328	foot	ft
millimetre	mm	0.0394	inch	in
Area				
hectare	ha	2.471	acre	
square kilometre	km²	0.3861	square mile	sq ml
square metre	m²	10.764	square foot	sq ft
square metre	m²	1550	square inch	sq in
square centimetre	cm²	0.155	square inch	sq in
Volume				
cubic metre	m³	35.336	cubic foot	cu ft
cubic metre	m³	1.308	cubic yard	cu yd
cubic centimetre	cm³	0.061	cubic inch	cu in
cubic centimetre	cm³	0.0338	fluid ounce	fl oz
Liquid volume				
litre	l	0.0013	cubic yard	cu yd
litre	l	61.02	cubic inch	cu in
litre	l	0.22	Imperial gallon	gal
litre	l	0.2642	US gallon	US gal
litre	l	1.7596	pint	pt
Mass				
metric tonne	t	0.984	long ton	lg ton
metric tonne	t	1.102	short ton	sh ton
kilogram	kg	2.205	pound, avoirdupois	lb
gram	g or gr	0.0353	ounce, avoirdupois	oz

Unit mass

kilograms/cubic metre	kg/m³	0.062	pounds/cubic foot	lbs/cu ft
kilograms/cubic metre	kg/m³	1.686	pounds/cubic yard	lbs/cu yd
tonnes/cubic metre	t/m³	1692	pound/cubic yard	lbs/cu yd
kilograms/sq centimetre	kg/cm²	14.225	pounds/square inch	lbs/sq in
kilogram-metre	kg.m	7.233	foot-pound	ft-lb

Force

meganewton	MN	9.3197	tons force	tonf
kilonewton	kN	225	pounds force	lbf
newton	N	0.225	pounds force	lbf

Pressure and stress

meganewton per square metre	MN/m²	9.3197	tons force/square foot	tonf/ft²
kilopascal	kPa	0.145	pounds/square inch	psi
bar		14.5	pounds/square inch	psi
kilograme metre	kgm	7.2307	foot-pound	ft-lb

Energy

kilocalorie	kcal	3.968	British thermal unit	Btu
metric horsepower	CV	0.9863	horse power	hp
kilowatt	kW	1.341	horse power	hp

Speed

kilometres/hour	km/h	0.621	miles/hour	mph

CONVERSION FACTORS - IMPERIAL TO METRIC

Multiply Imperial	Unit	By	To obtain metric	Unit
Length				
statute mile	ml	1.609	kilometre	km
yard	yd	0.9144	metre	m
foot	ft	0.3048	centimetre	cm
inch	in	25.4	millimetre	mm
Area				
acre	acre	0.4047	hectare	ha
square mile	sq ml	2.59	square kilometre	km²
square foot	sq ft	0.0929	square metre	m²
square inch	sq in	0.0006	square metre	m²
square inch	sq in	6.4516	square centimetre	cm²
Volume				
cubic foot	cu ft	0.0283	cubic metre	m³
cubic yard	cu yd	0.7645	cubic metre	m³
cubic inch	cu in	16.387	cubic centimetre	cm³
fluid ounce	fl oz	29.57	cubic centimetre	cm³
Liquid volume				
cubic yard	cu yd	764.55	litre	l
cubic inch	cu in	0.0164	litre	l
Imperial gallon	gal	4.5464	litre	l
US gallon	US gal	3.785	litre	l
US gallon	US gal	0.833	Imperial gallon	gal
pint	pt	0.5683	litre	l
Mass				
long ton	lg ton	1.016	metric tonne	tonne
short ton	sh ton	0.907	metric tonne	tonne
pound	lb	0.4536	kilogram	kg
ounce	oz	28.35	gram	g

Unit mass

pounds/cubic foot	lb/ cu ft	16.018	kilogram's/cubic metre	kg/m³
pounds/cubic yard	lb/cu yd	0.5933	cubic/cubic metre	kg/m³
pounds/cubic yard	lb/cu yd	0.0006	tonnes/cubic metre	t/m³
foot-pound	ft-lb	0.1383	kilogram-metre	kg.m

Force

tons force	tonf	0.1073	meganewton	MN
pounds force	lbf	0.0045	kilonewton	kN
pounds force	lbf	4.45	newton	N

Pressure and stress

pounds/square inch	psi	0.1073	kilogram/sq. centimetre	kg/cm²
pounds/square inch	psi	6.89	kilopascal	kPa
pounds/square inch	psi	0.0689	bar	
foot-pound	ft-lb	0.1383	kilogram metre	kgm

Energy

British Thermal Unit	Btu	0.252	kilocalorie	kcal
horsepower (hp)	hp	1.014	metric horsepower	CV
horsepower (hp)	hp	0.7457	kilowatt	kW

Speed

miles/hour	mph	1.61	kilometres/hour	km/h

CONVERSION TABLES – cont'd

Length

Millimetre	mm	1 in	=	25.4 cm	1 mm =	0.0394 in
Centimetre	cm	1 in	=	2.54 cm	1 cm =	0.3937 in
Metre	m	1 ft	=	0.3048 m	1 m =	3.2808 ft
		1 yd	=	0.9144 m	1 m =	1.0936 yd
Kilometre	km	1 mile	=	1.6093 km	1 km =	0.6214 mile

Note:
- 1 cm = 10 mm
- 1 m = 100 cm
- 1 km = 1,000 m
- 1 ft = 12 in
- 1 yd = 3 ft
- 1 mile = 1,760 yd

Area

Square millimetre	mm²	1 in²	=	645.2 mm²	1 mm² =	0.0016 in²
Square centimetre	cm²	1 in²	=	6.4516 cm²	1 cm² =	1.1550 in²
Square metre	m²	1 ft²	=	0.0929 m²	1 m² =	10.764 ft²
		1 yd²	=	0.8361 m²	1 m² =	1.1960 yd²
Square Kilometre	km²	1 mile2	=	2.590 km²	1 km² =	0.3861 mile2

Note:
- 1 cm² = 100 m²
- 1 m² = 10,000 cm²
- 1 km² = 100 hectares
- 1 ft² = 144 in²
- 1 yd² = 9 ft²
- 1 mile² = 640 acres
- 1 acre = 4,840 yd²

Volume

Cubic Centimetre	cm³	1 cm³	=	0.0610 in³	1 in³ =	16.387 cm³
Cubic Decimetre	dm³	1 dm³	=	0.0353 ft³	1 ft³ =	28.329 dm³
Cubic metre	m³	1 m³	=	35.3147 ft³	1 ft³ =	0.0283 m³
		1 m³	=	1.3080 yd³	1 yd³ =	0.7646 m³
Litre	L	1 L	=	1.76 pint	1 pint =	0.5683 L
			=	2.113 US pt		0.4733 US L

Note:
- 1 dm³ = 1,000 cm³
- 1 m³ = 1,000 dm³
- 1 L = 1 dm³
- 1 HL = 100 L
- 1 ft³ = 1.728 in³
- 1 yd³ = 27 ft³
- 1 pint = 20 fl oz
- 1 gal = 8 pints

Mass

Milligram	mg	1 mg	=	0.0154	grain	1 grain	=	64.935 mg
Gram	g	1 g	=	0.0353	oz	1 oz	=	28.35 g
Kilogram	kg	1 kg	=	2.2046	lb	1 lb	=	0.4536 kg
Tonne	t	1 t	=	0.9842	ton	1 ton	=	1.016 t

Note:

1 g	=	1,000 mg	1 oz	=	437.5 grains
1 kg	=	1000 g	1 lb	=	16 oz
1 t	=	1,000 kg	1 stone	=	14 lb
			1 cwt	=	112 lb
			1 ton	=	20 cwt

Force

Newton	N	1 lbf	=	4.448 N	1 kgf	=	9.807 N
Kilonewton	kN	1 lbf	=	0.00448 kN	1 ton f	=	9.964 kN
Meganewton	MN	100 tonf	=	0.9964 MN			

Pressure and stress

Kilonewton per square metre	kN/m^2	1 lbf/in^2	=	6.895 kN/m^2
		1 bar	=	100 kN/m^2
Meganewton per square metre	MN/m^2	1 $tonf/ft^2$	=	107.3 kN/m^2 = 0.1073 MN/m^2
		1 kgf/cm^2	=	98.07 kN/m^2
		1 lbf/ft^2	=	0.04788 kN/m^2

Temperature

Degree Celcius °C

$$°C = \frac{5 \times (°F - 32)}{9} \qquad °F = \frac{(9 \times °C)}{5} + 32$$

CONVERSION TABLES – cont'd

Metric Equivalents

1 km	=	1000 m
1 m	=	100 cm
1 cm	=	10 mm
1 km²	=	100 ha
1 ha	=	10,000 m²
1 m²	=	10,000 cm²
1 cm²	=	100 mm²
1 m³	=	1,000 litres
1 litre	=	1,000 cm³
1 metric tonne	=	1,000 kg
1 quintal	=	100 kg
1 N	=	0.10197 kg
1 kg	=	1000 g
1 g	=	1000 mg
1 bar	=	14.504 psi
1 cal	=	427 kg.m
1 cal	=	0.0016 cv.h
torque unit	=	0.00116 kw.h
1 CV	=	75 kg.m/s
1 kg/cm²	=	0.97 atmosph

English Unit Equivalents

1 mile	=	1760 yd
1 yd	=	3 ft
1 ft	=	12 in
1 sq mile	=	640 acres
1 acre	=	43,560 sq ft
1 sq ft	=	144 sq in
1 cu ft	=	7.48 gal liq
1 gal	=	231 cu in
	=	4 quarts liq
1 quart	=	32 fl oz
1 fl oz	=	1.80 cu in
	=	437.5 grains
1 stone	=	14 lb
1 cwt	=	112 lb
1 sh ton	=	2000 lb
1 lg ton	=	2240 lb
	=	20 cwt
1 lb	=	16 oz, avdp
1 Btu	=	778 ft lb
	=	0.000393 hph
	=	0.000293 kwh
1 hp	=	550 ft-lb/sec
1 atmosph	=	14.7 lb/in^2

CONVERSION TABLES - cont'd

Power Units		
kW	=	Kilowatt
HP	=	Horsepower
CV	=	Cheval Vapeur (Steam Horsepower)
	=	French designation for Metric Horsepower
PS	=	Pferderstarke (Horsepower)
	=	German designation of Metric Horsepower
1 HP	=	1.014 CV = 1.014 PS
	=	0.7457 kW
1 PS	=	1 CV = 0.9863 HP
	=	0.7355 kW
1 kW	=	1.341 HP
	=	1.359 CV
	=	1.359 PS

SPEED CONVERSION

km/h	m/min	Mph	fpm
1	16.7	0.6	54.7
2	33.3	1.2	109.4
3	50.0	1.9	164.0
4	66.7	2.5	218.7
5	83.3	3.1	273.4
6	100.0	3.7	328.1
7	116.7	4.3	382.8
8	133.3	5.0	437.4
9	150.0	5.6	492.1
10	166.7	6.2	546.8
11	183.3	6.8	601.5
12	200.0	7.5	656.2
13	216.7	8.1	710.8
14	233.3	8.7	765.5
15	250.0	9.3	820.2
16	266.7	9.9	874.9
17	283.3	10.6	929.6
18	300.0	11.2	984.3
19	316.7	11.8	1038.9
20	333.3	12.4	1093.6
21	350.0	13.0	1148.3
22	366.7	13.7	1203.0
23	383.3	14.3	1257.7
24	400.0	14.9	1312.3
25	416.7	15.5	1367.0
26	433.3	16.2	1421.7
27	450.0	16.8	1476.4
28	466.7	17.4	1531.1
29	483.3	18.0	1585.7
30	500.0	18.6	1640.4
31	516.7	19.3	1695.1
32	533.3	19.9	1749.8
33	550.0	20.5	1804.5
34	566.7	21.1	1859.1
35	583.3	21.7	1913.8

SPEED CONVERSION – cont'd

km/h	m/min	Mph	fpm
36	600.0	22.4	1968.5
37	616.7	23.0	2023.2
38	633.3	23.6	2077.9
39	650.0	24.2	2132.5
40	666.7	24.9	2187.2
41	683.3	25.5	2241.9
42	700.0	26.1	2296.6
43	716.7	26.7	2351.3
44	733.3	27.3	2405.9
45	750.0	28.0	2460.6
46	766.7	28.6	2515.3
47	783.3	29.2	2570.0
48	800.0	29.8	2624.7
49	816.7	30.4	2679.4
50	833.3	31.1	2734.0

FORMULAE

Two dimensional figures

Figure	Diagram of figure	Surface area	Perimeter
Square		a^2	$4a$
Rectangle		ab	$2(a+b)$
Triangle		$\frac{1}{2}ch$ $\frac{1}{2}ab \sin C$ $\sqrt{s(s-a)(s-b)(s-c)}$ where $s = \frac{1}{2}(a+b+c)$	$a+b+c = 2s$
Circle		πr^2 $\frac{1}{4}\pi d^2$ where $2r = d$	$2\pi r$ πd
Parallelogram		ah	$2(a+b)$
Trapezium		$\frac{1}{2}h(a+b)$	$a+b+c+d$
Ellipse		Approximately πab	$\pi(a+b)$
Hexagon		$2.6 \times a^2$	
Octagon		$4.83 \times a^2$	
Sector of circle		$\frac{1}{2}rb$ or $\frac{q}{360}\pi r^2$ note: b = angle $\frac{q}{360} \times 2\pi r$	
Segment of a circle		$S - T$ where S = area of sector T = area of triangle	
Bellmouth		$\frac{3}{14} \times r^2$	

FORMULAE - continued

Three dimensional figures

Figure	Diagram of figure	Surface area	Volume
Cube		$6a^2$	a^3
Cuboid/ rectangular block		$2(ab + ac + bc)$	abc
Prism/ triangular block		$bd + hc + dc + ad$	$\frac{1}{2}hcd$ $\frac{1}{2}ab \sin C \, d$ $d\sqrt{\{s(s-a)(s-b)(s-c)\}}$ where $s = \frac{1}{2}(a + b + c)$
Cylinder		$2\pi rh + 2\pi r^2$ $\pi dh + \frac{1}{2}\pi d^2$	$\pi r^2 h$ $\frac{1}{4}\pi d^2 h$
Sphere		$4\pi r^2$	$\frac{4}{3}\pi r^3$
Segment of sphere		$2\pi Rh$	$\frac{1}{6}\pi h(3r^2 + h^2)$ $\frac{1}{3}\pi h^2(3R - h)$
Pyramid		$(a + b)l + ab$	$\frac{1}{3}abh$
Frustrum of a pyramid		$l(a+b+c+d) + \sqrt{(ab+cd)}$ [regular figure only]	$\frac{h}{3}(ab + cd + \sqrt{abcd})$
Cone		$\pi rl + \pi r^2$ $\frac{1}{2}\pi dh + \frac{1}{4}\pi d^2$	$\frac{1}{3}\pi r^2 h$ $\frac{1}{12}\pi d^2 h$
Frustrum of a cone		$\pi r^2 + \pi R^2 + \pi h(R + r)$	$\frac{1}{3}\pi h(R^2 + Rr + r^2)$

FORMULAE

Formula	Description
Pythagoras theorem	$A^2 = B^2 + C^2$ where A is the hypotenuse of a right-angled triangle and B and C are the two adjacent sides

Simpsons Rule	The Area is divided into an even number of strips of equal width, and therefore has an odd number of ordinates at the division points
	
	area = $\dfrac{S(A + 2B + 4C)}{3}$
	
	where S = common interval (strip width)
	           A = sum of first and last ordinates
	           B = sum of remaining odd ordinates
	           C = sum of the even ordinates
	
	The Volume can be calculated by the same formula, but by substituting the area of each co-ordinate rather than its length.

Trapezoidal Rule	A given trench is divided into two equal sections, giving three ordinates, the first, the middle and the last.
	
	volume = $\dfrac{S \times (A + B + 2C)}{2}$
	
	where S = width of the strips
	           A = area of the first section
	           B = area of the last section
	           C = area of the rest of the sections

Prismoidal Rule	A given trench is divided into two equal sections, giving three ordinates, the first, the middle and the last.
	
	volume = $\dfrac{L \times (A + 4B + C)}{6}$
	
	where L = total length of trench
	           A = area of the first section
	           B = area of the middle section
	           C = area of the last section

EARTHWORK

Weights of typical materials handled by excavators.
The weight of the material is that of the state in its natural bed and includes moisture.
Adjustments should be made to allow for loose or compacted states.

Material	kg/m³	lb/cu yd
Adobe	1914	3230
Ashes	610	1030
Asphalt, rock	2400	4050
Basalt	2933	4950
Bauxite: alum ore	2619	4420
Borax	1730	2920
Caliche	1440	2430
Carnotite	2459	4150
Cement	1600	2700
Chalk (hard)	2406	4060
Cinders	759	1280
Clay: dry	1908	3220
Clay: wet	1985	3350
Coal: bituminouos	1351	2280
Coke	510	860
Conglomerate	2204	3720
Dolomite	2886	4870
Earth: dry	1796	3030
Earth: moist	1997	3370
Earth: wet	1742	2940
Feldspar	2613	4410
Felsite	2495	4210
Fluorite	3093	5220
Gabbro	3093	5220
Gneiss	2696	4550
Granite	2690	4540
Gravel, dry	1790	3020
Gypsum	2418	4080

Material	kg/m³	lb/cu yd
Lignite broken	1244	2100
Limestone	2596	4380
Magnesite, magnesium ore	2993	5050
Marble	2679	4520
Marl	2216	3740
Peat	700	1180
Potash	2193	3700
Pumice	640	1080
Quartz	2584	4360
Rhyolite	2400	4050
Sand: dry	1707	2880
Sand: wet	1831	3090
Sand and gravel - dry	1790	3020
- wet	2092	3530
Sandstone	2412	4070
Schist	2684	4530
Shale	2637	4450
Slag (blast)	2868	4840
Slate	2667	4500
Taconite	3182	5370
Topsoil	1440	2430
Trachyte	2400	4050
Traprock	2791	4710

Material	kg/m³	lb/cu yd
Snow - dry	130	220
- wet	509.61	860
Water	1000	62
Quarry waste	1438	90
Hardcore (consolidated)	1928	120

EARTHWORK – cont'd

Transport Capacities

Type of vehicle	Capacity of vehicle	
	Payload	Heaped capacity
Wheelbarrow	150	0.10
1 tonne dumper	1250	1.00
2.5 tonne dumper	4000	2.50
Articulated dump truck (Volvo A20 6x4)	18500	11.00
Articulated dump truck (Volvo A35 6x6)	32000	19.00
Large capacity rear dumper (Euclid R35)	35000	22.00
Large capacity rear dumper (Euclid R85)	85000	50.00

Bulkage of soils (After excavation)

Type of soil	Approximate bulking of 1 m^3 after excavation
Vegetable soil and loam	25 - 30%
Soft clay	30 - 40%
Stiff clay	10 - 20%
Gravel	20 - 25%
Sand	40 - 50%
Chalk	40 - 50%
Rock, weathered	30 - 40%
Rock, unweathered	50 - 60%

Shrinkage of materials (On being deposited)

Type of soil	Approximate bulking of 1 m^3 after excavation
Clay	10%
Gravel	8%
Gravel and sand	9%
Loam and light sandy soils	12%
Loose vegetable soils	15%

Voids in material used as sub bases or beddings

Material	m³ of voids/m³
Alluvium	0.37
River grit	0.29
Quarry sand	0.24
Shingle	0.37
Gravel	0.39
Broken stone	0.45
Broken bricks	0.42

Angles of repose

Type of soil		degrees
Clay	- dry	30
	- damp, well drained	45
	- wet	15 - 20
Earth	- dry	30
	- damp	45
Gravel	- moist	48
Sand	- dry or moist	35
	- wet	25
Loam		40

Slopes and angles

Ratio of base to height	Angle in degrees
5 : 1	11
4 : 1	14
3 : 1	18
2 : 1	27
1½ : 1	34
1 : 1	45
1 : 1½	56
1 : 2	63
1 : 3	72
1 : 4	76
1 : 5	79

EARTHWORK – cont'd

Grades (In Degrees and Percents)

Degrees	Percent	Degrees	Percent
1	1.8	24	44.5
2	3.5	25	46.6
3	5.2	26	48.8
4	7.0	27	51.0
5	8.8	28	53.2
6	10.5	29	55.4
7	12.3	30	57.7
8	14.0	31	60.0
9	15.8	32	62.5
10	17.6	33	64.9
11	19.4	34	67.4
12	21.3	35	70.0
13	23.1	36	72.7
14	24.9	37	75.4
15	26.8	38	78.1
16	28.7	39	81.0
17	30.6	40	83.9
18	32.5	41	86.9
19	34.4	42	90.0
20	36.4	43	93.3
21	38.4	44	96.6
22	40.4	45	100.0

Bearing powers

Ground conditions		Bearing Power		
		km/mý	lb/iný	metric t/m²
Rock (broken)		483	70	50
Rock (solid)		2,415	350	240
Clay,	dry or hard	380	55	40
	medium dry	190	27	20
	soft or wet	100	14	10
Gravel,	cemented	760	110	80
Sand,	compacted	380	55	40
	clean dry	190	27	20
Swamp and alluvial soils		48	7	5

Earthwork support

Maximum depth of excavation in various soils without the use of earthwork support:

Ground conditions	Feet (ft)	Metres (m)
Compact soil	12	3.66
Drained loam	6	1.83
Dry sand	1	0.3
Gravelly earth	2	0.61
Ordinary earth	3	0.91
Stiff clay	10	3.05

It is important to note that the above table should only be used as a guide.
Each case must be taken on its merits and, as the limited distances given above are approached, careful watch must be kept for the slightest signs of caving in.

CONCRETE WORK

Weights of concrete and concrete elements

Type of Material	kg/m³	lb/cu ft
Ordinary concrete (dense aggregates)		
Non-reinforced plain or mass concrete		
Nominal weight	2305	144
Aggregate - limestone	2162 to 2407	135 to 150
- gravel	2244 to 2407	140 to 150
- broken brick	2000 (av)	125 (av)
- other crushed stone	2326 to 2489	145 to 155
Reinforced concrete		
Nominal weight	2407	150
Reinforcement - 1%	2305 to 2468	144 to 154
- 2%	2356 to 2519	147 to 157
- 4%	2448 to 2703	153 to 163
Special concretes		
Heavy concrete		
Aggregates - barytes, magnetite	3210 (min)	200 (min)
steel shot, punchings	5280	330
Lean mixes		
Dry-lean (gravel aggregate)	2244	140
Soil-cement (normal mix)	1601	100

Type of material		kg/m² per mm thick	lb/sq ft per inch thick
Ordinary concrete (dense aggregates)			
Solid slabs (floors, walls etc.)			
Thickness:	75 mm or 3 in	184	37.5
	100 mm or 4 in	245	50
	150 mm or 6 in	378	75
	250 mm or 10 in	612	125
	300 mm or 12 in	734	150
Ribbed slabs			
Thickness:	125 mm or 5 in	204	42
	150 mm or 6 in	219	45
	225 mm or 9 in	281	57
	300 mm or 12 in	342	70
Special concretes			
Finishes etc			
Rendering, screed etc			
Granolithic, terrazzo		1928 to 2401	10 to 12.5
Glass-block (hollow) concrete		1734 (approx)	9 (approx)
Prestressed concrete		Weights as for reinforced concrete (upper limits)	
Air-entrained concrete		Weights as for plain or reinforced concrete	

CONCRETE WORK – cont'd

Average weight of aggregates

Materials	Voids %	Weight kg/m³
Sand	39	1660
Gravel 10 - 20 mm	45	1440
Gravel 35 - 75 mm	42	1555
Crushed stone	50	1330
Crushed granite (over 15 mm)	50	1345
(n.e. 15 mm)	47	1440
'All-in' ballast	32	1800 - 2000

Material	kg/m³	lb/cu yd
Vermiculite (aggregate)	64-80	108-135

Material	kg/m³	lb/cu ft
All-in aggregate	1999	125

Common mixes (per m³)

Recommended mix	Class of work suitable for: -	Cement (kg)	Sand (kg)	Coarse aggregate (kg)	No 50kg bags cement per m³ of combined aggregate
1:3:6	Roughest type of mass concrete such as footings, road haunching over 300mm thick	208	905	1509	4.00
1:2.5:5	Mass concrete of better class than 1:3:6 such as bases for machinery, walls below ground etc.	249	881	1474	5.00
1:2:4	Most ordinary uses of concrete, such as mass walls above ground, road slabs etc. and general reinforced concrete work	304	889	1431	6.00
1:1.5:3	Watertight floors, pavements and walls, tanks, pits, steps, paths, surface of 2 course roads, reinforced concrete where where extra strength is required	371	801	1336	7.50
1:1:2	Work of thin section such as fence posts and small precast work	511	720	1206	10.50

CONCRETE WORK – cont'd

Prescribed mixes for ordinary structural concrete

Weights of cement and total dry aggregates in kg to produce approximately one cubic metre of fully compacted concrete together with the percentages by weight of fine aggregate in total dry aggregates.

Conc. grade	Nominal max. size of aggregate (mm)	40		20		14		10	
	Workability	Med.	High	Med.	High	Med.	High	Med.	High
	Limits to slump that may be expected (mm)	50-100	100-150	25-75	75-125	10-50	50-100	10-25	25-50
7	Cement (kg)	180	200	210	230	-	-	-	-
	Total aggregate (kg)	1950	1850	1900	1800	-	-	-	-
	Fine aggregate (%)	30-45	30-45	35-50	35-50	-	-	-	-
10	Cement (kg)	210	230	240	260	-	-	-	-
	Total aggregate (kg)	1900	1850	1850	1800	-	-	-	-
	Fine aggregate (%)	30-45	30-45	35-50	35-50	-	-	-	-
15	Cement (kg)	250	270	280	310	-	-	-	-
	Total aggregate (kg)	1850	1800	1800	1750	-	-	-	-
	Fine aggregate (%)	30-45	30-45	35-50	35-50	-	-	-	-
20	Cement (kg)	300	320	320	350	340	380	360	410
	Total aggregate (kg)	1850	1750	1800	1750	1750	1700	1750	1650
	Sand Zone 1 (%)	35	40	40	45	45	50	50	55
	Sand Zone 2 (%)	30	35	35	40	40	45	45	50
	Sand Zone 3 (%)	30	30	30	35	35	40	40	45
25	Cement (kg)	340	360	360	390	380	420	400	450
	Total aggregate (kg)	1800	1750	1750	1700	1700	1650	1700	1600
	Sand Zone 1 (%)	35	40	40	45	45	50	50	55
	Sand Zone 2 (%)	30	35	35	40	40	45	45	50
	Sand Zone 3 (%)	30	30	30	35	35	40	40	45
30	Cement (kg)	370	390	400	430	430	470	460	510
	Total aggregate (kg)	1750	1700	1700	1650	1700	1600	1650	1550
	Sand Zone 1 (%)	35	40	40	45	45	50	50	55
	Sand Zone 2 (%)	30	35	35	40	40	45	45	50
	Sand Zone 3 (%)	30	30	30	35	35	40	40	45

Weights of bar reinforcement

Nominal sizes (mm)	Cross-sectional area (mm²)	Mass kg/m	Length of bar m/tonne
6	28.27	0.222	4505
8	50.27	0.395	2534
10	78.54	0.617	1622
12	113.10	0.888	1126
16	201.06	1.578	634
20	314.16	2.466	405
25	490.87	3.853	260
32	804.25	6.313	158
40	1265.64	9.865	101
50	1963.50	15.413	65

CONCRETE WORK – cont'd

Weights of bars (at specific spacings)

Weights of metric bars in kilogrammes per square metre

Size (mm)	Spacing of bars in millimetres									
	75	100	125	150	175	200	225	250	275	300
6	2.96	2.220	1.776	1.480	1.27	1.110	0.99	0.89	0.81	0 74
8	5.26	3.95	3.16	2.63	2.26	1.97	1.75	1.58	1.44	1.32
10	8.22	6.17	4.93	4.11	3.52	3.08	2.74	2.47	2.24	2.06
12	11.84	8.88	7.10	5.92	5.07	4.44	3.95	3.55	3.23	2 96
16	21.04	15.78	12.63	10.52	9.02	7.89	7.02	6.31	5.74	5 26
20	32.88	24.66	19.73	16.44	14.09	12.33	10.96	9.87	8.97	8.22
25	51.38	38.53	30.83	25.69	22.02	19.27	17.13	15.41	14.01	12.84
32	84.18	63.13	50.51	42.09	36.08	31.57	28.06	25.25	22.96	21.04
40	131.53	98.65	78.92	65.76	56.37	49.32	43.84	39.46	35.87	32.88
50	205.51	154.13	123.31	102.76	88.08	77.07	68.50	61.65	56.05	51.38

Basic weight of steelwork taken as 7850 kg/m³

Basic weight of bar reinforcement per metre run = 0.00785 kg/mm²

The value of PI has been taken as 3.141592654

Fabric reinforcement

Preferred range of designated fabric types and stock sheet sizes

Fabric reference	Longitudinal wires			Cross wires			Mass
	Nominal wire size (mm)	Pitch (mm)	Area (mm/m²)	Nominal wire size (mm)	Pitch (mm)	Area (mm/m²)	(kg/m²)
Square mesh							
A393	10	200	393	10	200	393	6.16
A252	8	200	252	8	200	252	3.95
A193	7	200	193	7	200	193	3.02
A142	6	200	142	6	200	142	2.22
A98	5	200	98	5	200	98	1.54
Structural mesh							
B1131	12	100	1131	8	200	252	10.90
B785	10	100	785	8	200	252	8.14
B503	8	100	503	8	200	252	5.93
B385	7	100	385	7	200	193	4.53
B283	6	100	283	7	200	193	3.73
B196	5	100	196	7	200	193	3.05
Long mesh							
C785	10	100	785	6	400	70.8	6.72
C636	9	100	636	6	400	70.8	5.55
C503	8	100	503	5	400	49.00	4.34
C385	7	100	385	5	400	49.00	3.41
C283	6	100	283	5	400	49.00	2.61
Wrapping mesh							
D98	5	200	98	5	200	98	1.54
D49	2.5	100	49	2.5	100	49	0.77
Stock sheet size	Length 4.8 m			Width 2.4 m		Sheet area 11.52m²	

CONCRETE WORK – cont'd

Wire

SWG	6g	5g	4g	3g	2g	1g	1/0g	2/0g	3/0g	4/0g	5/0g
diameter											
in	0.192	0.212	0.232	0.252	0.276	0.300	0.324	0.348	0.372	0.400	0.432
mm	4.9	5.4	5.9	6.4	7.0	7.6	8.2	8.8	9.5	0.2	1.0
area											
in^2	0.029	0.035	0.042	0.050	0.060	0.071	0.082	0.095	0.109	0.126	0.146
mm^2	19	23	27	32	39	46	53	61	70	81	95

Average weight kg/m³ of steelwork reinforcement in concrete for various building elements

	kg/m³ concrete
Substructure	
Pile caps	110 - 150
Tie beams	130 - 170
Ground beams	230 - 330
Bases	90 - 130
Footings	70 - 110
Retaining walls	110 - 150
Superstructure	
Slabs - one way	75 - 125
Slabs - two way	65 - 135
Plate slab	95 - 135
Cantilevered slab	90 - 130
Ribbed floors	80 - 120
Columns	200 - 300
Beams	250 - 350
Stairs	130 - 170
Walls - normal	30 - 70
Walls - wind	50 - 90

Note: For exposed elements add the following % :

Walls 50%, Beams 100%, Columns 15%

CONCRETE WORK – cont'd

Formwork stripping times

Normal Curing Periods

Conditions under which concrete is maturing	Minimum periods of protection for different types of cement						
	Number of days (where the average surface temperature of the concrete exceeds 10°C during the whole period)			Equivalent maturity (degree hours) calculated as the age of the concrete in hours multiplied by the number of degrees Celsius by which the average surface temperature of the concrete exceeds -10°C		
	Other	SRPC	OPC or RHPC	Other	SRPC	OPC or RHPC
1. Hot weather or drying winds	7	4	3	3500	2000	1500
2. Conditions not covered by 1	4	3	2	2000	1500	1000

KEY

OPC - Ordinary Portland Cement

RHPC - Rapid-hardening Portland cement

SRPC - Sulphate-resisting Portland cement

Minimum period before striking formwork

	Minimum period before striking		
	Surface temperature of concrete		
	16 °C	17 °C	t °C(0-25)
Vertical formwork to columns, walls and large beams	12 hours	18 hours	$\dfrac{300}{t+10}$ hours
Soffit formwork to slabs	4 days	6 days	$\dfrac{100}{t+10}$ days
Props to slabs	10 days	15 days	$\dfrac{250}{t+10}$ days
Soffit formwork to beams	9 days	14 days	$\dfrac{230}{t+10}$ days
Props to beams	14 days	21 days	$\dfrac{360}{t+10}$ days

MASONRY

Weights of bricks and blocks

Walls and components of walls	kg/m² per mm thick	lb/sq ft per inch thick
Blockwork		
Hollow clay blocks; average)	1.15	6
Common clay blocks	1.90	10
Brickwork		
Engineering clay bricks	2.30	12
Refactory bricks	1.15	6
Sand-lime (and similar) bricks	2.02	10.5

Weights of stones

Type of stone	kg/m3	lb/cu ft
Natural stone (solid)		
Granite	2560 to 2927	160 to 183
Limestone - Bath stone	2081	130
- Marble	2723	170
- Portland stone	2244	140
Sandstone	2244 to 2407	140 to 150
Slate	2880	180
Stone rubble (packed)	2244	140

Quantities of bricks and mortar

Materials per m² of wall:		
Thickness	**No. of Bricks**	**Mortar m³**
Half brick (112.5 mm)	58	0.022
One brick (225 mm)	116	0.055
Cavity, both skins (275 mm)	116	0.045
1.5 brick (337 mm)	174	0.074
Mass brickwork per m³	464	0.36

Mortar mixes: Quantities of Dry Materials

	Imperial cu yd			Metric m³		
Mix	Cement cwts	Lime cwts	Sand cu yds	Cement tonnes	Lime tonnes	Sand cu m
1:3	7.0	-	1.04	0.54	-	1.10
1:4	6.3	-	1.10	0.40	-	1.20
1:1:6	3.9	1.6	1.10	0.27	0.13	1.10
1:2:9	2.6	2.1	1.10	0.20	0.15	1.20
0:1:3	-	3.3	1.10	-	0.27	1.00

MASONRY – cont'd

Mortar mixes for various uses

Mix	Use
1:3	Construction designed to withstand heavy loads in all seasons
1:1:6	Normal construction not designed for heavy loads. Sheltered and moderate conditions in spring and summer. Work above d:p:c - sand, lime bricks, clay blocks etc.
1:2:9	Internal partitions with blocks which have high drying shrinkage, pumicblocks, etc. any periods
0:1:3	Hydraulic lime only should be used in this mix and may be used for construction not designed for heavy loads and above d:p:c spring and summer.

Quantities of bricks and mortar required per m² of walling

Description	Unit	Nr of bricks required	Mortar required (m³)		
			No frogs	Single frogs	Double frogs
Standard bricks					
Brick size					
215 x 102.5 x 50 mm					
half brick wall (103 mm)					
(103 mm)	m²	72	0.022	0.027	0.032
2 x half brick cavity wall					
(270 mm)	m²	144	0.044	0.054	0.064
one brick wall					
(215 mm)	m²	144	0 052	0 064	0.076
one and a half brick wall					
(328 mm)	m²	216	0.073	0.091	0.108
mass brickwork	m³	576	0.347	0.413	0.480
Brick size					
215 x 102.5 x 65 mm					
half brick wall	m²	58	0.019	0.022	0.026
(103 mm)					
2 x half brick cavity wall					
(270 mm)	m²	116	0.038	0 045	0.055
one brick wall					
(215 mm)	m²	116	0 046	0.055	0.064
one and a half brick wall					
(328 mm)	m²	174	0.063	0.074	0.088
mass brickwork	m³	464	0.307	0 360	0.413
Metric modular bricks		**Perforated**			
Brick co-ordinating size					
200 x 100 x 75 mm					
90 mm thick	m²	67	0.016	0.019	
190 mm thick	m²	133	0.042	0.048	
290 mm thick	m²	200	0.068	0.078	
Brick co-ordinating size					
200 x 100 x 100 mm					
90 mm thick	m²	50	0.013	0.016	
190 mm thick	m²	100	0.036	0.041	
290 mm thick	m²	150	0.059	0.067	
Brick co-ordinating size					
300 x 100 x 75 mm					
90 mm thick	m²	33	-	0.015	
Brick co-ordinating size					
00 x 100 x 100 mm					
90 mm thick	m²	44	0.015	0 018	

Note: Assuming 10 mm deep joints.

MASONRY – cont'd

Mortar required per m² blockwork (9.88 blocks/m²)

Wall thickness	75	90	100	125	140	190	215
Mortar m³/m²	0.005	0.006	0.007	0.008	0.009	0.013	0.014

Mortar Mixes

Mortar Group	Cement: lime: sand	Masonry cement: sand	Cement: sand with plasticiser
1	1 : 0-0.25:3		
2	1 : 0.5 :4-4.5	1 : 2.5-3.5	1 : 3-4
3	1 : 1:5-6	1 : 4-5	1 : 5-6
4	1 : 2:8-9	1 : 5.5-6.5	1 : 7-8
5	1 : 3:10-12	1 : 6.5-7	1 : 8

Group 1: strong inflexible mortar
Group 5: weak but flexible.

All mixes within a group are of approximately similar strength.
Frost resistance increases with the use of plasticisers.
Cement:lime:sand mixes give the strongest bond and greatest resistance to rain penetration.
Masonry cement equals ordinary Portland cement plus a fine neutral mineral filler and an air entraining agent.

Calcium Silicate Bricks

Type	Strength	Location
Class 2 crushing strength	14.0N/mm2	not suitable for walls
Class 3	20.5N/mm2	walls above dpc
Class 4	27.5N/mm2	cappings and copings
Class 5	34.5N/mm2	retaining walls
Class 6	41.5N/mm2	walls below ground
Class 7	48.5N/mm2	walls below ground

The Class 7 calcium silicate bricks are therefore equal in strength to Class B bricks.
Calcium silicate bricks are not suitable for DPCs

Durability of Bricks

FL	Frost resistant with low salt content
FN	Frost resistant with normal salt content
ML	Moderately frost resistant with low salt content
MN	Moderately frost resistant with normal salt content

Brickwork Dimensions

No. of Horizontal Bricks	Dimensions mm	No. of Vertical courses	No. of Vertical courses
1/2	112.5	1	75
1	225.0	2	150
1 1/2	337.5	3	225
2	450.0	4	300
2 1/2	562.5	5	375
3	675.0	6	450
3 1/2	787.5	7	525
4	900.0	8	600
4 1/2	1012.5	9	675
5	1125.0	10	750
5 1/2	1237.5	11	825
6	1350.0	12	900
6 1/2	1462.5	13	975
7	1575.0	14	1050
7 1/2	1687.5	15	1125
8	1800.0	16	1200
8 1/2	1912.5	17	1275
9	2025.0	18	1350
9 1/2	2137.5	19	1425
10	2250.0	20	1500
20	4500.0	24	1575
40	9000.0	28	2100
50	11250.0	32	2400
60	13500.0	36	2700
75	16875.0	40	3000

MASONRY – cont'd

Standard available block sizes			
Block	Length x height		
	Co-ordinating size	Work size	Thicknesses (work size)
A	400 x 100	390 x 90	(75, 90, 100, 140 &
	400 x 200	440 x 190	(190 mm
	450 x 225	440 x 215	(75, 90, 100, 140,
			(190 & 215 mm
B	400 x 100	390 x 90	(75, 90, 100, 140 &
	400 x 200	390 x 190	(190 mm
	450 x 200	440 x 190	(
	450 x 225	440 x 215	(75, 90, 100, 140,
	450 x 300	440 x 290	(190 & 215 mm
	600 x 200	590 x 190	(
	600 x 225	590 x 215	(
C	400 x 200	390 x 190	(
	450 x 200	440 x 190	(
	450 x 225	440 x 215	(60 & 75 mm
	450 x 300	440 x 290	(
	600 x 200	590 x 190	(
	600 x 225	590 x 215	(

TIMBER

Weights of timber

Material	kg/m³	lb/cu ft
General	806 (avg)	50 (avg)
Douglas fir	479	30
Yellow pine, spruce	479	30
Pitch pine	673	42
Larch, elm	561	35
Oak (English)	724 to 959	45 to 60
Teak	643 to 877	40 to 55
Jarrah	959	60
Greenheart	1040 to 1204	65 to 75
Quebracho	1285	80

Material	kg/m² per mm thickness	lb/sq ft per inch thickness
Wooden boarding and blocks		
Softwood	0.48	2.5
Hardwood	0.76	4
Hardboard	1.06	5.5
Chipboard	0.76	4
Plywood	0.62	3.25
Blockboard	0.48	2.5
Fibreboard	0.29	1.5
Wood-wool	0.58	3
Plasterboard	0.96	5
Weather boarding	0.35	1.8

TIMBER

Conversion tables (for sawn timber only)

Inches >	Millimetres	Feet >	Metres
1	25	1	0.300
2	50	2	0.600
3	75	3	0.900
4	100	4	1.200
5	125	5	1.500
6	150	6	1.800
7	175	7	2.100
8	200	8	2.400
9	225	9	2.700
10	250	10	3.000
11	275	11	3.300
12	300	12	3.600
13	325	13	3.900
14	350	14	4.200
15	375	15	4.500
16	400	16	4.800
17	425	17	5.100
18	450	18	5.400
19	475	19	5.700
20	500	20	6.000
21	525	21	6.300
22	550	22	6.600
23	575	23	6.900
24	600	24	7.200

Planed softwood

The finished end section size of planed timber is usually 3/16" less than the original size from which it is produced. This however varies slightly dependant upon availability of material and origin of species used.

Standard (timber) to cubic metres and cubic metres to standards (timber)

m³	m³/Standards	Standard
4.672	1	0.214
9.344	2	0.428
14.017	3	0.642
18.689	4	0.856
23.361	5	1.070
28.033	6	1.284
32.706	7	1.498
37.378	8	1.712
42.05	9	1.926
46.722	10	2.140
93.445	20	4.281
140.167	30	6.421
186.890	40	8.561
233.612	50	10.702
280.335	60	12.842
327.057	70	14.982
373.779	80	17.122
420.502	90	19.263
467.224	100	21.403

1 cu metre = 35.3148
1 cu ft = 0.028317 cu metres
1 std = 4.67227 cu metres

TIMBER

Standards (timber) to cubic metres and cubic metres to standards (timber)

1 cu metre = 35.3148 cu ft = 0.21403 std

1 cu ft = 0.028317 cu metres

1 std = 4.67227 cu metres

Basic sizes of sawn softwood available (cross sectional areas)

Thickness (mm)	Width (mm)								
	75	100	125	150	175	200	225	250	300
16	x	x	x	x					
19	x	x	x	x					
22	x	x	x	x					
25	x	x	x	x	x	x	x	x	x
32	x	x	x	x	x	x	x	x	x
36	x	x	x	x					
38	x	x	x	x	x	x	x		
44	x	x	x	x	x	x	x	x	x
47*	x	x	x	x	x	x	x	x	x
50	x	x	x	x	x	x	x	x	x
63	x	x	x	x	x	x	x		
75		x	x	x	x	x	x	x	x
100		x		x		x		x	x
150				x		x			x
200						x			
250								x	
300									x

* This range of widths for 47 mm thickness will usually be found to be available in construction quality only.

Note: The smaller sizes below 100 mm thick and 250 mm width are normally but not exclusively of European origin. Sizes beyond this are usually of North and South American origin.

TIMBER

Basic lengths of sawn softwood available (metres)

1.80	2.10	3.00	4.20	5.10	6.00	7.20
2.40	3.30	4.50	5.40	6.30		
2.70	3.60	4.80	5.70	6.60		
	3.90			6.90		

Note: Lengths of 6.00 m and over will generally only be available from North American species and may have to be recut from larger sizes.

Reductions from basic size to finished size of timber by planing of two opposed faces

Purpose	15 - 35 mm	36 - 100 mm	101 - 150 mm	over 150 mm
a) constructional timber	3 mm	3 mm	5 mm	6 mm
b) matching interlocking boards	4 mm	4 mm	6 mm	6 mm
c) wood trim not specified in BS 584	5 mm	7 mm	7 mm	9 mm
d) joinery and cabinet work	7 mm	9 mm	11 mm	13 mm

Note: The reduction of width or depth is overall the extreme size and is exclusive of any reduction of the face by the machining of a tongue or lap joints.

Weights of metalwork

Material	kg/m³	lb/cu ft
Metals, steel construction, etc		
Iron		
- cast	7207	450
- wrought	7687	480
- ore - general	2407	150
- (crushed) Swedish	3682	230
Steel	7854	490
Copper		
- cast	8731	545
- wrought	8945	558
Brass	8497	530
Bronze	8945	558
Aluminium	2774	173
Lead	11322	707
Zinc (rolled)	7140	446

	g/mm² per metre	lb/sq ft per foot
Steel bars	7.85	3.4

Structural steelwork	Net weight of member @ 7854 kg/m³
rivetted	+ 10% for cleats, rivets, bolts, etc
welded	+ 1.25% to 2.5% for welds, etc
Rolled sections	
beams	+ 2.5%
stanchions	+ 5% (extra for caps and bases)
Plate	
web girders	+ 10% for rivets or welds, stiffeners, etc

	kg/m	lb/ft
Steel stairs : industrial type		
1 m or 3ft wide	84	56
Steel tubes		
50 mm or 2 in bore	5 to 6	3 to 4
Gas piping		
20 mm or 3/4 in	2	1¼

KERBS/EDGINGS/CHANNELS

Precast Concrete Kerbs to BS 7263
Straight kerb units: length from 450 to 915 mm

150mm high x 125mm thick
  bullnosed   type BN
  half battered   type HB3

255mm high x 125mm thick
  45 degree splayed   type SP
  half battered   type HB2

305mm high x 150mm thick
  half battered   type HB1

Quadrant kerb units

150 mm high x 305 and 455 mm radius to match	type BN	type QBN
150 mm high x 305 and 455 mm radius to match	type HB2, HB3	type QHB
150 mm high x 305 and 455 mm radius to match	type SP	type QSP
255 mm high x 305 and 455 mm radius to match	type BN	type QBN
255 mm high x 305 and 455 mm radius to match	type HB2, HB3	type QHB
225 mm high x 305 and 455 mm radius to match	type SP	type QSP

Angle kerb units
  305 x 305 x 225 mm high x 125 mm thick
    bullnosed external angle   type XA
    splayed external angle to match type SP   type XA
    bullnosed internal angle   type IA
    splayed internal angle to match type SP   type IA

Channels
  255 mm wide x 125 mm high flat   type CS1
  150 mm wide x 125 mm high flat type   CS2
  255 mm wide x 125 mm high dished   type CD

Transition kerb units

from kerb type SP to HB	left handed	type TL
	right handed	type TR
from kerb type BN to HB	left handed	type DL1
	right handed	type DR1
from kerb type BN to SP	left handed	type DL2
	right handed	type DR2

Radial kerbs and channels

All profiles of kerbs and channels	
External radius	**Internal radius**
1000 mm	3000
2000	4500
3000	6000
4500	7500
6000	9000
7500	1050
9000	1200
1050	
1200	

Precast Concrete Edgings to BS 7263

Round top type ER	Flat top type EF	Bullnosed top type EBN
150 x 50 mm	150 x 50 mm	150 x 50 mm
200 x 50	200 x 50	200 x 50
250 x 50	250 x 50	250 x 50

BASES

Cement Bound Material for Bases and Sub-bases

CBM1: very carefully graded aggregate from 37.5 - 75ym, with a 7-day strength of 4.5N/mm2

CBM2: same range of aggregate as CBM1 but with more tolerance in each size of ggregate with a 7-day strength of 7.0N/mm2

CBM3: crushed natural aggregate or blastfurnace slag, graded from 37.5mm - 150ym for 40mm aggregate, and from 20 - 75ym for 20mm aggregate, with a 7-day strength of 10N/mm2

CBM4: crushed natural aggregate or blastfurnace slag, graded from 37.5mm - 150ym for 40mm aggregate, and from 20 - 75ym for 20mm aggregate, with a 7-day strength of 15N/mm2

INTERLOCKING BRICK/BLOCK ROADS/PAVINGS

Sizes of Precast Concrete Paving Blocks to BS 6717: Part 1

Type R blocks
200 x 100 x 60 mm
200 x 100 x 65
200 x 100 x 80
200 x 100 x 100

Type S
Any shape within a 295 mm space

Sizes of Clay Brick Pavers to BS 6677: Part 1
200 x 100 x 50 mm thick
200 x 100 x 65
210 x 105 x 50
210 x 105 x 65
215 x 102.5 x 50
215 x 102.5 x 65

Type PA: 3 kN
Footpaths and pedestrian areas, private driveways, car parks, light vehicle traffic and over-run.

Type PB: 7 kN
Residential roads, lorry parks, factory yards, docks, petrol station forecourts, hardstandings, bus stations.

PAVING AND SURFACING

Weights and sizes of paving and surfacing

Description of Item		Quantity per tonne
Paving 50 mm thick	900 x 600 mm	15
Paving 50 mm thick	750 x 600 mm	18
Paving 50 mm thick	600 x 600 mm	23
Paving 50 mm thick	450 x 600 mm	30
Paving 38 mm thick	600 x 600 mm	30
Path edging	914 x 50 x 150 mm	60
Kerb (including radius and tapers)	125 x 254 x 914 mm	15
Kerb (including radius and tapers)	125 x 150 x 914 mm	25
Square channel	125 x 254 x 914 mm	15
Dished channel	125 x 254 x 914 mm	15
Quadrants	300 x 300 x 254 mm	19
Quadrants	450 x 450 x 254 mm	12
Quadrants	300 x 300 x 150 mm	30
Internal angles	300 x 300 x 254 mm	30
Fluted pavement channel	255 x 75 x 914 mm	25
Corner stones	300 x 300 mm	80
Corner stones	360 x 360 mm	60
Cable covers	914 x 175 mm	55
Gulley kerbs	220 x 220 x 150 mm	60
Gulley kerbs	220 x 200 x 75 mm	120

Material	kg/m³	lb/cu yd
Tarmacadam	2306	3891
Macadam (waterbound)	2563	4325
Vermiculite (aggregate)	64-80	108-135
Terracotta	2114	3568
Cork - compressed	388	24

	kg/m²	lb/sq ft
Clay floor tiles, 12.7mm	27.3	5.6
Pavement lights	122	25
Damp proof course	5	1

	kg/m² per mm thickness	lb/sq ft per inch thickness
Paving Slabs (stone)	2.3	12
Granite setts	2.88	15
Asphalt	2.30	12
Rubber flooring	1.68	9
Poly-vinylchloride	1.94 (avg)	10 (avg)

PAVING AND SURFACING - cont'd

Coverage (m²) per cubic metre of materials used as sub bases or capping layers

Consolidated thickness laid in (mm)	Square metre coverage		
	Gravel	Sand	Hardcore
50	15.80	16.50	-
75	10.50	11.00	-
100	7.92	8.20	7.42
125	6.34	6.60	5.90
150	5.28	5.50	4.95
175	-	-	4.23
200	-	-	3.71
225	-	-	3.30
300	-	-	2.47

Approximate rate of spreads

Average thickness of course mm	Description	Approximate rate of spread			
		Open Textured		Dense, Medium & Fine Textured	
		kg/m²	m²/t	kg/m²	m²/t
35	14 mm open textured or dense wearing course	60-75	13-17	70-85	12-14
40	20 mm open textured or dense base course	70-85	12-14	80-100	10-12
45	20 mm open textured or dense base course	80-100	10-12	95-100	9-10
50	20 mm open textured or dense, or 28 mm dense base course	85-110	9-12	110-120	8-9
60	28 mm dense base course, 40 mm open textured of dense base course or 40 mm single course as base course		8-10	130-150	7-8
65	28 mm dense base course, 40 mm open textured or dense base course or 40 mm single course	100-135	7-10	140-160	6-7
75	40 mm single course, 40 mm open textured or dense base course, 40 mm dense roadbase	120-150	7-8	165-185	5-6
100	40 mm dense base course or roadbase	-	-	220-240	4-4.5

Surface Dressing Roads: Coverage (m²) per tonne of Material

Size in mm	Sand	Granite chips	Gravel	Limestone Chips
Sand	168	-	-	-
3	-	148	152	165
6	-	130	133	144
9	-	111	114	123
13	-	85	87	95
19	-	68	71	78

Sizes of Flags to BS 7263

Reference	Nominal Size	Thickness
A	600 x 450 mm	50 and 63 mm
B	600 x 600	50 and 63
C	600 x 750	50 and 63
D	600 x 900	50 and 63
E	450 x 450	50 and 70 chamfered top surface
F	400 x 400	50 and 65 chamfered top surface
G	300 x 300	50 and 60 chamfered top surface

Sizes of Natural Stone Setts to BS 435

Width		Length		Depth
100 mm	x	100 mm	x	100 mm
75	x	150 to 250	x	125
75	x	150 to 250	x	150
100	x	150 to 250	x	100
100	x	150 to 250	x	150

SPORTS
Sizes of Sports Areas

Sizes in <u>metres</u> given include clearances:

Association football			114	x	72
	Junior		108	x	58
	International		100 - 110	x	64 - 75
Football	American	Pitch	109.80	x	48.80
		Overall	118.94	x	57.94
	Australian Rules	Overall	135 - 185	x	110 - 155
	Canadian	Overall	145.74	x	59.47
	Gaelic		128 - 146.4	x	76.8 - 91.50
Handball			91 - 110	x	55 - 65
Hurling			137	x	82
Rugby	Union pitch		56	x	81
	League pitch		134	x	80
Hockey pitch			100.5	x	61
Men's lacrosse pitch			106	x	61
Women's lacrosse pitch			110	x	60
Target archery ground			150	x	50
Archery (Clout)			7.3m firing area		
			Range 109.728 (Women), 146.304 (Men).		
			182.88 (Normal range)		
400m Running Track			115.61 bend length x 2		
6 lanes			84.39 straight length x 2		
		Overall	176.91 long	x	92.52 wide
Baseball		Overall	60m	x	70
Basketball			14.0	x	26.0
Camogie			91 - 110	x	54 - 68

Tables and Memoranda

Discus and Hammer	Safety cage 2.74m square	
	Landing area 45 arc (65° safety) 70 m radius	
Javelin	Runway	36.5 x 4.27
	Landing area	80 - 95 x 48
Jump High	Running area	38.8 x 19
	Landing area	5 x 4
Long	Runway	45 x 1.22
	Landing area	9 x 2.750
Triple	Runway	45 x 1.22
	Landing area	7.3 x 2.75
Korfball		90 x 40
Netball		15.25 x 30.48
Pole Vault	Runway	45 x 1.22
	Landing area	5 x 5
Polo		275 x 183
Rounders	Overall	19 x 17
Shot Putt	Base	2.135 dia
	Landing area	65° arc, 25m radius from base
Shinty		128 -183 x 64 - 91.5
Tennis	Court	23.77 x 10.97
	Overall minimum	36.27 x 18.29
Tug-of-war		46 x 5

SEEDING/TURFING AND PLANTING

BS 3882: 1994 Topsoil Quality

Topsoil grade	Properties
Premium	natural topsoil, high fertility, loamy texture, good soil structure, suitable for intensive cultivation
General Purpose	natural or manufactured topsoil of lesser quality than Premium, suitable for agriculture or amenity landscape, may need fertilizer or soil structure improvement.
Economy	selected subsoil, natural mineral deposit such as river silt or greensand. The grade comprises two subgrades; "Low clay" and "High clay" which is more liable to compaction in handling. This grade is suitable for low production agricultural land and amenity woodland or conservation planting areas.

Forms of Trees to BS 3936: 1992

Standards: shall be clear with substantially straight stems. Grafted and budded trees shall have no more than a slight bend at the union. Standards shall be designated as Half, Extra light, Light, Standard, Selected standard, Heavy, and Extra heavy.

Sizes of Standards

Heavy standard	12-14 cm girth x 3.50 to 5.00 m high
Extra Heavy standard	14-16 cm girth x 4.25 to 5.00 m high
Extra Heavy standard	16-18 cm girth x 4.25 to 6.00 m high
Extra Heavy standard	18-20 cm girth x 5.00 to 6.00 m high

Semi-mature trees: between 6.0m and 12.0 m tall with a girth of 20 to 75 cm at 1.0 m above ground.

Feathered trees: shall have a defined upright central leader, with stem furnished with evenly spread and balanced lateral shoots down to or near the ground.

Whips: shall be without significant feather growth as determined by visual inspection.

Multi-stemmed trees: shall have two or more main stems at, near, above or below ground.

Seedlings grown from seed and not transplanted shall be specified when ordered for sale as:

1+0 one year old seedling		
2+0 two year old seedling		
1+1 one year seed bed,	one year transplanted	= two year old seedling
1+2 one year seed bed,	two years transplanted	= three year old seedling
2+1 two year seed bed,	one year transplanted	= three year old seedling
1u1 two years seed bed,	undercut after 1 year	= two year old seedling
2u2 four years seed bed,	undercut after 2 years	= four year old seedling

Cuttings. The age of cuttings (plants grown from shoots, stems, or roots of the mother plant) shall be specified when ordered for sale. The height of transplants and undercut seedlings/cuttings (which have been transplanted or undercut at least once) shall be stated in centimetres. The number of growing seasons before and after transplanting or undercutting shall be stated.

0+1	one year cutting
0+2	two year cutting
0+1+1	one year cutting bed, one year transplanted = two year old seedling
0+1+2	one year cutting bed, two years transplanted = three year old seedling

SEEDING/TURFING AND PLANTING

Grass Cutting Capacities in m2 per Hour

Speed mph	Width Of Cut in metres												
	0.5	0.7	1.0	1.2	1.5	1.7	2.0	2.0	2.1	2.5	2.8	3.0	3.4
1.0	724	1127	1529	1931	2334	2736	3138	3219	3380	4023	4506	4828	5472
1.5	1086	1690	2293	2897	3500	4104	4707	4828	5069	6035	6759	7242	8208
2.0	1448	2253	3058	3862	4667	5472	6276	6437	6759	8047	9012	9656	10944
2.5	1811	2816	3822	4828	5834	6840	7846	8047	8449	10058	11265	12070	13679
3.0	2173	3380	4587	5794	7001	8208	9415	9656	10139	12070	13518	14484	16415
3.5	2535	3943	5351	6759	8167	9576	10984	11265	11829	14082	15772	16898	19151
4.0	2897	4506	6115	7725	9334	10944	12553	12875	13518	16093	18025	19312	21887
4.5	3259	5069	6880	8690	10501	12311	14122	14484	15208	18105	20278	21726	24623
5.0	3621	5633	7644	9656	11668	13679	15691	16093	16898	20117	22531	24140	27359
5.5	3983	6196	8409	10622	12834	15047	17260	17703	18588	22128	24784	26554	30095
6.0	4345	6759	9173	11587	14001	16415	18829	19312	20278	24140	27037	28968	32831
6.5	4707	7322	9938	12553	15168	17783	20398	20921	21967	26152	29290	31382	35566
7.0	5069	7886	10702	13518	16335	19151	21967	22531	23657	28163	31543	33796	38302

Number of plants per m2 : For plants planted on an evenly spaced grid
Planting distances

mm	0.10	0.15	0.20	0.25	0.35	0.40	0.45	0.50	0.60	0.75	0.90	1.00	1.20	1.50
0.10	100.00	66.67	50.00	40.00	28.57	25.00	22.22	20.00	16.67	13.33	11.11	10.00	8.33	6.67
0.15	66.67	44.44	33.33	26.67	19.05	16.67	14.81	13.33	11.11	8.89	7.41	6.67	5.56	4.44
0.20	50.00	33.33	25.00	20.00	14.29	12.50	11.11	10.00	8.33	6.67	5.56	5.00	4.17	3.33
0.25	40.00	26.67	20.00	16.00	11.43	10.00	8.89	8.00	6.67	5.33	4.44	4.00	3.33	2.67
0.35	28.57	19.05	14.29	11.43	8.16	7.14	6.35	5.71	4.76	3.81	3.17	2.86	2.38	1.90
0.40	25.00	16.67	12.50	10.00	7.14	6.25	5.56	5.00	4.17	3.33	2.78	2.50	2.08	1.67
0.45	22.22	14.81	11.11	8.89	6.35	5.56	4.94	4.44	3.70	2.96	2.47	2.22	1.85	1.48
0.50	20.00	13.33	10.00	8.00	5.71	5.00	4.44	4.00	3.33	2.67	2.22	2.00	1.67	1.33
0.60	16.67	11.11	8.33	6.67	4.76	4.17	3.70	3.33	2.78	2.22	1.85	1.67	1.39	1.11
0.75	13.33	8.89	6.67	5.33	3.81	3.33	2.96	2.67	2.22	1.78	1.48	1.33	1.11	0.89
0.90	11.11	7.41	5.56	4.44	3.17	2.78	2.47	2.22	1.85	1.48	1.23	1.11	0.93	0.74
1.00	10.00	6.67	5.00	4.00	2.86	2.50	2.22	2.00	1.67	1.33	1.11	1.00	0.83	0.67
1.20	8.33	5.56	4.17	3.33	2.38	2.08	1.85	1.67	1.39	1.11	0.93	0.83	0.69	0.56
1.50	6.67	4.44	3.33	2.67	1.90	1.67	1.48	1.33	1.11	0.89	0.74	0.67	0.56	0.44

Grass Clippings Wet: Based on 3.5 m3 /ton

Annual Kg/100 m2	Average 20 cuts Kg/100m2	m2 /tonne	m2 /m3
32.0	1.6	61162.1	214067.3

Nr of Cuts	22	20	18	16	12	4
Kg/cut	1.45	1.60	1.78	2.00	2.67	8.00

Area capacity of 3 tonne vehicle per load						
m2	206250	187500	168750	150000	112500	37500

Load m3	100 m2 units / m3 of vehicle space					
1	196.4	178.6	160.7	142.9	107.1	35.7
2	392.9	357.1	321.4	285.7	214.3	71.4
3	589.3	535.7	482.1	428.6	321.4	107.1
4	785.7	714.3	642.9	571.4	428.6	142.9
5	982.1	892.9	803.6	714.3	535.7	178.6

FENCING AND GATES

Types of Preservative to BS 5589:1989

Creosote (tar oil) can be "factory" applied
- by pressure to BS 144: pts 1&2
- by immersion to BS 144: pt 1
- by hot and cold open tank to BS 144: pts 1&2

Copper/chromium/arsenic (CCA)
Organic solvent (OS)
- by full cell process to BS 4072 pts 1&2
- by double vacuum (vacvac) to BS 5707 pts 1&3
- by immersion to BS 5057 pts 1&3

Pentachlorophenol (PCP)
- by heavy oil double vacuum to BS 5705 pts 2&3

Boron diffusion process (treated with disodium octaborate to BWPA Manual 1986.
Note: Boron is used on green timber at source and the timber is suppled dry.

Cleft Chestnut Pale Fences to BS 1722:Part 4:1986

Pales	Pale spacing	Wire lines	
900 mm long	75 mm	2	temporary protection
1050	75 or 100	2	light protective fences
1200	75	3	perimeter fences
1350	75	3	perimeter fences
1500	50	3	narrow perimeter fences
1800	50	3	light security fences

Close-boarded Fences to BS 1722 :Pt 5: 1986

Close-boarded fences 1.05 to 1.8m high
Type BCR (recessed) or BCM (morticed) with concrete posts 140 x 115 mm tapered and Type BW with timber posts:

Palisade Fences to BS 1722:pt 6:1986.

Wooden palisade fences
Type WPC with concrete posts 140 x 115 mm tapered and Type WPW with timber posts.

For both types of fence:
- Height of fence 1050 mm: two rails
- Height of fence 1200 mm: two rails
- Height of fence 1500 mm: three rails
- Height of fence 1650 mm: three rails
- Height of fence 1800 mm: three rails
- Height of fence 1800 mm: three rails

Post and Rail Fences to BS 1722: part 7

Wooden post and rail fences
Type MPR 11/3 morticed rails and Type SPR 11/3 nailed rails
Height to top of rail 1100 mm
Rails: three rails 87 mm 38 mm

Type MPR 11/4 morticed rails and Type SPR 11/4 nailed rails
Height to top of rail 1100 mm
Rails: four rails 87 mm 38 mm.

Type MPR 13/4 morticed rails and Type SPR 13/4 nailed rails
Height to top of rail 1300 mm
Rail spacing 250 mm, 250 mm, and 225 mm from top
Rails: four rails 87 mm 38 mm.

Steel Posts to BS 1722: Part 1.

Rolled steel angle iron posts for chain link fencing:

Posts	Fence height	Strut	Straining post
1500 x 40 x 40 x 5 mm	900 mm	1500 x 40 x 40 x 5 mm	1500 x 50 x 50 x 6 mm
1800 x 40 x 40 x 5 mm	1200 mm	1800 x 40 x 40 x 5 mm	1800 x 50 x 50 x 6 mm
2000 x 45 x 45 x 5 mm	1400 mm	2000 x 45 x 45 x 5 mm	2000 x 60 x 60 x 6 mm
2600 x 45 x 45 x 5 mm	1800 mm	2600 x 45 x 45 x 5 mm	2600 x 60 x 60 x 6 mm
3000 x 50 x 50 x 6 mm with arms	1800 mm	2600 x 45 x 45 x 5 mm	3000 x 60 x 60 x 6 mm

Concrete Posts to BS 1722: Part 1.

Concrete posts for chain link fencing:

Posts and straining posts	Fence height	Strut
1570 mm 100 x 100 mm	900 mm	1500 mm x 75 x 75 mm
1870 mm 125 x 125 mm	1200 mm	1830 mm x 100 x 75 mm
2070 mm 125 x 125 mm	1400 mm	1980 mm x 100 x 75 mm
2620 mm 125 x 125 mm	1800 mm	2590 mm x 100 x 85 mm
3040 mm 125 x 125 mm	1800 mm	2590 mm x 100 x 85 mm (with arms)

Rolled Steel Angle Posts to BS 1722: Part 2.

Rolled steel angle posts for rectangular wire mesh (field) fencing

Posts	Fence height	Strut	Straining post
1200 x 40 x 40 x 5 mm	600 mm	1200 x 75 x 75 mm	1350 x 100 x 100 mm
1400 x 40 x 40 x 5 mm	800 mm	1400 x 75 x 75 mm	1550 x 100 x 100 mm
1500 x 40 x 40 x 5 mm	900 mm	1500 x 75 x 75 mm	1650 x 100 x 100 mm
1600 x 40 x 40 x 5 mm	1000 mm	1600 x 75 x 75 mm	1750 x 100 x 100 mm
1750 x 40 x 40 x 5 mm	1150 mm	1750 x 75 x 100 mm	1900 x 125 x 125 mm

Concrete Posts to BS 1722: Part 2.

Concrete posts for rectangular wire mesh (field) fencing

Posts	Fence height	Strut	Straining post
1270 x 100 x 100 mm	600 mm	1200 x 75 x 75 mm	1420 x 100 x 100 mm
1470 x 100 x 100 mm	800 mm	1350 x 75 x 75 mm	1620 x 100 x 100 mm
1570 x 100 x 100 mm	900 mm	1500 x 75 x 75 mm	1720 x 100 x 100 mm
1670 x 100 x 100 mm	600 mm	1650 x 75 x 75 mm	1820 x 100 x 100 mm
1820 x 125 x 125 mm	1150 mm	1830 x 75 x 100 mm	1970 x 125 x 125 mm

Cleft Chestnut Pale Fences to BS 1722: part 4: 1986

Timber Posts to BS 1722: Part 2.

Timber posts for wire mesh and hexagonal wire netting fences.
Round timber for general fences

Posts	Fence height	Strut	Straining post
1300 x 65 mm dia.	600 mm	1200 x 80 mm dia	1450 x 100 mm dia
1500 x 65 mm dia	800 mm	1400 x 80 mm dia	1650 x 100 mm dia
1600 x 65 mm dia.	900 mm	1500 x 80 mm dia	1750 x 100 mm dia
1700 x 65 mm dia.	1050 mm	1600 x 80 mm dia	1850 x 100 mm dia
1800 x 65 mm dia.	1150 mm	1750 x 80 mm dia	2000 x 120 mm dia

Squared timber for general fences

Posts	Fence height	Strut	Straining post
1300 x 75 x 75 mm	600 mm	1200 x 75 x 75 mm	1450 x 100 x 100 mm
1500 x 75 x 75 mm	800 mm	1400 x 75 x 75 mm	1650 x 100 x 100 mm
1600 x 75 x 75 mm	900 mm	1500 x 75 x 75 mm	1750 x 100 x 100 mm
1700 x 75 x 75 mm	1050 mm	1600 x 75 x 75 mm	1850 x 100 x 100 mm
1800 x 75 x 75 mm	1150 mm	1750 x 75 x 75 mm	2000 x 125 x 100 mm

Steel Fences to BS 1722: Pt 9: 1992

Mild steel fences: round or square verticals; flat standards and horizontals.
Tops of vertical bars may be bow-top, blunt, or pointed

Round or square bar railings.

	Fence height	Top/bottom rails and flat posts	Vertical bars
Light	1000 mm	40 x 10 mm 450 mm in ground	12 mm dia at 115 mm cs
	1200 mm	40 x 10 mm 550 mm in ground	12 mm dia at 115 mm cs
	1400 mm	40 x 10 mm 550 mm in ground	12 mm dia at 115 mm cs
Light	1000 mm	40 x 10 mm 450 mm in ground	16 mm dia at 120 mm cs
	1200 mm	40 x 10 mm 550 mm in ground	16 mm dia at 120 mm cs
	1400 mm	40 x 10 mm 550 mm in ground	16 mm dia at 120 mm cs
Medium	1200 mm	50 x 10 mm 550 mm in ground	20 mm dia at 125 mm cs
	1400 mm	50 x 10 mm 550 mm in ground	20 mm dia at 125 mm cs
	1600 mm	50 x 10 mm 600 mm in ground	22 mm dia at 145 mm cs
	1800 mm	50 x 10 mm 600 mm in ground	22 mm dia at 145 mm cs
Heavy	1600 mm	50 x 10 mm 600 mm in ground	22 mm dia at 145 mm cs
	1800 mm	50 x 10 mm 600 mm in ground	22 mm dia at 145 mm cs
	2000 mm	50 x 10 mm 600 mm in ground	22 mm dia at 145 mm cs
	2200 mm	50 x 10 mm 600 mm in ground	22 mm dia at 145 mm cs

Timber Field Gates to BS 3470: 1975

Gates made to this standard are designed to open one way only.
All timber gates are 1100 mm high.
Width over stiles 2400, 2700, 3000, 3300, 3600, and 4200 mm.
Gates over 4200 mm should be made in two leaves.

Steel Field Gates to BS 3470: 1975

Heavy duty: width over stiles 2400, 3000, 3600 and 4500 mm
Light duty: width over stiles 2400, 3000, and 3600 mm
All steel gates are 1100 mm high.

Domestic Front Entrance Gates to BS 4092:part 1: 1966

Metal gates: Single gates are 900 mm high minimum, 900 mm, 1000 mm and 1100 mm wide

Domestic Front Entrance Gates to BS 4092:part 2: 1966.

Wooden gates: All rails shall be tenoned into the stiles
Single gates are 840 mm high minimum, 801 mm and 1020 mm wide
Double gates are 840 mm high minimum, 2130, 2340 and 2640 mm wide

---

Timber Bridle Gates to BS 5709:1979 (Horse Or Hunting Gates)

Gates open one way only
Minimum width between posts   1525 mm
Minimum height   1100 mm

Timber Kissing Gates to BS 5709:1979

Minimum width   700 mm
Minimum height   1000 mm
Minimum distance between shutting posts   600 mm
Minimum clearance at mid-point   600 mm

Metal Kissing Gates to BS 5709:1979

Sizes are the same as those for timber kissing gates.
Maximum gaps between rails 120 mm.

Categories of Pedestrian Guard Rail to BS 3049:1976

Class A for normal use
Class B where vandalism is expected
Class C where crowd pressure is likely

DRAINAGE

Weights and dimensions - vitrified clay pipes

Product	Nominal diameter (mm)	Effective length (mm)	BS 65 limits of tolerance		Crushing Strength (kN/m)	Weight	
			min (mm)	max (mm)		kg/pipe	kg/m
Supersleve	100	1600	96	105	35.00	14.71	9.19
	150	1750	146	158	35.00	29.24	16.71
Hepsleve	225	1850	221	236	28.00	84.03	45.42
	300	2500	295	313	34.00	193.05	77.22
	150	1500	146	158	22.00	37.04	24.69
Hepseal	225	1750	221	236	28.00	85.47	48.84
	300	2500	295	313	34.00	204.08	81.63
	400	2500	394	414	44.00	357.14	142.86
	450	2500	444	464	44.00	454.55	181.63
	500	2500	494	514	48.00	555.56	222.22
	600	2500	591	615	57.00	796.23	307.69
	700	3000	689	719	67.00	1111.11	370.45
	800	3000	788	822	72.00	1351.35	450.45
Hepline	100	1600	95	107	22.00	14.71	9.19
	150	1750	145	160	22.00	29.24	16.71
	225	1850	219	239	28.00	84.03	45.42
	300	1850	292	317	34.00	142.86	77.22
Hepduct (Conduit)	90	1500	-	-	28.00	12.05	8.03
	100	1600	-	-	28.00	14.71	9.19
	125	1750	-	-	28.00	20.73	11.84
	150	1750	-	-	28.00	29.24	16.71
	225	1850	-	-	28.00	84.03	45.42
	300	1850	-	-	34.00	142.86	77.22

Weights and dimensions - vitrified clay pipes

Nominal internal diameter (mm)	Nominal wall thickness (mm)	Approximate weight kg/m
150	25	45
225	29	71
300	32	122
375	35	162
450	38	191
600	48	317
750	54	454
900	60	616
1200	76	912
1500	89	1458
1800	102	1884
2100	127	2619

Wall thickness, weights and pipe lengths vary, depending on type of pipe required.

The particulars shown above represent a selection of available diameters and are applicable to strength class 1 pipes with flexible rubber ring joints.

Tubes with Ogee joints are also available.

DRAINAGE - cont'd

Weights and dimensions - PVC-U pipes

	Nominal size	Mean outside diameter (mm)		Wall thickness (mm)	Weight kg/m
		min	max		
Standard pipes					
	82.4	82.4	82.7	3.2	1.2
	110.0	110.0	110.4	3.2	1.6
	160.0	160.0	160.6	4.1	3.0
	200.0	200.0	200.6	4.9	4.6
	250.0	250.0	250.7	6.1	7.2
Perforated pipes					
- heavy grade	As above	As above	As above	As above	As above
- thin wall	82.4	82.4	82.7	1.7	-
	110.0	110.0	110.4	2.2	-
	160.0	160.0	160.6	3.2	-

Width of trenches required for various diameters of pipes

Pipe diameter (mm)	Trench n.e. 1.5 m deep (mm)	Trench over 1.5 m deep (mm)
n.e.100	450	600
100-150	500	650
150-225	600	750
225-300	650	800
300-400	750	900
400-450	900	1050
450-600	1100	1300

DRAINAGE BELOW GROUND AND LAND DRAINAGE

Flow of Water Which Can Be Carried by Various Sizes of Pipe

Clay or concrete pipes

Pipe size	Gradient of pipeline							
	1:10	1:20	1:30	1:40	1:50	1:60	1:80	1:100
	Flow in litres per second							
DN 100 15.0	8.5	6.8	5.8	5.2	4.7	4.0	3.5	
DN 150 28.0	19.0	16.0	14.0	12.0	11.0	9.1	8.0	
DN 225 140.0	95.0	76.0	66.0	58.0	53.0	46.0	40.0	

Plastic pipes

Pipe size	Gradient of pipeline							
	1:10	1:20	1:30	1:40	1:50	1:60	1:80	1:100
	Flow in litres per second							
82.4mm i/dia	12.0	8.5	6.8	5.8	5.2	4.7	4.0	3.5
110mm i/dia	28.0	19.0	16.0	14.0	12.0	11.0	9.1	8.0
160mm i/dia	76.0	53.0	43.0	37.0	33.0	29.0	25.0	22.0
200mm i/dia	140.0	95.0	76.0	66.0	58.0	53.0	46.0	40.0

Vitrified (Perforated) Clay Pipes and Fittings to BS En 295-5 1994

Length not specified

75 mm bore	250 mm bore	600 mm bore
100	300	700
125	350	800
150	400	1000
200	450	1200
225	500	

Pre-cast Concrete Pipes: Pre-stressed Non-pressure Pipes and Fittings: Flexible Joints to BS 5911 :Pt.103: 1994

Rationalized metric nominal sizes: 450, 500

Length: 500 - 1000 by 100 increments
           1000 - 2200 by 200 increments
           2200 - 2800 by 300 increments

Angles: length: 450 - 600 angles 45, 22.5, 11.25 °
                   600 or more angles 22.5, 11.25 °

Tables and Memoranda 463

Pre-cast Concrete Pipes: Un-reinforced and Circular Manholes and Soakaways to BS 5911 :Pt.200: 1994

Nominal Sizes:

Shafts: 675, 900 mm

Chambers: 900, 1050, 1200, 1350, 1500, 1800, 2100, 2400, 2700, 3000 mm.

Large chambers: To have either tapered reducing rings or a flat reducing slab in order to accept the standard cover.

Ring depths:
1. 300 - 1200 mm by 300 mm increments except for bottom slab and rings below cover slab, these are by 150 mm increments.
2. 250 - 1000 mm by 250 mm increments except for bottom slab and rings below cover slab, these are by 125 mm increments.

Access hole: 750 x 750 mm for DN 1050 chamber
1200 x 675 mm for DN 1350 chamber

Calculation of Soakaway Depth

The following formula determines the depth of concrete ring soakaway that would be required for draining given amounts of water.

$$h = \frac{4ar}{3\pi D^2}$$

h = depth of the chamber below the invert pipe
A = The are to be drained
r = The hourly rate of rainfall (50 mm per hour)
π = pi
D = internal diameter of the soakaway

This table shows the depth of chambers in each ring size which would be required to contain the volume of water specified. These allow a recommended storage capacity of 1/3 (one third of the hourly rain fall figure).

Table showing required depth of concrete ring chambers in metres

AREA m2 Ring Size	50	100	150	200	300	400	500
0.9	1.31	2.62	3.93	5.24	7.86	10.48	13.10
1.1	0.96	1.92	2.89	3.85	5.77	7.70	9.62
1.2	0.74	1.47	2.21	2.95	4.42	5.89	7.37
1.4	0.58	1.16	1.75	2.33	3.49	4.66	5.82
1.5	0.47	0.94	1.41	1.89	2.83	3.77	4.72
1.8	0.33	0.65	0.98	1.31	1.96	2.62	3.27
2.1	0.24	0.48	0.72	0.96	1.44	1.92	2.41
2.4	0.18	0.37	0.55	0.74	1.11	1.47	1.84
2.7	0.15	0.29	0.44	0.58	0.87	1.16	1.46
3.0	0.12	0.24	0.35	0.47	0.71	0.94	1.18

Pre-cast Concrete Inspection Chambers and Gullies to BS 5911 :Pt.230: 1994

Nominal sizes: 375 diameter, 750, 900 mm deep
                 450 diameter, 750, 900, 1050, 1200 mm deep

Depths: from the top for trapped or un-trapped units:
                     centre of outlet 300 mm
                     invert (bottom) of the outlet pipe 400 mm

Depth of water seal for trapped gullies:
                 85 mm, rodding eye int. diam. 100 mm

Cover slab: 65 mm min.

Ductile Iron Pipes to BS En 598 : 1995

Type K9 with flexible joints should be used for surface water drainage.
5500 mm or 8000 mm long

80 mm bore	400 mm bore	1000 mm bore
100	450	1100
150	500	1200
200	600	1400
250	700	1600
300	800	
350	900	

Bedding Flexible Pipes: Pvc-u Or Ductile Iron

Type 1	=	100mm fill below pipe, 300mm above pipe: single size material
Type 2	=	100mm fill below pipe, 300mm above pipe: single size or graded material
Type 3	=	100mm fill below pipe, 75mm above pipe with concrete protective slab over
Type 4	=	100mm fill below pipe, fill laid level with top of pipe
Type 5	=	200mm fill below pipe, fill laid level with top of pipe
Concrete	=	25mm sand blinding to bottom of trench, pipe supported on chocks, 100mm concrete under the pipe, 150mm concrete over the pipe.

Bedding Rigid Pipes: Clay Or Concrete

(for vitrified clay pipes the manufacturer should be consulted)

Class D:	Pipe laid on natural ground with cut-outs for joints, soil screened to remove stones over 40mm and returned over pipe to 150mm min depth. Suitable for firm ground with trenches trimmed by hand.
Class N:	Pipe laid on 50mm granular material of graded aggregate to Table 4 of BS 882, or 10mm aggregate to Table 6 of BS 882, or as dug light soil (not clay) screened to remove stones over 10mm. Suitable for machine dug trenches.
Class B:	As Class N, but with granular bedding extending half way up the pipe diameter.
Class F:	Pipe laid on 100mm granular fill to BS 882 below pipe, minimum 150mm granular fill above pipe: single size material. Suitable for machine dug trenches.
Class A:	Concrete 100mm thick under the pipe extending half way up the pipe, backfilled with the appropriate class of fill. Used where there is only a very shallow fall to the drain. Class A bedding allows the pipes to be laid to an exact gradient.
Concrete surround:	25mm sand blinding to bottom of trench, pipe supported on chocks, 100mm concrete under the pipe, 150mm concrete over the pipe. It is preferable to bed pipes under slabs or wall in granular material.

PIPED SUPPLY SYSTEMS

Identification of Service Tubes From Utility to Dwellings

Utility	Colour	Size	Depth
British Telecom	grey	54 mm od	450 mm
Electricity	black	38 mm od	450 mm
Gas	yellow	42 mm od rigid	450 mm
		60 mm od convoluted	
Water	may be blue	(normally untubed)	750 mm

ELECTRICAL SUPPLY/POWER/LIGHTING SYSTEMS

Electrical Insulation Class En 60.598 BS 4533

Class 1:   luminaires comply with class 1 (I) earthed electrical requirements
Class 2:   luminaires comply with class 2 (II) double insulated electrical requirements
Class 3:   luminaires comply with class 3 (III) electrical requirements

Protection to Light Fittings

BS EN 60529:1992 Classification for degrees of protection provided by enclosures.
(IP Code - International or ingress Protection)

1st characteristic: against ingress of solid foreign objects.

The figure
2   indicates that fingers cannot enter
3   that a 2.5 mm diameter probe cannot enter
4   that a 1.0 mm diameter probe cannot enter
5   the fitting is dust proof (no dust around live parts)
6   the fitting is dust tight (no dust entry)

2nd characteristic: ingress of water with harmful effects:

The figure
0   indicates unprotected
1   vertically dripping water cannot enter
2   water dripping 15° (tilt) cannot enter
3   spraying water cannot enter
4   splashing water cannot enter
5   jetting water cannot enter
6   powerful jetting water cannot enter
7   proof against temporary immersion
8   proof against continuous immersion

Optional additional codes: A-D protects against access to hazardous parts;

H   High voltage apparatus
M   fitting was in motion during water test
S   fitting was static during water test
W   protects against weather

Marking code arrangement: (example) IPX5S = IP (International or Ingress Protection);
X (denotes omission of firstcharacteristic);
5 = jetting;
S = static during water test.

LANDFILL TAX

Waste liable at the lower rate

Group	Description of material	Conditions	
1	Rocks and soils	Naturally occurring	includes clay, sand, gravel, sandstone, limestone, crushed stone, china clay, construction stone, stone from the demolition of buildings or structures, slate, topsoil, peat, silt and dredgings glass includes fritted enamel, but excludes glass fibre and glass reinforced plastics
2	Ceramic or concrete materials		ceramics includes bricks, bricks and mortar, tiles, clay ware, pottery, china and refractories concrete includes reinforced concrete, concrete blocks, breeze blocks and aircrete blocks, but excludes concrete plant washings
3	Minerals	Processed or prepared, not used	moulding sands excludes sands containing organic binders clays includes moulding clays and clay absorbents, including Fuller's earth and bentonite man-made mineral fibres includes glass fibres, but excludes glass-reinforced plastic and asbestos silica, mica and mineral abrasives
4	Furnace slags		vitrified wastes and residues from thermal processing of minerals where, in either case, the residue is both fused and insoluble slag from waste incineration
5	Ash		comprises only bottom ash and fly ash from wood, coal or waste combustion excludes fly ash from municipal, clinical, and hazardous waste incinerators and sewage sludge incinerators
6	Low activity inorganic compound		comprises only titanium dioxide, calcium carbonate, magnesium carbonate, magnesium oxide, magnesium hydroxide, iron oxide, ferric hydroxide, aluminium oxide, aluminium hydroxide & zirconium dioxide
7	Calcium sulphate	Disposed of either at a site not licensed to take putrescible waste or in a containment cell which takes only calcium sulphate	includes gypsum and calcium sulphate based plasters, but excludes plasterboard
8	Calcium hydroxide and brine	Deposited in brine cavity	
9	Water	Containing other qualifying material in suspension	

Volume to weight conversion factors

Waste category	Typical waste types	Cubic metres to tonne - multiply by:	Cubic yards to tonne - multiply by:
Inactive or inert waste	Largely water insoluble and non or very slowly biodegradable: e.g. sand, subsoil, concrete, bricks, mineral fibres, fibreglass etc.	1.5	1.15

SPON'S PRICEBOOKS 2003

Spon's Architects' & Builders' Price Book 2003
Davis Langdon & Everest

The most detailed professionally relevant source of construction price information currently available.

New features for 2003 include:
- A new section on the Aggregates Tax.
- A new section on Part L of the Building Regulations and its impact upon costs.

HB & CD-ROM ♦ £115.00
0-415-30116-5 ♦ 1056pp

Spon's Mechanical & Electrical Services Price Book 2003
Mott Green & Wall

Still the only annual services engineering price book available.

New features for 2003 include:
- Wide range of building types for both elemental and all-in m2 rates.
- The electrical section is now in line with the CAWS that SMM7 follow.

HB & CD-ROM ♦ £115.00
0-415-30122-X ♦ 704pp

Receive a **free cd-rom when you order any Spon 2003 Pricebook** to help you produce tender documents, customise data, perform keyword searches and simple calculations.

Spon's Landscape & External Works Price Book 2003
Davis Langdon & Everest

The only comprehensive source of information for detailed landscape and external works cost.

New features for 2003 include:
- Notes on the aggregate tax with examples of the impact of the tax on measured works.
- Costs of external wall block work, special bricks, clay drainage and much more!

HB & CD-ROM ♦ £85.00
0-415-30120-3 ♦ 512pp

Spon's Civil Engineering & Highway Works Price Book 2003
Davis Langdon & Everest

More than just a price book, it is a comprehensive, work manual that all those in the civil engineering, surveying and construction business will find it hard to work without.

New features for 2003 include:
- A revision and expansion of both the Outputs and the Tables and Memoranda sections with more useful data.

HB & CD-ROM ♦ £120.00
0-415-30118-1 ♦ 704pp

To Order: Tel: +44 (0) 8700 768853, or +44 (0) 1264 343071 Fax: +44 (0) 1264 343005, or Post: Spon Press Customer Services, Thomson Publishing Services, Cheriton House, Andover, Hants, SP10 5BE, UK Email: book.orders@tandf.co.uk

For a complete listing of all our titles visit
www.sponpress.com

Spon Press
Taylor & Francis Group

ESSENTIAL READING FROM SPON PRESS

The Regeneration of Public Parks

Edited by Jan Woudstra and Ken Fieldhouse, University of Sheffield, UK

'I feel this book will be welcome in the library of every landscape designer, every park manager and every garden historian. I would recommend it to anyone working in the landscape, urban design and parks profession and it is essential reading for anyone with an interest or active involvement in the care and refurbishment of public parks.' - *Journal of Environmental Planning and Management*

The Regeneration of Public Parks highlights the new enthusiasm for Britain's public parks and provides a valuable overview of the public park design, sustainability, management and regeneration. It will be essential to all those involved with or interested in public parks. The expanding interest in public parks in the UK has been encouraged by the grants of the Heritage Lottery Fund (Urban Parks Programme), but the renaissance is an international phenomenon, fired by general concerns about the urban environment.

The book has been prepared by a team of experts involved with, and responsible for public parks, all are specialists in their fields. The first part of the book deals with general issues relating to public parks, while the second part relates to specific design elements within the parks and their management: ironwork, walks and paved surfaces, planting and bedding, lakes and water bodies.

June 2000: 276x219: 200pp
132 b+w photos; 8 pages colour illustrations
Pb: 0-419-25900-7: £30.00

To Order: Tel: +44 (0) 8700 768853, or +44 (0) 1264 343071 Fax: +44 (0) 1264 343005, or Post: Spon Press Customer Services, Thomson Publishing Services, Cheriton House, Andover, Hants, SP10 5BE, UK Email: book.orders@tandf.co.uk

For a complete listing of all our titles visit:
www.sponpress.com

Spon Press
Taylor & Francis Group

Index

Access Covers and Frames 298, 299
Aeration 253
Aftercare 211, 214, 247, 248
    approximate estimates 371
Agricultural Drainage 302-306, 308
    approximate estimates 385
Anchoring 121, 235-236
Anchors 111, 130-132, 258, 274-279
Aquatic Planting 245
    approximate estimates 393
Aquatic Plants 224, 245
Artificial Grass 191
    approximate estimates 363
Artificial Surfaces 190
Asphalt 114, 116, 172, 174, 175
Backfilling Materials 308
Ball Stop Fencing 268
Balusters 152, 269, 270
    approximate estimates 382, 394
Bar Railings 269
Bark Mulch 215
Bark Surfaces 177
Barriers 236, 263, 273, 274, 283
Bases 150, 169, 170, 274, 275, 282
    approximate estimates 353
Bedding 161, 256, 301
Beds 118, 142, 208, 255, 256, 292, 294
Blinding 142, 154, 317
Blockwork 148
Bollards 280, 282-285, 327, 328
    approximate estimates 384, 395
Bound Aggregates 177
Boundary Fencing 259-262
Brick Copings 147, 148
Brick Paving 179
    approximate estimates 353
Brick/Block Walling 145
Brickwork 135, 137, 146, 147, 293
Building Fabric Sundries 163
Buildings 126
Bulb Planting 245
Bulbs 225, 226, 245, 255
    approximate estimates 375-377

Bulkhead Fittings 322
Butt Joints 178-181, 184, 185, 274, 306

Car Parks 175
    approximate estimates 349, 350
Carcassing Metal 152
Cast Stone 113, 126, 149, 150, 153, 318
Cattle Grids 273
Cement 141, 144, 145, 160
Cement Mortar 146, 147
Cement:Sand 160, 187
Ceramic Tiles 160
Chambers 159, 292, 293, 296
Channels 166, 168
    approximate estimates 363
Chippings 160, 175, 176
Cladding 113, 153
Classical Temple 126
Clay
    brick pavings 180
    channels 293, 294
    drainage 118
    gullies 296, 297
    pipes 119, 295 306, 307
    tiles 160
Clear Finishing 161, 162
Clearing Site Vegetation 135
Cleft Rail Fencing 265
Climbers 219
Climbing Equipment 288, 289
Close Boarded Fencing 265, 266
Clothes Line Fittings 291
Cobble Paviors 184
Cold Water 314
Compacting 118, 121, 139, 163, 294, 313
Completion 187
Concrete 120, 135-137, 140-143, 160, 164-171, 311, 312
    blocks 113, 148, 167, 178, 188
    bollards 283
    fencing 267
    grass concrete 188, 189
        approximate estimates 358
Concrete
    gullies 296, 301
    in situ concrete 140-143, 170, 292, 311
        approximate estimates 344
    outfalls 311

Index

Concrete cont'd
   paviors 184
   pipes 301
   reinforced concrete 135-137
      approximate estimates 342, 363
   revetments 111, 130
   screeds 160
   setts 166, 184
      approximate estimates 358
Copings 148
Coshh 123
   approximate estimates 337
Crash Barriers 274
Crib 127
Crib Walls 127
Cricket Squares 202, 209
Cutting 208, 211, 255
   approximate estimates 369, 370
   curved cutting 180
   grass cutting 212, 251, 252
      approximate estimates 371
   hedges 250, 257, 258
   raking cutting 180
   shrubs 246
Cycle holders 182, 188
Damp Proof Courses 151
Damp Proofing 154, 155
Decking 152
   approximate estimates 394
Detailed 123, 159
Directional Signage 282
Disposal 112, 120, 137, 138, 311
Disposal Systems 292-313
Ditching 299-301
Drainage
   medium 156
   layer 157, 158
   channel(s) 159, 169
   system(s) 168, 299, 302
   chambers 292
   below ground 292-298
   land drainage 299-312
   mole drainage 301
   slits 302
   agricultural drainage 302-308
   calculation table 302
Dry Stone Walling 148
Dwarf Walls 149
Earthwork Support 118, 119, 163, 292, 294, 295, 306
Edge Maintenance 254

Edge Restraints 179, 184, 274
   approximate estimates 348, 351
Edging Boards 176, 192
Edgings 164-166, 168, 178, 180, 186, 187, 192, 317
   approximate estimates 347, 348
Electrical accessories 333
Embankments 129-131
Employer's requirements 124
Erosion Control Mats 129, 130, 133
Excavation and Path Preparation 175
Expansion Joints 171, 188
Fencing 118, 124, 164-291
   approximate estimates 337, 365, 378-383
Fertilizers 193, 199, 200, 214-216, 231, 245, 247-250
Field Gates 271, 272
Filling 112, 134, 136, 139, 169, 176
   approximate estimates 391
Filling to Make Up Levels 112, 136, 139, 169, 176
Finishing Coat 161
Flagpoles 289, 290
Flexible Sheet Materials 132, 133
Flexible Sheet Tanking 155
Folly 126
Forestry 249, 250
Forestry planting 249, 250
Formwork 143, 164, 171, 292
   approximate estimates 357
Fountain Kits 319
Fountains 315, 316, 318, 319
Fungicide 192, 194
Gabions 111, 128, 131
   approximate estimates 342
Gates 258, 260, 262-264, 266, 268, 270-273
Glacial Boulders 149
Grading 129, 139, 196, 229
Granite Paving 185
   approximate estimates 357
Granite Setts 166, 167, 185
   approximate estimates 347
Granite Walls 148
Granolithic Paving 160
Grass Blocks 116, 179
Grass Concrete 128, 131, 188, 189
   approximate estimates 358
Grass Cutting 212, 251, 252
   approximate estimates 371
Grass Reinforcement 132

Index

Grass Maintenance Materials 192, 194

Grass Seed 132-134, 188, 201, 202, 204, 206
   approximate estimates 359, 365
Grasscrete 131, 188
   approximate estimates 342, 350
Green Roof Systems 156, 158
Grilles 286
   approximate estimates 377
Grit Bins 277
Groundcover Planting 245
Groundwork 127-139
   approximate estimates 339, 340
Growth Retardant 248
Guards 241, 242, 254, 263, 286
Gullies 168, 296-298, 301
Handrails 152
   approximate estimates 394
Hardcore 112, 136, 169
   approximate estimates 343, 348, 384
Hardcore Blinding 132
Harrowing 207, 213, 253
Haunching 140, 164-168, 179, 184
   approximate estimates 347
Health and Safety 123
   approximate estimates 337
Hedge 242, 250, 257, 258
   approximate estimates 372
Hedge Cutting 257, 258
Herbaceous and Groundcover Planting 245
Herbaceous Plants 245, 255
Herbicides 170, 176, 217, 230, 231, 246, 248, 249
Hose 207, 314, 315
In Situ Concrete 142
   approximate estimates 344
Industrially Grown Turf 209, 210
Intercepting Traps 294
Interlocking Units 116, 128, 179
Intruder Guards 263
Irrigation 159, 191, 236, 257, 314
   approximate estimates 392
Irrigation Systems 236
   approximate estimates 363
Kerb Drainage Systems 168
Kerbs 164-168, 296, 298
   approximate estimates 347, 348
Kidney Flint Cobbles 187
   approximate estimates 358
Kiln Dried Sand Joints 180, 184

   approximate estimates 361
Lamps 320, 331, 332
Land Drainage 299-312
Landscape Maintenance 251-258
Lanterns 320, 321, 328-332
Leaf Clearance 212, 254
Leaky pipe 314, 315
   approximate estimates 392
Liability 155
   approximate estimates 371
Lichen 250
Lifebuoy Stations 277
Lighting Columns 330, 332, 333
Lime Granules 201
Lintels 151
Liquid Applied Tanking 154
Liquid Sod 208
Litter and Grit Bins 277
Litter Clearance 254
Maintenance 194
   approximate estimates 371
Maintenance Operations 211, 214, 247, 248
Manhole Covers 168, 298
Market Prices of
   aquatic plants 224
   backfilling materials 308
   bulbs 225, 226
   containers 281
   grass maintenance materials 192, 194
   grass seed 201, 202, 204
   lamps 320
   materials 114
   mulching materials 215
   planting materials 215, 216
   pre-seeding materials 192
   sand pits 289
   surfacing materials 189
   topsoil 112, 134
   trees 224
   trees, shrubs and plants 217, 218, 220, 222
   wild flora seed mixtures 204
Marking 175, 213
Marking Car Parks 175
Marking Out Pitches 213
Masonry 113, 129, 144-151
Mastic Asphalt Tanking 154
Metal Framed Seats 278
Method Statements 123
Mild Steel Railings 270

Mixes 140, 145, 197, 253

Mole Drainage 301
    approximate estimates 389
Mosaic 288
Movement of Materials 179
Moving Equipment 287
Mulch 215, 236, 237, 246, 247
    approximate estimates 375
Mulching Materials 215
Mulching of Tree Pits 236
Mushroom Compost 215, 232, 233, 244
Natural Stone 113, 136, 148, 153, 165, 166, 185
    approximate estimates 355, 356
Natural Stone Rubble 148
Office 125, 190
Oil Paint 161
Operations After Planting 246
Ornamental Garden Buildings 126
Ornamental Masonry 149, 150
Ornamental Pools 317, 318
Outfalls 307, 311
Painting/Clear Finishing 161, 162
Parking Posts 284
    approximate estimates 384
Path Preparation 175
Paths 175
    firepaths 116, 179
        approximate estimates 359
    footpaths 172, 174
        approximate estimates 362
Pavilion 126
Paving Blocks 167, 178, 188, 274, 298
Paving(s) 114, 116, 136, 142, 160, 164-291,
Paviors 116, 136, 179, 180, 184, 298
    approximate estimates 361
Pedestrian Guard Rails 272
Pesticides 200
Picnic Benches 280
Pier Caps 150, 151
Piers 146, 147, 149, 150
    approximate estimates 345
Pipe Laying 306
Piped Ditching 301

Pipes 119, 295, 296, 301, 302, 306-310, 312, 314
    approximate estimates 385
Pipeways 163
Pits 135, 137,163, 289, 292

Plant Containers 281, 282
Planted Areas 247, 255, 256
Planters 236, 255, 256
    approximate estimates 377, 384
Planting Materials 215, 216
Planting Operations 230, 232, 234
Plants 217-225, 243-245
    approximate estimates 373-376, 394
Plastered/rendered/roughcast 160
Play Sculptures 288
Play structures 288
Playground Equipment 191, 286-288
Playgrounds 191, 192
    approximate estimates 363-365
Ponds 155, 315, 317
    approximate estimates 393
Pool Surrounds 318
Pools 317, 318
Precast Concrete 189, 284, 291, 292
    approximate estimates 342, 343, 353, 358, 379, 390
    block edgings 178
    bollards 283
    channels 167-169
        approximate estimates 347
    cover slabs 312
    edging units 165
    flags 274
        approximate estimates 354
    kerbs, channels, edgings 164
        approximate estimates 347
    litter bins 275
    paving blocks 178, 274
    paving slabs 136
    pavings 181-183
    plant containers 282
    sills, lintels, copings, features 151
    vehicular paving 178
Prefabricated Buildings 126
Prefabricated Timber Unit Decking 152
Programmes 124
    approximate estimates 337
Protection 124, 156-158, 228, 241, 242, 285
    approximate estimates 363
Protective Fencing 228, 258
Pruning Shrubs 254
Quarry Tiles 160, 161
Rabbit netting 261
Radial Paving 182, 183, 185
Railings 161, 269, 270
    approximate estimates 382

Reinforced In Situ Concrete 142, 164, 170, 292
  approximate estimates 342, 348
Reinforced Turf 117, 210
  approximate estimates 370
Reinforcement 126, 132, 143, 171, 292
Removing Trees 134
Rendering 154, 296, 298
Resin Bound Macadam 175
Resources 155
Retaining Walls 127-130, 132
  approximate estimates 343
Risk Assessments 123
Road Repairs 175
Rockery Stone 149
Rolling 139, 176, 177, 196, 207, 208, 211, 213, 229, 253
Root Barriers 156, 236
Roundabouts 287
Rubble walling 148
Safety Channels 168
Safety Surfacing 288
  approximate estimates 364
Sand 144, 160, 308
  approximate estimates 361
  Amsterdam tree sand 117, 235
  blinding 131, 169, 176, 318
    approximate estimates 359-361, 393
  pits 289
  Playsand 189
  reject sand 118, 294
  sharp sand 118, 144, 294
    approximate estimates 348, 351
  slitting 302
Scarifying 213, 253
Screeds 157, 160
Screen Fencing 266, 267
Seats and Benches 277
Secure Storage 125
Security 124, 263, 273, 285
  approximate estimates 382
Security Fencing 263
  approximate estimates 382
Seed Market Prices of 201-204
  approximate estimates 368-370
Seed Mixtures 204, 210
  approximate estimates 343
Seeded Areas 207

Seesaws 288
Setting Out 124, 129, 213, 242, 243, 256
Setts 166, 167, 184, 185
  approximate estimates 357, 358
Shrub Planting 216, 243, 244
Shrubs 127, 217-220, 222, 228, 244, 246, 254
  approximate estimates 338, 375, 376, 384
Signage 282
Signs 273, 324
  approximate estimates 378
Sika 154
Sills 151
Silt Box 168
Silt Pits 292
Site Accommodation 124
Site Preparation 134, 136
Site Protection 228
Slate 185
Slides 287
Slip Layers 156
Slip Resistant 160
Slotted Grating 169
Soakaways 120-122, 311, 312
  approximate estimates 391
Soil
  conditioners 197, 198, 244, 245
  preparation 192
  stabilization 127, 128, 130-134
  treatments 199, 200
Sports Areas 190
Sports Equipment 290
Sports Pitches 202
  approximate estimates 365
Spot Bedding 189
Spring Equipment 289
Stakes 235, 241, 244, 245, 254, 255, 261, 263
Standpipes 314
Step Irons 120, 121, 293, 312
Stepping Stones 182, 185
Steps 141, 327
Stiles and Kissing Gates 272
Stopcocks 314
Street Area Floodlighting 320 - 332
Street Furniture 273-290
  approximate estimates 384
Stump Grinding 135
Subsoiling 138, 195, 228
  approximate estimates 367
Sundries 163, 171

Index

Surface 114, 172, 175-177, 179, 190, 232, 233
   approximate estimates 363-365
   erosion 111, 130
   finishes 160-162
   treatment 111, 117, 131, 139, 169, 176, 177, 183, 235, 253, 292
Surfacing Materials 189
Suspension(s) 329, 330
Swings 287
Synthaprufe 154
Tanking 154, 155
Temple 126
Temporary 125, 128, 130, 228, 255
   approximate estimates 337
Temporary roads 125
Tender and Contract Documents 123
Tennis Courts 190, 202, 209, 290
Tensar 128-130, 132, 210, 267
   approximate estimates 341
Terram 121, 133, 175, 306, 313
   approximate estimates 384
Ties 146, 235, 254
Tiles 160, 161, 185, 191, 210
Timber
   bollards 285
   decking 152
      approximate estimates 394
   edgings 192
      approximate estimates 347
   fencing 264-266
   gates 271
   log retaining walls 132
      approximate estimates 344
   outdoor seats 279, 280
   plant containers 281
   trellis 267
Timbercrib 127
Toilet Facilities 125
   approximate estimates 337
Top Dressing 192, 194, 214
Topiary 243
Topsoil 112, 133-137, 192, 196, 198, 199, 208, 211, 214, 215, 229-231, 234, 294, 308, 316-318
   approximate estimates 393, 394, 366
Transformers 324, 326, 333
Transplants 216, 220, 221, 236, 242
Transport 208
Traps 119, 294-296
Tree Grilles 286

Tree Guards 241, 242, 254
Trees 134, 212, 216-218, 220, 222-224, 228, 235-243, 250, 254
   approximate estimates 338
Trellis 267
Trenches 117-119, 122, 132, 135, 137, 163, 164, 175, 235, 294, 295, 303, 313, 316
   approximate estimates 341
Trenchless Drainage System 302
Trip Rails 264
   approximate estimates 383
Tubes 250
Turf
   aeration 213, 253
   areas 253
      approximate estimates 393
   industrially grown 210
   management 208
   preparation of turf beds 208
   systems 117
Turfing 192-214
   approximate estimates 366-370
Unclimbable Fencing 124
Underwater Lighting 326
Unreinforced Concrete 170
Vehicle Crash Barriers 274
Vehicular Paving 178
   approximate estimates 352
Walkways 125
Wall Brackets 320, 321
Walls 113, 145-150, 154, 155, 160
Water Features 315, 316, 318
   approximate estimates 393
Waterfall Construction 317
Waterproof Rendering 154
Waterproofing 154-159
Watertanks 319
Watertight Floors 141
Weedkillers 200, 201
Wild Flora Seed 204
Wild Flora Seed Mixtures 204
Willow 132
Willow Walling 132
Windbreak Fencing 267
Window Boxes 281
Work to Existing Planting 250
York Stone 166, 185, 186
   approximate estimates 354-356
York Stone cladding 153

CD-ROM Single-User Licence Agreement

We welcome you as a user of this Spon Press CD-ROM and hope that you find it a useful and valuable tool. Please read this document carefully. **This is a legal agreement** between you (hereinafter referred to as the "Licensee") and Taylor and Francis Books Ltd., under the imprint of Spon Press (the "Publisher"), which defines the terms under which you may use the Product. **By breaking the seal and opening the package containing the CD-ROM you agree to these terms and conditions outlined herein. If you do not agree to these terms you must return the Product to your supplier intact, with the seal on the CD case unbroken.**

1. Definition of the Product

The product which is the subject of this Agreement, *Spon's Landscape and External Works Price Book on CD-ROM* (the "Product") consists of:

1.1 Underlying data comprised in the product (the "Data")
1.2 A compilation of the Data (the "Database")
1.3 Software (the "Software") for accessing and using the Database
1.4 A CD-ROM disk (the "CD-ROM")

2. Commencement and Licence

2.1 This Agreement commences upon the breaking open of the package containing the CD-ROM by the Licensee (the "Commencement Date").
2.2 This is a licence agreement (the "Agreement") for the use of the Product by the Licensee, and not an agreement for sale.
2.3 The Publisher licenses the Licensee on a non-exclusive and non-transferable basis to use the Product on condition that the Licensee complies with this Agreement. The Licensee acknowledges that it is only permitted to use the Product in accordance with this Agreement.

3. Installation and Use

3.1 The Licensee may provide access to the Product for individual study in the following manner: The Licensee may install the Product on a secure local area network on a single site for use by one user. For more than one user or for a wide area network or consortium, use is only permissible with the express permission of the Publisher in writing and requires the payment of the appropriate fee as specified by the Publisher, and signature by the Licensee of a separate multi-user licence agreement.
3.2 The Licensee shall be responsible for installing the Product and for the effectiveness of such installation.
3.3 Text from the Product may be incorporated in a coursepack. Such use is only permissible with the express permission of the Publisher in writing and requires the payment of the appropriate fee as specified by the Publisher and signature of a separate licence agreement.

4. Permitted Activities

4.1 The Licensee shall be entitled:
4.1.1 to use the Product for its own internal purposes;
4.1.2 to download onto electronic, magnetic, optical or similar storage medium reasonable portions of the Database provided that the purpose of the Licensee is to undertake internal research or study and provided that such storage is temporary;
4.1.3 to make a copy of the Database and/or the Software for back-up/archival/disaster recovery purposes.
4.2 The Licensee acknowledges that its rights to use the Product are strictly as set out in this Agreement, and all other uses (whether expressly mentioned in Clause 5 below or not) are prohibited.

5. Prohibited Activities

The following are prohibited without the express permission of the Publisher:

5.1 The commercial exploitation of any part of the Product.
5.2 The rental, loan (free or for money or money's worth) or hire purchase of the product, save with the express consent of the Publisher.
5.3 Any activity which raises the reasonable prospect of impeding the Publisher's ability or opportunities to market the Product.
5.4 Any networking, physical or electronic distribution or dissemination of the product save as expressly permitted by this Agreement.
5.5 Any reverse engineering, decompilation, disassembly or other alteration of the Product save in accordance with applicable national laws.
5.6 The right to create any derivative product or service from the Product save as expressly provided for in this Agreement.
5.7 Any alteration, amendment, modification or deletion from the Product, whether for the purposes of error correction or otherwise.

6. General Responsibilities of the Licensee

6.1 The Licensee will take all reasonable steps to ensure that the Product is used in accordance with the terms and conditions of this Agreement.

6.2 The Licensee acknowledges that damages may not be a sufficient remedy for the Publisher in the event of breach of this Agreement by the Licensee, and that an injunction may be appropriate.

6.3 The Licensee undertakes to keep the Product safe and to use its best endeavours to ensure that the product does not fall into the hands of third parties, whether as a result of theft or otherwise.

6.4 Where information of a confidential nature relating to the product or the business affairs of the Publisher comes into the possession of the Licensee pursuant to this Agreement (or otherwise), the Licensee agrees to use such information solely for the purposes of this Agreement, and under no circumstances to disclose any element of the information to any third party save strictly as permitted under this Agreement. For the avoidance of doubt, the Licensee's obligations under this sub-clause 6.4 shall survive termination of this Agreement.

7. Warrant and Liability

7.1 The Publisher warrants that it has the authority to enter into this Agreement, and that it has secured all rights and permissions necessary to enable the Licensee to use the Product in accordance with this Agreement.

7.2 The Publisher warrants that the CD-ROM as supplied on the Commencement Date shall be free of defects in materials and workmanship, and undertakes to replace any defective CD-ROM within 28 days of notice of such defect being received provided such notice is received within 30 days of such supply. As an alternative to replacement, the Publisher agrees fully to refund the Licensee in such circumstances, if the Licensee so requests, provided that the Licensee returns the Product to the Publisher. The provisions of this sub-clause 7.2 do not apply where the defect results from an accident or from misuse of the product by the Licensee.

7.3 Sub-clause 7.2 sets out the sole and exclusive remedy of the Licensee in relation to defects in the CD-ROM.

7.4 The Publisher and the Licensee acknowledge that the Publisher supplies the Product on an "as is" basis. The Publisher gives no warranties:

7.4.1 that the Product satisfies the individual requirements of the Licensee; or

7.4.2 that the Product is otherwise fit for the Licensee's purpose; or

7.4.3 that the Data are accurate or complete or free of errors or omissions; or

7.4.4 that the Product is compatible with the Licensee's hardware equipment and software operating environment.

7.5 The Publisher hereby disclaims all warranties and conditions, express or implied, which are not stated above.

7.6 Nothing in this Clause 7 limits the Publisher's liability to the Licensee in the event of death or personal injury resulting from the Publisher's negligence.

7.7 The Publisher hereby excludes liability for loss of revenue, reputation, business, profits, or for indirect or consequential losses, irrespective of whether the Publisher was advised by the Licensee of the potential of such losses.

7.8 The Licensee acknowledges the merit of independently verifying Data prior to taking any decisions of material significance (commercial or otherwise) based on such data. It is agreed that the Publisher shall not be liable for any losses which result from the Licensee placing reliance on the Data or on the Database, under any circumstances.

7.9 Subject to sub-clause 7.6 above, the Publisher's liability under this Agreement shall be limited to the purchase price.

8. Intellectual Property Rights

8.1 Nothing in this Agreement affects the ownership of copyright or other intellectual property rights in the Data, the Database or the Software.

8.2 The Licensee agrees to display the Publisher's copyright notice in the manner described in the Product.

8.3 The Licensee hereby agrees to abide by copyright and similar notice requirements required by the Publisher, details of which are as follows:
" © 2003 Spon Press. All rights reserved. All materials in *Spon's Landscape and External Works Price Book on CD-ROM* are copyright protected. © 2001 Adobe Systems Incorporated. All rights reserved. No such materials may be used, displayed, modified, adapted, distributed, transmitted, transferred, published or otherwise reproduced in any form or by any means now or hereafter developed other than strictly in accordance with the terms of the licence agreement enclosed with the CD-ROM. However, text and images may be printed and copied for research and private study within the preset program limitations. Please note the copyright notice above, and that any text or images printed or copied must credit the source."

8.4 This Product contains material proprietary to and copyrighted by the Publisher and others. Except for the licence granted herein, all rights, title and interest in the Product, in all languages, formats and media throughout the world, including all copyrights therein, are and remain the property of the Publisher or other copyright owners identified in the Product.

9. Non-assignment

This Agreement and the licence contained within it may not be assigned to any other person or entity without the written consent of the Publisher.

10. Termination and Consequences of Termination

10.1 The Publisher shall have the right to terminate this Agreement if:

10.1.1	the Licensee is in material breach of this Agreement and fails to remedy such breach (where capable of remedy) within 14 days of a written notice from the Publisher requiring it to do so; or
10.1.2	the Licensee becomes insolvent, becomes subject to receivership, liquidation or similar external administration; or
10.1.3	the Licensee ceases to operate in business.
10.2	The Licensee shall have the right to terminate this Agreement for any reason upon two months' written notice. The Licensee shall not be entitled to any refund for payments made under this Agreement prior to termination under this sub-clause 10.2.
10.3	Termination by either of the parties is without prejudice to any other rights or remedies under the general law to which they may be entitled, or which survive such termination (including rights of the Publisher under sub-clause 6.4 above).
10.4	Upon termination of this Agreement, or expiry of its terms, the Licensee must:
10.4.1	destroy all back up copies of the product; and
10.4.2	return the Product to the Publisher.

11. General

11.1	*Compliance with export provisions*
	The Publisher hereby agrees to comply fully with all relevant export laws and regulations of the United Kingdom to ensure that the Product is not exported, directly or indirectly, in violation of English law.
11.2	*Force majeure*
	The parties accept no responsibility for breaches of this Agreement occurring as a result of circumstances beyond their control.
11.3	*No waiver*
	Any failure or delay by either party to exercise or enforce any right conferred by this Agreement shall not be deemed to be a waiver of such right.
11.4	*Entire agreement*
	This Agreement represents the entire agreement between the Publisher and the Licensee concerning the Product. The terms of this Agreement supersede all prior purchase orders, written terms and conditions, written or verbal representations, advertising or statements relating in any way to the Product.
11.5	*Severability*
	If any provision of this Agreement is found to be invalid or unenforceable by a court of law of competent jurisdiction, such a finding shall not affect the other provisions of this Agreement and all provisions of this Agreement unaffected by such a finding shall remain in full force and effect.
11.6	*Variations*
	This Agreement may only be varied in writing by means of variation signed in writing by both parties.
11.7	*Notices*
	All notices to be delivered to: Spon Press, an imprint of Taylor & Francis Books Ltd., 11 New Fetter Lane, London EC4P 4EE, UK.
11.8	*Governing law*
	This Agreement is governed by English law and the parties hereby agree that any dispute arising under this Agreement shall be subject to the jurisdiction of the English courts.

If you have any queries about the terms of this licence, please contact:

Spon's Price Books
Spon Press
an imprint of Taylor & Francis Books Ltd.
11 New Fetter Lane
London EC4P 4EE
United Kingdom
Tel: +44 (0) 20 7583 9855
Fax: +44 (0) 20 7842 2298
www.sponpress.com

CD-ROM Installation Instructions

System requirements

Minimum

- 66 MHz processor
- 12 MB of RAM
- 10 MB available hard disk space
- Quad speed CD-ROM drive
- Microsoft Windows 95/98/2000/NT
- VGA or SVGA monitor (256 colours)
- Mouse

Recommended

- 133 MHz (or better) processor
- 16 MB of RAM
- 10 MB available hard disk space or more
- 12x speed CD-ROM drive
- Microsoft Windows 95/98/2000/NT
- SVGA monitor (256 colours) or better
- Mouse

Microsoft ® is a registered trademark and Windows™ is a trademark of the Microsoft Corporation.

Installation

How to install *Spon's Landscape and External Works Price Book 2003 CD-ROM*

Windows 95/98/2000/NT

Spon's Landscape and External Works Price Book 2003 CD-ROM should run automatically when inserted into the CD-ROM drive. If it fails to run, follow the instructions below.

- Click the **Start** button and choose **Run**.
- Click the **Browse** button.
- Select your CD-ROM drive.
- Select the Setup file [setup.exe] then click **Open**.
- Click the OK button.
- Follow the instructions on screen.
- The installation process will create a folder containing an icon for *Spon's Landscape and External Works Price Book 2003 CD-ROM* and also an icon on your desk top.

How to run the *Spon's Landscape and External Works Price Book 2003 CD-ROM*

- Double click the icon (from the folder or desktop) installed by the Setup program.
- Follow the instructions on screen.

For further help with using *Spon's Landscape and External Works Price Book 2003 CD-ROM* please visit our website www.pricebooks.co.uk

© COPYRIGHT ALL RIGHTS RESERVED

All materials in *Spon's Landscape and External Works Price Book 2003 CD-ROM* are copyright protected. No such materials may be used, displayed, modified, adapted, distributed, transmitted, transferred, published or otherwise reproduced in any form or by any means now or hereafter developed other than strictly in accordance with the terms of the above licence agreement.

The software used in *Spon's Landscape and External Works Price Book 2003 CD-ROM* is furnished under a licence agreement. The software may be used only in accordance with the terms of the licence agreement.